OIL-IN-WATER NANOSIZED EMULSIONS FOR DRUG DELIVERY AND TARGETING

AF794953

OIL-IN-WATER NANOSIZED EMULSIONS FOR DRUG DELIVERY AND TARGETING

TAMILVANAN SHUNMUGAPERUMAL
Department of Pharmaceutics
National Institute of Pharmaceutical Education
and Research (NIPER)
Guwahati, Assam, India

WILEY

This edition first published 2021
© 2021 John Wiley & Sons, Inc.

All rights reserved. No part of this publication may be reproduced, stored in a retrieval system, or transmitted, in any form or by any means, electronic, mechanical, photocopying, recording or otherwise, except as permitted by law. Advice on how to obtain permission to reuse material from this title is available at http://www.wiley.com/go/permissions.

The right of Tamilvanan Shunmugaperumal to be identified as the author of this work has been asserted in accordance with law.

Registered Office
John Wiley & Sons, Inc., 111 River Street, Hoboken, NJ 07030, USA

Editorial Office
111 River Street, Hoboken, NJ 07030, USA

For details of our global editorial offices, customer services, and more information about Wiley products visit us at www.wiley.com.

Wiley also publishes its books in a variety of electronic formats and by print-on-demand. Some content that appears in standard print versions of this book may not be available in other formats.

Limit of Liability/Disclaimer of Warranty
In view of ongoing research, equipment modifications, changes in governmental regulations, and the constant flow of information relating to the use of experimental reagents, equipment, and devices, the reader is urged to review and evaluate the information provided in the package insert or instructions for each chemical, piece of equipment, reagent, or device for, among other things, any changes in the instructions or indication of usage and for added warnings and precautions. While the publisher and authors have used their best efforts in preparing this work, they make no representations or warranties with respect to the accuracy or completeness of the contents of this work and specifically disclaim all warranties, including without limitation any implied warranties of merchantability or fitness for a particular purpose. No warranty may be created or extended by sales representatives, written sales materials or promotional statements for this work. The fact that an organization, website, or product is referred to in this work as a citation and/or potential source of further information does not mean that the publisher and authors endorse the information or services the organization, website, or product may provide or recommendations it may make. This work is sold with the understanding that the publisher is not engaged in rendering professional services. The advice and strategies contained herein may not be suitable for your situation. You should consult with a specialist where appropriate. Further, readers should be aware that websites listed in this work may have changed or disappeared between when this work was written and when it is read. Neither the publisher nor authors shall be liable for any loss of profit or any other commercial damages, including but not limited to special, incidental, consequential, or other damages.

Library of Congress Cataloging-in-Publication Data

Names: Shunmugaperumal, Tamilvanan, author.
Title: Oil-in-water nanosized emulsions for drug delivery and targeting / Tamilvanan Shunmugaperumal.
Description: First edition. | Hoboken : Wiley, [2021] | Includes bibliographical references and index.
Identifiers: LCCN 2021019779 (print) | LCCN 2021019780 (ebook) | ISBN 9781119585220 (hardback) | ISBN 9781119585329 (adobe pdf) | ISBN 9781119585251 (epub)
Subjects: LCSH: Drug delivery systems. | Emulsions. | Oils and fats–Solubility. | Nanomedicine.
Classification: LCC RS199.5 .S58 2021 (print) | LCC RS199.5 (ebook) | DDC 615.1/9–dc23
LC record available at https://lccn.loc.gov/2021019779
LC ebook record available at https://lccn.loc.gov/202101978

Cover Design: Wiley
Cover Image: Courtesy of Tamilvanan Shunmugaperumal

Set in 10/12pt Times Ten by SPi Global, Pondicherry, India

10 9 8 7 6 5 4 3 2 1

MIX
Paper from responsible sources
FSC www.fsc.org FSC® C013604

I dedicate this book to my wife, Suriaprabha Marchen for her support, my son, Arunachalam Tamilvanan for his play-time sacrifice with me, my director, Dr. USN Murty for his encouragement and sustained help, and all coauthors for their enthusiasm toward achieving the vision of this book.

I dedicate this book to my father (late) Shunmugaperumal and mother Pethammal. I also dedicate this book to Prof. Simon Benita, School of Pharmacy, Hebrew University of Jerusalem, Israel.

CONTENTS

LIST OF CONTRIBUTORS

Abhinab Goswami, Research scholar, Department of Pharmaceutics, National Institute of Pharmaceutical Education and Research (NIPER), Guwahati, Assam, India

Oly Katari, Research scholar, Department of Pharmaceutics, National Institute of Pharmaceutical Education and Research (NIPER), Guwahati, Assam, India

Datta Maroti Pawde, Research scholar, Department of Pharmaceutics, National Institute of Pharmaceutical Education and Research (NIPER), Guwahati, Assam, India

Syed Nazrin Ruhina Rahman, Research scholar, Department of Pharmaceutics, National Institute of Pharmaceutical Education and Research (NIPER), Guwahati, Assam, India

Tamilvanan Shunmugaperumal, Department of Pharmaceutics, National Institute of Pharmaceutical Education and Research (NIPER), Guwahati, Assam, India

FOREWORD

It gives me an immense pleasure to write foreword for a book written by my colleague. This book provides comprehensive information concerning the oil-in-water nanosized emulsions as drug delivery carrier and drug targeting purpose. Very importantly, the generations of nanosized emulsions are updated according to the present-day research focus, especially for reachable and unreachable organs/parts of the human body and thus, permitting the target place-precised drug delivery and/or improved drug bioabsorption. A dedicated section covered in this book under heading, "Biofate of Nanosized Emulsions," after their administration via different routes (dermal, nasal, ocular and parenteral) is a subject matter of interest for researchers. I strongly feel that this book will serve as a guide for formulation scientists working on oil-in-water nanosized emulsions in academic and industry levels.

DR. USN MURTY
Director
National Institute of Pharmaceutical Education and Research (NIPER),
Guwahati, Assam, India

PREFACE

Over the years, so many new chemical entities (NCE) are constantly being discovered/designed/synthesized and even sometimes few NCE claim to be "molecular breakthrough" showing additional/exceptional better therapeutic activities in managing/treating single or multiple ailments. The so-called molecular breakthrough NCE, in most of the times, are unable to provide the intended therapeutic application as they were thought initially. Thus, a question has come to find out the answer "why the sudden lacuna that prevents the successful development of NCE from laboratory to market (L-to-M)." In the present scenario, the L-to-M consists of multiple stages that include laboratory-scale preformulation studies, systematic optimization of a formula to incorporate the NCE into the selected formulation, NCE-incorporated formulation characterization and safety assessment at non/preclinical levels, clinical trials using the optimized NCE-laden formulation including the toxicity assessment at multicentric levels, patenting and regulatory approval processes, market assessment, targeting and positioning of final product, etc. Although each stages of product development require considerable/significant attention and care, the preformulation and formulation development stages become a cutting-edge decision-making stage where the formulation development scientist will decide whether to allow the NCE-laden formulation to proceed for further stages or to send back the NCE to the previous NCE design/discovery/synthesis stage. Furthermore, the cutting-edge decision-making stage is very important in industrial point-of-view to avoid any further utilization of manpower, money, and time on a particular molecular breakthrough NCE.

Ranging from 40 to 70%, the molecular breakthrough NCE possess exceptionally water-insoluble characteristics and therefore compelling the formulation developers to look into or to go for any suitable formulation that may fit for the selected NCE rather than simply trying to make aqueous/nonaqueous-based solution-type formulation. One of the amazing drug delivery carrier systems that suit for such NCE is oil-in-water (o/w)-type nanosized emulsions. The o/w nanosized emulsions possess the success stories for taking the active pharmaceutical ingredients (APIs) into desirable target organs of human body, particularly in terms of non-parenteral (especially ocular) and parenteral routes. The in-or-new-born internal structures observed in the dispersed oil droplets of o/w nanosized emulsions are the welcome additions in emulsion technology wherein two or three API molecules can easily be accommodated simultaneously during the emulsification time itself. Such recent observations of in-or-new-born internal structures in the o/w nanosized emulsions include "Cerberus and Janus" and corresponding names given to the emulsions are Cerberus emulsions and Janus emulsions. These types of new emulsions, although at their rudimentary stage of development, are capable of escalating the o/w nanosized emulsion's utility in multiple directions viz., drug solubility enhancers, dual/triple drug-loading vehicle, imaging and theragnostic purposes, etc.

This book provides very comprehensively the details of o/w nanosized emulsions at the grass-root level so that the formulation scientist can understand how to select the proper excipients such as emulsifier combination, oil or oil combination, and other ingredients to systematically optimize the emulsion formula matching with ICH guidelines Q8(R2), Q9, and Q10, to characterize the emulsions and perform safety assessment, to monitor the biofate of nanosized emulsions, etc. All these abovesaid issues are detailed in this book with few case studies wherever possible.

Chapter 1 starts with terminology confusion prevailing about emulsion in medical and pharmaceutical fields. The purpose of this section is to clarify the terminology needed to indicate authentically this drug delivery/targeting system and finally gives reasons for judicious selection of the term "nanosized emulsion." It will also provide generations of oil-in-water nanosized emulsions so far developed with short justification followed by a brief description regarding the purpose and contents of the book.

Chapter 2 introduces the quality-by-design (QbD) approach applied onto the emulsion optimization during the preformulation studies. The effect of amount of NCE or lipophilic drug, quantity of excipients (oils, emulsifiers, and other excipients) onto the drug incorporation patterns, final particle size distribution of dispersed oil droplets of the emulsion, stability of final emulsion over different temperature conditions, etc., are the subject of interest discussed in this chapter.

Chapter 3 provides an overview of how nanosized emulsions are serially or systematically characterized to meet the requirements against physical, chemical, biological, and safety points of view. Safety evaluations using animal or human volunteers and cell-culture models are separately discussed in terms of the requirements needed by the regulatory agencies for allowing the emulsions to undergo clinical trials and then, commercial usage.

Chapter 4 shows how changes in terms of formulation excipients were being constantly made on the nanosized emulsions over the last few decades. To understand better this changeover process, the o/w nanosized emulsions are classified based on the generations with decade-wise gap.

Chapter 5 arranges the various issues relevant to the o/w nanosized emulsions to implicate in parenteral and ocular drug delivering systems. Understanding the mechanisms of interactions between emulsion droplets and plasma protein components helps the formulation developers to design long-circulating stealth emulsions for parenteral drug delivery and drug targeting purposes. Similar attention is also being given to the consequences of nanosized emulsions following ocular topical instillation or intraocular injection. The published reports in conjunction with these points are covered thoroughly in this chapter.

Chapter 6 starts with a narrative on how the emulsion surfaces can possibly be decorated with different functional molecules for the purpose of extracting a multifunctional activity in a single drug delivery and drug targeting system. Various medical applications of emulsions achieved/obtained through different administration routes are discussed in detail in this chapter.

Chapter 7 intends to provide an overview of some selected and miscellaneous uses of nanosized emulsions to interface with the recent application.

Chapter 8 describes the various steps ranging from laboratory level manufacturing together with safety aspects evaluation in animal and human eyes to industrial scaleup and then successful commercialization of cyclosporin A-loaded nanosized emulsions. Other case study included in this chapter is fish oil-based emulsions.

With these collective information, this book serves as a guide for emulsion formulation developers working in academic environment and at the industrial level.

Tamilvanan Shunmugaperumal

CHAPTER 1

INTRODUCTION: AN OVERVIEW OF NANOSIZED EMULSIONS

TAMILVANAN SHUNMUGAPERUMAL

Department of Pharmaceutics, National Institute of Pharmaceutical Education and Research (NIPER), Guwahati, Assam, India

Oil-in-Water Nanosized Emulsions for Drug Delivery and Targeting,
First Edition. Tamilvanan Shunmugaperumal.
© 2021 John Wiley & Sons, Inc. Published 2021 by John Wiley & Sons, Inc.

EXPANSION OF ABBREVIATIONS

API	active pharmaceutical ingredients
BCS	biopharmaceutics classification system
DCS	developability classification system
GSE	general solubility equation
LSW	Lifshitz and Slezov and Wagner theory
MS	metastable
NPs	nanoparticles
O/W	oil-in-water
T_m	melting point
USFDA	United States Food and Drug Administration
W/O	water-in-oil

1.1. INTRODUCTION

Inadequacy is the term that influences in every corner of modern scientific world. Starting from clothes, food, water, and shelter for covering the basic needs of living to affordable medicines for managing any ailments, the inadequacy plays major roles to judge the standard living of every country in the world. In the present scenario of democratic countries like India, USA, etc., the political party which acquires the adequate number of parliamentary members either elected by highest number of people votes in the constituency or after election shuffling with opposite party, is able to form the government for ruling the country another 3 or 5 years. Here too the adequacy means a lot to form the government and the inadequacy leads the party to sit opposite to ruling party. Coming to the affordable medicines, the presence of adequate amount of active pharmaceutical ingredient (API) in a dosage form becomes the prime importance to elicit the desired and often required pharmacological activity following administration of the same into human body. But achieving the adequate amount of API in the particular dosage form requires a lot of formulation knowledge input. How to manufacture the dosage form with adequate amount of API? The answer to this query primarily depends on solubility of API followed by its permeability, molecular size, therapeutic index, etc. If the water solubility of API is low, then, the enhancement of API's solubility becomes a major concern in pharmaceutical industries for converting the inadequate API aqueous solubility amount into adequate category. This type of converting the API amount from inadequate to adequate category needs the tremendous supports from formulation scientist who has the experience to solve this concern. The lucrative way adapted routinely in the pharmaceutical industry for deleting the word inadequate from the API aqueous solubility is just to combine the API with a suitable nanosized API delivery system or just to pulverize or micronize the API into nano-level size ranges. This book accentuates the use of one of the nanosized API delivery system to erase the inadequate aqueous solubility from API. Before proceeding with the nanosized API delivery system, a brief history of nanotechnology is being described below.

1.1.1. Nanotechnology: Definition

Pulverized API having nanometer size ranged particles and intact API-loaded nanoparticles (either preformed or forming *in situ*) are considered one of the most-prevalent improvement methods which have been used to overcome the problem of API's poor solubility and, thus, bioavailability, as well as to achieve targeted API delivery. Having said the importance of pulverized API as nanoparticles (NPs) or API-entrapped nanoparticulate system, there is no single definition of what a NP is. This might be because of the highly multidisciplinary nature of nanotechnology. The term "nanotechnology" was first used by Norio Taniguchi in 1974, at the University of Tokyo, Japan, for any material in the nanometer size range (Taniguchi 1974) and the materials somehow handled by human beings in their everyday life. According to the United States Food and Drug Administration (USFDA), materials are classified as being in the nanoscale range if they have at least one dimension at the size range of approximately 1–100 nm. However, the characteristic properties of nanoscale such as solubility, light scattering and surface effects are predictable and even these properties are indeed continuous characteristics of the bulk materials (Donaldson and Poland 2013), the definitions of "nanomaterial" based on size are often inconsistent and the upper end of the nanoscale at 100 nm is an arbitrary cut-off (Boverhof et al. 2015). Thus, the 100-nm limit is often considered constraining and, according to a more-inclusive definition, particles <1,000 nm in each dimension (submicron particles) is designated as NPs (Keck and Müller 2006). The latter more-inclusive definition is particularly applicable in the pharmaceutical field since the particle size in the nanometer range can lead to increased dissolution rates because of the increase in surface area and increased saturation solubility (Junghanns and Müller 2008). In this respect, the generation of nanometer range particles from the API molecule itself is coming under NPs with a nomenclature of API nanocrystals or API nanosuspensions. On the other hand, encapsulating the API into a preformed or *in situ* forming nanosized API delivery carrier is also grouped under NPs. Now the question will arise, at an industrial environment, which type of API is going for API nanocrystals/nanosuspensions formation and which type of API needs to go as cargo in a nanosized carrier? To clarify this confusion, the following discussion is devoted wherein the poorly soluble API is grouped into two categories.

1.1.1.1. Poorly Water-Soluble Grease Ball and Brick Dust Molecules It is now well known that the poor API solubility is an issue for approximately 70–90% of new APIs (Merisko-Liversidge and Liversidge 2011; Müller and

Keck 2012). According to Lipinski (2005) there are two different classes are available within the poorly soluble APIs viz. (i) grease ball and (ii) brick dust molecules.

Hydrophobic APIs have a limited capacity to interact with the water phase, in accordance with "*similia similibus solvuntur*" (like dissolves like), and these APIs are solubility-limited by poor hydration. The poorly soluble APIs restricted in solubility by poor hydration are described in the popular scientific jargon as "grease ball" molecules, due to their high hydrophobicity and lack of interaction with water. Although they possess poor water solubility (insolubility is probably due to the salvation extreme), the grease ball molecules are easily soluble to lipids or oils and therefore these molecules can plausibly be formulated into lipid-or oil-based formulations. In contrast, the brick dust molecules are poorly soluble due to crystal packing interactions being insoluble to both aqueous solvents and lipids or oils. So, the aqueous solubility of brick dust molecules could be enhanced by the formation of amorphous materials or particle size decrease, e.g., nanocrystallization. Thus, formulating APIs as nanocrystals should be mainly used as a solubility enhancement formulation approach to brick dust molecules rather than to grease balls.

In other words, Yalkowsky and coworkers established the General Solubility Equation (GSE), in which the solubility of a compound is expressed as a function of the melting point (T_m) and its lipophilicity (in the form of the octanol-water partition coefficient, $\log K_{o/w}$ or simply the $\log P$ value) (Jain and Yalkowsky 2001). The GSE states the following relationship:

$$\log S_0 = 0.5 - 0.01(T_m - 25) - \log P \quad (1.1)$$

where S_0 is the intrinsic solubility, i.e., the solubility of the non-ionised (neutral species).

Specifically, "grease balls" represent highly lipophilic compounds ($\log K_{o/w} > 3$), which are poorly hydrated and their solubility is solvation limited, whereas "brick dust" compounds display lower lipophilicity and higher melting points ($T_m > 200°C$) and their solubility is limited by the strong intermolecular bonds within the crystal (Bergström et al. 2007), therefore, they said to have solid-state-limited solubility. Collectively, the APIs with solvation-limited solubility are lipophilic, relatively large molecules and lack conjugated systems. Most of these grease ball molecules have been developed as oral dosage forms, however, the developed dosage forms typically include several different excipients that may improve dissolution (disintegration and dispersion) and solubilization. This indicates that an extensive API development process would be required to bring these grease ball molecules to the market. In contrast, the APIs with solid-state-limited solubility (brick dust molecules) are often flat, typically with an extended ring structure and display high aromaticity. These molecular features are important for forming a more stable crystal lattice. Hence, brick dust molecules have been found to benefit from formulation approaches such as particle size reduction and amorphization, whereas grease balls can be formulated as lipid or oil-based formulations (Tuomela et al. 2016). Table 1.1 shows the various properties of brick dust and grease ball APIs.

It should be noted, however, that while compounds described as solvation-limited (grease ball) often display intrinsic solubility values in the lower nanomolar scale, none of the compounds included in the solid-state-limited (brick dust) category had intrinsic solubility values even in this lower nanomolar scale (Wassvik et al. 2008). Interestingly, griseofulvin which has a solubility of approximately 15 μM was the least soluble compound found in the dataset published by Wassvik et al. (2008). Two fundamental explanations may be given for why this is the case. First of all, after using the GSE relationship [Eq. (1.1)] along with the knowledge that compounds with a log *P* value of 2 or less will tend to have solid-state-dependent solubility, the estimated or expected intrinsic solubility value of compounds having extraordinarily high melting points (of about 325°C) would be about 30 μM. Secondly, most of the solely solubility-limited compounds because of their solid-state are terminated relatively early in the API development process. So, in the quest of affordable APIs for therapeutic activity, the grease ball molecule looks competitively better than the brick dust molecule during the API or dosage form development process.

TABLE 1.1. Typical Differences Between Grease Ball and Brick Dust Molecules

Grease Ball API	Brick Dust API
Highly lipophilic compound with high log *P* (>3 or 4) and a low melting point (<200°C)	Compound with high log *P* (<2) and high melting point (>200°C)
Poorly soluble compounds restricted in solubility by poor hydration are described as grease ball molecule	Compound with strong intermolecular bonds and/or complex interaction patterns with large number of interaction points between the molecules in the crystal lattice which often shows a limited capacity to dissociate from the solid form. This type of compounds are called as brick dust (stone-like)
These compound cannot form bonds with water molecules, thus their solubility is limited by the solvation process	The solubility of compounds in water is restricted due to strong intermolecular bonds within the crystal structure
Usual formulation approaches do not work, solubility enhancement through the use of a polar promoiety may prove useful	If the molecule has brick dust nature, a polar promoiety may work as this strategy which might disrupt the intermolecular interactions that led to the high crystallinity
Grease ball APIs are the candidates for entrapment into various lipid- and oil-based nanoformulations	Brick dust fraction that dissolves neither in oil nor in water cannot be administered as self-emulsifying API delivery system

Extracted from Bergström and Larsson (2018).

Preparing the pulverized API suspended in aqueous or non-aqueous medium (is termed as API nanosuspensions) is being suggested as a universal delivery approach for those group of orally administrable APIs that fall into class II (low solubility and high permeability) and class IV (low solubility and low permeability) of the Biopharmaceutics Classification system (BCS) (Keck and Müller 2006; Shegokar and Müller 2010). Another elegant way proposed by Butler and Dressman (2010) to classify the API molecules is the Developability Classification System (DCS) as this way of categorizing the API molecules is in a more biorelevant manner. According to the DCS, the API molecules can further be sub-categorized into two types to distinguish between dissolution rate-limited (DCS Class IIa) and solubility-limited (DCS Class IIb) API molecules (as shown in Flowchart 1.1). More importantly, the intrinsic solubility and the related intraluminal API concentration for API molecules belonging to Class IIb and IV are too low to achieve sufficient flux over the epithelial membrane. Therefore, the API molecules belonging to DCS Class IIb and IV often utilize the complexation or other formulation approaches based on solid-state modifications and even these approaches might be preferable compared with nanocrystals or nanosuspensions (Chen et al. 2017; Möschwitzer 2013; Shah et al. 2016).

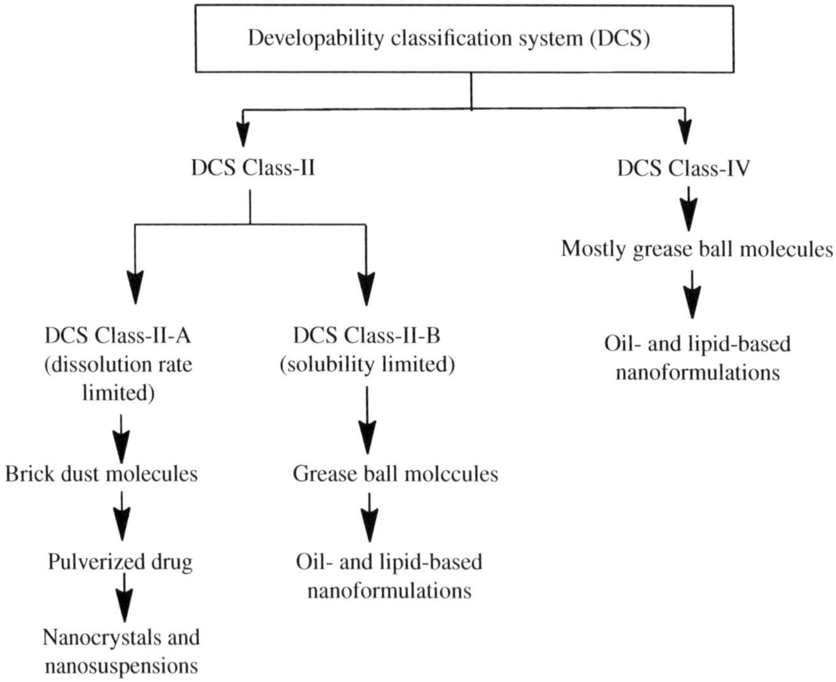

Flowchart 1.1. API sub-categorization.

After understanding the clear-cut difference between the grease ball and brick dust API molecules, the oil-based heterogeneous, dispersion (liquid-retentive) system, like emulsion, is the main centre of focal point to solve the solubility (and thus the intestinal permeability) problems of grease ball API molecules.

1.1.2. Nanosized Emulsions

An emulsion is a biphasic liquid preparation consisting of two immiscible liquids, one of which (the dispersed phase) is finely and uniformly dispersed as globules throughout the second phase (the continuous phase) (Barkat et al. 2011). If the amount of oil phase is significantly low when compared to the amount of water phase, then, the final emulsion is termed as oil-in-water (o/w) emulsion. Conversely, when the amount of water phase is significantly lower than the oil phase, the resulting emulsion system appears to be somewhat more viscous and is called as water-in-oil (w/o) emulsion. Both o/w and w/o type of emulsion systems are stabilized against the aggregation, coalescence and separation of dispersed oil or water phase by the addition of a third component called as emulsifying agent or emulgent or emulgator. In fact, the therapeutic emulsions are being stabilized by the addition of more than one emulgent molecules in order to prevent random collision of-and then the coalescence of-dispersed oil or water phase of the o/w or w/o emulsion. Mixing of appropriate amounts of oil, water and emulgent leads to the formation of an emulsion and this whole process is being named as emulsification. Apart from the amount of dispersed oil or water phase which will determine the type of final emulsion formed (whether o/w or w/o), the amount of single or multiple emulgent molecule added during the emulsification process will obviously control the type of emulsion produced. In addition, the size-reduction equipments such as high-energy or low-energy homogenizer used to mix the oil and water phases along with single or multiple emulgent molecules will also direct the final emulsion produced in terms of mean size of the dispersed phase in the final emulsion. Interestingly, both high-and low-energy homogenizers will generate emulsions with nano-range droplets particle sizes of the dispersed phase.

The emulsion generated by the low-energy or spontaneous emulsification process is termed as microemulsion. On the other hand, the emulsion produced by means of the high-energy homogenizers is called as nanoemulsion or nanosized emulsion. Again there are few basic differences between microemulsion and nanosized emulsion. Firstly, the microemulsion possesses the dispersed phase droplet diameter value well below the range of 100 nm, typically even below 10 nm as well. But the dispersed phase droplet diameter of the nanosized emulsion lies above 100 nm that too in between 300 to 500 nm level. Secondly, the notable difference between these two emulsions is being their physical appearances. While the microemulsion creates solution like transparent appearance, the nanosized emulsion looks like slightly bluish-coloured white milk.

TABLE 1.2. Typical Differences Between Microemulsion and Oil-in-Water Nanosized Emulsion

Microemulsion	Oil-in-Water Nanosized Emulsion
Emulsifier concentration is high, i.e., 30–40%	Emulsifier concentration is comparatively less, preferably 2–5% and can be increased up to 10%
It is a thermodynamically stable isotropic mixture	It is thermodynamically unstable but kinetically stable for some time with the help of emulsifier molecule's coverage onto the dispersed oil droplet surface
Particle size ranges below 10 nm or below 100 nm	Particle size ranges from 200 to 700 nm
It is toxic for parenteral application but oral dosage form is possible. Any water-soluble API can be used	All routes of application are possible

Thirdly, a moderately higher amount of single or multiple emulgent molecules is used for producing the microemulsion when compared to the emulsifier molecules' amount which is being utilized to prepare the nanosized emulsion. Table 1.2 depicts the basic differences between microemulsion and o/w nanosized emulsion. In spite of these basic differences between microemulsion and o/w nanosized emulsion, these two terms (micro- and nano-emulsions) are, however, used interchangeably in many medical and pharmaceutical literatures.

1.1.2.1. Nanosized Emulsions: Prime Candidates for Nanoparticle Engineering As discussed above, emulsions exist as two types often reported as being the same, thermodynamically stable (microemulsion) and thermodynamically unstable or metastable (MS) varieties (nanosized emulsion), the prevailing form is a function of surface tension that is related to dispersion entropy. The factors that are critical in long-term product stability are presented in flowchart form (Flowchart 1.2). The MS or nanosized emulsions are considerably more susceptible to the destabilization phenomena than thermodynamically stable variety. Destabilization does occur due to aggregation or coalescence of dispersed oil droplets of o/w nanosized emulsion and Ostwald ripening phenomena. Whereas the aggregation or coalescence destabilization phenomena are related to the mechanics of the droplet surface, the Ostwald ripening describes the change of an inhomogeneous structure or size in the dispersed oil droplets of emulsion over time. The presence of inhomogeneous sizes in the dispersed oil droplets allow the smaller particles to dissolve and deposit on the larger particles in order to reach a more thermodynamically stable state wherein the surface to area ratio is minimized, i.e., the formation of larger-sized dispersed oil droplets at the expense or unison of smaller-sized oil droplets. This is

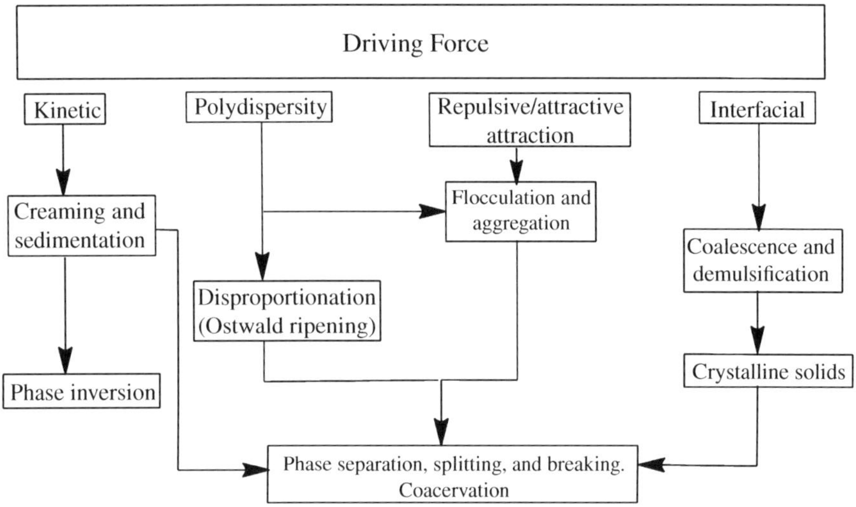

Flowchart 1.2. Critical factors to determine long-term product stability.

accentuated by the smaller particle dimensions because of the increased internal Laplace pressure that is related to droplet surface curvature. The Ostwald ripening is otherwise termed as particle coarsening. This causes interfaces with low curvature to grow at the expense of interfaces with high curvature and the average size-scale of the domains to increase in size with time. The driving force for this process is the minimization of the total interfacial energy of the system. Coarsening typically occurs under conditions where the volume fractions of the phases are nearly at their equilibrium values. The coarsening process has a profound effect on the properties of materials. For example, the Ostwald ripening or particle coarsening impacts directly on continuous phase solubility of the dispersed phase components and encapsulated API.

The manufacturers of new colloidal products (nanosized emulsions) have to be aware of the best manufacturing techniques, limitations of producing new products and efficacy of the final administered product (Margreiter 2002). Because emulsions that are easy to produce on a laboratory-scale can be much more difficult to mass produce. The lab-designed product does not always come with an acceptable range of built-in invulnerability to variable storage conditions and viable shelf life of the final marketed product. Worryingly, MS nanosized emulsion droplets are notable for the size-related improved stability and with smaller sizes the formulated products show better medium-term stability (Medlicott et al. 2004). They are also worth considering for regular use since they have rapid API release and viable API partition coefficients (Chen et al. 2004), say from interior to interface and thus to bulk, which is primarily because of the exceptionally large surface area of nanospheres. Furthermore, adaptation is possible as the surface fluidity of interface emulsifiers can be selectively modified. In terms of API delivery, nanosized emulsions have been reported to have a particular favourable predisposition for vascular wall and capillaries (Mizushima 1996).

The o/w nanosized emulsions are nanometric-sized emulsions, typically exhibiting diameters of up to 500 nm or often quoted as being 400–800 nm. However, particles in this range tend to be thermodynamically unstable non-self forming requiring mechanical energy input and are opaque. Nanosized emulsions are also frequently known as miniemulsions, fine-dispersed emulsions, submicron emulsions and so forth, but are all characterized by a great stability in suspension due to their very small size, essentially the consequence of significant steric stabilization between droplets, which goes to explain why the Ostwald ripening is the only adapted droplet destabilization process (detailed below).

1.1.2.2. Ostwald Ripening-Adapted Droplet Destabilization Process for the Greater Stability of Nanosized Emulsions The main particularity of nanosized emulsions, making them prime candidates for nanoparticle engineering, is their great stability of droplet suspension. A kinetic stability that lasts for months, stability against dilution or even against temperature changes, totally unlike the (thermodynamically stable) microemulsions. Emulsions are thermodynamically unstable systems, due to the free energy of emulsion formation (ΔG_f) greater than zero. The large positive interfacial energy term ($\lambda\Delta A$) outweighs the entropy of droplet formation ($T\Delta S_f$), also positive. The terms λ and ΔA, respectively represent the surface tension and the surface area gained with emulsification. Emulsion instability is therefore induced by the positive sign of ΔG_f [Eq. (1.2)].

$$\Delta G_f - \lambda\Delta A\, T\Delta S_f \tag{1.2}$$

Accordingly, the physical destabilization of emulsions is related to the spontaneous trend towards a minimal interfacial area between the two immiscible phases. Therefore, a minimization of interfacial area is attained by two mechanisms:

1. Flocculation followed mostly by coalescence, and
2. Ostwald ripening.

In nanosized emulsion systems, flocculation is naturally prevented by steric stabilization, essentially due to the sub-micrometric droplet size. In short (Napper 1983; Tadros 1982; Tadros et al. 2004), when interfacial droplet layers overlap, steric repulsion occurs, from two main origins. The first one is the unfavorable mixing of the stabilizing chain of the adsorbed layer, depending on the interfacial density, interfacial layer thickness δ, and Flory–Huggins parameter $\chi_{1,2}$ (which reflects the interactions between the interfacial layer and solvent).

The second one is the reduction of the configurational entropy, due to the bending stress of the chains, which occurs when inter-droplet distance h becomes lower than δ.

Generally, the sum of the energies of interaction U_T adopts a typical shape of systems wherein molecules repel and particles attract each other, showing a weak minimum, around $h = 2\delta$, and a very rapid increase below this value (see Fig. 1.1 for illustration). The depth of the minimum $|U_0|$ will induce predispositions for coagulate in the colloidal suspension, that is to say, it is intimately linked to the stability of the suspension. $|U_0|$ is shown to be dependent on the particle radius, the Hamaker constant A, and the adsorbed layer thickness δ, with the result that the higher the δ/r ratio, the lower the value of $|U_0|$. Now, in the case of nanoemulsion droplets, δ/r becomes very high in comparison with macroemulsions, which in the end totally inhibits its ability to coagulate. On the other hand, it is worth noting that the small droplet sizes also induce stabilization against sedimentation or creaming, in so far as the droplets are solely under the influence of the Brownian motion.

Taking all this into account, the destabilization of nanosized emulsions is due only to a mass transfer phenomenon between the droplets through the bulk phase, well described in the literature (Taylor 1998) as Ostwald ripening in emulsions. At the origin of this destabilization process, the differences, however slight, of the droplet radius induce differences in chemical potential of the material within the drops. The reduction of free energy in the emulsion will result in the decrease of the interfacial area, and therefore in the growth of the bigger emulsion droplets at the expense of the smaller ones. The dispersed phase migrates through the bulk from the smaller droplets to the bigger ones, owing to the higher solubility in the bulk of the smaller droplets. Ostwald ripening is initiated and will increase throughout the process. As an illustration and under the assumption that only one component composes the dispersed

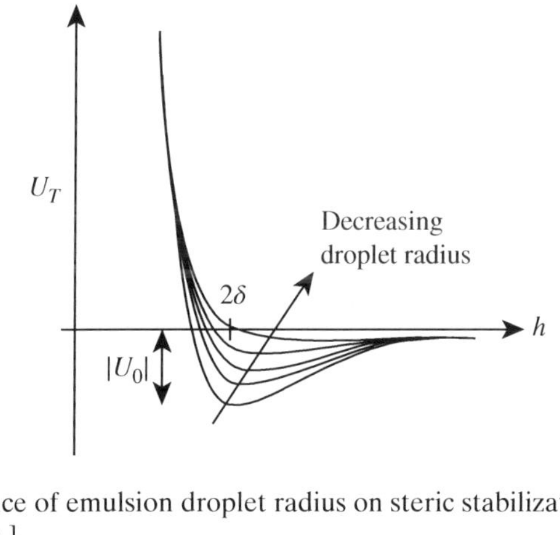

Figure 1.1. Influence of emulsion droplet radius on steric stabilization. [Adapted from Anton et al. (2008).]

phase, the solubility, $C(\Gamma)$, of the dispersed material throughout the dispersion medium is expressed as a function of the droplet radius r, from the Kelvin equation (Skinner and Sambles 1972) [Eq. (1.3)],

$$C(\Gamma) = C_\alpha \exp\left(\frac{2\gamma M}{\rho RT\Gamma}\right) \tag{1.3}$$

where C_∞ is the bulk solubility of the dispersed phase, M its molar mass, and ρ its density.

In most studies, the follow-up of Ostwald ripening as the temporal evolution of the droplet diameter still remains well fitted, even under the approximations involved in Eq. (1.3). In addition, the literature provides a number of theories dealing with calculations of the rate of ripening, such as the most famous (and complete) given by Lifshitz and Slezov (1959, 1961) and Wagner (1961), the so called LSW theory. Besides the consideration of Eq. (1.3), the diffusion of dispersed materials through the continuous medium is assumed to be diffusion-controlled, i.e., crossing the interface with ease. Details on LSW theories are fully developed and discussed in the literature (Dunning 1973; Kahlweit 1975; Taylor 1998) leading to the commonly used expression of the ageing rate, ω, in Eq. (1.4),

$$\omega = \frac{\mathrm{d}r_\mathrm{c}^3}{\mathrm{d}t} = \frac{8\mathrm{D}C_\alpha \gamma M}{9\rho^2 RT} \tag{1.4}$$

where r_c is the critical radius of the system at any given time, at the frontier between the growth and decrease of the droplets. Consequently, Ostwald ripening is reflected by a linear relationship between the cube radius and time.

In processes involved in nanoparticle engineering, i.e., for multi-component emulsion droplets, by adding monomer, polymer, or simply surfactant or co-surfactant, the above approximation is surpassed. The rate of ripening can be reduced by several orders of magnitude when the additive has a substantially lower solubility in the bulk phase than the main component of the droplet. This phenomenon has been widely studied (Buscall et al. 1979; Davis and Smith 1973; Davis et al. 1981; Higuchi and Misra 1962; Kabalnov et al. 1985, 1987; Smith and Davis 1973; Taylor and Ottewill 1994), since it appears to be an efficient method to reduce the Ostwald ripening rate, even when using small amounts of additives. In short, it is explained by the difference of solubility in the continuous phase between the dispersed phase noted (1) and the additive (2), less soluble in this example. The first step remains similar to the ripening without additives, since only the component (1) diffuses from the smaller to the larger droplets, due to the higher chemical potential of the materials within the smaller drops. Gradually, the chemical potential in the larger droplets increases due to the presence of the component (2), until the diffusion process of (1) is stopped. Equilibrium is reached between the two opposing effects and the limiting process becomes the diffusion of the less soluble additive (2), significantly reducing the ripening rate and the nanoemulsion destabilization.

A final remark, which may be of importance here, concerns the influence on the nanosized emulsion destabilization of layer density and structure in the interfacial zone. Indeed, up to now it has been considered that Ostwald ripening is a diffusion-controlled process, but this assumption does not take into account the fact that surfactants, polymeric emulsifiers or stabilizers can create a thick steric barrier at the droplet interface (Goldberg and Higuchi 1969; Yotsuyanagi et al. 1973). As a consequence, the diffusion of the inner material of the droplets may be slowed down, reducing the ripening rate. The substantial difference in stability between nanoemulsions and nanocapsules (another colloidal API delivery system having polymeric outer shells covered on the dispersed oil droplets) for instance, appears essentially from such details.

Before proceeding into Chapter 2, a brief description concerning classification of nanosized emulsions is presented below.

1.1.2.3. Classification of Oil-in-Water Nanosized Emulsions Purely based on the emulsifier combinations used to stabilize the dispersed oil droplets of the emulsions, the o/w nanosized emulsions can be classified into three types (Fig. 1.2). First type includes emulsions prepared using the emulsifiers that are having the capacity to assemble at the o/w interface and able to produce a minus (negative) charge in the vicinity of dispersed oil droplets of the emulsions. The emulsions thus formed are termed as anionic or negatively-charged emulsions. The emulsions made with the inclusion of emulsifiers that are assembled at the o/w interface and competent to confer a plus (positive) charge in the vicinity of dispersed oil droplets of the emulsions are called as cationic or positively-charged emulsions. The literature suggests that neither triglycerides nor phospholipidic emulsifier's components of the conventional or anionic

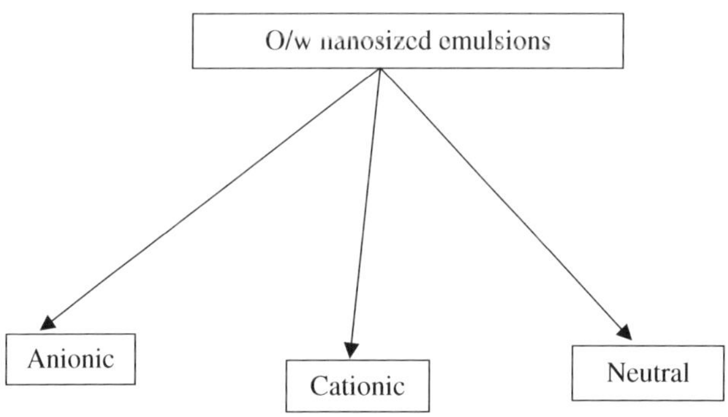

Figure 1.2. Classification of oil-in-water (o/w) nanosized emulsions based on emulsifier molecules.

emulsions are able to significantly sustain the incorporated lipophilic API release in simulated or real physiological environments under sink conditions. Therefore, in an attempt to prolong and/or optimize the API release, cationic lipid or polysaccharide emulsifiers are added to the emulsions to elicit mucoadhesion with anionic ocular tissues by an electrostatic adhesion. Indeed, cationic emulsions prepared on the basis of stearylamine, oleylamine and chitosan can serve this purpose. It was initially believed and now has become clearer from many reports in the literature that an occurrence of electrostatic attraction between the cationic emulsified droplets and anionic cellular moieties of the ocular and topical skin surface tissues enhance the bioavailability of emulsions containing lipophilic APIs (Lallemand et al. 2003; Piemi et al. 1999; Tamilvanan et al. 2002; Vandamme 2002). There is another type of emulsions that are neutral in terms of the charge on the dispersed droplets. These are instead stabilized through steric effects exerted by the emulsifier molecule present in the emulsion formulation.

According to Capek (2004), the stability of the electrostatically- and sterically-stabilized o/w nanosized emulsions can be controlled by the charge of the electrical double layer and the thickness of the droplet surface layer formed by non-ionic emulsifier, respectively. In spite of the similarities between electrostatically- and sterically-stabilized emulsions, there are large differences in the partitioning of molecules of ionic and non-ionic emulsifiers between the oil and water phases and the thickness of the interfacial layers at the droplet surface (Capek 2004). The thin interfacial layer (the electrical double layer) at the surface of electrostatically stabilized droplets does not create any steric barrier for mass transfer. This may not necessarily be true for the thick interfacial layer formed by a non-ionic emulsifier. The sterically-stabilized oil droplets, however, can favor the transfer of materials within the intermediate agglomerates. Hence, the stability of electrosterically-stabilized emulsion (δ_k) is controlled by the ratio of the thickness of the non-ionic emulsifier adsorption layer (δ) to the thickness of the electrical double layer (k^{-1}) around the oil droplets (Capek 2004).

$$\left[\delta / k^{-1}\right] = (\delta_k) \tag{1.5}$$

1.2. CONCLUSION

This section says that although the o/w nanosized emulsions belong to MS category in terms of stability aspects, many competing forces actually determine the stability of emulsions. The group of API molecules suitable to be incorporated into the o/w nanosized emulsions is carefully revived by interpreting the physicochemical properties (molecular size and structure, melting point, log P value, etc.) of individual APIs along with their solubility and permeability characteristics.

REFERENCES

Anton, N., Benoit, J.-P., and Saulnier, P. (2008), Design and production of nanoparticles formulated from nano-emulsion templates—a review, *J. Control. Release*, 128(3), 185–199. doi:10.1016/j.jconrel.2008.02.007

Barkat, A.K., Naveed, A., Haji, M.S.K. et al. (2011), Basics of pharmaceutical emulsions: a review, *Afr. J. Pharm. Pharmacol.*, 5 (25), 2715–2725. doi:10.5897/AJPP11.698

Bergström, C.A.S. and Larsson, P. (2018), Computational prediction of drug solubility in water-based systems: qualitative and quantitative approaches used in the current drug discovery and development setting, *Int. J. Pharm.*, 540 (1–2), 185–193. doi:10.1016/j.ijpharm.2018.01.044

Bergström, C.A.S., Wassvik, C.M., Johansson, K. et al. (2007), Poorly soluble marketed drugs display solvation limited solubility, *J. Med. Chem.*, 50 (23), 5858–5862. doi:10.1021/jm0706416

Boverhof, D.R., Bramante, C.M., Butala, J.H. et al. (2015), Comparative assessment of nanomaterial definitions and safety evaluation considerations, *Regul. Toxicol. Pharmacol.*, 73 (1), 137–150. doi:10.1016/j.yrtph.2015.06.001

Buscall, R., Davis, S.S., and Potts, D.C. (1979), The effect of long-chain alkanes on the stability of oil-in-water emulsions. The significance of Ostwald ripening, *Colloid Polym. Sci.*, 257, 636–644. doi:10.1007/BF01548833

Butler, J.M. and Dressman, J.B. (2010), The developability classification system: application of biopharmaceutics concepts to formulation development, *J. Pharm. Sci.*, 99 (12), 4940–4954. doi:10.1002/jps.22217

Capek, I. (2004), Degradation of kinetically-stable o/w emulsions, *Adv. Colloid Interf. Sci.*, 107, 125–155. doi:10.1016/S0001-8686(03)00115-5

Chen, H., Chang, X., Weng, T. et al. (2004), A study of microemulsion systems for transdermal delivery of triptolide, *J. Control. Release*, 98 (3), 427–436. doi:10.1016/j.jconrel.2004.06.001

Chen, M.-L., John, M., Lee, S.L., and Tyner, K.M. (2017), Development considerations for nanocrystal drug products, *AAPS J.*, 19 (3), 642–651. doi:10.1208/s12248-017-0064-x

Davis, S.S., Round, H.P., and Purewal, T.S. (1981), Ostwald ripening and the stability of emulsion systems: an explanation for the effect of an added third component, *J. Colloid Interf. Sci.*, 80 (2), 508–511. doi:10.1016/0021-9797(81)90210-1

Davis, S.S. and Smith, A. (1973), *Emulsion Theory and Practice*, Academic Press, London.

Donaldson, K. and Poland, C.A. (2013), Nanotoxicity: challenging the myth of nano-specific toxicity, *Curr. Opin. Biotechnol.*, 24 (4), 724–734. doi:10.1016/j.copbio.2013.05.003

Dunning, W.J. (1973), Particle growth in suspensions, No. 38 in SCI monograph, Academic Press, London.

Goldberg, A.H. and Higuchi, W.I. (1969), Mechanisms of interphase transport. II. Theoretical considerations and experimental evaluation of interfacially controlled transport in solubilized systems, *J. Pharm. Sci.*, 58 (11), 1341–1352. doi:10.1002/jps.2600581109

Higuchi, W.I. and Misra, J. (1962), Physical degradation of emulsions via the molecular diffusion route and the possible prevention thereof, *J. Pharm. Sci.*, 51, 459–466. doi:10.1002/jps.2600510514

Jain, N. and Yalkowsky, S.H. (2001), Estimation of the aqueous solubility I: application to organic nonelectrolytes, *J. Pharm. Sci.*, 90(2), 234–252. doi:10.1002/1520-6017(200102)90:2%3C234::aid-jps14%3E3.0.co;2-v

Junghanns, J.U.A.H. and Müller, R.H. (2008), Nanocrystal technology, drug delivery and clinical applications, *Int. J. Nanomed.*, 3 (3), 295–309. doi:10.2147/ijn.s595

Kabalnov, A.S., Pertsov, A.V., Aprosin, Y.D. et al. (1985), Influence of nature and composition of disperse phase on stability of direct emulsions against transcondensation, *Kolloidn. Zh.* (Engl. Transl.), 46, 965–967.

Kabalnov, A.S., Pertsov, A.V., and Shchukin, E.D. (1987), Ostwald ripening in two-component disperse phase systems: application to emulsion stability, *Colloids Surf.*, 24, 19–32. doi:10.1016/0166-6622(87)80258-5

Kahlweit, M. (1975), Ostwald ripening of precipitates, *Adv. Colloid Interf. Sci.*, 5, 1–35. doi:10.1016/0001-8686(75)85001-9

Keck, C.M. and Müller, R.H. (2006), Drug nanocrystals of poorly soluble drugs produced by high pressure homogenisation, *Eur. J. Pharm. Biopharm.*, 62 (1), 3–16. doi:10.1016/j.ejpb.2005.05.009

Lallemand, F., Felt-Baeyens, O., Basseghir, K. et al. (2003), Cyclosporine A delivery to the eye: a pharmaceutical challenge, *Eur. J. Pharm. Biopharm.*, 56, 307–318. doi:10.1016/S0939-6411(03)00138-3

Lifshitz, I.M. and Slezov, V.V. (1959), Kinetics of diffusive decomposition of supersaturated solid solutions, *Soviet Phys. J. Exp. Theor. Phys.*, 35 (8), 331.

Lifshitz, I.M. and Slezov, V.V. (1961), The kinetics of precipitation from supersaturated solid solutions, *J. Phys. Chem. Solids*, 19 (1–2), 35–50. doi:10.1016/0022-3697(61)90054-3

Lipinski, C.A. (2005), Solubility in water and DMSO: issues and potential solutions, in: Borchardt, R., Kerns, E., Lipinski, C., Thakker, D., and Wang, B., Eds., *Pharmaceutical Profiling in Drug Discovery for Lead Selection*, AAPS Press, Arlington, pp. 93–125.

Margreiter, R. (2002), Efficacy and safety of tacrolimus compared with ciclosporin microemulsion in renal transplantation: a randomised multicentre study, *The Lancet*, 359 (9308), 741–746. doi:10.1016/s0140-6736(02)07875-3

Medlicott, N.J., Waldron, N.A., and Foster, T.P. (2004), Sustained release veterinary parenteral products, *Adv. Drug Deliv. Rev.*, 56 (10), 1345–1365. doi:10.1016/j.addr.2004.02.005

Merisko-Liversidge, E. and Liversidge, G.G. (2011), Nanosizing for oral and parenteral drug delivery: a perspective on formulating poorly-water soluble compounds using wet media milling technology, *Adv. Drug Deliv. Rev.*, 63 (6), 427–440. doi:10.1016/j.addr.2010.12.007

Mizushima, Y. (1996), Lipid microspheres (lipid emulsions) as a drug carrier—an overview, *Adv. Drug Deliv. Rev.*, 20 (2–3), 113–115. doi:10.1016/0169-409X(95)00114-M

Möschwitzer, J.P. (2013), Drug nanocrystals in the commercial pharmaceutical development process, *Int. J. Pharm.*, 453 (1), 142–156. doi:10.1016/j.ijpharm.2012.09.034

Müller, R.H. and Keck, C.M. (2012), Twenty years of drug nanocrystals: where are we, and where do we go?, *Eur. J. Pharm. Biopharm.*, 80 (1), 1–3. doi:10.1016/j.ejpb.2011.09.012

Napper, D.H. (1983), *Polymeric Stabilisation of Colloidal Dispersions*, Academic Press, London.

Piemi, M.P., Korner, D., Benita, S. et al. (1999), Positively and negatively charged submicron emulsions for enhanced topical delivery of antifungal drugs, *J. Control. Release*, 58, 177–187. doi:10.1016/s0168-3659(98)00156-4

Shah, D.A., Murdande, S.B., and Dave, R.H. (2016), A review: pharmaceutical and pharmacokinetic aspect of nanocrystalline suspensions, *J. Pharm. Sci.*, 105 (1), 10–24. doi:10.1002/jps.24694

Shegokar, R. and Müller, R.H. (2010), Nanocrystals: industrially feasible multifunctional formulation technology for poorly soluble actives, *Int. J. Pharm.*, 399 (1–2), 129–139. doi:10.1016/j.ijpharm.2010.07.044

Skinner, L.M. and Sambles, J.R. (1972), Kelvin equation—a review, *J. Aerosol. Sci.*, 3 (3), 199–210. doi:10.1016/0021-8502(72)90158-9

Smith, A. and Davis, S.S. (1973), Proceedings: the role of molecular diffusion in the bulk stability of o-w hydrocarbon emulsions, *J. Pharm. Pharmacol.*, 25 (Suppl), 117.

Tadros, T.F. (1982), Polymer adsorption and colloid stability, in: *The Effect of Polymer on Dispersion Properties*, Academic Press, London.

Tadros, T.F., Izquierdo, P., Esquena, J. et al. (2004), Formation and stability of nanoemulsions, *Adv. Colloid Interf. Sci.*, 108–109, 303–318. doi:10.1016/j.cis.2003.10.023

Tamilvanan, S., Gursoy, R.N., and Benita, S. (2002), Emulsion-based delivery systems for enhanced drug absorption, *Pharm. Technol.*, 131, 156–161.

Taniguchi, N. (1974), On the basic concept of nanotechnology, in: *Proceedings of the International Conference on Production Engineering*, Tokyo, Part-II, Japan Society of Precision Engineering, Japan, pp. 18–23.

Taylor, P. (1998), Ostwald ripening in emulsions, *Adv. Colloid.Interf. Sci.*, 75 (2), 107–163. doi:10.1016/S0001-8686(98)00035-9

Taylor, P. and Ottewill, R.H. (1994), Ostwald ripening in O/W miniemulsions formed by the dilution of O/W microemulsions, in: *Progress in Colloid and Polymer Science*, Springer, Berlin, Germany, pp. 199–203.

Tuomela, A., Hirvonen, J., and Peltonen, L. (2016), Stabilizing agents for drug nanocrystals: effect on bioavailability, *Pharmaceutics*, 8 (2), 1–18. doi:10.3390/pharmaceutics8020016

Vandamme, T.F. (2002), Microemulsions as ocular drug delivery systems: recent developments and future challenges, *Prog. Retin. Eye Res.*, 21, 15–34. doi:10.1016/s1350-9462(01)00017-9

Wagner, C. (1961), Theorie der alterung von niederschlägen durch umlösen (ostwaldreifung), *Z. Elektrochem.*, 65 (7–8), 581–591. doi:10.1002/bbpc.19610650704

Wassvik, C.M., Holmen, A.G., and Draheim, R. (2008), Molecular characteristics for solid-state limited solubility, *J. Med. Chem.*, 51 (10), 3035–3039. doi:10.1021/jm701587d

Yotsuyanagi, T., Higuchi, W.I., and Ghanem, A.H. (1973), Theoretical treatment of diffusional transport into and through an oil-water emulsion with an interfacial barrier at the oil-water interface, *J. Pharm. Sci.*, 62, 40–43. doi:10.1002/jps.2600620106

CHAPTER 2

FORMULATION DEVELOPMENT OF OIL-IN-WATER NANOSIZED EMULSIONS

SYED NAZRIN RUHINA RAHMAN* AND TAMILVANAN SHUNMUGAPERUMAL

Department of Pharmaceutics, National Institute of Pharmaceutical Education and Research (NIPER), Guwahati, Assam, India

*The funding provided by Department of Biotechnology, Govt. of India to Syed Nazrin Ruhina Rahman is greatly acknowledged.

Oil-in-Water Nanosized Emulsions for Drug Delivery and Targeting,
First Edition. Tamilvanan Shunmugaperumal.
© 2021 John Wiley & Sons, Inc. Published 2021 by John Wiley & Sons, Inc.

EXPANSION OF ABBREVIATIONS

apo	apolipoprotein
AI	artificial intelligence
API	active pharmaceutical ingredient
CCD	central composite design
CKC	cetalkonium chloride
CMAs	critical material attributes
CMC	chemistry, manufacturing, and control
CPPs	critical process parameters
CQAs	critical quality attributes
CTAB	cetyltrimethyl ammonium bromide
DBP	di-butyl-phthalate
2DEHPA	bis(2-ethylhexyl) hydrogen phosphate
DMPC	dimyristoylphosphatidylcholine
DMPE	dimyristoylphosphatidylethanolamine
DOTAP	1,2-dioleoyl-sn-glycero-3-trimethylammoniumpropane
2D PAGE	two-dimensional polyacrylamide gel electrophoresis
DPPC	dipalmitoylphosphatidylcholine
ELM	emulsion liquid membranes
FbD	Formulation by Design
FMEA	failure mode effect analysis
HLB	hydrophilic–lipophilic balance
IC	impression cytology
ICH	International Council for Harmonisation
IVCM	*in vivo* confocal microscopy
MA	material attribute
MCT	medium-chain triglycerides
ML	machine learning

MPS	mean particle size
NCE	new chemical entity
$NPEO_{10}$	nonylphenol-poly (ethylene oxide)
OFAT	one-factor-at-a-time
O/W	oil-in-water
PBCA	poly(*n*-butylcyanoacrylate)
PC	phosphatidyl choline
6-PC	*n*-hexanoyl lysolecithin
PDI	polydispersity index
PEO	polyoxyethylene
PFOB	perfluorooctyl bromide
PP	processing parameter
PPO	polyoxypropylene
TPGS	tocopheryl polyethylene glycol 1000 succinate
QAC	quaternary ammonium compound
QbD	Quality by Design
QbT	Quality by Testing
QRM	quality risk management
QTPP	quality target product profile
REM	risk estimation matrix
RPN	risk priority number
SDS	sodium dodecyl sulfate
ZP	zeta potential

2.1. INTRODUCTION

Therapeutically, the oil-in-water (o/w) nanosized emulsions are used mainly as delivery carriers for lipophilic active pharmaceutical ingredient (API) molecules that show pharmacological activities after administration via parenteral, ocular, and transdermal routes. Furthermore, the o/w nanosized emulsions having anionic, cationic, or neutral charged dispersed oil droplets can be made especially by changing the emulsifiers so that the first step of engineered droplet surfaces could be obtained to extract multifunctional activities. The second step of engineered droplet surfaces in emulsions usually attains by decorating the droplet surface with anchoring or homing moiety either by conjugation or simple adsorption reaction. By combining both the surface charge optimization and engineered droplet surfaces, the o/w nanosized emulsions are indeed in recent years useful for API delivery and/or targeting to otherwise inaccessible internal organs of the human body (Tamilvanan 2009).

In contrast to microemulsions, the o/w nanosized emulsions are thermodynamically instable, which can significantly reduce their applicability. As discussed in Chapter 1, the thermodynamic instability behavior of emulsions includes phenomena such as flocculation, coalescence, creaming, and Ostwald ripening. The physical instability of emulsions is due to the spontaneous tendency toward a minimal interfacial area and hence a minimal surface free energy between the dispersed oil phase and the aqueous continuous/dispersion medium. Minimizing the interfacial area is mainly achieved by two mechanisms: first, coagulation possibly followed by coalescence and second, by Ostwald ripening. Coalescence is often considered as the most important destabilization mechanism leading to coursing of dispersions and can be prevented by a careful choice of stabilizers or emulsifiers. The molecular diffusion of solubilizate (Ostwald ripening), however, will occur as soon as curved interfaces are present. During Ostwald ripening, the molecular diffusion of the lipophilic components occurs from small particles to larger particles due to the difference in escaping tendency (vapor pressure) between the two. Ostwald ripening is thus directly proportional to the solubility of the lipophilic component in the dispersion medium as well as to the particle size distribution. Various physical factors such as density difference between the dispersion medium and dispersed oil droplets, strong hydrodynamic agitation, interdroplet interaction, and the droplet interfacial viscoelasticity can affect or perturb the colloidal stability of emulsion droplets and thus their shelf life and therapeutic utility. Therefore, the stabilization of emulsion droplets is in many cases very important and desirable. There are different ways of stabilization of the o/w nanosized emulsions, all of which are related to the modification of the interface between the dispersion medium and emulsified oil droplets. Depending on the intended medical/therapeutic application of nanosized emulsions, various kinds of emulsifier molecules ranging from small surfactants or surface-active polymers to polylayered interfacial coatings produced by multicomponent emulsifier films are considered.

This chapter initially starts with the different excipients or ingredients used in the emulsion preparation followed by a short overview on the lipophilic APIs' incorporation pattern into the o/w nanosized emulsions. One more important section included in this chapter is how to optimize a formula for ensuring a quality emulsion formulation. This section introduces a case study that shows the Quality by Design (QbD) approach applied onto the emulsions to optimize a formula during preformulation studies. The effect of the amount of new chemical entity (NCE) or lipophilic API on the physical stability of emulsions, quantity of excipients (oils, emulsifiers, and other excipients) onto the API incorporation patterns, final particle size distribution of dispersed oil droplets of the emulsion on different storage temperature conditions, etc., will be the subject of interest to be discussed in this chapter. It is to be informed to readers that the discussion of majority of the traditional excipients (oils, emulsifiers, and other excipients) can also be found from Chapter 4 wherein the classification of emulsions based on their generation appears.

2.2. FDA-APPROVED OILS, EMULSIFIERS, AND AUXILIARY OR MISCELLANEOUS EXCIPIENTS

A comprehensive and non-exhaustive list of different excipients used to make the o/w nanosized emulsions are shown in Table 2.1. Interestingly, the conventional or traditional excipients such as oils and emulsifiers are divided into two categories. The oils are classified according to their origin or source, i.e., animal, mineral, and plant. Based on the charges, the emulsifiers are majorly grouped as amphoteric, anionic, cationic, neutral, and nonionic. The emulsifiers are again separated into two groups depending on the specific activity they produce due to their charge presence. In one group, the amphoteric, anionic, neutral, and

TABLE 2.1. Comprehensive and Non-exhaustive List of Different Conventional or Traditional Excipients Used to Make the O/W Nanosized Emulsions

Excipients	Selected Examples
Oils	*Animal origin:* Lanolin, squalene (shark liver oil) *Mineral origin:* Paraffin light, paraffin oil, silicone oil, vaseline *Plant origin:* Arachis oil, castor oil, corn oil, glycerol monostearate, medium-chain monoglycerides, medium-chain triglycerides, olive oil, sesame oil, soyabean oil, etc.
Emulsifiers: Amphoteric, anionic, neutral and nonionic	*Oil soluble:* Cholesterol, cremophor RH, phospholipids (lipoid E 80) including phosphatidylcholine, phosphatidylethanolamine, sphingomyelin, etc. *Water soluble:* Miranol C 2 M (disodium cocoamphodiacetate) Miranol MHT [(1*H*-imidazolium, 4,5-dihydro-1-(carboxymethyl)-1-(2-hydroxyethyl)-2-undecyl-, hydrogen sulfate (salt), monosodium salt)] Poloxamer [(poly(ethylene glycol)-block-poly(propylene glycol)-block-poly(ethylene glycol)] 188 and 407 Polysorbate/Tween 20 {2-[2-[3,4-bis(2-hydroxyethoxy) oxolan-2-yl]-2-(2-hydroxyethoxy)ethoxy]ethyl dodecanoate} Polysorbate/Tween 80 {2-[2-[3,5-bis(2-hydroxyethoxy) oxolan-2-yl]-2-(2-hydroxyethoxy)ethoxy]ethyl (*E*)-octadec-9-enoate} Transcutol P (diethylene glycol monoethyl ether) Tyloxapol [4-(1,1,3,3-tetramethylbutyl)phenol polymer with formaldehyde and oxirane] TPGS (tocopheryl polyethylene glycol succinate)
Emulsifiers: Cationic	*Lipid:* DMPE, DOTAP, oleylamine, stearylamine *Polysaccharide:* Chitosan
Miscellaneous	α-Tocopherol, EDTA, glycerin, methylparaben, propylparaben, sorbitol, thiomersal, xylitol

nonionic emulsifiers are gathered while the cationic emulsifier is kept alone in other group. All the emulsifiers without the cationic emulsifier is again subgrouped based on their solubility, whether lipid/oil or water, and hence the emulsifiers are grouped into two categories, i.e., oil soluble and water soluble. The cationic emulsifier is classified into two based on its chemical nature, i.e., carbohydrates or lipids, and hence the cationic polysaccharides and cationic lipids.

2.2.1. Issues Related to Oil Selection to Make the O/W Nanosized Emulsions for Medical Application

In general, the o/w nanosized emulsion should be formulated with compatible vehicles and additives. The components of internal (dispersed oil droplets) and external (aqueous continuous medium) phases of nanosized emulsions should be chosen to confer enhanced solubility and stability to the incorporated therapeutically active lipophilic API. In principle, the function of selected excipients should not only be limited to improve the overall physical stability of the emulsions or enhance the API's solubility but also they should contribute to influence the biofate or therapeutic index of the incorporated API after administration via parenteral, ocular, percutaneous, and nasal routes. The general considerations concerning excipient selection and optimum concentrations mainly in relation to the oils, emulsifiers, and miscellaneous excipients are presented comprehensively below.

At the time of preformulation studies, the solubility data of APIs need to be generated in the different oil selected, that too either alone or in combination of one oil with other oils in a definite ratio. Two more important points should be kept in mind before selecting the oil or oil combination. First point is compatibility of the oil or oil combination not only with other formulation excipients but also with the site of applications, i.e., ocular/skin tissues. Second point is related to auto- or self-oxidation potential of oil or oil combination during and after emulsion manufacturing, because oils are triglycerides and are prone to auto- or self-oxidation over the emulsion's processing and storage time periods. To minimize or to eliminate some extent the auto- or self-oxidation, the additional excipient recommended to include in the emulsion formula is antioxidants, especially α-tocopherol in the concentration range from 0.001 to 0.002% w/w.

The final oil-phase concentration in emulsions meant for ocular use is now widely accepted to be at or below 5% w/w taking into account that the emulsion must be kept in a low-viscosity range of between 2 and 3 centipoises, which also is the optimal viscosity for ocular preparations (Lee and Robinson 1986). This viscosity range may be suitable to take the emulsion into the syringe (i.e., syringeability) for parenteral application into human body. However, for all other medical uses, the amount of oil may be varied but generally is within 5–20% w/w. Sometimes, a mixture of oils may be employed to facilitate API solubilization in the oil phase. Jumaa and Müller (1998, 1999) reported the effect of mixing castor oil with medium-chain triglycerides (MCT) on the viscosity of castor oil. Mixing of castor oil with MCT at a ratio of 1 : 1 (w/w) led to a decrease in the viscosity of castor oil and simultaneously to a decrease in the interfacial tension of the oil phase. The presence of free fatty acids in the castor oil (may be as an impurity) can act as coemulsifiers to lower the interfacial tension between dispersed oil droplets and continuous aqueous medium and subsequently to produce a more stable emulsion formulation in comparison with the other oil phases. In addition to the digestible oils from the family of triglycerides, including soybean oil, sesame seed oil, cottonseed oil, and safflower oil, which are routinely used for making medical emulsions, alternative biocompatible ingredients such as α-tocopherol and/or other tocols were also investigated for API delivery purposes via o/w emulsions (Constantinides et al. 2004, 2006). But the emulsions formed from tocols are often considered as microemulsion systems with few exceptions (Constantinides et al. 2004, 2006). Very recently, it was shown that playing with oil combinations could generate the dispersed oil droplets with bi-compartmental structure possessing different polarity and thus paving the strategy of dual API loading in the o/w macro- and nano-sized emulsions (Puri et al. 2019). The details of this type of emulsions are discussed in Chapter 7.

2.2.2. Issues Related to Emulsifier Selection to Stabilize the O/W Nanosized Emulsions for Medical Application

Not only the chemical nature of emulsifiers but also their concentration used determine the type of emulsion produced. For example, the spontaneously forming thermodynamically stable microemulsion systems require a high emulsifier/surfactant concentration [usually at 20% and higher (w/w)] along with an alkanol component. But the kinetically stable nanosized emulsions can be prepared by using relatively lower surfactant concentrations. For example, a 20% o/w nanosized emulsion may only require a surfactant concentration of 1–5%. The kinetic stability of the nanosized emulsions can be achieved by creating a barrier at the oil–water interface, protecting the emulsion from breakage (Capek 2004). These barriers may be of electrostatic or steric nature and prevent emulsion droplets from direct contact. The most common way to stabilize emulsions is by surfactant adsorbed at the interface between the dispersed oil droplets and continuous aqueous dispersion medium. Surfactant adsorption layers do not only reduce the interfacial tension but can also provide an electrical charge to the emulsion droplets (ionic surfactants) or create the strong steric barrier via bulky molecular groups directed toward the dispersion medium (nonionic surfactants).

Traditionally, lecithins or phospholipids are the emulsifiers of choice to produce o/w nanosized emulsions, because the phospholipid emulsifier molecule structure is more or less similar to the endogenous phospholipids, which build the cells and/tissues. However, additional emulsifiers preferably dissolved in the aqueous phase are usually included in the emulsion composition. A typical example of the aqueous soluble emulsifiers are nonionic surfactants (e.g., Tween 20), which are preferred because they are less irritant than their ionic counterparts. The nonionic block copolymer of polyoxyethylene-polyoxypropylene (PEO-PPO), Pluronics F68 (Poloxamer 188) is included to stabilize the emulsion through strong steric repulsion. However, surfactants such as Miranol MHT (lauroamphodiacetate and sodium tridecethsulfate) and Miranol C_2M (cocoamphodiacetate) were also used in earlier ophthalmic emulsions (Muchtar and Benita 1994). It should be added that commercially available cyclosporin A-loaded anionic emulsion (Restasis®) contains only polysorbate 80 and carbomer 1,342 at alkaline pH to stabilize the anionic emulsion. To prepare a cationic emulsion, cationic lipids (stearyl-and oleyl-amines) or polysaccharides (chitosan) are added to the formulation. Strikingly, a stable emulsion prepared based on chitosan–lecithin combination was also reported (Ogawa et al. 2003). Conversely, a cationic emulsion based on an association of poloxamer 188 and chitosan without the incorporation of lecithin was prepared and also demonstrated adequate stability (Calvo et al. 1997; Jumaa and Müller 1999). Similarly, a report from our group also indicated the stability of oil droplets through the cation conferring chitosan along with poloxamer 188 as a mixed emulsifier (Tamilvanan et al. 2010). Since the free fatty acid generating phospholipid emulsifier molecule is omitted from the nanosized emulsion system, the stable nanosized emulsion produced from chitosan–poloxamer emulsifier combination would significantly reduce the generation of microclimate acidic pH in the vicinity of oil phase, oil–water interface, and water phase of the emulsion (Tamilvanan et al. 2010). These non-phospholipid-based emulsions should therefore pave the way to incorporate the acid-labile molecules like therapeutic peptides and proteins, and to delineate the scope of applying lyophilization process for the development of a solid or dry emulsion. With the addition of suitable cryo- or lyo-protectant at optimum concentration, the preparation of lyophilized solid dry-powder form of non-phospholipid-based o/w nanosized emulsions is possible in recent years. Figure 2.1 shows the o/w nanosized emulsions in liquid form (before lyophilization) and solid-dry powder form after the addition of different cryo- or lyo-protectant molecules.

Oil-in-water emulsion compositions based on α-tocopherol (or α-tocopherol derivative) as the disperse phase has been described in a patent granted to Dumex (Sonne 2015). Interestingly, the emulsifying agent used to make tocol-based emulsions are restricted to vitamin E TPGS (D-alpha-tocopheryl polyethylene glycol 1000 succinate) taking into consideration of toxicological issues. According to a patent by Nakajima et al. (2003), functional emulsions for use in food, APIs, and cosmetics were reported. These emulsions were stabilized with various span products such as Span 80, Span 40, etc. There is a group of emulsions that were not prepared by using the traditional anionic, cationic, and nonionic surfactant molecules. These emulsions do contain the oil

Figure 2.1. Freeze-dried emulsions using different cryo- or lyo-protectants and reconstitution of freeze-dried powder into nanosized emulsion. (See color insert.)

or oil combination and therefore they can be termed as "surfactant-free emulsions." Another group of colloidal dispersions whose final appearance is white similar to the traditional emulsions but these dispersions do not contain both surfactants and oil or oil combination. Taking the physical appearance (white color) into consideration, these colloidal dispersions get the term "surfactant- and oil-free emulsions." Both of these two emulsions (surfactant-free and surfactant- and oil-free) are briefly discussed in Chapter 7.

2.2.3. Importance of Charge-Stabilized Nanosized Emulsions

At present, emulsions stabilized by positively charged, cationic surfactants are most often used as colloidal API carriers (Tamilvanan 2004). Kim et al. (2005) used an emulsion of squalene in water stabilized by the cationic surfactant 1,2-dioleoyl-sn-glycero-3-trimethylammoniumpropane (DOTAP), which facilitated gene transfer in biological fluid even in the presence of 90% serum in the dispersion medium. The emulsion droplets play the role of mucosal gene carriers and can form stable complexes with DNAs. Here, the DNA was incorporated at the end of emulsion preparation by the *de novo* method. Compared with liposomal carriers, cationic emulsions demonstrated a 200-fold increase in transfectional efficacy in both lungs and tissues (Kim et al. 2003, 2005). The nature of oil selected as the dispersed phase is another important factor that can affect the applicability of such emulsions for transfection. Three different oils were used for the disperse phase: soybean oil, linseed oil, and squalene (Kim et al. 2003). The transfection activities of the nanosized emulsion carriers in the presence of serum followed the order squalene > soybean oil > linseed oil, and the squalene emulsions were also most stable. From these data, the authors concluded that stability of a carrier system is a necessary requirement to form stable complexes with DNA, and this stability determines the *in vivo* transfection.

It is known from the literatures that the interaction between cationic liposomes and polyanionic macromolecules like DNA is dependent on ± ratio, and at the ratio of maximum transfection there is a major aggregation leading to destabilization of formulation or desorption of DNA from the formulation (Liu et al. 1997). Furthermore, Simberg et al. (2003) suggest that an understanding of the interplay between lipoplex composition, its interaction with serum, hemodynamics, and target tissue properties (susceptibility to transfection) could explain the biodistribution and efficient *in vivo* transfection following intravenous administration of cationic lipid-DNA complexes (lipoplexes) into mouse. However, it is interesting to see what could happen when the cationic nanosized emulsion is applied to *in vitro* cell culture models in the presence of serum. The serum stability of emulsion/DNA complex was reported (Yi et al. 2000). Further studies are, however, necessary to be carried out to understand clearly the origin of the serum stability of this emulsion. In addition, the transfection efficiency of this emulsion was not affected by time up to 2 h post-emulsion/DNA complex formation. This means that the o/w cationic nanosized emulsion allows the experimenter to have a wider time window to work within during transfection study.

The o/w nanosized emulsions stabilized by both cationic and anionic lipidic emulsifiers were investigated in order to compare the degree of binding and uptake by specific cells that over-expressed tumor receptors (Goldstein et al. 2007a). Immunoemulsions were prepared by conjugating an antibody to the surfactant molecule via a hydrophobic linker and then the antibody-conjugated surfactant was used to make the emulsion by the *de novo* method. The anionic stabilized emulsions showed decreased stability leading to phase separation after 20 days of storage. The reduced stability of anionic immunoemulsion could be attributed to the rapid decrease of the zeta-potential caused by the positively charged conjugated antibody and consequently, due to a lower electrostatic repulsion between the colloidal droplets (Goldstein et al. 2007b). On the other hand, immunoemulsions stabilized by both anionic and cationic emulsifiers exhibited a multifold increase in cell binding in contrast to the emulsions without antibodies.

The cationic o/w nanosized emulsions were also found to be effective vehicles to improve the skin permeability of incorporated lipophilic molecules in dermatological applications (Yilmaz and Borchert 2005). Because epithelial cells of the skin carry a negative surface charge, they show high selectivity and permeability to positively charged solutes. Thus, positively charged nanosized emulsions are promising systems for enhancing the skin permeability for APIs included in the colloidal droplets. The authors also showed that ceramides could be successfully delivered in a transdermal route by means of nanosized emulsions stabilized by a positively charged interfacial layer of the naturally occurring molecule, phytosphingosine. Other applications of nanosized emulsions as carriers, stabilized by ionic surfactants, in the pharmaceutical and cosmetic fields have been briefly reviewed by Solans et al. (2005) and Tamilvanan (2008).

Anionic phospholipids are also commonly utilized for the stabilization of API-carrying nanosized emulsion droplets both individually and in binary mixtures (Trotta et al. 2002). Soybean lecithin and modified phospholipid, *n*-hexanoyl lysolecithin (6-PC), alone and as 1 : 1 mixtures were used as stabilizers of MCT droplets in water (Trotta et al. 2002). Although individual uncharged phospholipids provide emulsion droplets, a moderate negative charge for stabilization, mixed phospholipids produce much more stable emulsions and a large negative zeta-potential value. A possible explanation for this phenomenon is related to the increased incorporation of polar compounds from the soya lecithin into the mixed interfacial film when 6-PC is present. This interfacial film acts as a stabilizer by forming a high energy barrier that repels adjacent droplets and leads to the formation of stabilized emulsified droplets. The stability of the emulsion did not noticeably change, even in the presence of the model destabilizing API, indomethacin, demonstrating the high potential for such mixed emulsifiers for the formulation of colloidal API delivery systems (Trotta et al. 2002). Lysolecithin has one fatty acid ester chain removed from the glycerol backbone, in addition, lysolecithin is toxic (destroys RBC cell membranes). Furthermore, although the role of phospholipids is essential for the stability of the emulsions, possible cataractogenic effects due to the phosphatidyl choline (PC) and, basically, to a derivative of the same, lysophosphatidyl, have been described by different authors (Cotlier et al. 1975; Kador and Kinoshita 1978).

A new class of surface-active dialkyl maleates can be utilized for emulsion polymerization (Abele et al. 1997). Here, the emulsion droplets of monomeric maleates are self-stabilized and simultaneously serve as liquid "reactive storage carriers." Three types of head group in the dialkyl maleates were studied—nonionic, cationic, and zwitterionic with different lengths of hydrophobic alkyl chain. Cationic and zwitterionic dialkyl maleates with the longest alkyl chains -$C_{16}H_{33}$ and $C_{17}H_{35}$ provided the best stability for o/w nanosized emulsions. When compared with the data obtained for the well-known nonionic surfactant nonylphenol-poly (ethylene oxide) ($NPEO_{10}$) and the cationic cetyltrimethyl ammonium bromide (CTAB), an excellent stabilizing capacity especially for the cationic maleates can be stated. Whereas nonionic dialkyl maleates show almost the same emulsifying ability and stability as $NPEO_{10}$, the cationic derivatives of these novel surfactants are more effective in stabilization than the traditional CTAB.

Sometimes anionic surfactants are especially added to emulsion droplets for the stabilization of "reactive storage carriers" subjected to further chemical transformation. Sodium dodecyl sulfate (SDS) was utilized to stabilize miniemulsion droplets, which in the subsequent step, were polymerized and formed poly(*n*-butylcyanoacrylate) (PBCA) nanoparticles, suitable for targeting API delivery to specific cells (Weiss et al. 2007). It is worth mentioning that SDS is predominantly used to achieve required miniemulsion stability (Landfester 2006). In some cases, however, cationic surfactants are also used in miniemulsion formulations, which were reported first in the late seventies of the 20th century. In general, however, stability of miniemulsions does not depend on the sign of the surfactant charge and is mainly determined by the surfactant coverage of the reactive carriers (miniemulsion droplets). The same factor is also crucial for the size of miniemulsion droplets after steady-state miniemulsions are obtained (Landfester 2006).

2.2.4. Importance of Neutral-Charged (Sterically Stabilized) Nanosized Emulsions

In many cases, however, greater emulsion stability can be achieved without imparting a significant surface charge to the emulsion droplets, by means of steric stabilization (Capek 2004). Nonionic surfactants possessing bulky hydrophilic groups like PEO protruding into the dispersion media decrease coalescence arising from droplet collisions. Another contribution to the steric stabilization of emulsions by nonionic surfactants is provided by the close packing of PEO chains at the droplet surface. The compact packing of PEO chains at the droplet surface creates steric stabilization because little or no interpenetration of PEO chains on different droplet surfaces occurs due to entropic repulsion (Dale et al. 2006). Large head groups carrying simultaneously charges of opposite signs, such as in zwitterionic surfactants, can cause similar effects. In polar dispersion media of low-to-medium ionic strength, these groups are, as a rule, strongly solvated (hydrated in the most common case of H_2O) (Yaseen et al. 2006). Voluminous and on an average almost non-charged hydration shells, surrounding the emulsion droplet, possess a significant steric rigidity and can also effectively stabilize emulsions. There are, however, only a few examples in the literature that use zwitterionic surfactants as effective emulsion stabilizers. For example, lecithin was used for the stabilization of perfluorooctyl bromide (PFOB) in water emulsions, to be used as oxygen-carrying system in a bio-artificial liver device (Moolman et al. 2004). The Sauter mean diameter of 0.2 µm PFOB emulsion droplet in water was obtained by high-pressure homogenization. The emulsion was stable for several months even at a volume fraction of 20%. Nonionic surfactants are more often used for emulsion stabilization than zwitterionic phospholipids because they are synthetically manufactured, can be well defined analytically, and have significantly less batch-to-batch variation than naturally occurring (egg yolk, soybean) lecithins.

The nonionic surfactant Span-83 was used for stabilizing water droplets in oil to form a reactive storage carrier for the synthesis of calcium carbonate nanoparticles by means of a two membrane system (Hu et al. 2004). Firstly, an aqueous emulsion was prepared in kerosene stabilized by 0.02596 wt% Span and containing CO_3^{-2} ions in the droplets of the dispersed phase. The oil phase contained also a 0.02792 M solution of bis(2-ethylhexyl) hydrogen phosphate (2DEHPA), a well-known molecular carrier for the transportation of metal ions across emulsion liquid membranes (ELM). In the second stage, a $CaCl_2$ aqueous solution-filled dialysis tube was placed into the o/w emulsion and due to the reaction between CO_3^{-2} and Ca^{2+} ions in the aqueous droplets, $CaCO_3$ nanoparticles were obtained. Similarly, ZnS nanoparticles were prepared in inverse water-oil-emulsion (Naskar et al. 2006). The stabilization of emulsions was provided by the addition of 5 wt% of Span 80 or Span 20, respectively, to the oil phase (cyclohexane). The dispersed phase contained a mixture of zinc acetate and thioacetamide, which react upon heating to form ZnS. The authors demonstrated that for the preparation of ZnS nanoparticles, the use of Span 20 was more favorable because of the smaller emulsion droplet size and therefore lower and more homogeneous size of the final particles. Another advantage was the higher stability of Span 20 against hydrolysis as compared to Span 80.

In general, fulfilling both stabilization mechanisms (smaller droplet size and lesser susceptibility of surfactant toward chemical degradation) simultaneously leads not only to the highest emulsion stability but also to lesser sensitivity to changes in the external conditions such as pH, ionic strength, and temperature. Therefore, the use of mixtures of different classes of surfactants for emulsion stabilization is frequently the most effective solution in many practical cases.

2.2.5. Advantages of Nanosized Emulsions Stabilized by Mixed or Multicomponent Emulsifier Molecules

Sometimes mixtures of natural zwitterionic surfactants used for emulsion stabilization contain small amounts of polar compounds that can be incorporated into the adsorption layer and lead to a modest droplet charge, which additionally stabilizes the emulsion (Tamilvanan 2008). Surface layer with charged natural admixtures reported by Trotta et al. (2002) is only a particular case of a very large class of emulsion-stabilizing systems based on a tailored application of ionic–zwitterionic surfactant mixtures. Mixtures of dipalmitoylphosphatidylcholine (DPPC) and homologues and dimyristoylphosphatidylethanolamine (DMPE) phospholipids were utilized by Ishii and Nii (2005) for stabilizing model API-carrying o/w nanosized emulsions. In contrast to the data, the main stability factor was found to be the optimal average hydrophilic–lipophilic balance (HLB) value of the stabilizers' mixture, defined similarly for nonionic surfactants (Trotta et al. 2002). For example, emulsions prepared with mixtures of dimyristoylphosphatidylcholine (DMPC, zwitterionic) and DMPE behaved similarly to emulsions prepared by DMPC alone. This fact was explained by the equivalence of HLB values for both surfactants used, regardless of their ionic nature. However, the ionic character of a surfactant like DMPE (and therefore the charge of respective emulsion droplets) can be affected by the pH of the dispersion medium.

The o/w nanosized emulsions stabilized by mixed ionic/nonionic surfactants revealed very high physical stability and were found to be most appropriate for dermatological applications as a ceramide-carrying colloidal system (Yilmaz and Borchert 2005). Greater emulsion stability was achieved by the combination of the nonionic steric stabilizer Tween 80 and the phospholipid co-stabilizers phytosphingosine and phosphatidylethanolamine.

2.2.6. "Stealth" Property of Nanosized Emulsions: In Vitro Demonstrations

Anionic emulsion formulations capture apolipoproteins (apo) along with other plasma proteins within minutes after an infusion in human blood, facilitating their fast elimination. In contrast, cationic emulsions reveal a much longer retention time in the plasma. Moreover, cationic colloidal carriers can promote the penetration of therapeutic agents into cell surfaces possibly via an endocytotic mechanism (Calvo et al. 1997). To improve the API targeting efficacy of colloidal carriers of anionic emulsions and to further prolong the circulating effect of the cationic emulsions, a mixed stabilizer film at the oil–water droplet interface composed of nonionic Poloxamer 188 and ionic lipoid E80 and stearylamine/oleylamine was created by combining the effects of electrostatic and steric barriers at the oil–water interface (Tamilvanan et al. 2005). In order to prove this concept, surface (charge)-modified o/w nanosized emulsions (cationic and anionic) were prepared following the well-established combined emulsification techniques (*de novo*) and these two emulsions were characterized for their droplet size distribution and surface charge. Marketed lipofundin MCT 10%, deoxycholic acid-based anionic emulsions, oleylamine-/stearylamine-based cationic emulsions, and oleic acid-based anionic emulsions were selected in this study. The effect of these emulsions on *in vitro* adsorption of plasma proteins was investigated by means of two-dimensional polyacrylamide gel electrophoresis (2D PAGE).

Competition between steric repulsion by poloxamer and electrostatic attraction by ionic components led to the sensitive adsorption of small molecular weight proteins like apo, albumins by the surface of colloidal carriers whereas all emulsion droplets were effectively shielded from the adsorption of larger proteins like immunoglobulins, fibrinogen, etc., enhancing the shelf life of emulsion formulations in the blood (Fig. 2.2). The apoA-I along with apoA-IV have been suggested to modulate the distribution of apoE between the different lipoprotein particles in the blood and thereby affect their clearance (Tamilvanan et al. 2005). In addition, the attachment of apoE would greatly alter the *in vivo* distribution of fat emulsions since this protein is a ligand for the apoE-specific receptors on the liver parenchymal cells. The higher the preferential adsorption of apoA-I onto the cationic emulsion droplets, the more intensified the displacement/redistribution of apoE would, therefore, be expected to occur on these types of cationic emulsion formulations in the blood (Tamilvanan et al. 2005). Indeed, the ratio of apoA-I to apoA-IV was very close to 1 for

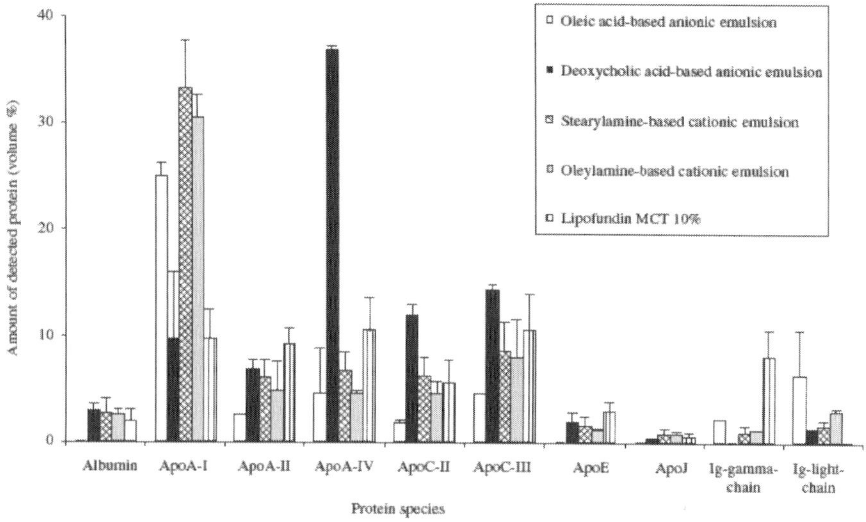

Figure 2.2. Amount of major proteins on the 2-D gels of plasma proteins adsorbed on emulsions with negative or positive surface charge in comparison with Lipofundin MCT 10%. [Taken with permission from Elsevier (Tamilvanan et al. 2005).] (See color insert.)

Lipofundin® MCT 10% whereas it was about 0.26 for deoxycholic acid-based anionic emulsion and above 5 for oleic acid-based anionic emulsion and both cationic emulsions. This indicates that emulsions having similar surface/interfacial charge imparted by different anion-forming stabilizers (oleic or deoxycholic acids) exhibited markedly different protein adsorption patterns.

2.2.7. Advantages of Stabilizers in Nanosized Emulsions

In some practical cases where emulsions are applied under real, sometimes quite harsh conditions, such as high temperature or shear stresses, the stabilization of colloidal carriers by conventional surfactants (ionic as well as nonionic) can appear to be insufficient to keep the initially acceptable emulsion properties intact. For ionic surfactants as stabilizers, the stabilization mechanism based on the electrical double layer fails at high amounts of electrolyte. The situation for nonionic surfactants as stabilizers is only slightly better because their molecules are typically not strongly adsorbed (Tadros et al. 2004; Tadros 2006). Therefore, the most effective way to stabilize emulsions by creation of a protective adsorption layer is the use of amphiphilic macromolecular compounds or polymeric surfactants (Tadros et al. 2004). In contrast to the commonly used monomolecular surfactants, polymeric surfactants do not only adsorb much stronger but also retain this ability under high electrolyte concentration or/and high temperature (Tadros et al. 2004; Tadros 2006). This is made possible due to a special molecular design of surface active polymers that have in their structure anchor groups responsible for strong adsorption at the interface and stabilizing groups protruding from the interface into the dispersion medium and forming a bulky layer with thicknesses of several nanometers.

Most often used stabilizers for the preparation of emulsions, in the fields of agrochemicals, pharmaceuticals, and personal care products, are either block or graft copolymers. In block copolymers, the hydrophobic blocks reside at the surface or even partly penetrate in the oil droplet, making trains or short loops whereas the hydrophilic blocks protrude in the dispersion medium as loops or tails providing steric stabilization (Benichou et al. 2004). As examples, PEO-PPO-PEO triblock copolymer (commercially available as "Pluronics") or PPO-PEO-PPO can be mentioned. Triblock copolymers are, however, not the most efficient stabilizers because the PPO chain is not hydrophobic enough to attach strongly at the o/w interface (Benichou et al. 2004). The surface activity of these polymeric surfactants is rather the result of a rejective anchoring or negative enthalpic energy change of the PPO group because of its low solubility in water and most oils. Alternative and more efficient graft copolymers consist of a polymeric backbone attached to the interface and several chains dangling into the continuous phase and forming at the interface a "brush" structure.

A typical example of commercial graft was described (Jumaa and Müller 2002). Here, mixtures of polyoxyethylene-660-12-hydroxystearate (Solutol HS15) with the anionic lipid composition. Lipoid S75 was employed to enhance the long term as well as accelerated (by freezing and centrifugation) stability of o/w nanosized emulsions. Emulsion stabilized by phospholipids displayed a stable behavior after autoclaving and centrifugation but de-emulsified after freezing. In contrast, emulsions prepared only with Solutol HS15 demonstrated a significant change in particle size after autoclaving. The best results were obtained using a stabilizer mixture revealing a combination of electrostatic stabilization mechanism typical for the anionic phospholipids and the steric stabilization mechanism originating from nonionic polymeric surfactant. The combination of stabilization mechanisms improved the emulsion's stability, compared to the emulsion's stability prepared using only the individual surfactants.

The commercially available cationic block-copolymer Eudragit E100 was utilized as both emulsion stabilizer and solidifying agent upon further drying (Cui et al. 2007). Due to the specific properties of Eudragit E100, no surfactant or organic solvent additives were employed in order to fulfill common ecological, toxicological, and manufacturing safety requirements during the preparation of a redispersible dry emulsion.

The naturally occurring polymers cyclodextrins are commonly used in pharmaceutical aqueous formulations for inclusion of less soluble or instable APIs. Their use, however, is complicated by poor interaction with biological membranes. To overcome this difficulty, the cyclodextrins were chemically modified, including grafting of fatty acid chains to the hydroxyl groups in order to provide the optimum surface activity to the polymer (Memisoglu et al. 2002). Upon addition of Miglyol to an aqueous phase, nanocapsules with an oily interior were formed spontaneously. These nanocapsules had an average size of approximately 300 nm and revealed an excellent long-term stability for at least 5 months. Interestingly, the most appropriate properties for nanocapsule formation were observed for cyclodextrin derivatives with linear C_6 chains. Compounds possessing longer aliphatic chains caused the formation of large polydisperse aggregates and were therefore unsuitable for encapsulation. The preparation of nanocapsules with modified cyclodextrins without the addition of surfactants is a promising tool for the development of novel API delivery systems. A similar approach of adding cyclodextrin instead of surfactant molecules to stabilize the o/w nanosized emulsions is currently receiving much attention (as briefly discussed in Chapter 7). Furthermore, a variety of current approaches to increase emulsion stability with help of emulsifier molecules have been summarized in a recent review by Grigoriev and Miller (2009).

2.2.8. Miscellaneous Additives

Additives other than antioxidants such as preservatives (e.g., benzalkonium chloride, chlorocresol, and parabens) are included in emulsions to prevent microbial spoilage of multidose medical emulsions. α-Tocopherol is a good example of an antioxidant used to obtain a desirable stabilized emulsion under prolonged storage conditions. The presence of components of natural origin such as lecithin or oils with high calorific potential renders the emulsion a good medium to promote microbial growth when it is packaged in multidose containers. Pharmaceutical products when distributed into multidose containers, especially for parenteral and ocular administrations, should be properly preserved against microbial contamination and proliferation during storage under normal conditions and proper use. Incorporation of preservatives in single-dose vials is also a common procedure if filtration is used as a sterilization method (Tamilvanan 2008). Sznitowska et al. (2002) studied the physicochemical compatibility between the lecithin-stabilized emulsion and 12 antimicrobial agents over 2 years of storage at room temperature. Preliminary physicochemical screening results indicated that the addition of chlorocresol, phenol, benzyl alcohol, thiomersal, chlorhexidine gluconate, and bronopol should be avoided due to the occurrence of an unfavorable pH change followed by coalescence of lecithin-stabilized droplets of the emulsion.

Furthermore, the efficacy of antimicrobial preservation was assessed using the challenge test according to the method described by the European Pharmacopoeia.

Despite good physicochemical compatibility, neither parabens nor benzalkonium chloride showed satisfactory antibacterial efficacy in the emulsion against the tested microorganisms and consequently were not suitable for preservation. Therefore, higher concentrations of antimicrobial agents or their combinations may be required for efficient preservation of the lecithin-stabilized emulsion probably because of unfavorable phase partitioning of the added antimicrobials within the different internal structures of the emulsion (Sznitowska et al. 2002). This finding clearly indicates that the possible electrostatic attraction between the negatively charged lipid moieties of the mixed emulsifying film formed around the anionic emulsified oil droplets (Abele et al. 1997) and the quaternary cationic ammonium groups of the preservative is not the plausible cause for the reduced activity of the benzalkonium chloride. Thus, the possible intercalation of this surfactant in either the cationic or anionic interfacial mixed emulsifying film is likely to occur, preventing benzalkonium chloride from eliciting its adequate preservative action (Sznitowska et al. 2002).

For the development of cationic emulsions in ophthalmology, the use of quaternary ammonium compound (QAC) for their cationic property rather than their preservative effect is being considered. Therefore, Liang et al. (2008) suggested the use of lipophilic cetalkonium chloride (CKC), one of the longest alkyl-chain components, as a cationic agent in ophthalmic emulsions. With a highly lipophilic QAC, the distribution between the oil and aqueous phases of the emulsion is modified because of the affinity toward the oil phase, further favoring the cationic agent role over the preservative role. Using *in vivo* confocal microscopy (IVCM) for *in vivo* tissues images and impression cytology (IC) for *ex vivo* epithelium inflammatory marker expression in correlation with standard immunohistology for deep infiltration and apoptosis, the same author assessed the toxicological effects of benzalkonium chloride/CKC emulsion/solution formulations on the ocular surface of rabbits following multiple topical applications (Liang et al. 2008). These *in vivo* and *ex vivo* experimental approaches demonstrated that ocular surface toxicity was reduced by using an emulsion instead of a traditional solution and that a CKC emulsion was safe for future ocular administration.

Overall, it is preferable to formulate nanosized emulsions devoid of preservative agents and fill it in sterile single-dose packaging units (manufactured using heat blow-fill-seal technology) to prevent potential contamination due to repeated use of multidose packaging. It should be pointed out that the two available ocular emulsion products (Refresh Endura® and Restasis®, Allergan, Irvine, CA, USA) on the market are preservative free and packed in single-use vials only. Currently there is no commercial parenteral emulsion that contains preservatives and research concerning the problem of preservation of nanosized emulsion is very limited (Jumaa et al. 2002; Pongcharoenkiat et al. 2002; Han and Washington 2005; Swietlikowska and Sznitowska 2006).

2.3. CURRENT AND NEAR FUTURE DIRECTION

2.3.1. Colloidal Particle-Stabilized Emulsions

In addition to surfactants, polymers, and biomolecules, colloidal particles have long been recognized to stabilize droplets, e.g., as in Pickering emulsions (Pickering 1907). Examples of various colloidal particle types used to stabilize emulsions include $BaSO_4$, crystalline ferric oxide, carbon black, bentonite, kaolinite clay, latex, and silica particles. The effectiveness of emulsion stabilization by particles depends on the particle size, shape, concentration, and wettability (expressed as the contact angle at the three-phase boundary and is equivalent to the HLB of a surfactant), as well as the level of particle aggregation as controlled by particle–particle interaction (Aveyard et al. 2003). It is proposed that the conventional model for emulsion stabilization by solid particles assumes the formation of a "densely packed" layer at the oil–water interface, which prevents droplet coalescence by a steric barrier mechanism. If charged, particles may give rise to electrostatic repulsion, which further enhances emulsion stability.

Many commercial products based on emulsions (for cosmetic and food uses) include both surfactants and particles, hence the characterization and mechanisms of stabilization for emulsions with mixed interfacial layers are of great importance (Binks et al. 2007a, b). These droplet/surfactant/particle systems are also of interest in designing low-surfactant emulsion systems with enhanced stability (Midmore 1998). Tambe and Sharma (1993) reported the improvement in the stability of emulsions stabilized with calcium carbonate particles following the addition of stearic acid. This synergistic effect was attributed to the adsorption of surfactant on solid particles resulting in the change in particle wettability. A strong emulsification synergy in stabilizing o/w emulsions between hydrophilic colloidal silica (Ludox) and nonionic PEO surfactants added from the water phase has been demonstrated by Midmore (1998). The synergy was attributed to the POE micellar adsorption or POE chain bridging between silica particles, which resulted in better interfacial adsorption. Synergistic stabilization of o/w emulsions by a mixture of nonionic surfactants and hydrophilic silica nanoparticles has recently been studied by Binks et al. (2007a, b) highlighting the importance of preparation protocol. Attachment energy of a silica particle at the oil–water interface was correlated to emulsion stability. Velev et al. (1996) reported that the adsorption of lysine onto negatively charged latex particles enabled interfacial adsorption due to the reduction of hydration forces and electrostatic repulsion between droplets and particles. The influence of droplet–particle electrostatic interaction on emulsion stability was studied by Binks and Catherine (2005) and Lan et al. (2007). Addition of cationic surfactant CTAB to negatively charged hydrophilic silica dispersions as the aqueous phase of emulsions resulted in enhanced stability at low CTAB concentrations, i.e., the region where CTAB is preferably adsorbed onto silica surface and promoted interfacial adsorption due to neutralization of surface charge and optimization of particle hydrophobicity. This synergistic emulsion stabilization comes from three sources: CTAB (1) hydrophobizes silica nanoparticles by charge neutralization, (2) promotes nanoparticle aggregation, and (3) reduces interfacial tension. However, when the CTAB concentration is such so that both droplets and silica are positively charged, interfacial adsorption and emulsion stability are lost and thus emulsion stability is dramatically reduced.

Simovic and Prestidge (2007) showed that nanoparticle layers significantly influence the release kinetics of a model lipophilic API [di-butyl-phthalate (DBP)] from polydimethylsiloxane o/w emulsions; either sustained or enhanced release can be achieved depending on the nanoparticle layer structure and API loading level. Nanoparticle layers can be engineered to facilitate a range of release behaviors and offer great potential in the delivery of poorly soluble APIs. Particle-stabilized emulsions have been extensively studied in terms of stabilization mechanisms (Binks and Lumsdon 2000; Binks and Catherine 2005), synergy with common emulsifiers (Lan et al. 2007; Binks et al. 2007a, b), and interfacial properties (Simovic and Prestidge 2003, 2008). However, few reports have focused on their carrier properties as dermal delivery vehicles, e.g., penetration and targeting skin layers. Recently, the influence of nanoparticle coating of submicron (nanosized) o/w emulsion droplets on the *in vitro* release and dermal delivery characteristics, with particular emphasis on potential controlled release and targeted skin delivery of all-trans-retinol, was reported by Eskandar et al. (2009). MCT o/w emulsions have been stabilized with mixed interfacial layers composed of lecithin or oleylamine and hydrophilic silica nanoparticles using a simple cold high-pressure homogenization technique. These emulsion-based hybrid API delivery systems showed improved topical delivery of all-trans-retinol; nanoparticle layers significantly improved the performance of o/w emulsions as encapsulation and delivery systems for all-trans-retinol. Therefore, emulsion-based hybrid API delivery systems should have a potential for future directions.

2.4. LIPOPHILIC API INCORPORATION PATTERN INTO NANOSIZED EMULSIONS

There are four different approaches to incorporate lipophilic APIs or heat labile molecules into the oil phase or at the o/w interface of the nanosized emulsions, namely (Tamilvanan and Benita 2004),

(A) extemporaneous API addition,
(B) de novo emulsion preparation,
(C) an interfacial incorporation approach, which includes the recently developed SolEmul® technology, and
(D) incorporation of antibodies, DNA protein, oligonucleotide, or heat labile molecules.

2.4.1. Extemporaneous API Addition

Cohen et al. (1996), when looking for a new galenic presentation form for amphotericin B with better ocular tolerance over the commercially available Fungizone® eye drops, incorporated the API directly into the preformed 20%

emulsion, Intralipid®. However, after addition of the solid API particles or API solution, several physical changes such as phase separation, precipitation, or creaming may occur thus limiting such practices in o/w nanosized emulsion preparations. Therefore, ocular active lipophilic agents are not normally incorporated into the emulsions by this extemporaneous addition method.

2.4.2. *De Novo* Emulsion Preparation

In principle, the lipophilic API molecules (thermostable) should however be incorporated by a *de novo* process as described earlier. Thus, the API is initially solubilized or dispersed together with an emulsifier in suitable single-oil or oil mixture by means of heating. The water phase containing the osmotic agent with or without an additional emulsifier is also heated and mixed with the oil phase by means of high-speed mixers. Further homogenization takes place to obtain the needed small droplet size range of the emulsion. A terminal sterilization by filtration, or steam, then follows. The emulsion thus formed contains most of the API molecules within its oil phase or its oil–water interface. This is a generally accepted and standard method to prepare lipophilic API-loaded nanosized emulsions for parenteral, ocular, percutaneous, and nasal uses, as illustrated in Fig. 2.3. This process is normally carried out under aseptic conditions and nitrogen or argon atmosphere to prevent both contamination and potential oxidation of sensitive excipients.

2.4.3. Interfacial Incorporation Approach

Since many APIs of commercial interest generally have a solubility that is too low in FDA-approved oils, Lance et al. (1995) proposed a method to incorporate such APIs into the interfacial o/w layer of the emulsion droplets. This can be achieved by initially dissolving the API along with the phospholipid (emulsifier) in an organic solvent, instead of in the oil.

Following the solvent evaporation, the obtained phospholipids/API co-mixture is used in the de novo production of the emulsions (Davis and Washington 1988). However, this approach suffers from possible API nanocrystal formation and from the use of organic solvent during the emulsion preparation process. To overcome such drawbacks, a novel SolEmul® technology was developed in which an additional high-speed homogenization step is included to mix the API with emulsion. The API particles are micronized to the nanosize range prior to incorporation into the emulsions. By this technique, adequate amounts of lipophilic APIs can be substantially incorporated into the lipophilic core or intercalated between the selected emulsifier molecular films at the o/w interface of the emulsions. The APIs reported to have been incorporated by this novel approach are amphotericin B, carbamazepine, and itraconazole (Buttle et al. 2002; Müller and Schmidt 2002; Akkar and Müller 2003a, b). However, it should be emphasized that all the lipophilic API molecules that have been

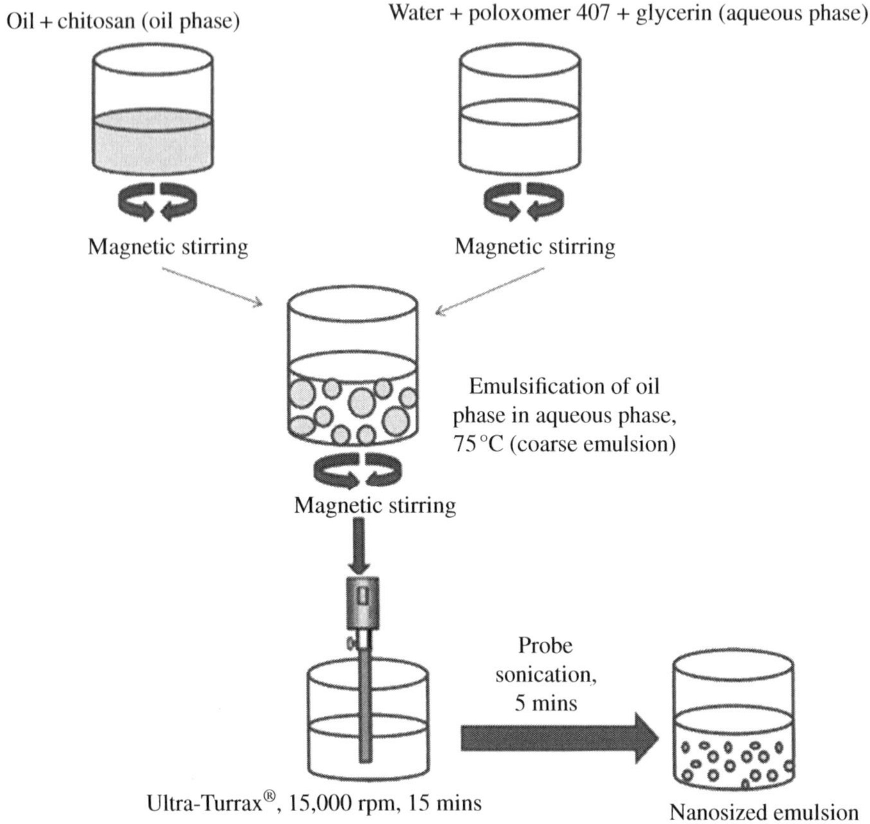

Figure 2.3. Schematic diagram for preparation of nanosized emulsion. (See color insert.)

incorporated into the emulsions by SolEmul® technology are meant only for parenteral use (Buttle et al. 2002; Müller and Schmidt 2002; Akkar and Müller 2003a, b) and so far no ocular, nasal, and topical active agents have been incorporated by this approach although there is no regulatory reason to exclude this technical improvement when designing emulsion formulations for these applications.

2.4.4. Incorporation of Antibodies, DNA Protein, Oligonucleotide, or Heat Labile Molecules

Both extemporaneous API addition (method A) into the preformed emulsion and *de novo* emulsion preparation (method B) are useful for the incorporation of heat labile molecules into the o/w nanosized emulsions. For example, cyclosporin A (peptidic molecule) was successfully incorporated without API degradation into the emulsion by following the *de novo* method (Tamilvanan et al.

2001). The extemporaneous addition of the solid API or API previously solubilized in another solvent or oil to the o/w nanosized emulsions is not a favored approach in technology wise as it might compromise the integrity of the emulsion. However, since therapeutic DNA and single-stranded oligos or siRNA are water soluble due to their polyanionic character, the aqueous solution of these compounds need to be added directly to the o/w cationic nanosized emulsion in order to interact electrostatically with the cationic emulsion droplets and thus associate/link superficially at the oil–water interface of the emulsion (Teixeira et al. 1999; Tamilvanan 2004; Hagigit et al. 2010). During *in vivo* condition when administered via parenteral and ocular routes, the release of the DNA and oligos from the associated emulsion droplet surfaces should therefore initially be dependent solely on the affinity between the physiological anions of the biological fluid and cationic surface of the emulsion droplets. The mono- and di-valent anions containing biological fluid available in parenteral route is plasma and in ocular topical route is tear fluid, aqueous humor, and vitreous. Moreover, these biofluids contain multitude of macromolecules and nucleases. There is a possibility that endogenous negatively charged biofluid's components could dissociate the DNA and oligos from cationic emulsion. It is noteworthy to conduct during the preformulation development stages an *in vitro* release study for therapeutic DNA and oligos-containing cationic nanoemulsion in these biological fluids and this type of study could be considered as an indicator for the strength of the interaction occurred between DNA or oligo and the emulsion (Hagigit et al. 2008). Interestingly, the stability of oligos (a 17-base oligonucleotide, partially phosphorothioated) was validated using a gel-electrophoresis method. After incorporating the oligos into the cationic nanosized emulsion as well as during *in vitro* experiments of oligos-containing emulsion in vitreous fluid at different time periods, the emulsions were phase separated by Triton X-100 and then the degradation of oligos was also monitored following the same validated gel-electrophoresis method (Hagigit et al. 2008). No appearance of new band was seen in comparison to the standard aqueous oligos solution. This result indicates that the oligos did not undergo degradation against the conditions applied to prepare a sterile emulsion.

In order to bring the nanosized emulsion closer to otherwise inaccessible pathological target tissues, homing devices/ligands such as antibodies and cell recognition proteins are usually linked somehow onto the particle surfaces. Various methods have been employed to couple ligands to the surface of the nanosized emulsions with reactive groups. These can be divided into covalent and noncovalent couplings. Noncovalent binding by simple physical association of targeting ligands to the nanocarrier surface has the advantage of eliminating the use of rigorous, destructive reaction agents. Common covalent coupling methods involve formation of a disulfide bond, cross-linking between two primary amines, reaction between a carboxylic acid group and primary amine, reaction between maleimide and thiol, reaction between hydrazide and aldehyde, and reaction between a primary amine and free aldehyde (Nobs et al. 2004). For antibody-conjugated anionic emulsions, the reaction of the carboxyl derivative of the coemulsifier molecule with free amine groups of the antibody and disulfide bond formation between coemulsifier derivative and reduced antibody were the two reported conjugation techniques so far (Song et al. 1996; Lundberg et al. 1999, 2004). However, by the formation of a thioether bond between the free maleimide reactive group already localized at the o/w interface of the emulsion oil droplets and a reduced monoclonal antibody, the antibody-tethered cationic emulsion was developed for active targeting to tumor cells (Goldstein et al. 2005, 2007a, b).

2.5. QbD APPROACH TO OPTIMIZE EMULSION

The QbD paradigm, as clearly delineated in the International Council for Harmonisation (ICH) Q8(R2), Q9, and Q10 guidelines, indicates the necessity to follow initial risk assessment studies, factor screening studies, and then the optimization of any pharmaceutical products not only for improving the final product quality but also to provide regulatory flexibility for the industry to improve their manufacturing processes [ICH Q9 guideline 2005; ICH Q10 guideline 2008; ICH Q8(R2) guideline 2009]. The guideline ICH Q8(R2) described a QbD-based approach of Formulation by Design (FbD). Whereas the ICH Q9 promulgated the need of Quality Risk Management (QRM), the ICH Q10 signified the way of obtaining the quality final products. The typical steps required for the new API products during their formulation development stage are shown in Flowchart 2.1.

Like other dosage forms, the o/w nanosized emulsions do contain many raw materials, manufacturing process, and testing of in-process materials and final products. At the very outset, the selection of each raw materials and their proper amount is one of the prime steps. Following the raw materials' assortment, the manufacturing process is to be selected from the myriad of available size-reduction machineries. In many cases, these two (raw materials and manufacturing process) constraints are often called as independent variables or critical process parameters (CPPs). Studying the influence of CPPs on the so-called dependent or response variables [also known as critical quality attributes (CQAs)] will eventually determine the final product quality and also provide flexibility for the industry to advance their manufacturing processes for meeting the stringent regulatory procedures. Hence, the studies relating the influence of CPPs on CQAs look in a diverse manner that depends on the changes of regulatory procedures over the time periods. Initially, it was thought that the product quality along with its performance, *in vitro* and *in vivo*, could be safeguarded by performing the traditional quality by testing (QbT) approach usually via one-factor-at-a-time (OFAT) experimental approach. In the QbT approach, the quality of both API and its product is mainly ensured by testing

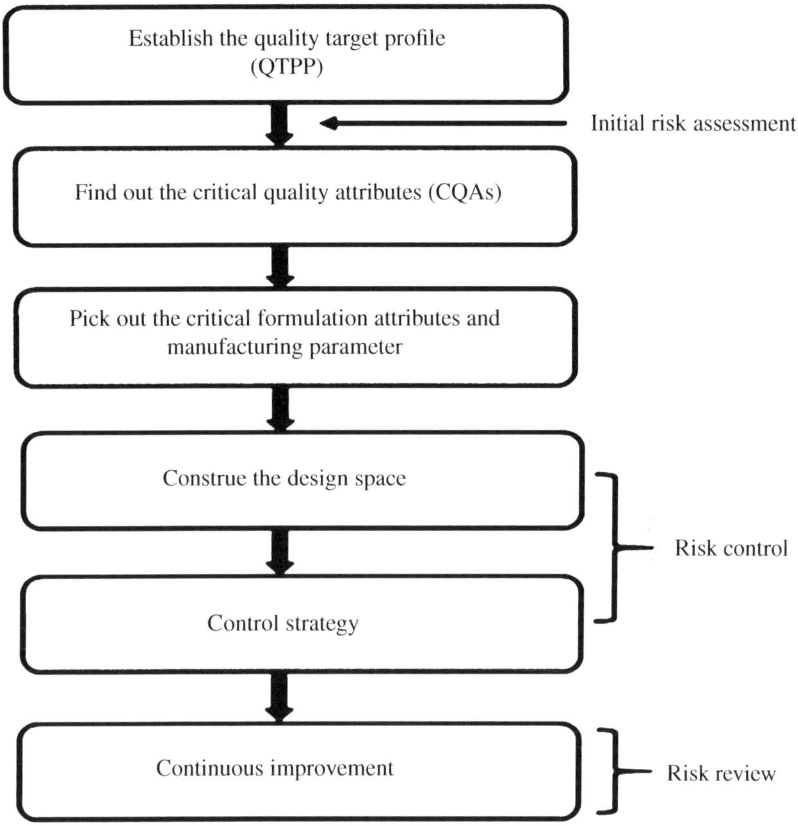

Flowchart 2.1. Typical steps involved for the new drug products during their formulation development stage as per the quality by design (QbD) approach of formulation by design (FbD).

of raw materials, a fixed manufacturing process, and testing of in-process materials and end product (Yu 2008).

This traditional framework has certain drawbacks. Any minor changes made in input materials and processes (including equipment) for anticipated variability are empirical and addressed via the OFAT experimental approach. This development practice is not cost-effective and results in incomplete product and process understandings, which in turn leads to restrictive (or fixed) manufacturing processes that are unable to compensate for the regular variability in input materials, processes, manufacturing equipment, and laboratory instrumentation (Debevec et al. 2018). As mentioned earlier, the QbT approach also requires extensive testing to comply with restrictive FDA-approved specifications (Yu 2008).

For these reasons, traditional industry practices are often extraordinarily expensive and time consuming and present an overwhelming burden to regulatory agencies for reviewing multiple chemistry, manufacturing, and control (CMC) supplements. Another major drawback of the traditional approach is that it does not differentiate products with regard to complexity (for example, APIs with high potency and/or narrow therapeutic indices) and associated level of quality and health risks to manufacturers and consumers, respectively (Yu 2008). These drawbacks of "minimal" or "traditional approach" and need for higher regulatory flexibility via an "enhanced approach" prompted a paradigm shift in the industry practice [ICH Q8(R2) guideline 2009].

The need for transition from traditional QbT to an enhanced approach was formally communicated through an ICH Q8 guidance published in May 2006, which emphasized that "quality cannot be tested into products, rather it should be built into products by design" (FDA Guidance for Industry 2006).

These innovative frameworks are fully reflected in current regulatory guidance on QbD and PAT [FDA Guidance for Industry 2004; ICH Q8(R2) guideline 2009] and are encouraged for industry practice.

Both QbD and PAT share common goals of providing a rapid and science- and risk-based road map for product development and economically effective strategies for process monitoring and analytical testing. The QbD strategy involves an end-to-end integration of six key elements, which are quality target product profile (QTPP), risk assessments related to process and product design, DOEs, design space, control strategy, and continuous process improvement. Each of these elements is essential for product development, manufacturing, and quality assurance. Through parallel evolution of QbD and PAT, it is now well established that developing a robust formulation and manufacturing process requires a thorough understanding of inter-relationships between material attributes (MAs), processing parameters (PPs), and dependent product quality attributes at each stage of product development. In this regard, pharmaceutical, chemical, and engineering industries, research institutions, and regulatory agencies worldwide have significantly contributed through joint collaborations, workshops, conference presentations, and publications. More recently, a 5-year pilot QbD program (March 2011–April 2016) was launched between the FDA and the EMEA (EMA 2017).

In this pilot program, the subject matter experts from both agencies had thoroughly exchanged their viewpoints to allow a joint evaluation of QbD elements in an effort to harmonize agencies' expectation for regulatory submissions worldwide. As a result of this pilot program, both the FDA and the EMEA have mutually agreed on several pertinent topics and level of details required in a regulatory submission (EMA 2014).

These include risk assessments related to process and product design, DOEs, and design space, which are briefly summarized below. In QbD-based product development, the MAs and PPs are initially identified from prior scientific knowledge using risk assessment techniques. One of commonly utilized risk assessment tools is failure mode effect analysis (FMEA). In this approach, a risk priority number (RPN) is calculated for each factor that can impact the CQAs of the product. The RPN is calculated by multiplying numerical rankings (e.g., 1–5 or 1–10) ascribed to three risk components (severity of harm, probability of occurrence, and detectability) associated with each factor, i.e.,

$$\text{RPN} = \text{Severity} \times \text{Occurence} \times \text{Detectability} \quad (2.1)$$

The RPN values can be presented in a tabular format or as a Pareto plot for a quantitative display of relative risk rank order. It needs to be emphasized that all the MAs and PPs that can potentially affect CQAs are considered as part of the initial risk analysis; however, only a subset of these attributes and parameters are selected for development studies as warranted by the outcome of risk analysis (Badawy et al. 2016).

This QbD element has originated from the Pareto principle. As a rule of thumb, about 80% of the problems originate from roughly 20% of the factors identified, that is why Pareto concept is sometimes referred to as 80/20 rule (Orloff 2011).

Subsequently, the combinations and interactions of identified subset of MAs and PPs are studied as part of the product development through DOE, which is regarded as a "toolkit" component of a QbD approach. The collective outcome from formulation and process design DOEs is utilized in finalizing the list of critical material attributes (CMAs) and CPPs, thereby establishing a design space and an overall control strategy. Based on current ICH Q8(R2), a design space is defined as the "multidimensional combination and interaction of input variables (e.g., material attributes) and process parameters that have been demonstrated to provide assurance of quality" [ICH Q8(R2) guideline 2009].

A generalized scheme representing the evolution of QbD and multidimensional combination and interaction of critical input variables (CMAs and CPPs) on CQAs of an output product is given in Flowchart 2.2. CMAs of input materials may include physical, chemical, or microbiological attributes of APIs, excipients, and other components (e.g., purified water, solvent) that are used in formulating a quality product.

2.5.1. Case Study for Optimizing Systematically a Formula to Make O/W Nanosized Emulsions

The case study initially starts with the risk assessment plan for o/w nanosized emulsions by utilizing Ishikawa fish-bone diagram and RPN score.

2.5.1.1. Initial Quality Risk Assessment Studies To find out the influence of CMAs and/or CPPs (independent variables) on CQAs (dependent variables) of topical ophthalmic emulsions, the initial risk assessment studies were

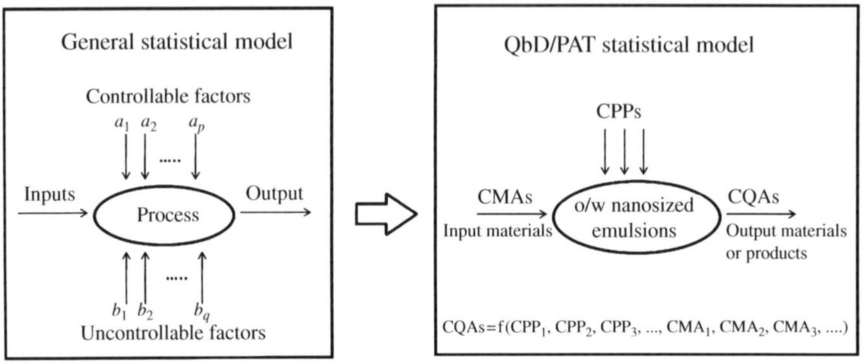

Flowchart 2.2. Evolution of QbD and multidimensional combination and interactions of critical input variables (CMAs and CPPs) on critical response variables (CQAs) for the preparation of o/w nanosized emulsions. [Adapted from Montgomery (2013) and Yu et al. (2014).]

performed. By employing the Minitab 18 software (M/s Minitab Inc., Philadelphia, PA, USA), an Ishikawa fish-bone diagram was constructed to ascertain the potential cause–effect relationship among the product and process variables. Prioritization studies were carried out to select the CMAs/CPPs with high risk by constructing the Risk Estimation Matrix (REM) for qualitative analysis of risk by assigning low-, medium-, and high risk(s) levels to each MA and/or PP of topical ophthalmic emulsions (Table 2.2). Furthermore, the quantitative estimation of risk(s) and detection of the plausibility of failure modes associated with the emulsions were assessed with the help of the FMEA (Table 2.3). The rank order scores, ranging between 1 and 10 each, were allotted to the CMAs/CPPs (independent variables) for indicating severity, detectability, and occurrence of risks. The FMEA defines the RPN according to the formula already shown in Eq. (2.1) (Fahmy et al. 2012).

The parameter D is the ease that a failure mode can be detected because the more detectible a failure mode is, the less risk it presents to product quality. For D, the rank 1 is considered as easily detectable, 5 as moderately detectable, and 10 as hard to detect. The parameter O is the occurrence probability or the likelihood of an event occurring. For O, the rank 1 is considered as unlikely to occur, 5 as 50 : 50 chance of occurring, and 10 as likely to occur. The parameter S is a measure of how severe of an effect a given failure mode would cause. For S, the rank 1 is considered as no effect, 5 as moderate effect, and 10 as severe effect. Using this procedure, the REM carried out for qualitative analysis of risk associated with each MA and/or PP.

The Ishikawa fish-bone diagram constructed for topical ophthalmic emulsions is simply portraying the cause–effect relationship among the factors that potentially affect the final product CQAs (shown in Fig. 2.4). The parameters

TABLE 2.2. Risk Estimation Matrix (REM) for Qualitative Analysis of Risk Constructed After Assigning Low, Medium, and High Risk(s) Levels to Each Material Attributes and Process Parameters of Topical Ophthalmic Emulsions

Critical Quality Attributes (CQAs)	REM for Qualitative Analysis of Risk Assigned to						
	Volume of Castor Oil	Amount of Chitosan	Amount of Poloxamer 407	Premixing Time	Homogenization Time	Homogenization Speed	Probe Sonication Time
Mean particle size	High	Medium	Medium	Medium	High	High	Medium
Polydispersity index	Low	Low	Medium	Medium	Medium	High	Medium
Zeta potential	Low	High	Low	Low	Low	Low	Low

TABLE 2.3. Summary of Failure Mode and Effect Analysis (FMEA) Demonstrating Risk Priority Number (RPN) Scores for Various Materials and Process Variables Affecting the Critical Quality Attributes (CQAs) Such As Mean Particle Size (MPS), Polydispersity Index (PDI), and Zeta Potential (ZP)

Failure Modes	Detection (D)	Occurrence (O)	Severity (S)	RPN (=DOS)	Consequences on CQAs
Volume of castor oil (ml)	5	5	8	200	MPS
Amount of chitosan (mg)	6	5	7	210	MPS and ZP
Amount of poloxamer (mg)	5	5	5	125	MPS and PDI
Premixing time (min)	5	7	7	245	MPS and PDI
Homogenization time (min)	5	4	6	120	MPS and PDI
Homogenization speed (rpm)	5	6	7	210	MPS and PDI
Probe sonication time (min)	2	3	3	18	MPS and PDI

outlined in the Ishikawa fish-bone diagram assisted in the identification of the failure modes, i.e., the modes through which a system, process step, or piece of equipment might fail. Table 2.2 illustrates the REM carried out for qualitative analysis of risk associated with each MA and/or PP. The REM suggested that factors such as amount of chitosan, speed and time of homogenization, and volume of castor oil were found to be high risk, while the factors like the

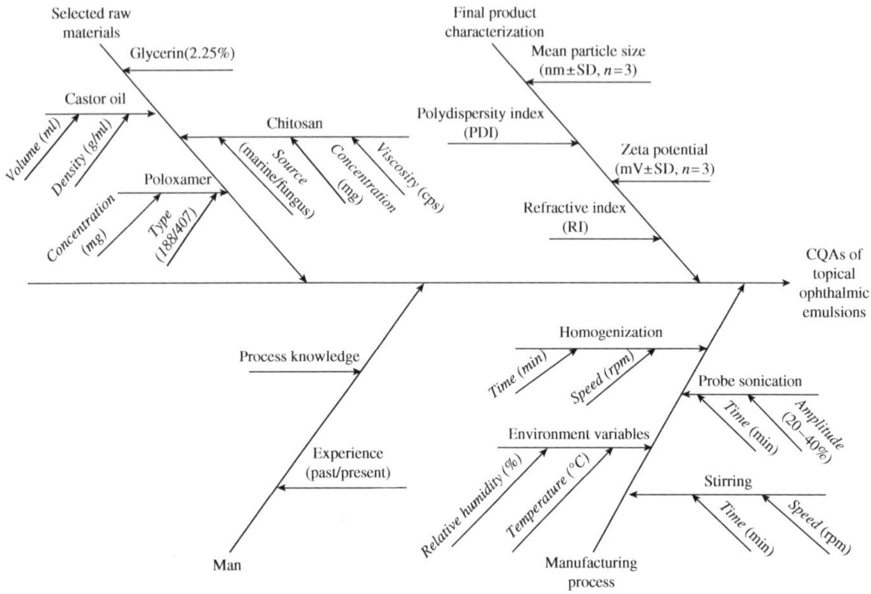

Figure 2.4. Ishikawa fish-bone diagram made with the help of Minitab 18 software showcasing the potential cause–effect relationship among the product and process variables for topical ocular emulsions.

amount of poloxamer, times for premixing, and probe sonication were associated with medium risk. An in-house exercise employing extensive brain storming among the diverse research group members as well as the existing literature reports were used for prioritizing the factors or parameters and allotting the scores for RPN computation. Using FMEA, the modes of failure can be prioritized for risk management purposes according to their seriousness of their consequences (effects), how rottenly they occur, and how easily they can be detected. Through this information, the variables that are needed to be further studied and controlled were found out or short-listed. In addition, the process of doing the FMEA analysis within a larger organization facilitates systematic gathering of current knowledge inside the organization. Furthermore, with the help of knowledge management system, the FMEA analysis allows the information on risk to be stored for future use. In this manner, it is important for the larger organization in which the turnover results in the loss of institutional memory. The outcomes of FMEA analysis are to define the RPN and further computation of RPN.

Based on the REM analysis, an elaborative risk assessment was carried out by assigning ordinal scores to each MA and PP. Table 2.3 displays the details of MAs and PPs employed during FMEA analysis together with their computed RPN scores, which collectively explain their effect and plausible/possible consequences on CQAs of topical ocular emulsions. To discriminate the high-risk factors against the low-risk factors, a critical cutoff value of RPN that was fixed for ocular emulsions is 100 or above. With lone exception of probe sonication time (which shows the RPN scores of 18 only), the factors such as amount of chitosan, amount of poloxamer, homogenization time, homogenization speed, premixing time, and volume of castor oil possessed the high RPN scores. The factors associated with high RPN scores were finally subjected to factor screening study by employing the Taguchi design.

Table 2.4 shows the selective list of various designs used for optimization and screening of CPPs of o/w nanosized emulsions.

2.5.1.2. *Factor Screening Study* The principles of factor sparsity are applied for the factor screening studies in which only a few of the factors among the numerous ones are identified to explain the major proportion of the experimental variation in the final product (Negi et al. 2015). Whereas the active or influential variables were responsible for the major variability, remaining all other factors were termed as inactive, less influential, or simply the noise variables. Based on Eq. (2.1), the CMAs/CPPs that produced high RPN values were finally subjected to factor screening studies to quantitatively estimate the

TABLE 2.4. Selective List of Various Designs Used for Optimization and Screening of CPPs of O/W Nanosized Emulsion [As Per Design-Expert® (version 11.1.0.1, Stat-Ease Inc., Minneapolis, MN, USA) Software]

Design for	
Optimization	Screening
Box-Behnken	**Miscellaneous**
Central composite	• Irregular res V
• Face centered	• Plackett–Burman
• Orthogonal quadratic	• Taguchi OA
• Practical ($k>5$)	**Randomized**
• Spherical	• Min-run characterize
• Rotatable ($k<6$)	• Min-run screen
Miscellaneous	• Multilevel categoric
• 3-Level fractional	• Optimal (custom)
• Hybrid	• Regular two-level
• Pentagonal	**Split-plot**
• Hexagonal	• Multilevel categoric
Mixture	• Optimal (custom)
• Simplex lattice	• Regular two-level
• Simplex centroid	
Split-plot	
• Central composite	
• Optimal (custom)	
Supersaturated	

risk associated with the formulation and process variables of topical ophthalmic emulsions. A 7-factor 8-run Taguchi design was employed for screening the formulation and process variables of the ophthalmic emulsions. The design matrix was prepared using Design-Expert® (version 11.1.0.1, Stat-Ease Inc., Minneapolis, MN, USA) software. Table 2.5 enlists the Taguchi design matrix selected for the preparation of topical ophthalmic emulsions along with the description of their respective low and high levels. A total of eight trial formulations were thus prepared as per the screening design and evaluated for the identified/selected CQAs [mean particle size (MPS), polydispersity index (PDI), and zeta potential (ZP)]. The analysis of the design-generated data was performed by fitting it to the first-order linear model obviating the interaction effect(s), while analyzing coefficients for each of the factors. The Half-normal and Pareto charts were used for quantitatively identifying the effect of each MAs/PPs on the selected CQAs for screening.

2.5.1.3. Factor Screening Studies by Taguchi Design To establish the orthogonal effects between formulation or process variables and selected CQAs, the Taguchi design is mainly employed. This design helps not only in filtering the most significant critical variables from the insignificant ones but

TABLE 2.5. Taguchi Design Matrix Portraying the Layout of Various Experimental Runs for Factor Screening of Topical Ophthalmic Emulsions

Critical Material Attributes (CMAs) and Critical Process Parameters (CPPs, also Called as Independent Variables) with Their Code	Levels		Critical Quality Attributes (CQAs, also Called as Dependent Variables)
	Low (−1)	High (+1)	
A: Castor oil (ml)	1	2	
B: Chitosan (mg)	6	18	
C: Poloxamer (mg)	75	100	Mean particle size (MPS, nm)
D: Premixing time (min)	10	15	
E: Homogenization time (min)	15	20	Polydispersity index (PDI)
F: Homogenization speed (min)	15,000	17,000	Zeta potential (ZP, mV)
G: Probe sonication time (min)	5	10	

Run	*A*	*B*	*C*	*D*	*E*	*F*	*G*
1	1	−1	1	−1	1	−1	1
2	−1	1	1	−1	−1	1	1
3	−1	−1	−1	−1	−1	−1	−1
4	1	−1	1	1	−1	1	−1
5	1	1	−1	−1	1	1	−1
6	−1	1	1	1	1	−1	−1
7	−1	−1	−1	1	1	1	1
8	1	1	−1	1	−1	−1	1

also in optimizing further the formula for the development of topical ophthalmic emulsions. Figure 2.5a–c depict the Pareto charts for screening of influential formulation and process variables as per Taguchi design using selected CQAs. The standard *t* limit and Bonferroni limit are depicted as a black line and a red line, respectively, in the Pareto chart. Similarly, the positive and negative effects of each formulation and process variables on each CQAs are also depicted using yellow and blue box in the Pareto chart, respectively.

If a particular formulation and process variables showed an effect that exceeds the standard *t* limit in the Pareto chart, then the variables produced a significant effect on the CQAs. On the other side, any of the formulation and process variables showing the effect that is lower than the standard *t* limit will be considered to produce a nonsignificant influence on the CQAs. In a similar manner, the significances of the formulation and process variables are also determined by the Bonferroni limit in the Pareto chart.

The formulation and process variable such as homogenization speed was found to exceed either *t* value limit or Bonferroni limit for the MPS (Fig. 2.5a). This indicates that this formulation and process variable might produce the most significant influence on MPS. Although the homogenization speed was found to produce a significant influence on MPS, it did not influence the PDI and ZP values. Taking the insignificant influence of homogenization speed on PDI and ZP into consideration, the homogenization speed was kept at a constant value of 15,000 rpm. For PDI and ZP, none of the formulation and process variables has shown significant influence (Fig. 2.5b, c) as they did not cross the *t* value limit in the Pareto chart. However, the premixing time deliberately omitted and considered as dummy factor since this variable involves in the initial dispersion/mixing of oil and water phases during the emulsion preparation step. The formulation and process variables such as premixing time, homogenization time, homogenization speed, and probe sonication time were decided to fix at constant values of 10 min, 15 min, 15,000 rpm, and 5 min, respectively, due to their insignificant influence on the selected three CQAs of the topical ophthalmic emulsions. By taking our previous experience in making of topical ophthalmic emulsions into consideration, the formulation and process variables such as the amounts of castor oil, chitosan, and poloxamer were, however, selected as critical formulation and process variables that were ultimately needed to be further optimized using the face-centered CCD.

2.5.1.4. Formulation Optimization by QbD: Experimental Design To identify the effect of (Taguchi design-) screened CPPs (independent variables) such as castor oil concentration (1–2 ml, A), chitosan concentration (6–18 mg, B), and poloxamer concentration (75–100 mg, C) on the three CQAs (dependent or response variables) like MPS (R_1), PDI (R_2), and ZP (R_3), a three-factor face-centered central composite design (CCD) was used. A total of 20 experiments, including 6 replications of central points (the leverage value of 0.1182 and α value of 1) together with 8 factorial points and 6 axial points (shown in

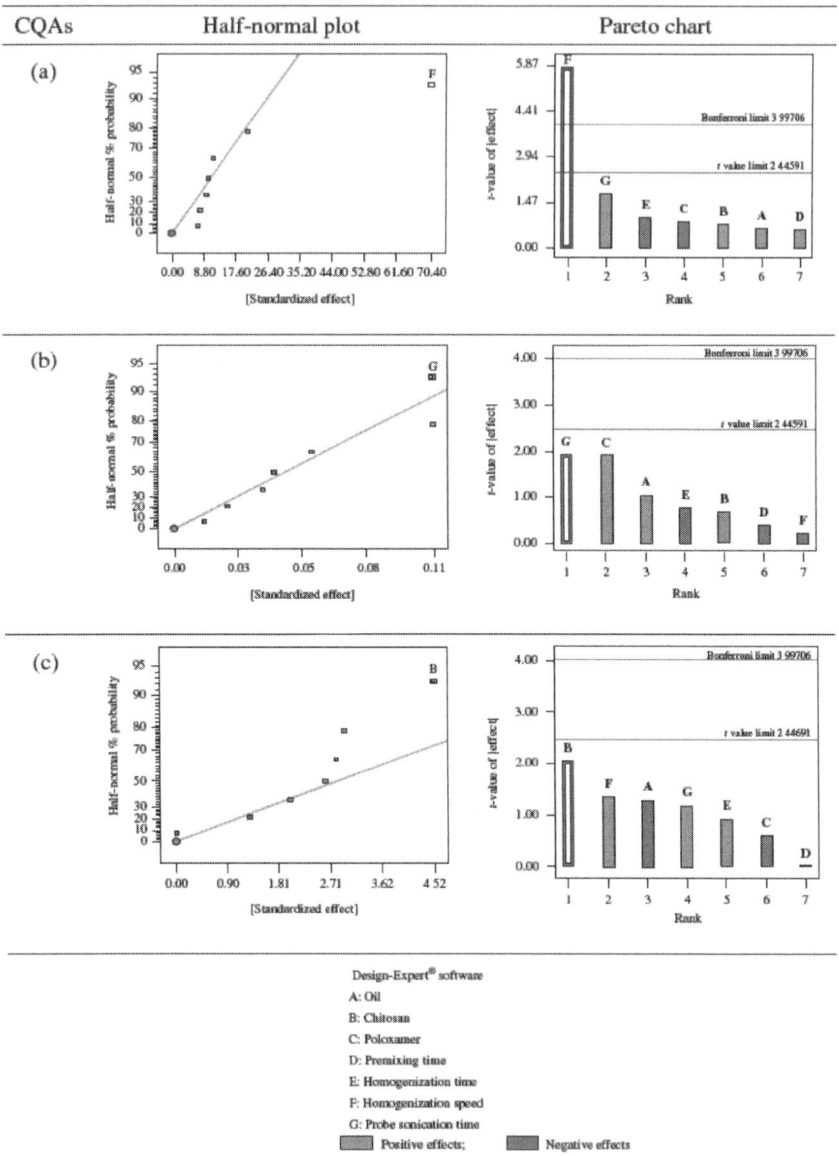

Figure 2.5. Half-normal and Pareto charts for screening of influential formulation and process variables as per Taguchi design using selected critical quality attributes (CQAs): (a) mean particle size, (b) polydispersity index, and (c) zeta potential. (See color insert.)

Table 2.6), were run using Design-Expert® software. While the leverage value indicates the potential for a design point (axial, central, or factorial) for influencing the model fit, the α value determines the geometry of the design region and for face-centered CCD, it is equal to 1 defining a square design geometry.

TABLE 2.6. Independent and Dependent (Response) Variables: Face-Centered Central Composite Design (CCD) Scheme

Run	Type	Critical Process Parameters (CPPs, also Called as Independent Variables)			Critical Quality Attributes (CQAs, also Called as Dependent Variables)		
		Castor Oil (ml)	Chitosan (mg)	Poloxamer (mg)	MPS (nm)	PDI	ZP (mV)
1	Axial	2	12	87.5	676.4	0.655	27.6
2	Center	1.5	12	87.5	609.8	0.655	27.2
3	Factorial	1	18	75	247.6	0.376	24.6
4	Axial	1.5	12	75	396.4	0.275	35.5
5	Axial	1.5	18	87.5	449.4	0.661	29.1
6	Center	1.5	12	87.5	609.8	0.655	27.2
7	Factorial	2	6	100	417.1	0.62	17.1
8	Factorial	2	18	100	438.1	0.469	23.2
9	Factorial	2	6	75	390.5	0.438	28.5
10	Center	1.5	12	87.5	609.8	0.655	27.2
11	Axial	1.5	12	100	441.5	0.488	18.8
12	Axial	1	12	87.5	374.2	0.435	33.6
13	Axial	1.5	6	87.5	333.9	0.408	30.6
14	Center	1.5	12	87.5	609.8	0.655	27.2
15	Factorial	1	6	75	240.1	0.287	17.9
16	Factorial	1	6	100	447.6	0.676	15.6
17	Center	1.5	12	87.5	609.8	0.655	27.2
18	Factorial	1	18	100	453.4	0.62	25.1
19	Factorial	2	18	75	492.8	0.537	20.9
20	Center	1.5	12	87.5	609.8	0.655	27.2

MPS, mean particle size; PDI, polydispersity index; ZP, zeta potential.

Each row indicates an experiment while the columns present the composition and the critical quality attributes (CQAs) of the non-phospholipid-based topical ophthalmic emulsions.

Table 2.7 summarizes the screened CPPs and their levels. The values of CQAs were fed into the software and second-order mathematical model was used to study the factor–response (CPPs–CQAs) relationship. An appropriate polynomial model was chosen based on the significant terms ($p < 0.005$, ANOVA), the least significant lack of fit, coefficient of variance, the multiple correlation coefficient, and adjusted multiple correlation coefficient provided by Design-Expert® software. The center point was repeated six times to determine the repeatability of the method.

The search for design space and optimized emulsion formula was carried out by numerical and graphical techniques using the software. The interaction effect of the CPPs on the CQAs was visualized using the response surfaces and 3D contour plots of the fitted polynomial regression equation. Furthermore, the quantitative comparison between the theoretical prediction and obtained

TABLE 2.7. Low and High Levels of Critical Material Attributes (CMAs) and Critical Process Parameters (CPPs, also Called as Independent Variables) Along with the Goals of Critical Quality Attributes (CQAs, also Called as Dependent Variables) Used for Making Face-Centered Central Composite Design (CCD) Throughout the Initial Pilot Screening Study to Prepare Topical Ophthalmic Emulsions

Screened CMAs and CPPs	Quantity	Levels	
		Low	High
X_1: Castor oil (ml)	1, 1.5, 2	1	2
X_2: Chitosan (mg)	6, 12, 18	6	18
X_3: Poloxamer (mg)	75, 87.5, 100	75	100
CsA (% w/w)	*0.05 or 0.1*		
Glycerin (g)	*2.25*		
Double distilled water (ml)	*Upto 100*		
Selected CQAs		Goal	
R_1: Mean particle size		Minimize	
R_2: Polydispersity index		Minimize	
R_3: Zeta potential		Maximize	

Note: Italics numbers and texts indicate constant quantity of independent variables while the quantity of other independent variables were altered.

experimental values was used by plotting the predicted versus observed value curve to validate/verify the chosen face-centered CCD model. The predicted error is the difference between the experimental value and the predicted value divided by predicted value.

2.5.1.5. *Systematic Optimization Using Face-Centered CCD* A Box–Wilson central composite design, commonly termed as CCD, is frequently utilized to build a second-order polynomial for the response variables (CQAs) in RSM without using a complete full factorial design of experiments. There are two main varieties of CCD, namely, face-centered CCD and rotatable CCD. Due to its simplicity, regions of interest, and operability, the face-centered CCD was chosen in the present study. As per the face-centered CCD, the optimal composition and the experimental conditions to prepare the topical ophthalmic emulsions were fixed (Table 2.7). Although the goal of ZP is fixed at maximum (Table 2.7) by considering the previous observation that a stable cationic nanosized emulsion should contain the ZP value that ranges from +25 to +45 mV (Tamilvanan et al. 2010), the stable stability of emulsion is not increasing with the increase of ZP's values. With the help of polynomial regression equation, the factor–response relationship was examined for the response function (Yi) using the generalized response surface model [as given in Eq. (2.2)]. In Eq. (2.2), the terms X_1, X_2, and X_3 indicate the three different factors (CPPs, independent variables) such as amounts of castor oil, chitosan, and poloxamer, respectively. While the term a_0 represents intercept (*a* constant), the linear (first-order), quadratic (second-order), and interactive polynomial coefficients are given as a_1, a_2, and a_3 (in general, a_i), a_{11}, a_{22}, and a_{33} (in general, a_{ii}), and a_{12}, a_{13}, and a_{23} (in general, a_{ij}), respectively.

$$Y_i = a_0 + a_1X_1 + a_2X_2 + a_3X_3 + a_{12}X_1X_2 + a_{13}X_1X_3 + a_{23}X_2X_3 + a_{11}X_1^2 + a_{22}X_2^2 + a_{33}X_3^2 \quad (2.2)$$

The estimated model equations (both coded and actual), regression coefficients, R^2, adjusted R^2, regression (P-value and F-value), and standard deviation related to the effect of the three CPPs (independent variables) are presented in Table 2.8. Although the actual model equation contains the levels specified in the original units for each CPP, this equation should not be used to determine the relative impact of each CPP (factor). Therefore, the coded model equation is used for identifying the relative impact of the CPPs by comparing the regression coefficient values of each factor. Nevertheless, a positive value in the regression equation represents an effect that favors optimization due to synergistic effect, while a negative value indicates an inverse relationship or antagonistic effect between the factor and the responses (Woitiski et al. 2009). It should be mentioned that nonsignificant ($p<0.005$) linear terms (main CCD effect) were included in the final reduced model if quadratic or interaction terms containing these variables were found to be significant ($p<0.05$). In the present study, the response surface analysis demonstrated that the second-order polynomial used for MPS has a higher coefficient of determination ($R^2 = 0.8186$) compared to ZP value ($R^2 = 0.6525$) and PDI value ($R^2 = 0.7188$). The obtained coefficient of determination showed that >82% of the response variation of the MPS, PDI, and ZP values could be described by response surface models as the function of the CPPs. It was observed that the lack of fit had no indication of significant ($p<0.05$) for the final reduced model, therefore providing the satisfactory fitness of the RSM to the significant independent variables (factors) effect. From Table 2.8, it was observed that two independent variables (A and B) exhibited a positive effect on the response of ZP (R_3). Both MPS (R_1) and PDI (R_2) showed a positive effect for all of the tested three independent variables (A, B, and C). The interaction coefficients with more than one factor, or higher order terms in the regression equation, represent the interaction between terms or the quadratic relationship, respectively, which suggest a nonlinear relationship between factors and responses (Motwani et al. 2008). Therefore, a different degree of response than it is originally predicted by regression equation may be expected from the independent variables as they are varied at different levels or more than one factor or variables is changed simultaneously (Woitiski et al. 2009). In the present face-centered CCD modeling, all the responses (MPS, PDI, and ZP or R_1, R_2, and R_3) were affected by the interaction of independent variables and hence displaying a quadratic relationship. The interaction effects between B and C was favorable for response R_3 (showing a positive regression

TABLE 2.8. Estimated Coded and Actual Model Equations Along with ANOVA of Regression Coefficients for Face-Centered Central Composite Design (CCD) Effects (Main or Linear, Quadratic and Interaction) Against the Critical Quality Attributes (CQAs, Dependent Variables) to Determine the Best Fitted Quadratic Equation

CQAs	Model Equations	R^2	Adjusted R^2
Mean particle size (R_1)	**Coded:** $R_1 = 569.86 + 65.20 \times A + 25.21 \times B + 43.03 \times C + 13.75 \times AB - 55.18 \times AC - 10.38 \times BC + 15.35 \times A^2 - 118.30 \times B^2 - 91.00 \times C^2$ **Actual:** $PS = -5992.56500 + 663.59545 \times a + 88.29447 \times b + 120.25931 \times c + 4.58333 \times a \times b - 8.82800 \times a \times c - 0.138333 \times b \times c + 61.41818 \times a^2 - 3.28598 \times b^2 - 0.582371 \times c^2$	0.8186	0.6553
Polydispersity index (R_2)	**Coded:** $R_2 = 0.6038 + 0.0325 \times A + 0.0234 \times B + 0.0960 \times C - 0.0106 \times AB - 0.0649 \times AC - 0.0494 \times BC + 0.0180 \times A^2 + 0.0075 \times B^2 + 0.1455 \times C^2$ **Actual:** $PDI = -9.27017 + 0.800295 \times a + 0.061847 \times b + 0.194161 \times c - 0.003542 \times a \times b - 0.010380 \times a \times c - 0.000658 \times b \times c + 0.071818 \times a^2 + 0.000207 \times b^2 - 0.000931 \times c^2$	0.7188	0.4658
Zeta potential (R_3)	**Coded:** $R_3 = 28.89 + 0.0500 \times A + 1.32 \times B - 2.76 \times C - 2.21 \times AB - 0.9125 \times AC + 2.06 \times BC - 0.8136 \times A^2 - 1.56 \times B^2 - 4.26 \times C^2$ **Actual:** $ZP = -180.64250 + 31.48864 \times a - 0.037576 \times b + 4.44347 \times c - 0.737500 \times a \times b - 0.146000 \times a \times c + 0.027500 \times b \times c - 3.25455 \times a^2 - 0.043434 \times b^2 - 0.027287 \times c^2$	0.6525	0.3397

CCD effects	**Variables**	R_1		R_2		R_3	
		***F*-value**	***P*-value**	***F*-value**	***P*-value**	***F*-value**	***P*-value**
Linear term	A	7.61	0.0202	1.09	0.3219	0.0013	0.9714
	B	1.14	0.3111	0.5629	0.4704	0.9391	0.3554
	C	3.32	0.0986	9.47	0.0117	4.11	0.0702
Interaction term	AB	0.2709	0.6141	0.0928	0.7668	2.11	0.1769
	AC	4.36	0.0633	3.46	0.0924	0.3590	0.5624
	BC	0.1542	0.7028	2.01	0.1872	1.83	0.2054
Quadratic term	A^2	0.1161	0.7403	0.0911	0.7689	0.0981	0.7605
	B^2	6.89	0.0254	0.0157	0.9027	0.3624	0.5606
	C^2	4.08	0.0711	5.99	0.0344	2.69	0.1317

Note: *A*/*a*: Castor oil, *B*/*b*: Chitosan, *C*/*c*: Poloxamer.

coefficient value of 2.06) and non-favorable for R_1 and R_2 (showing negative regression coefficient values). The interaction effects between A and B was favorable for response R_1 (showing a positive regression coefficient value of 13.75) and non-favorable for R_2 and R_3 (showing negative regression coefficient values). However, it was observed that the interaction between A and C had an inverse effect for all three responses variables studied. Noticeably, quadratic effects (A^2, B^2, and C^2) were also seen between all three independent variables and all of the studied three response variables (MPS, PDI, and ZP or R_1, R_2, and R_3). The highest and positive quadratic effect for all three independent variables was noticed for the MPS response variable R_1 (showing a positive regression coefficient value of 15.35). Similarly, the highest and negative quadratic effect was also noted for the MPS response variable R_1 (showing a negative regression coefficient −118.30).

The coefficient significance of the quadratic polynomial models was evaluated by using Analysis of Variance (ANOVA). For any of the terms in the models, a large F-value and a small p-value indicated more significant effect on the respective response variables (Joglekar 1987). Table 2.8 shows the effect of independent variables on the variation of the physicochemical properties of topical ophthalmic emulsions. The variable that exhibited the largest and significant ($p < 0.05$) effect on the MPS of the emulsion for the linear term was castor oil amount. The other two variables (amounts of chitosan and poloxamer) showed insignificant effects. All the interaction terms showed insignificant effect on MPS. The quadratic terms of chitosan amount exhibited significant effect on MPS while other two terms showed insignificant effect.

The independent variables that most affect the ZP value of the emulsions for the linear term were poloxamer amount followed by the linear term of chitosan amount and castor oil amount but none of the linear terms has any significant effect. All the three interaction and quadratic effects have insignificant effect ($p > 0.05$) on the ZP value of the emulsions.

For the PDI, the linear and quadratic terms of poloxamer amount have the significant effect while all other linear, interaction, and quadratic terms have insignificant effect. Thus, it was indicated that in evaluating the response variation of PDI, it was important to consider the poloxamer amount.

2.5.1.8. Response Surface Plot Analysis The 3D response surface graphs for the selected CQAs (R_1:MPS, R_2:PDI, and R_3:ZP) of topical ophthalmic emulsions are shown in Fig. 2.6. To represent the 3D graphs for R_1 (MPS), there are three different plots designated as Fig. 2.6a–c. In a similar way, the 3D graphs for R_2 (PDI) and R_3 (ZP) are also represented, respectively, as Fig. 2.6d–f and Fig. 2.6g–i. From Fig. 2.6a and b, the MPS value increases with an increase in the amount of castor oil. Two possible explanations might be provided with this observation. The oil droplet disruption process during homogenization and probe sonication of coarse emulsion gets somewhat cumbersome due to the progressive increment of castor oil amount because of the resistance in oil

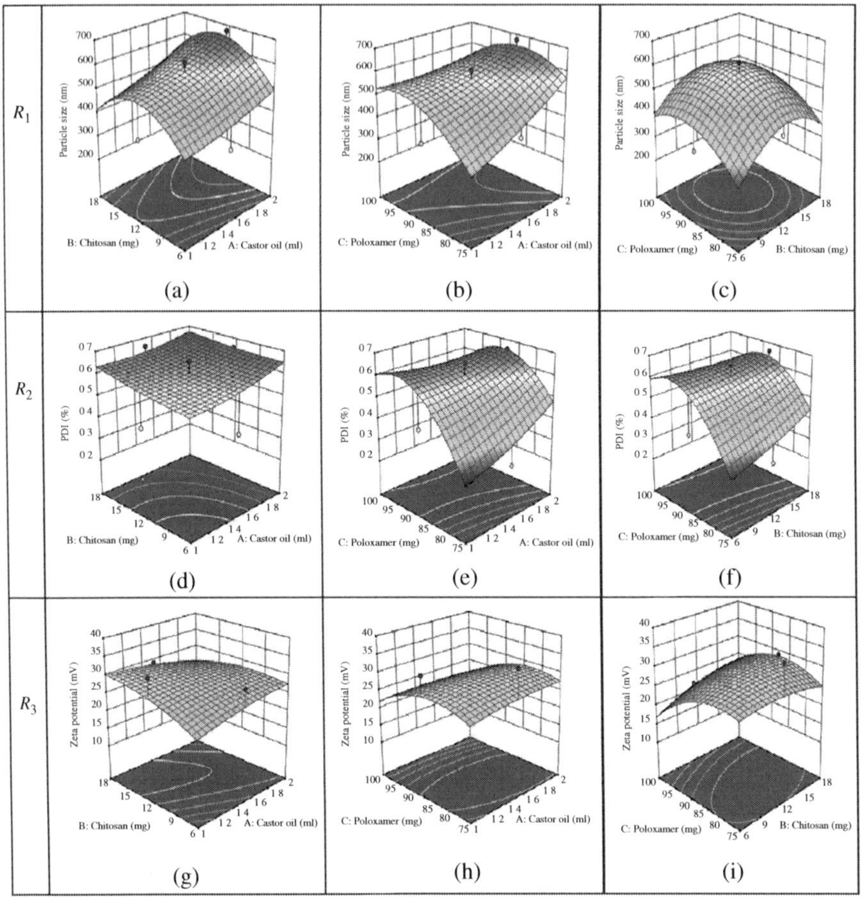

Figure 2.6. Response surface plots showing the interaction effects of castor oil and chitosan (a, d, g), castor oil and poloxamer (b, e, h), and chitosan and poloxamer (c, f, i) on response, mean particle size (R_1), polydispersity index (R_2), and zeta potential (R_3), respectively. (See color insert.)

flow and thus the diminution/restriction in the oil droplet breakup rate. The MPS value increase due to the increase of castor oil amount might also be attributed to the increased rates of oil droplets—droplets collision frequency especially at lower castor oil amount, which ultimately leads to the higher probability of coalescence of the smaller droplets and thus the bigger droplet formation at the expense of smaller droplets.

The presence of two different emulsifying agents (chitosan and poloxamer) either alone or in combination influenced the MPS in a biphasic manner. An initial increment until to reach a sharp break-point followed by a progressive decrement in the MPS value was noticed when these two emulsifying agents

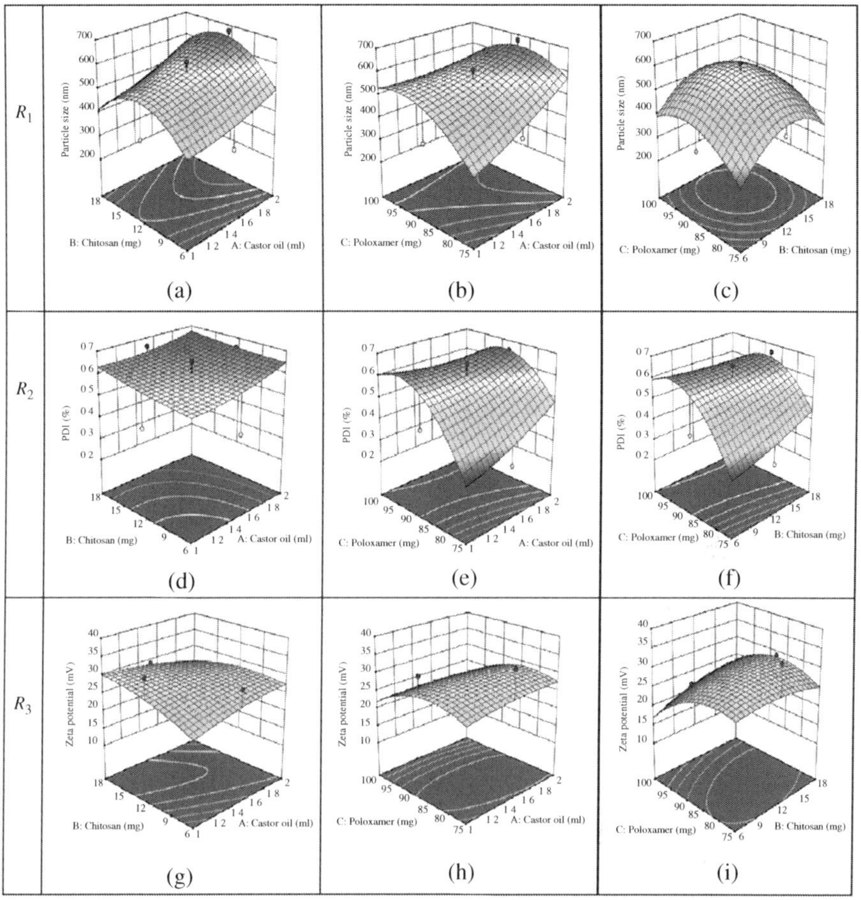

Figure 2.6. Response surface plots showing the interaction effects of castor oil and chitosan (a, d, g), castor oil and poloxamer (b, e, h), and chitosan and poloxamer (c, f, i) on response, mean particle size (R_1), polydispersity index (R_2), and zeta potential (R_3), respectively. (See color insert.)

flow and thus the diminution/restriction in the oil droplet breakup rate. The MPS value increase due to the increase of castor oil amount might also be attributed to the increased rates of oil droplets—droplets collision frequency especially at lower castor oil amount, which ultimately leads to the higher probability of coalescence of the smaller droplets and thus the bigger droplet formation at the expense of smaller droplets.

The presence of two different emulsifying agents (chitosan and poloxamer) either alone or in combination influenced the MPS in a biphasic manner. An initial increment until to reach a sharp break-point followed by a progressive decrement in the MPS value was noticed when these two emulsifying agents

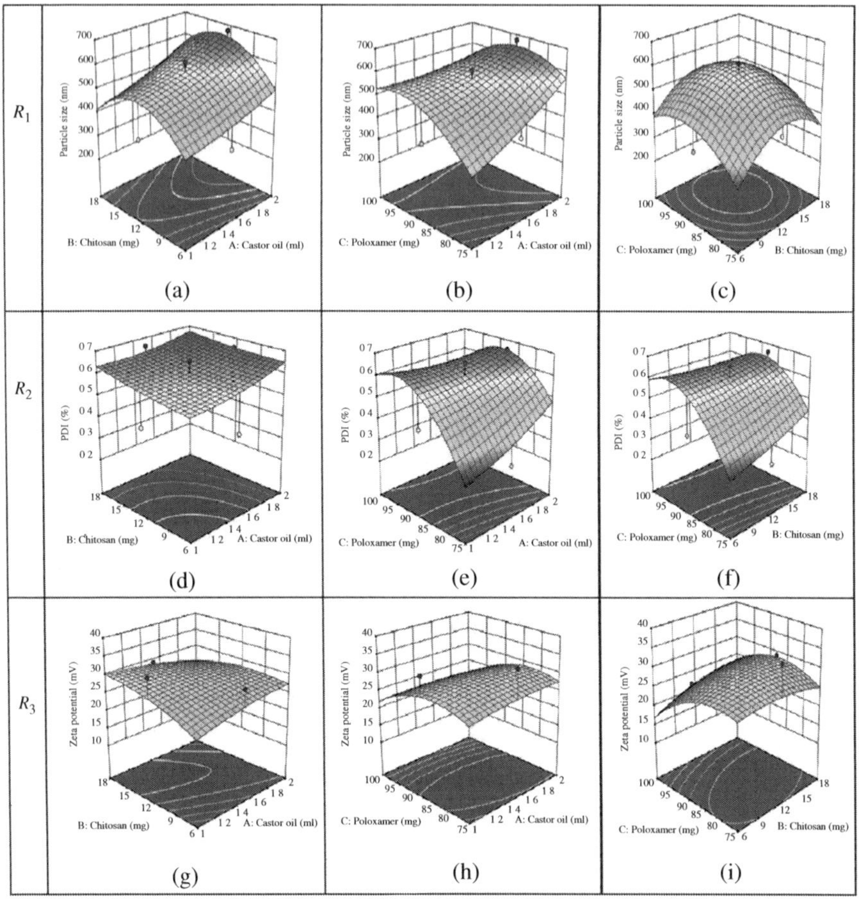

Figure 2.6. Response surface plots showing the interaction effects of castor oil and chitosan (a, d, g), castor oil and poloxamer (b, e, h), and chitosan and poloxamer (c, f, i) on response, mean particle size (R_1), polydispersity index (R_2), and zeta potential (R_3), respectively. (See color insert.)

flow and thus the diminution/restriction in the oil droplet breakup rate. The MPS value increase due to the increase of castor oil amount might also be attributed to the increased rates of oil droplets—droplets collision frequency especially at lower castor oil amount, which ultimately leads to the higher probability of coalescence of the smaller droplets and thus the bigger droplet formation at the expense of smaller droplets.

The presence of two different emulsifying agents (chitosan and poloxamer) either alone or in combination influenced the MPS in a biphasic manner. An initial increment until to reach a sharp break-point followed by a progressive decrement in the MPS value was noticed when these two emulsifying agents

interacted (Fig. 2.6c) as well as even in the occurrence of interaction between chitosan and castor oil (Fig. 2.6a) or poloxamer and castor oil (Fig. 2.6b). The initial particle size increment might be due to an inadequate amount of emulsifying agent(s) to form a mono- or multilayer film onto the dispersed oil droplets of the emulsion during emulsification (Jumaa and Müller 1998). Due to the attainment of the critical interfacial tension reduction at the vicinity of oil and water by the sufficient amount of emulsifying agent(s) resulted in the diminution of Laplace pressure, *p*, and thus the stress required for droplet deformation is reduced (Tang et al. 2012).

The PDI value is ranged from 0 to 1 wherein PDI < 0.1 indicates a nearly homogenous monodisperse formulation. The PDI value between 0.1 and 0.2 specifies a particle population with relatively narrow size distribution while the PDI > 0.3 designates polydispersity with variations in particle size without a distinct distribution shape (Müllar 1990; Cegnar et al. 2004). The experimental CPPs involved in the preparation of topical ophthalmic emulsions may influence the performance of CQAs (PDI). All of the studied CPPs (amounts of castor oil, chitosan, and poloxamer) produced a direct relationship with PDI values observed. An increase in CPPs amounts increases the PDI values (Fig. 2.6d–f). Higher is the amount of castor oil, lesser is the oil breakdown frequency during emulsification promoting the formation of larger sized oil droplets and thus the PDI to attain a higher value. Increasing the amounts of these two emulsifying agents (chitosan and poloxamer) either alone (Fig. 2.6d and e) or in combination (Fig. 2.6f) also increased the PDI values. For instance, the PDI value of >0.6 was noted when the interaction between chitosan and castor oil occurred (Fig. 2.6d). But the interaction between poloxamer and castor oil yielded the PDI value of ~0.6 (Fig. 2.6e). However, the combined interaction of two emulsifying agents produced the PDI value of <0.6 (Fig. 2.6f). These effects are possibly due to the progressive increase in the apparent viscosity of the emulsion, which ultimately provides a higher flow resistance in the batch emulsification process (Müllar 1990). In consequence, this condition increased the coalescence rate resulting in a large particle size to form. Moreover, the large particles with inadequate emulsifier film coverage tend to coalesce faster than small particles. This phenomenon contributed to the high PDI value. Figure 2.6g–i demonstrates that by increasing the chitosan or castor oil concentration, the ZP value increases. In contrast, a biphasic manner, i.e., an initial increase followed by decrease in the ZP values, was observed with an increase in poloxamer concentration (Fig. 2.6h and i). It should be added that the present topical ophthalmic emulsions were stabilized by both electrostatic and steric mechanisms due to the chitosan and poloxamer emulsifier combination. Whatever the strong repulsive Coulomb force occurred between the protonated chitosan molecules must be counterbalanced by the week van der Waals attraction forces or the steric hindrance effect of poloxamer. That is why the biphasic attitude was seen for the influence of poloxamer concentration on ZP values (Tamilvanan 2009).

2.5.1.7. Optimization of Responses for Formulation of CsA-Loaded Nanosized Emulsion By using graphical optimization, an overlay plot (Fig. 2.7g) showing the obtained design space (highlighted) was produced by the Design-Expert® software. From the overlay plot, the design space representing desired amounts of CPPs and the selected CQAs (R_1:MPS, R_2:PDI, and R_3:ZP) values of topical ophthalmic emulsions was found out. Within the design space, the optimized formula was selected that consists of minimum MPS and PDI values but maximum ZP value. The CPPs amounts found in the optimized formula are 1.39 ml castor oil, 6 mg chitosan, and 75 mg poloxamer. Similarly, the CQAs values established for the optimized formula within the design space are 260.173 nm, 0.275 and 25.85 mV, respectively, for MPS, PDI, and ZP.

To substantiate further the established optimized formula for topical ophthalmic emulsion within the design space, the predictability of chosen face-centered CCD model is at first corroborated by evaluating the randomly selected six different formulae along with the optimized formula for the actual CQAs values and comparing them with the predicted CQAs values. The diagnostic plot of actual versus predicted CQAs (R_1:MPS, R_2:PDI, and R_3:ZP) values are shown in Fig. 2.7a–f. Interestingly, the actual versus predicted plots for all of the CQAs were shown the r^2 value of greater than 0.9 indicating the establishment of the closeness between the values and hence conformed/justified the predictability of chosen model within the design space (Fig. 2.7b, d, and f). The value of randomness of scatter and deviations was found to be within ±4% in residual versus predicted value plot (Fig. 2.7a, c, and e). Furthermore, the chosen model also showed an overall mean percent error value of 0.30 ± 0.13%. In addition, an adequate precision should measure the signal-to-noise ratio, and its value of greater than 4 is desirable. As per the selected face-centered CCD model, the signal-to-noise ratio values shown by Design-Expert® software for MPS, PDI, and ZP were 8.701, 6.415, and 4.6524, respectively. This directly indicates that the adequate precision is produced and therefore this model can be used to navigate the design space to find out the optimized formula for topical ophthalmic emulsion.

2.6. CONCLUSION

While narrating the importance of selecting the components of emulsion like oil, emulsifying agents, tonicity-adjusting agent, etc., the emulsion formulator should also consider the application of Design Expert® software to optimize a formula for making the final emulsion. The non-exhaustive and selected list of excipients used to make emulsions suitable for regulatory approval is briefly discussed in this chapter in conjunction with a case study of how to optimize a formula by applying the ICH Q8(R2), Q9, and Q10 guidelines. The QbD approach using the Design Expert® software could also further be substantiated via artificial intelligence (AI) and machine learning (ML) as proposed recently by Ghate et al. (2019). The AI and ML approach applied on the

Critical quality attributes (CQAs)	Residual versus predicted plot	Actual versus predicted plot
R_1: MPS	(a) 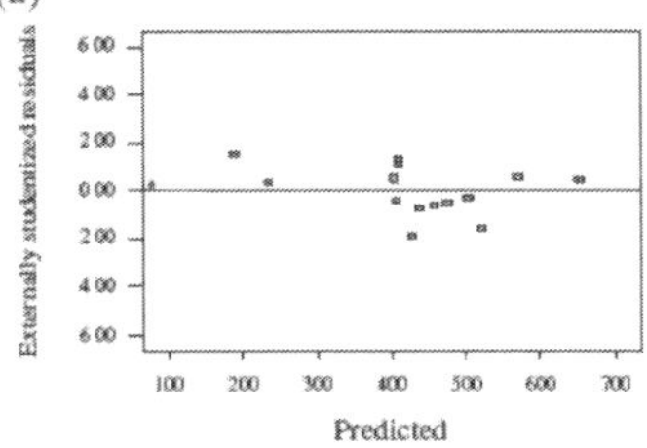	(b) 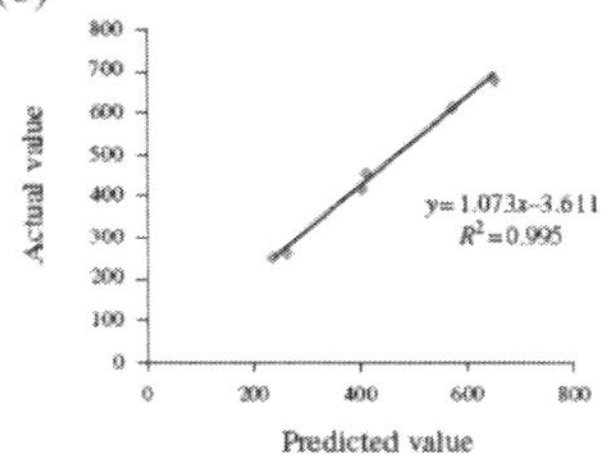
R_2: PDI	(c) 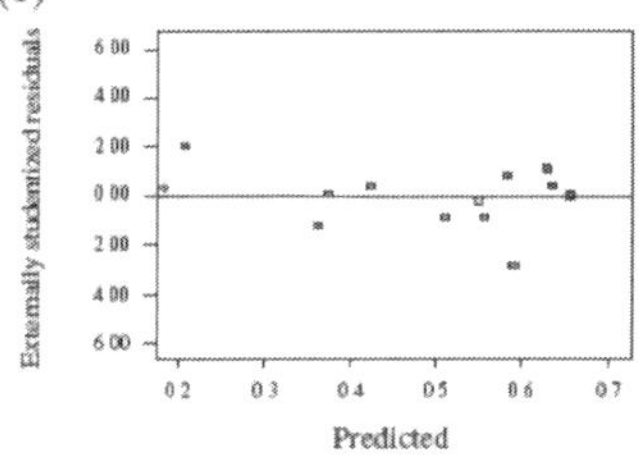	(d) 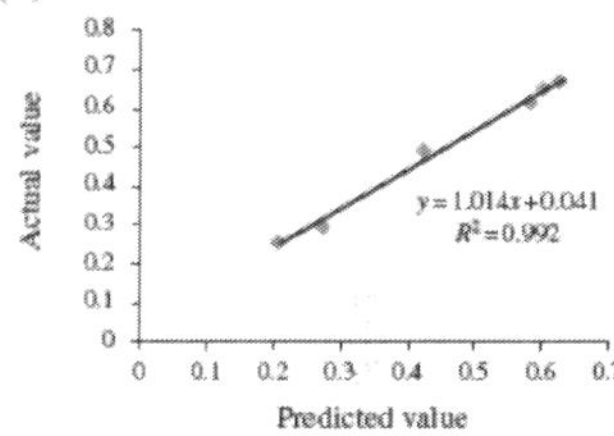
R_3: ZP	(e) 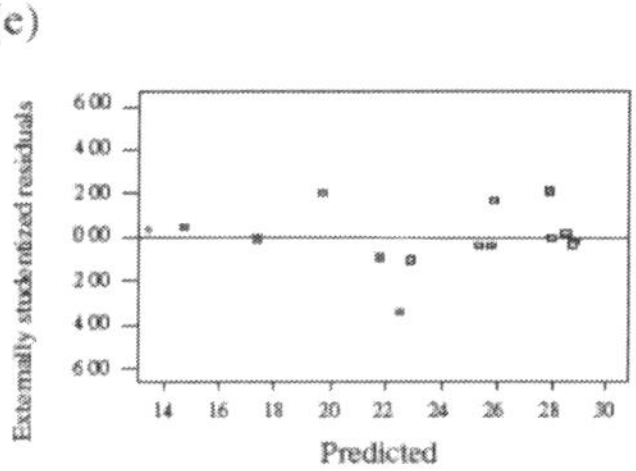	(f) 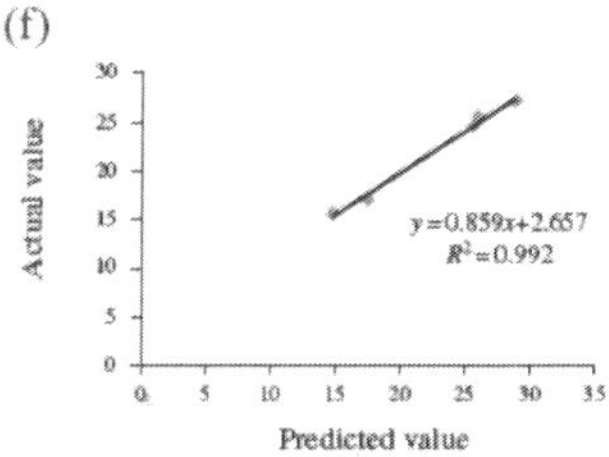
	(g) Overlay plot 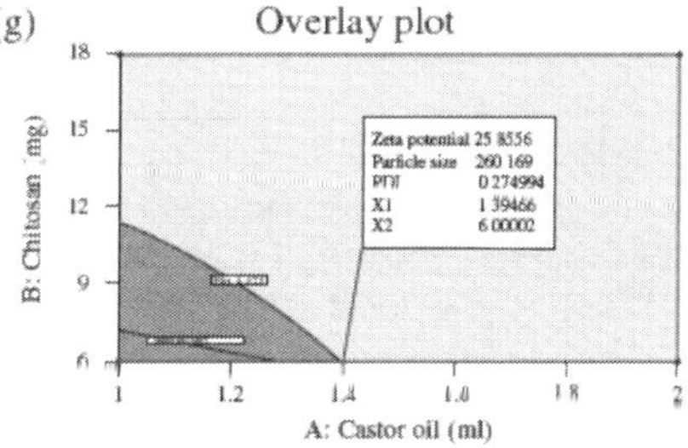	

Figure 2.7. Derived plots obtained from 3D-response surface model: residual versus predicted plot (a, c, e), actual versus predicted plot (b, d, f), and overlay plot (g). (See color insert.)

optimization and selection of emulsion formula is considered to reduce the amount of excipients, formulation preparation time, and thus the possible toxicity reduction due to high excipient concentration. Hence, the AI and ML approach is one of the welcome additions in the emulsion-making technology to get cleared rapidly the regulatory hurdles.

REFERENCES

Abele, S., Sjöberg, M., Hamaide, T. et al. (1997), Reactive surfactants in heterophase polymerization. 10. Characterization of the surface activity of new polymerizable surfactants derived from maleic anhydride, *Langmuir*, 13, 176–181. doi: 10.1021/la960577n

Akkar, A. and Müller, R.H. (2003a), Formulation of intravenous carbamazepine emulsions by SolEmuls® technology, *Eur. J. Pharm. Biopharm.*, 55, 305–312. doi:10.1016/s0939-6411(03)00028-6

Akkar, A. and Müller, R.H. (2003b), Intravenous itraconazole emulsions produced by SolEmuls technology, *Eur. J. Pharm. Biopharm.*, 56, 29–36. doi:10.1016/s0939-6411(03)00063-8

Aveyard, R., Binks, B.P., and Clint, J.H. (2003), Emulsions stabilized solely by colloidal particles, *Adv. Colloid Interf. Sci.*, 100 (102), 503–546. doi:10.1016/S0001-8686(02)00069-6

Badawy, S.I., Narang, A.S., LaMarche, K.R. et al. (2016), Integrated application of quality-by-design principles to drug product development: a case study of brivanib alaninate film-coated tablets, *J. Pharm. Sci.*, 105 (1), 168–181. doi:10.1016/j.xphs.2015.11.023

Benichou, A., Aserin, A., and Garti, N. (2004), Double emulsions stabilized with hybrids of natural polymers for entrapment and slow release of active matters, *Adv. Colloid Interf. Sci.*, 108 (109), 29–41. doi:10.1016/j.cis.2003.10.013

Binks, B.P. and Catherine, P. (2005), Nanoparticle silica-stabilised oil-in-water emulsions: improving emulsion stability, *Colloids Surf. A Physicochem. Eng. Aspects*, 253, 105–115. doi:10.1016/j.colsurfa.2004.10.116

Binks, B.P., Desforges, A., and Duff, D.G. (2007b), Synergistic stabilization of emulsions by a mixture of surface-active nanoparticles and surfactant, *Langmuir*, 23, 1098–1106. doi:10.1021/la062510y

Binks, B.P. and Lumsdon, S.O. (2000), Influence of particle wettability on the type and stability of surfactant-free emulsions, *Langmuir*, 16, 8622–8631. doi:10.1021/la000189s

Binks, B.P., Rodrigues, J.A., and Frith, W.J. (2007a), Synergistic interaction in emulsions stabilized by a mixture of silica nanoparticles and cationic surfactant, *Langmuir*, 23, 3626–3636. doi:10.1021/la0634600

Buttle, S., Schmidt, R.H., and Müller, R.H. (2002), Production of amphotericin B emulsions based on SolEmuls technology, in: *Fourth World Meeting on Pharmaceutics, Biopharmaceutics and Pharmaceutical Technology, Florence*, pp. 1535–1536.

Capek, I. (2004), Degradation of kinetically-stable o/w emulsions, *Adv. Colloid Interf. Sci.*, 107, 125–155. doi:10.1016/S0001-8686(03)00115-5

Cegnar, M., Kos, J., and Kristl, J. (2004), Cystatin incorporated in poly(lactide-co-glycolide) nanoparticles: development and fundamental studies on preservation of its activity, *Eur. J. Pharm. Sci.*, 22, 357–364. doi:10.1016/j.ejps.2004.04.003

Cohen, T., Sauvageon-Martre, H., Brossard, D. et al. (1996), Amphotericin B eye drops as a lipidic emulsion, *Int. J. Pharm.*, 137, 249–254. doi:10.1016/0378-5173(96)04473-0

Constantinides, P.P., Han, J., and Davis, S.S. (2006), Advances in the use of tocols as drug delivery vehicles, *Pharm. Res.*, 23, 243–255. doi:10.1007/s11095-005-9262-9

Constantinides, P.P., Tustian, A., and Kessler, D.R. (2004), Tocol emulsions for drug solubilization and parenteral delivery, *Adv. Drug Deliv. Rev.*, 56, 1243–1255. doi:10.1016/j.addr.2003.12.005

Cotlier, E., Baskin, M., and Kresca, L. (1975), Effects of lysophosphatidyl choline and phospholipase A on the lens, *Invest. Ophthalmol. Vis. Sci.*, 14, 697–701.

Cui, F., Wang, Y., Wang, J. et al. (2007), Preparation of redispersible dry emulsion using Eudragit E100 as both solid carrier and unique emulsifier, *Colloids Surf. A Physicochem. Eng. Aspects*, 307, 137–141. doi:10.1016/j.colsurfa.2007.05.013.

Calvo, P., Remuñá-López, C., Vila-Jato, J.L. et al. (1997), Development of positively charged colloidal drug carriers: chitosan-coated polyester nanocapsules and submicro-emulsions, *Colloid Polym. Sci.*, 275, 46–53. doi:10.1007/s003960050050

Davis, S.S. and Washington, C. (1988), Drug emulsion, European Patent 0,296, 845, A1.

Dale, P.J., Kijlstra, J., and Vincent, B. (2006), The temperature stability of single and mixed emulsions stabilized by nonionic surfactants, *Colloids Surf. A Physicochem. Eng. Aspects*, 291, 85–92. doi:10.1016/j.colsurfa.2006.06.016

Debevec, V., Srčič, S., and Horvat, M. (2018), Scientific, statistical, practical, and regulatory considerations in design space development, *Drug Dev. Ind. Pharm.*, 44 (3), 349–364. doi:10.1080/03639045.2017.1409755

EMA (2014), Questions and answers on level of detail in the regulatory submissions, EMA/59240/December 10, 2014.

EMA (2017), Report from the EMA-FDA QbD pilot program, EMA/213746/April 19, 2017.

Eskandar, N.G., Simovic, S., and Clive, A. (2009), Nanoparticle coated submicron emulsions: sustained in-vitro release and improved dermal delivery of all-trans-retinol, *Pharm. Res.*, 26, 1764–1775. doi:10.1007/s11095-009-9888-0

Fahmy, R. Kona, R. Dandu, R. et al. (2012), Quality by design I: application of failure mode effect analysis (FMEA) and Plackett-Burman design of experiments in the identification of "main factors" in the formulation and process design space for roller-compacted ciprofloxacin hydrochloride immediate-release tablets, *AAPS PharmSciTech.*, 13 (4), 1243–1254. doi: 10.1208/s12249-012-9844-x

FDA Guidance for Industry (2004) PAT-a framework for innovative pharmaceutical development, manufacturing, and quality assurance, (https://www.fda.gov/media/71012/download, Accessed on June 16, 2019).

FDA Guidance for Industry (2006), Q8 pharmaceutical development, (https://www.fda.gov/media/71524/download, Accessed on June 16, 2019).

Ghate, V.M., Kodoth, A.K., Raja, S. et al. (2019), Development of MART for the rapid production of nanostructured lipid carriers loaded with all-trans retinoic acid for dermal delivery, *AAPS PharmSciTech.*, 20, 162. doi:10.1208/s12249-019-1307-1

Goldstein, D., Gofrit, O., Nyska, A. et al. (2007a), Anti-HER2 cationic immunoemulsion as a potential targeted drug delivery system for the treatment of prostate cancer, *Cancer Res.*, 67, 269–275. doi:10.1158/0008-5472.CAN-06-2731

Goldstein, D., Nassar, T., Lambert, G. et al. (2005), The design and evaluation of a novel targeted drug delivery system using cationic emulsion-antibody conjugates, *J. Control. Release*, 108, 418–432. doi:10.1016/j.jconrel.2005.08.021

Goldstein, D., Sader, O., and Benita, S. (2007b), Influence of oil droplet surface charge on the performance of antibody-emulsion conjugates, *Biomed. Pharmacother.*, 61, 97–103. doi:10.1016/j.biopha.2006.08.005

Grigoriev, D.O. and Miller, R. (2009), Mono- and multilayer covered drops as carriers, *Curr. Opin. Colloid Interface Sci.*, 14, 48–59. doi:10.1016/j.cocis.2008.03.003

Hagigit, T., Abdulrazik, M., Orucov, F. et al. (2010), Topical and intravitreous administration of cationic nanoemulsions to deliver antisense oligonucleotides directed towards VEGF KDR receptors to the eye, *J. Control. Release*, 145 (3), 297–305. doi:10.1016/j.jconrel.2010.04.013

Hagigit, T., Nassar, T., Behar-Cohen, F. et al. (2008), The influence of cationic lipid type on in vitro release kinetics of antisense oligonucleotide from cationic nanoemulsions, *Eur. J. Pharm. Biopharm.*, 70 (1), 248–259. doi:10.1016/j.ejpb.2008.03.005

Han, J. and Washington, C. (2005), Partition of antimicrobial additives in an intravenous emulsion and their effect on emulsion physical stability, *Int. J. Pharm.*, 288 (2), 263–271. doi:10.1016/j.ijpharm.2004.10.002

Hu, Z., Deng, Y., and Sun, Q. (2004), Synthesis of precipitated calcium carbonate nanoparticles using a two-membrane system, *Colloid J.*, 66 (6), 745–750. doi:10.1007/s10595-005-0017-4

ICH (2005), Q9 quality risk management, (Available at http://www.ich.org/fileadmin/Public_Web_Site/ICH_Products/Guidelines/Quality/Q9/Step4/Q9_Guideline.pdf.) Accessed on August 1, 2019.

ICH (2008), Q10 pharmaceutical quality system, (Availabe at http://www.ich.org/fileadmin/Public_Web_Site/ICH_Products/Guidelines/Quality/Q10/Step4/Q10_Guideline.pdf.) Accessed on August 1, 2019.

ICH (2009), Q8(R2) pharmaceutical development. Part I: pharmaceutical development, and Part II: annex to pharmaceutical development, (Available at http://www.ich.org/fileadmin/Public_Web_Site/ICH_Products/Guidelines/Quality/Q8_R1/Step4/Q8_R2_Guideline.pdf.) Accessed on August 1, 2019.

Ishii, F. and Nii, T. (2005), Properties of various phospholipid mixtures as emulsifiers or dispersing agents in nanoparticle drug carrier preparation, *Colloids Surf. B Biointerfaces*, 41 (4), 257–262. doi:10.1016/j.colsurfb.2004.12.018

Joglekar, A.T. (1987), Product excellence through design of experiments, *Cereal Foods World*, 32, 857–868.

Jumaa, M., Furkert, F.H., and Müller, B.W. (2002), A new lipid emulsion formulation with high antimicrobial efficacy using chitosan, *Eur. J. Pharm. Biopharm.*, 53 (1), 115–123. doi:10.1016/s0939-6411(01)00191-6

Jumaa, M. and Müller, B.W. (1998), The effect of oil components and homogenization condition on the physicochemical properties and stability of parenteral fat emulsions, *Int. J. Pharm.*, 163 (s1–s2), 81–89. doi:10.1016/S0378-5173(97)00369-4

Jumaa, M. and Müller, B.W. (1999), Physicochemical properties of chitosan-lipid emulsions and their stability during the autoclaving process, *Int. J. Pharm.*, 183 (2), 175–184. doi:10.1016/s0378-5173(99)00086-1

Jumaa, M. and Müller, B.W. (2002), Parenteral emulsions stabilized with a mixture of phospholipids and PEG-660-12-hydroxy-stearate: evaluation of accelerated and long-term stability, *Eur. J. Pharm. Biopharm.*, 54 (2), 207–212. doi:10.1016/s0939-6411(02)00057-7

Kador, P.F. and Kinoshita, J.H. (1978), Phospholipid effects on the rat lens transport systems, *Exp. Eye Res.*, 26 (6), 657–665. doi:10.1016/0014-4835(78)90100-8

Kim, T.W., Chung, H., Kwon, I.C. et al. (2005), Airway gene transfer using cationic emulsion as a mucosal gene carrier, *J. Gen. Med.*, 7 (6), 749–758. doi:10.1002/jgm.711

Kim, Y.J., Kim, T.W., Chung, H. et al. (2003), The effects of serum on the stability and the transfection activity of the cationic lipid emulsion with various oils, *Int. J. Pharm.*, 252 (1–2), 241–252. doi:10.1016/s0378-5173(02)00676-2

Lan, Q., Yang, F., Zhang, S. et al. (2007), Synergistic effect of silica nanoparticle and cetyltrimethyl ammonium bromide on the stabilization of O/W emulsions, *Colloids*

Surf. A Physicochem. Eng. Aspects, 302 (1–3), 126–135. doi:10.1016/j.colsurfa.2007.02.010

Lance, M.R., Washington, C., and Davis, S.S. (1995), Structure and toxicity of amphotericin B/triglyceride emulsion formulations, *J. Antimicrob. Chemother.*, 36 (1), 119–128. doi:10.1093/jac/36.1.119

Landfester, K. (2006), Synthesis of colloidal particles in miniemulsions, *Annu. Rev. Mater. Res.*, 36, 231–279. doi:10.1146/annurev.matsci.36.032905.091025

Lee, V.H.L. and Robinson, J.R. (1986), Review: topical ocular drug delivery: recent developments and future challenges, *J. Ocul. Pharmacol.*, 2 (1), 67–108. doi:10.1089/jop.1986.2.67

Liang, H., Brignole-Baudouin, F., Rabinovich-Guilatt, L. et al. (2008), Reduction of quaternary ammonium-induced ocular surface toxicity by emulsions: an in vivo study in rabbits, *Mol. Vis.*, 14, 204–216. http://www.molvis.org/molvis/v14/a26

Liu, Y., Mounkes, L.C., Liggitt, H.D. et al. (1997), Factors influencing the efficiency of cationic liposome-mediated intravenous gene delivery, *Nat. Biotechnol.*, 15 (2), 167–173. doi:10.1038/nbt0297-167

Lundberg, B.B., Griffiths, G., and Hansen, H.J. (1999), Conjugation of an anti-B-cell lymphoma monoclonal antibody, LL2, to long-circulating drug-carrier lipid emulsions, *J. Pharm. Pharmacol.*, 51, 1099–1105. doi:10.1211/0022357991776787

Lundberg, B.B., Griffiths, G., and Hansen, H.J. (2004), Cellular association and cytotoxicity of anti-CD74-targeted lipid drug-carriers in B lymphoma cells, *J. Control. Release*, 94 (1), 155–161. doi:10.1016/j.jconrel.2003.09.016

Memisoglu, E., Bochot, A., Sen, M. et al. (2002), Amphiphilic β-cyclodextrins modified on the primary face: synthesis, characterization and evaluation of their potential as novel excipients in the preparation of nanocapsules, *J. Pharm. Sci.*, 91 (5), 1214–1224. doi:10.1002/jps.10105

Midmore, B.R. (1998), Synergy between silica and polyoxyethylene surfactants in the formation of o/w emulsions, *Colloids Surf. A Physicochem. Eng. Aspects*, 145 (1–3), 133–143. doi:10.1016/S0927-7757(98)00577-9

Montgomery, D.C. (2013), *Design and Analysis of Experiments*, 8th ed., Wiley, Hoboken, NJ; Chapter 1 (p. 3,19), Chapter 11 (p. 501), Chapter 14 (p. 605).

Moolman, F.S., Rolfes, H., van der Merwe, S.W. et al. (2004), Optimization of perfluorocarbon emulsion properties for enhancing oxygen mass transfer in a bio-artificial liver support system, *Biochem. Eng. J.*, 19 (3), 237–250. doi:10.1016/j.bej.2004.02.003

Motwani, S.K., Copra, S. Talegaonkar, S. et al. (2008), Chitosan-sodium alginate nanoparticles as submicroscopic reservoirs for ocular delivery: formulation, optimization and in vitro characterization, *J. Pharm. Biopharm.*, 68 (3), 513–525. doi:10.1016/j.ejpb.2007.09.009

Muchtar, S. and Benita, S. (1994), Emulsions as drug carriers for ophthalmic use, *Colloid Surf. A Physicochem. Eng. Aspects*, 91 (3), 181–190. doi:10.1016/0927-7757(94)02882-6

Müllar, R.H. (1990), *Colloidal Carriers for Controlled Drug Delivery and Targeting: Modification, Characterization and In Vivo Distribution*, Wiss Verl-Ges, Stuttgart, Germany.

Müller, R.H. and Schmidt, S. (2002). SolEmuls technology for i.v. emulsions of poorly soluble drugs: amphotericin B, in: *Fourth World Meeting on Pharamceutics, Biopharmaceutics and Pharmaceutical Technology, Florence*, pp. 1451–1452.

Nakajima, M., Nabetani, H., Ichikawa, S. et al. (2003), Functional emulsions, US Patent 6538019.

Naskar, M.K., Patra, A., and Chatterjee, M. (2006), Understanding the role of surfactants on the preparation of ZnS nanocrystals, *J. Colloid Interface Sci.*, 297 (1), 271–275. doi:10.1016/j.jcis.2005.10.057

Negi, P., Singh, B., Sharma, G. et al. (2015), Biocompatible lidocaine and prilocaine loaded-nanosized emulsion system for enhanced percutaneous absorption: QbD-based optimisation, dermatokinetics and in vivo evaluation, *J. Microencapsul.*, 32 (5), 419–431. doi:10.3109/02652048.2015.104651

Nobs, L., Buchegger, F., Gurny, R. et al. (2004), Current methods for attaching targeting ligands to liposomes and nanoparticles, *J. Pharm. Sci.*, 93 (8), 1980–1992. doi:10.1002/jps.20098

Ogawa, S., Decker, E.A., and McClements, D.J. (2003), Production and characterization of o/w emulsions containing cationic droplets stabilized by lecithin-chitosan membranes, *J. Agric. Food Chem.*, 51 (9), 2606–2812. doi:10.1021/jf020590f

Orloff, J.J. (2011), The promise and threat of quality risk management, *Pharm. Technol.*, 35 (2), 38–40.

Pickering, S.U. (1907), Emulsions, *J. Chem. Soc.*, 91, 2001–2021. doi:10.1039/CT9079102001

Pongcharoenkiat, N., Narsimhan, G., Lyons, R.T. et al. (2002), The effect of surface charge and partition coefficient on the chemical stability of solutes in o/w emulsions, *J. Pharm. Sci.*, 91 (2), 559–570. doi:10.1002/jps.10064

Puri, D., Khatik, G.L., and Tamilvanan, S. (2019), Studies on olive-and silicone-oils-based Janus macroemulsions containing ginger to manage primary dysmenorrheal pain, *Mater. Sci. Eng. C Mater. Biol. Appl.*, 100, 276–285. doi:10.1016/j.msec.2019.01.137

Q8(R2) (2009), ICH harmonized tripartite guideline for pharmaceutical development, current step 4 version, (Accessed on August 1, 2019).

Simberg, D., Weisman, S., Talmon, Y. et al. (2003), The role of organ vascularization and lipoplex-serum initial contact in intravenous murine lipofection, *J. Biol. Chem.*, 278, 39858–39865. doi:10.1074/jbc.M302232200

Simovic, S. and Prestidge, C.A. (2003), Adsorption of hydrophobic silica nanoparticles at the PDMS droplet-water interface, *Langmuir*, 19, 8364–8370. doi:10.1021/la0347197

Simovic, S. and Prestidge, C.A. (2007), Nanoparticle layers controlling drug release from emulsions, *Eur. J. Pharm. Biopharm.*, 67 (1), 39–47. doi:10.1016/j.ejpb.2007.01.011

Simovic, S. and Prestidge, C.A. (2008), Colloidosomes from controlled interaction of submicrometer triglyseride droplets and hydrophilic silica nanoparticles, *Langmuir*, 24, 7132–7137. doi:10.1021/la800862v

Song, Y.K., Liu, D., and Maruyama, K.Z. (1996), Antibody mediated lung targeting of long-circulating emulsions, *PDA J. Pharm. Sci. Technol.*, 50 (6), 372–377.

Sonne, M.R. (2015), Tocopherol compositions for delivery of biologically active agents, in: *USPTO Patent full-text and image database no. 6,193,985,* Dumex, Copenhagen, pp. 1–21.

Solans, C., Izquierdo, P., Nolla, J. et al. (2005), Nano-emulsions, *Curr. Opin. Colloid Interface Sci.*, 10 (3–4), 102–110. doi:10.1016/j.cocis.2005.06.004

Swietlikowska, D.W. and Sznitowska, M. (2006), Partitioning of parabens between phases of submicron emulsions stabilized with egg lecithin, *Int. J. Pharm.*, 312 (1–2), 174–178. doi:10.1016/j.ijpharm.2006.01.005

Sznitowska, M., Janicki, S., Dabrowska, E.A. et al. (2002), Physicochemical screening of antimicrobial agents as potential preservatives for submicron emulsions, *Eur. J. Pharm. Sci.*, 15 (5), 489–495. doi:10.1016/s0928-0987(02)00034-9

Tadros, T. (2006), Principles of emulsion stabilization with special reference to polymeric surfactants, *J. Cosmet. Sci.*, 57, 153–169.
Tadros, T., Vandamme, A., Levecke, B. et al. (2004), Stabilization of emulsions using polymeric surfactants based on inulin, *Adv. Colloid Interf. Sci.*, 108–109, 207–226. doi:10.1016/j.cis.2003.10.024
Tambe, D.E. and Sharma, M.M. (1993), Factors controlling the stability of colloid-stabilized emulsions: I. an experimental investigation, *J. Colloid Interface Sci.*, 157, 244–253. doi:10.1006/jcis.1993.1182
Tamilvanan, S. (2004), Oil-in-water lipid emulsions: implications for parenteral and ocular delivering systems, *Prog. Lipid Res.*, 43 (6), 489–533. doi:10.1016/j.plipres.2004.09.001
Tamilvanan, S. (2008), Oil-in-water nanosized emulsions: medical applications, in: Gad, S.C., Ed., *Pharmaceutical Manufacturing Handbook*, Chapter 7.4, John Wiley & Sons Publishers, Hoboken, NJ, pp. 1329–1368. doi:10.1002/9780470259818.ch34
Tamilvanan, S. (2009), Formulation of multifunctional oil-in-water nanosized emulsions for active and passive targeting of drugs to otherwise inaccessible internal organs of the human body, *Int. J. Pharm.*, 381 (1), 62–76. doi:10.1016/j.ijpharm.2009.08.001
Tamilvanan, S. and Benita, S. (2004), The potential of lipid emulsion for ocular delivery of lipophilic drugs, *Eur. J. Pharm. Biopharm.*, 58 (2), 357–368. doi:10.1016/j.ejpb.2004.03.033
Tamilvanan, S., Khoury, K., Gilhar, D. et al. (2001), Ocular delivery of cyclosporin A. I. Design and characterization of cyclosporin A-loaded positively-charged submicron emulsion, *STP Pharma Sci.*, 11, 421–426.
Tamilvanan, S., Kumar, B., Senthilkumar, S.R. et al. (2010), Stability assessment of injectable castor oil-based nanosized emulsion containing cationic droplets stabilized by poloxamer-chitosan emulsifier films, *AAPS PharmSciTech.*, 11 (2), 904–909. doi:10.1208/s12249-010-9455-3
Tamilvanan, S., Schmidt, S., Müller, R.H. et al. (2005), In vitro adsorption of plasma proteins onto the surface (charges) modified-submicron emulsions for intravenous administration, *Eur. J. Pharm. Biopharm.*, 59 (1), 1–7. doi:10.1016/j.ejpb.2004.07.001
Tang, S.Y., Manickam, S., Wei, T.K. et al. (2012), Formulation development and optimization of a novel cremophor EL-based nanoemulsion using ultrasound cavitation, *Ultrason. Sonochem.*, 19 (2), 330–345. doi:10.1016/j.ultsonch.2011.07.001
Teixeira, H., Dubernet, C., Puisieux, F. et al. (1999), Submicron cationic emulsions as a new delivery system for oligonucleotides, *Pharm. Res.*, 16, 30–36. doi:10.1023/A:1018806425667
Trotta, M., Pattarino, F., and Ignoni, T. (2002), Stability of drug-carrier emulsions containing phosphatidylcholine mixtures, *Eur. J. Pharm. Biopharm.*, 53 (2), 203–208. doi:10.1016/s0939-6411(01)00230-2
Velev, O.D., Furusawa, K., and Nagayama, K. (1996), Assembly of latex particles by using emulsion droplets as templates. 1. Microstructured hollow spheres, *Langmuir*, 12, 2374–2384. doi:10.1021/la9506786
Weiss, C.K., Lorenz, M.R., Landfester, K. et al. (2007), Cellular uptake behavior of unfunctionalized and functionalized PBCA particles prepared in a miniemulsion, *Macromol. Biosci.*, 7, 883–896. doi:10.1002/mabi.200700046
Woitiski, C.M. Veiga, F. Ribiero, A. et al. (2009), Design for optimization of nanoparticles integrating biomaterials for orally dose insulin, *Eur. J. Pharm. Biopharm.*, 73 (1), 25–33. doi:10.1016/j.ejpb.2009.06.002

Yaseen, M., Lu, J.R., Webster, J.R.P. et al. (2006), The structure of zwitterionic phosphocholine surfactant monolayers, *Langmuir*, 22, 5825–5832. doi:10.1021/la053316z

Yi, S.W., Yune, Y., Kim, T.W. et al. (2000), A cationic lipid emulsion/DNA complex as a physically stable and serum-resistant gene delivery system, *Pharm. Res.*, 17, 314–320. doi:10.1023/A:1007553106681

Yilmaz, H. and Borchert, H.H. (2005), Design of a phytosphingosine-containing, positively charged nanoemulsion as a colloidal carrier system for dermal application of ceramides, *Eur. J. Pharm. Biopharm.*, 60, 91–98. doi:10.1016/j.ejpb.2004.11.009

Yu, L.X. (2008), Pharmaceutical quality by design: product and process development, understanding, and control, *Pharm. Res.*, 25 (4), 781–791. doi:10.1007/s11095-007-9511-1

Yu, L.X., Amidon, G., Khan, M.A. et al. (2014), Understanding pharmaceutical quality by design, *AAPS J.*, 16 (4), 771–781. doi:10.1208/s12248-014-9598-3

CHAPTER 3

CHARACTERIZATION AND SAFETY ASSESSMENT OF OIL-IN-WATER NANOSIZED EMULSIONS

SYED NAZRIN RUHINA RAHMAN* AND TAMILVANAN SHUNMUGAPERUMAL

Department of Pharmaceutics, National Institute of Pharmaceutical Education and Research (NIPER), Guwahati, Assam, India

*The funding provided by Department of Biotechnology, Govt. of India to Syed Nazrin Ruhina Rahman is greatly acknowledged.

Oil-in-Water Nanosized Emulsions for Drug Delivery and Targeting, First Edition. Tamilvanan Shunmugaperumal.
© 2021 John Wiley & Sons, Inc. Published 2021 by John Wiley & Sons, Inc.

EXPANSION OF ABBREVATIONS

ACZM	acetazolamide
AFM	atomic force microscopy
API	active pharmaceutical ingredient
AZM	azithromycin
BSS	balanced salt solution
CAM	chorioallantoic membrane
CsA	cyclosporin A
CK	creatine kinase
CSF	cerebrospinal fluid
CXB	celecoxib
DE	dry eyes
DEE%	drug entrapment efficiency in percentage
DLS	dynamic light scattering
DLVO	Derjaguin, Landau, Verwey, and Overbeek
DMEM	Dulbecco's modified Eagle's medium
DP	degradation product
E	entering coefficient
EC	European Country
EDL	electric double layer
EIPA	5-(*N*-ethyl-*N*-isopropyl) amiloride
EU	European Union
FBS	fetal bovine serum
FCS	fetal calf serum
FRET	Forster resonance energy transfer
HET-CAM	Hen's Egg Test-Chorioallantoic Membrane
HPLC	high-performance liquid chromatography
ICH	International Council for Harmonization
IRAG	Interagency regulatory alternatives group

IOP	intraocular pressure
IS	irritation score
IUPAC	International Union of Pure and Applied Chemistry
IV	intravenously
IVIVC	*in vitro–in vivo* correlation
J_{ss}	steady state
LVET	low-volume eye test
MLS	multiple light scattering
MPS	mean particle size
MTT	3-(4,5-dimethylthiazol-2-yl)-2,5-diphenyltetrazolium bromide
NCs	nanocarrriers
NIBUT	noninvasive tear break-up time
NIST	National Institute of Standards and Technology
OD	optical density
OLI	ocular lesion index
O/w	oil-in-water
PCS	photon correlation spectroscopy
PBS	phosphate-buffered saline
PDI	polydispersity index
P_{app}	permeability coefficient
PAO	phenylarsine oxide
Psi	pounds per square inch
PSPD	position sensitive photodetector
PZC	point of zero charge
QY	quantum yield
RBC	red blood cell
RI	refractive index
SA	stearylamine
S	spreading coefficient
SC	stratum corneum
SDS	sodium dodecyl sulphate
SS	Sjögren's syndrome
STF	simulated tear fluid
TEM	transmission electron microscopy
TBUT	tear break-up-time
Turbiscan ags	Turbiscan automated ageing station
UC	ultracentrifugation
UF	ultrafiltration
UHPLC	ultra-high-performance liquid chromatography
$W_{1,2}$	work of adhesion
$W_{1,1}$	work of cohesion
ZP	zeta potential

3.1. INTRODUCTION

It is highly desirable to arrange serially or systematically the various *in vitro* and *ex vivo* characterization techniques reported so far in relation to the oil-in-water (o/w) nanosized emulsions. Very broadly, the serial or systematic arrangement of reported characterization techniques can be sorted out depending on the intended therapeutic application of the developed emulsion, stability of incorporated active pharmaceutical ingredient (API) molecules, stability of the oil and water phases itself, degradation/stability of the selected emulsifier molecules and single oil or oil combination, etc. Flowchart 3.1 displays the serial or systematic arrangement of reported characterization techniques along with the various evaluation techniques used to assess the safety of the o/w nanosized emulsions administered both via ophthalmic and parenteral routes. It should be added that few toxicity studies need to be performed for the emulsions to comply/fulfill the regulatory norms of different countries.

The objective of this chapter is to elaborate with experimental proofs concerning the characterization and safety assessment tests conducted for o/w nanosized emulsions at least in laboratory/pre- or non-clinical scale. Since most of the commercially available emulsions are mainly used for ophthalmic and parenteral administrations, the characterization and safety assessment studies included in this chapter are mostly relevant to the ophthalmic and parenteral emulsions. However, the o/w nanosized emulsions meant for other routes of administration could also be utilized for discussing the characterization and safety assessment studies.

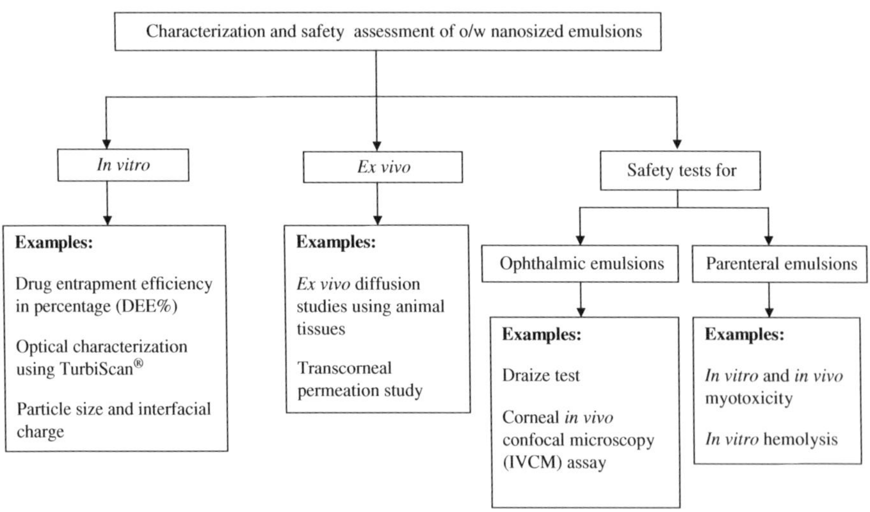

Flowchart 3.1. Non-exhaustive list of serially or systematically arranged various *in vitro* and *ex vivo* characterization techniques along with safety evaluation tests reported so far in relation to the o/w nanosized emulsions.

3.2. PHYSICAL AND VISUAL EVALUATIONS

3.2.1. Refractive Index

Theoretically, refractive index (RI) is the ratio of the velocity of light of a specified wavelength in the air to its velocity in the test samples. There are two definitions/explanations to determine the RI value of the test samples. Firstly, for the purpose of measuring the velocity of light of a specified wavelength, the RI can be defined as the sine of the angle of incidence divided by the sine of the angle of refraction. The absolute RI of air is 1.00027. Secondly, the absolute angle of refraction (relative to vacuum) can be calculated by dividing the RI relative to air by 1.00027. The ratio of the sines of the incident and refractive angles of light in the tested sample is equal to the ratio of light velocity to the velocity of light in a vacuum (that is why both definitions are correct). This equality is also referred to as Snell's law (Silla et al. 2019). Using this second definition (Snell's law), the RI is measured by using the formula as shown in Eq. (3.1),

$$n = \frac{c}{v} \tag{3.1}$$

where n = refractive index, c = speed of light in vacuum, and v = speed of light in medium.

Whether liquid or dispersion samples, the laboratory instrument used to accurately measure the RI was design/developed in the year 1874 by Ernst Abbe. Therefore, the instrument used for measuring the RI value of test samples is called as Abbe's refractometer. Determining the RI value of o/w nanosized emulsion correlates the acuteness of vision, especially the emulsion is administered via the topical ocular route. The RI value is also one of the useful parameters for matching the specific gravity and an osmotic pressure of cerebrospinal fluid (CSF) when the formulation is intrathecally administered. Therefore, if nanosized emulsion is also meant for intrathecal administration, then it is mandatory for the emulsion to be isobaric with that of the CSF as far as possible.

Using the Abbe's type refractometer at $25 \pm 0.5°C$, the RI value of nanosized emulsion can be determined. In fact, Tamilvanan and Kumar (2011) have prepared nanosized emulsions consisting of lecithin–oleic acid–poloxamer, poloxamer–chitosan, and lecithin–poloxamer emulsifier combinations to confer anionic, cationic, and neutral charges over dispersed oil droplets, respectively. Arachis oil was used as the oil phase to prepare all of these emulsions and the viscosity and RI of these emulsions were measured. Irrespective of the charges, the emulsions developed showed the RI value of 1.345, which is close to the RI value of water (1.332) indicating that the emulsion can be suitable for ocular use. A similar conclusion was made concerning the closeness of values of RI determined by Abbe's refractometer between nanosized emulsions and water (Baboota et al. 2007; Kumar 2014; Harika et al. 2015).

3.2.2. Viscosity

Since the emulsifier molecules are integral part of o/w nanosized emulsions to stabilize the dispersed oil droplets against aggregation, coalescence, and Ostwald ripening, the mechanism of action of these emulsifier molecules often related to reducing the o/w interfacial tension, providing steric hindrance to the dispersed oil droplets and somehow increasing the viscosity of emulsion. However, decreasing the amount of emulsifier molecules increases the interfacial tension between oil and water resulting in overall enhancement in the viscosity of emulsion due to the coalescence of dispersed oil droplets. Another most worrying fact in the nanosized emulsion is the influence of the environmental temperature or storage temperature conditions onto the overall viscosity of the emulsion. Hence, viscosity is one of the key thermophysical properties of nanosized emulsions in determining the physical stability of emulsions. In rheological term, the o/w nanosized emulsions simply belong to the category of pseudoplastic flow and therefore they are non-Newtonian fluid. For pseudoplastic flow, the exponential formula [Eq. (3.2)] has been used most frequently.

$$\eta = \frac{FN}{G} \tag{3.2}$$

The exponent N (Farrow's constant) rises as the flow becomes increasingly non-Newtonian. The terms F, G, and η represent, respectively, shearing stress, rate of shear, and viscosity coefficient. By the arrangement of the above, Eq. (3.3),

$$\log G = N \log F - \log \eta \tag{3.3}$$

A straight line relationship is obtained for many pseudoplastic systems when $\log G$ is plotted as a function of $\log F$ (Martin et al. 1964).

Various instruments are employed for measuring viscosity such as Brookfield viscometer, Ferranti-Shirley viscometer, Hoeppler falling ball viscometer, Ostwald viscometer, Stormer viscometer, and Ubbelohde capillary viscometer. An Ubbelohde capillary viscometer was used to measure the viscosity of the o/w nanosized emulsions (Tamilvanan and Kumar 2011). The developed emulsions consist of lecithin–oleic acid–poloxamer, poloxamer–chitosan, and lecithin–poloxamer emulsifier combinations to confer anionic, cationic, and neutral charges over dispersed oil droplets, respectively. Arachis oil was used as the oil phase to prepare all of these emulsions. The viscosity of these emulsions was measured and it was found to be 1.5 cps close to the viscosity values of water (0.894543) indicating that the emulsion can be suitable for administering via parenteral and topical ocular routes.

3.2.3. pH

As proposed by Danish biochemist Soren Sorensen in 1909, the expression related to the presence of hydrogen ion concentration in aqueous solutions is notified in terms of pH unit value. Sorensen defined the pH value as the negative of the logarithm of the concentration of hydrogen ions. For example, when water molecule dissociates, it yields a hydrogen ion and a hydroxide [Eq. (3.4)] (Petrucci et al. 2011).

$$H_2O \leftrightarrow H^+ + OH^- \tag{3.4}$$

The equilibrium equation yields the following formula for pH [Eq. (3.5)]:

$$pH = -\log_{10}\left[H^+\right] \tag{3.5}$$

Coming to the o/w nanosized emulsions, the pH of the emulsion usually fixes at close to neutral unit immediately after the autoclave-sterilized emulsions. However, the change in pH of the emulsion with time as a function of storage temperature was noticed by Abdulrazik et al. (2001). Temperature conditions of 4, 25, and 37°C were chosen to store the emulsions for checking temperature-induced pH change over the storage time period of 6 months. Figure 3.1 clearly shows the reduction of emulsion's pH at 25 and 37°C. A much more pronounced pH of the emulsion stored at 37°C was observed compared with the emulsion stored at 25°C. It is highly noteworthy to see the result of the emulsion stored at 4°C wherein the initial reduction of pH started only after 2 months. Even after initiation of pH reduction from 2 months onward, the progression in the reduction of pH was not so rapid in the following months as compared with the higher storage temperature conditions (25 and 37°C). The emulsion's pH was found to be decreased only by single unit over the storage time of 6 months at 4°C. Almost 1–4 units of pH values were found to decrease when the emulsion was stored at 37°C. The pH reduction was well correlated with accelerated oxidation and hydrolysis of the phospholipid emulsifier molecules present in the emulsion (Hansrani et al. 1983; Washington and Davis 1987). Neither creaming and coalescence (therefore, mean particle size, MPS) nor zeta potential (ZP) value reduction value was observed for the emulsion stored at 37°C even though the pH was decreased from a value of 7.5–3.5 over the period of 6 months storage (Abdulrazik et al. 2001). In recent years, the pH

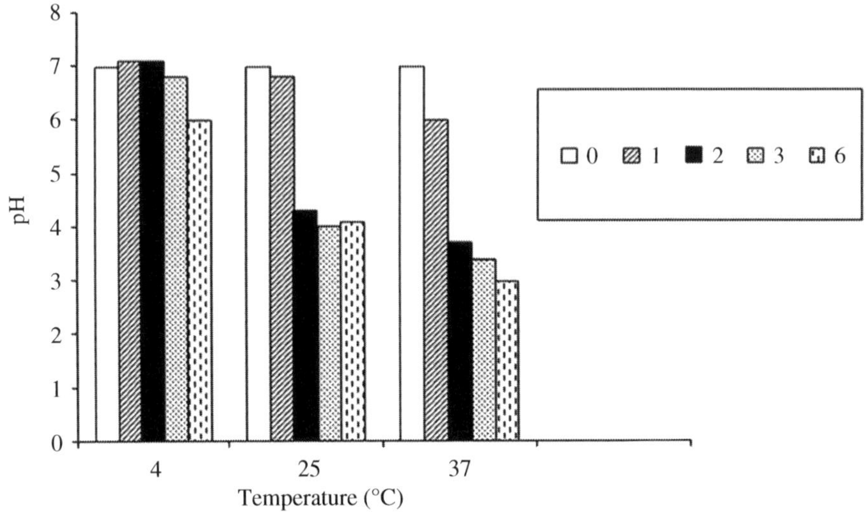

Figure 3.1. Influence of 6 months storage temperature (4, 25, and 37°C) conditions on the pH of oil-in-water nanosized emulsions.

change due to auto-oxidation of phospholipid emulsifier molecules is renamed as microclimate pH change.

During the initial emulsion development period at Second World War time, the various forms of phospholipid were ostensibly used as one of the emulsifier molecules to stabilize the pharmaceutical emulsions due to their structural closeness with endogenous lipids. However, even with the addition of antioxidant molecules like alpha-tocopherol, the phospholipid-stabilized emulsions showed the liberation of free fatty acid heavily over the stipulated storage time periods, thus leading to the microclimate pH change (Abdulrazik et al. 2001; Rabinovich-Guilatt et al. 2004). In addition, the microclimate pH change thus generated might be harmful to the emulsion-incorporated API molecules. To eradicate the generation of free fatty acid particularly from the phospholipid components and thus the microclimate acidic pH in the vicinity of oil phase, oil–water interface, and water phase of the emulsions, a non-phospholipid-based injectable nanosized emulsion containing cationic droplets stabilized by poloxamer–chitosan emulsifier film was prepared previously (Tamilvanan et al. 2010). Furthermore, it could be useful to develop the phospholipidless emulsion possessing the transient kinetic stability for at least over months. Even, the complete elimination of phospholipid from the emulsions unequivocally facilitated the development of solid-dry powder form of the liquid-retentive emulsion via lyophilization or nano-spray drying (unpublished results).

3.2.4. Creaming

Gravitational force plays an important role to determine/measure the stability of o/w nanosized emulsions. Depending on the density, the dispersed phase may travel either downward or upward direction. This movement of dispersed phase within the dispersion medium of emulsion acquires two different processes termed as creaming and sedimentation. The dispersed phase with high density moves downward to reach the bottom of the dispersion medium of the emulsion and gets the name sedimentation. The dispersed phase with lower density moves upward to reach the surface or top of the dispersion medium of the emulsion and catches the name creaming. Whatever it may be, excessive creaming and sedimentation results ultimately in the break of oil and water phases of the emulsion. Both creaming and sedimentation phenomena are depicted in Fig. 3.2.

Majority of the FDA-approved oils (animal, mineral, and vegetable based) used in the manufacturing of the o/w nanosized emulsions possess the density value either lower or close to the density value of water ($0.99777\,g\,ml^{-1}$ at 22°C). Thus, creaming is more common in o/w nanosized emulsions. However, in the case of emulsions with crystalline lipids, the density of oil phase increases during crystallization. Moreover, lipid droplets may be surrounded by thick and biopolymer coatings with higher density than oil. These types of emulsions have the tendency of both creaming and sedimentation. The rate of creaming is expressed in terms of creaming velocity, which can be calculated from Stokes' law [Eq. (3.6)].

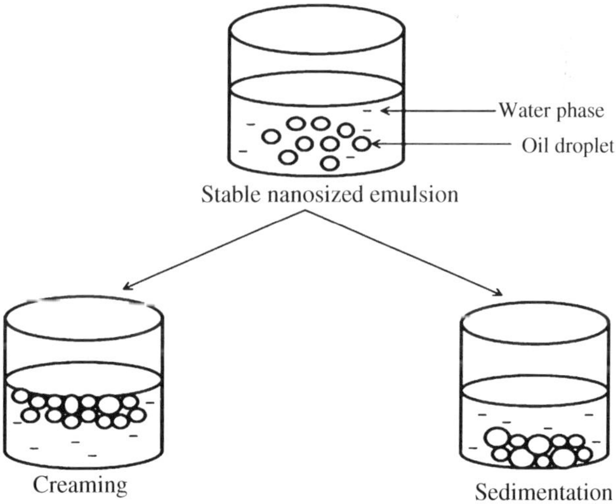

Figure 3.2. Density- and gravitational force-dependent/induced physical instabilities in o/w nanosized emulsions.

$$v = \frac{2gr^2\left(\rho_w - \rho_o\right)}{9\eta_w} \tag{3.6}$$

where v is velocity of the droplet, ρ_w and ρ_o are density of water and oil phase, respectively, r is radius of the droplet, η_w is viscosity of water phase, and g is acceleration due to gravity.

It can be seen from Stokes' law that creaming or sedimentation can be reduced by reducing the density difference between two phases, reducing droplet size, and increasing viscosity of the continuous phase. Generally, an emulsion can be assumed as stable for a low value of creaming velocity (e.g., $v < 1\,\text{mm}\,\text{day}^{-1}$) (Dickinson 1992; Akoh and Min 2002). Consequently, tiny droplets in nanosized emulsions can reduce gravitational separation. But the so-called *zig-zag* or tiny particle random motion (also named as Brownian motion) becomes prominent in the nanosized emulsions. Because, the Brownian motion significantly increases with decrease in droplet size. Interestingly, the gravitational separation (creaming or sedimentation) is dominant for large-size droplets and the Brownian motion is dominant for small-size droplets. Using the particle size distribution profile of o/w nanosized emulsions, McClements (2011) observed that creaming does not occur if the droplet size is below 10 nm. Though the interfacial layer consisting of single or multiple emulsifier molecular film significantly influences the properties of droplets, the presence of Brownian motion is highly inevitable. In addition, the density of emulsifier and volume surrounding the dispersed oil droplets also control the gravitational separation or instability of emulsion. Both these parameters are correlated to density of oil phase (ρ_o). Collectively, the density of oil phase is calculated by using the following Eq. (3.7).

$$\rho_o = \rho_s\varphi_s + \left(1 - \varphi_s\right)\rho_w \tag{3.7}$$

where ρ_s is density of emulsifier layer, ρ_w is density of water phase, and φ_s is volume fraction of emulsifier layer. The density of emulsifier layer is generally more than either oil or water phase. Thus, overall density of dispersed oil phase of the emulsion increases with volume fraction (φ_s) of emulsifier layer. This reduces the density difference between continuous and dispersed phase, resulting in the prevention of gravitational separation in nanosized emulsions. On the other hand, if the dispersed oil droplets are covered/surrounded with sufficiently thick and dense emulsifier film, then their density increases, which may result in sedimentation. Thus, by controlling the droplet size and thickness of the adsorbed emulsifier layer, it is possible to set the desired density for reducing creaming or sedimentation of droplets (Pathak 2017).

In practice, creaming or sedimentation of dispersed oil droplets of the emulsion can be measured at various time intervals according to the method reported by Krishna et al. (1998) as shown in the following Eq. (3.8):

$$CI = \frac{H_{aq}}{H_{tot}} \times 100 \quad (3.8)$$

where H_{aq} = height of the aqueous phase of emulsion, H_{tot} = total height of the emulsion.

From the above discussion, it is clear that emulsifier molecules play a vital role in determining or measuring the physical stability of the emulsions, especially creaming and sedimentation. However, as mentioned in Chapter 1, the emulsions can be classified into three types based upon the emulsifier combination used to prepare the autoclave-sterilized stable nanosized emulsions. Table 3.1 portrays the different types of emulsions along with their compositions. All these three different types emulsions were underwent for creaming index study.

About 9 ml of sterilized azithromycin (AZM)-loaded emulsions were taken in measuring cylinders and they were kept aside for the period of 90 days at room temperature (Tamilvanan et al. 2013). The creaming index (in centimeter) values were noted at the various time intervals (Table 3.2).

TABLE 3.1. Different Types of Emulsions Along with Their Compositions

Emulsion Types	Compositions
Cationic emulsion	Castor oil, azithromycin, chitosan, α-tocopherol, poloxamer 188, glycerin, and bi-distilled water
Anionic emulsion	Castor oil, azithromycin, lecithin, oleic acid, α-tocopherol, poloxamer 188, glycerin, and bi-distilled water
Neutral-charged emulsion	Castor oil, azithromycin, lecithin, α-tocopherol, poloxamer 188, glycerin, and bi-distilled water

TABLE 3.2. Creaming Index Values Observed for Different Emulsions Over the Period of 90 Days at Room Temperature

Time Intervals	Creaming Index (cm)		
	Non-phospholipid-Based Cationic Emulsion	Phospholipid-Based Anionic Emulsion	Neutral-Charged Emulsion
Initial	9.0	9.0	9.0
1 Hours	8.6	8.8	8.9
12 Hours	8.6	8.8	8.9
24 Hours	8.6	8.8	8.9
2 Days	8.6	8.8	7.9
4 Days	8.6	8.8	6.9
50 Days	8.4	6.4	4.6
90 Days	8.4	5.2	3.3

Only 0.6 cm reduction from the original volume of the two-phase system (9.0 cm) was noticed for the cationic emulsion. This indicates that the phase separation after 90 days was minimal for the cationic emulsion. However, the anionic emulsion showed 3.8 cm reduction and the neutral-charged emulsion had 5.7 cm reduction following 90 days stand-in position. Apparently, the emulsifier combinations used to stabilize the last two emulsions were unable to prevent the coalescence of emulsified oil droplets upon random collision leading to the phase-separation process.

3.2.5. Stability Testing

The stability testing for o/w nanosized emulsions is meant for the finished products (autoclave sterilized). The main aims of stability testing are to gain the information related to the performance of finished product after keeping it in various environmental conditions such as temperature, humidity, and light, and to establish a retest period for the API substance or a shelf life for the API product and recommended storage conditions.

The choice of test conditions defined in the International Council for Harmonization (ICH) guideline is based on an analysis of the effects of climatic conditions in the three regions of the European Country (EC), Japan, and the United States. The mean kinetic temperature in any part of the world can be derived from climatic data, and the world can be divided into four climatic zones, I–IV. This guideline addresses climatic zones I and II. The principle has been established that stability information generated in any one of the three regions of the EC, Japan, and the United States would be mutually acceptable to the other two regions, provided the information is consistent with this guideline and the labeling is in accord with national/regional requirements.

In general, an API should be evaluated under storage conditions (with appropriate tolerances) that test its thermal stability and, if applicable, its sensitivity to moisture. The storage conditions and the lengths of studies chosen should be sufficient to cover storage, shipment, and subsequent use. The long-term testing should cover a minimum of 12 months' duration on at least three primary batches at the time of submission and should be continued for a period of time sufficient to cover the proposed retest period. Additional data accumulated during the assessment period of the registration application should be submitted to the authorities if requested. Data from the accelerated storage condition and, if appropriate, from the intermediate storage condition can be used to evaluate the effect of short-term excursions outside the label storage conditions (such as that might occur during shipping).

Long term, accelerated, and, where appropriate, intermediate storage conditions for APIs are detailed in the sections below. The general case applies if the API or its product is not specifically covered by a subsequent section. Alternative storage conditions can be used if justified. Table 3.3 shows the ICH guidelines for the Stability testing protocol.

TABLE 3.3. Stability Testing Protocol as per ICH Guidelines

Study	Storage Condition	Minimum Time Period Covered by Data at Submission (Months)
Long term[a]	25°C ± 2°C/60% RH ± 5% RH or 30°C ± 2°C/65% RH ± 5% RH	12
Intermediate[b]	30°C ± 2°C/65% RH ± 5% RH	6
Accelerated	40°C ± 2°C/75% RH ± 5% RH	6

[a]It is up to the applicant to decide whether long-term stability studies are performed at 25 ± 2°C/60% RH ± 5% RH or 30°C ± 2°C/65% RH ± 5% RH.

[b]If 30°C ± 2°C/65% RH ± 5% RH is the long-term condition, there is no intermediate condition.

The autoclave-sterilization of the o/w nanosized emulsions itself is a very drastic stress condition because heating the emulsion up to 121°C and allowing the emulsion at 14–15 psi (pounds per square inch) for 15 min. These drastic stress conditions applied onto the emulsion will definitely disturb the physical stability of the nanosized emulsions. However, most of the emulsions developed and even commercialized did not show the physical instability problems following terminal sterilization, especially by autoclaving. Because, the dispersed oil droplets of o/w nanosized emulsions are basically stabilized by multiple numbers of emulsifier molecules that were added during the emulsion preparation step. However, the physical property values such as MPS, pH, and ZP of the nanosized emulsions will show definite changes. For example, the pH value may reduce to acidic unit due to the auto-oxidation of phospholipid emulsifier molecules or the sensitive fatty acids present in the oils, which were used to make the emulsions. Usually, the MPS of emulsions will get increment due to the coalescence or unison of small-sized oil droplets during the autoclave sterilization of emulsions. Although the ZP is the indication of charge present at the oil–water interface of the emulsions, especially the electronation/protonation of anion/cation-conferring emulsifier molecules, the ZP value might change by the pH variation that occurred due to the autoclave stressing conditions. Irrespective of the changes in physical property values (MPS, pH, and ZP), the emulsion after its terminal autoclave sterilization should show an acceptable or a desirable value, i.e., comparable values before and after the autoclave sterilization.

3.2.5.1. Stress Testing The stress testing is usually applied to the API molecules only. It helps to identify the likely API degradation products, which can, in turn, help establish the degradation pathways and the intrinsic stability of the molecule and validate the stability indicating power of the analytical procedures developed/used. The nature of the stress testing will depend on the individual API molecules and the type of final API product involved. Stress testing is likely to be carried out on a single batch of API molecules. It should include the effect of temperatures (in 10°C increments (e.g., 50, 60°C, ...) above that for accelerated testing), humidity (e.g., 75% RH or greater) where appropriate, oxidation, and photolysis on the API molecules. The testing should also evaluate the susceptibility of the API molecules to hydrolysis across a wide range of pH values when in solution or suspension. Photostability testing should be an integral part of stress testing. A case report is shown here to explain the importance of stress testing on an API molecule. The selected API molecule is cyclosporin A (CsA).

CsA is a major representative of a family of hydrophobic cyclic undecapeptides produced by the fungus *Tolypocladiuminflatum Gams* and other fungi imperfecti or via directed biosynthesis or chemical total synthesis. Over the past years, CsA has been successful when used topically in a variety of immune-mediated ocular surface phenomena like vernal conjunctivitis, dry eye syndrome, and the prevention of corneal allograft rejection (Lallemand et al. 2003). Figure 3.3 shows the structure of CsA.

CsA consists of International Union of Pure and Applied Chemistry (IUPAC)name:(3*S*,6*S*,9S,12*R*,15*S*,18*S*,21*S*,24*S*,30*S*,33*S*)-30-ethyl-33-[(*E*,1*R*,2*R*)-1-hydroxy-2-methylhex-4-enyl]-1,4,7,10,12,15,19,25,28-nonamethyl-6,9,18,

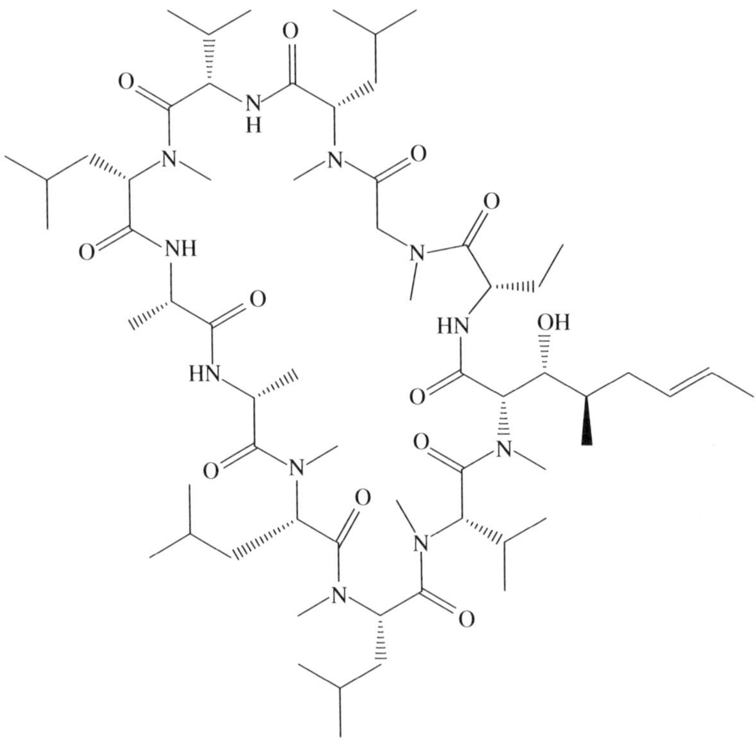

Figure 3.3. Structure of cyclosporine A. (Drawn by using ChemDraw Ultra 12.0.)

24-tetrakis(2-methylpropyl)-3,21-di(propan-2-yl)-1,4,7,10,13,16,19,22,25,28,31-undecazacyclotritriacontane-2,5,8,11,14,17,20,23,26,29,32-undecone. Its molecular formula is $C_{62}H_{111}N_{11}O_{12}$ and its molecular weight is 1,202.6 g mol^{-1}. It appears in white or almost white in color and available as crystalline form with white prismatic needles. The melting point of CsA is 148–151°C. It possesses the solubility in organic solvents such as methanol, ethanol, acetone, ether, and chloroform. It is also slightly soluble in saturated hydrocarbons and in water. The aqueous solubility of CsA does not exceed 25–50 μg ml^{-1} at 25°C. The partition coefficient (log *P*) value of CsA is 2.92 indicating its lipophilic nature. The CsA remains stable for at least 2 years when stored as a dry powder at 2–8°C in a dark place and for over 7 months at 40°C and 75% relative humidity (Czogalla 2009). Due its peptidic nature, the CsA molecular conformation is always prone to slight changes in the environmental conditions, which include pH, light, oxidation, etc. (Deshmukh and Sharma 2016). Based on this report, an intentional degradation study using CsA solution was conducted (unpublished results) and the results are briefly discussed below.

The 1 mg ml^{-1} CsA solution was made using acetonitrile. The selected stressor conditions to carry out the intentional degradation study are 0.1 N solutions of HCl and NaOH, 30% v/v H_2O_2, and sunlight for 30 min at ambient temperature. A 1 : 1 ratio of API and respective stressor solutions (HCl, NaOH, and H_2O_2) were made to represent acidic, basic, and oxidative degradations for treated CsA test samples. But for photolytic degradation study, 1 ml of API solution was taken in a dark brown color Eppendorf tube and the tube was kept directly under sunlight. Untreated CsA solution was taken as the control for all intentional degradation studies. Following the 30 min reaction time, samples were collected from both the treated and untreated samples using a syringe fitted with the 0.22 μ membrane filter and injected into the ultra-high-performance liquid chromatography (UHPLC) system (using the mobile phase consisting of 74% acetonitrile, 26% water, and 0.025% orthophosphoric acid) for measuring the CsA content and its possible degradation products (represented as DP_1 and DP_2) (Table 3.4).

From Table 3.4, it can be deduced that the untreated solution of CsA in acetonitrile did not show any DPs in UHPLC chromatogram. However, treating the solution of CsA in acetonitrile with the studied stress conditions resulted in the formation of two additional peaks (named as DP_1 and DP_2) in the UHPLC chromatogram. It should be of interest what could happen when the CsA is loaded in the o/w nanosized emulsion and subjected to the different stress conditions.

Fifty microliters of CsA (0.05 and 0.1% w/w)-loaded nanosized emulsions were taken in three different vials. The first vial was kept in an autoclave condition at 121°C and 15 psi pressure for 15 min. The second vial was placed in hot air oven condition at 70°C for 30 min while the third vial was set aside at a room temperature condition of 25°C for 30 min. Following the 30 min reaction

TABLE 3.4. Peak Area and UHPLC Chromatogram of Cyclosporin A (CsA) and Its Possible Degradation Products (DP_1 and DP_2) Observed in Treated Test Samples (1 mg ml^{-1} Solution of CsA in Acetonitrile) Following Acid, Alkali, Oxidation, and Sunlight Stress Conditions in Comparison with Untreated 1 mg ml^{-1} Solution of CsA in Acetonitrile

Untreated/Treated Samples	Inference	Peak Area (mAU·min) (Mean ± SD)	Chromatogram of CsA
Untreated/control sample	NA	1,338.3 ± 13.28 (CsA)	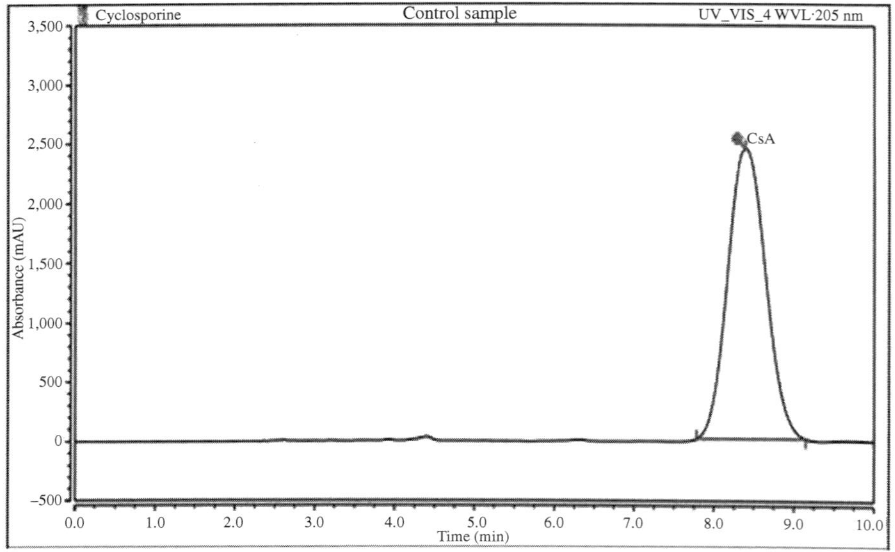

Acid (0.1 N HCl)	Degradation	13.06 ± 1.01 (DP_1) 1.305 ± 0.21 (DP_2) 617.82 ± 38.09 (CsA)	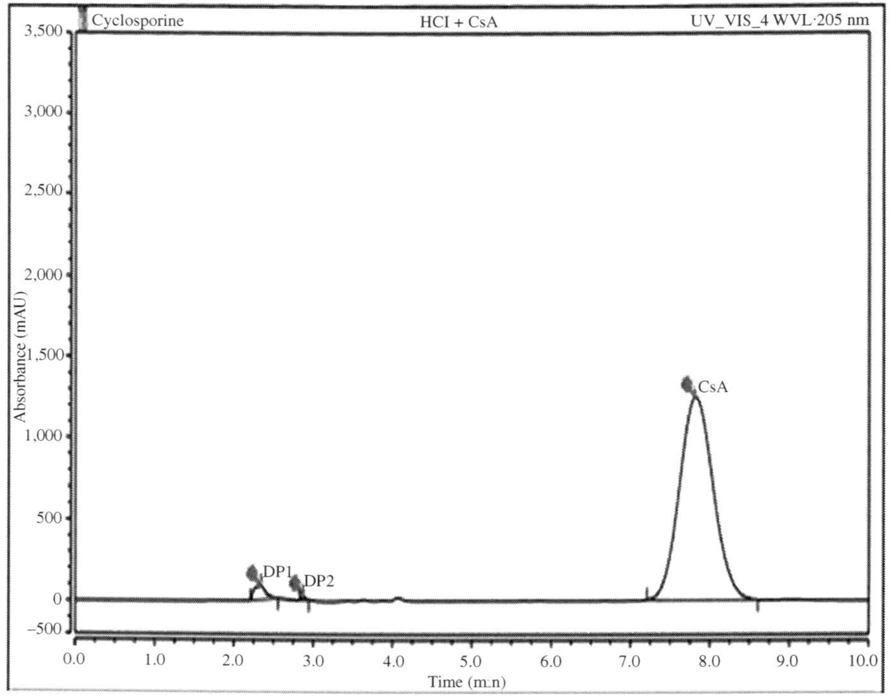

TABLE 3.4. ***Continued***

Untreated/Treated Samples	Inference	Peak Area (mAU·min) (Mean ± SD)	Chromatogram of CsA
Alkali (0.1 N NaOH)	Degradation	15.734 ± 0.73 (DP_1) 2.843 ± 0.43 (DP_2) 739.4 ± 26.99 (CsA)	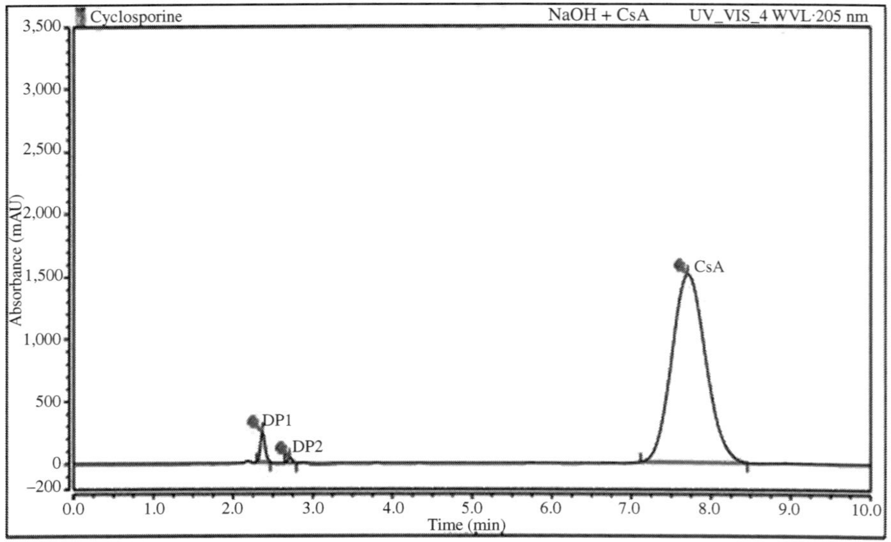

Hydrolysis (H_2O_2)	Degradation	260.1 ± 3.89 (DP_1) 62.78 ± 1.13 (DP_2) 539.94 ± 6.31 (CsA)	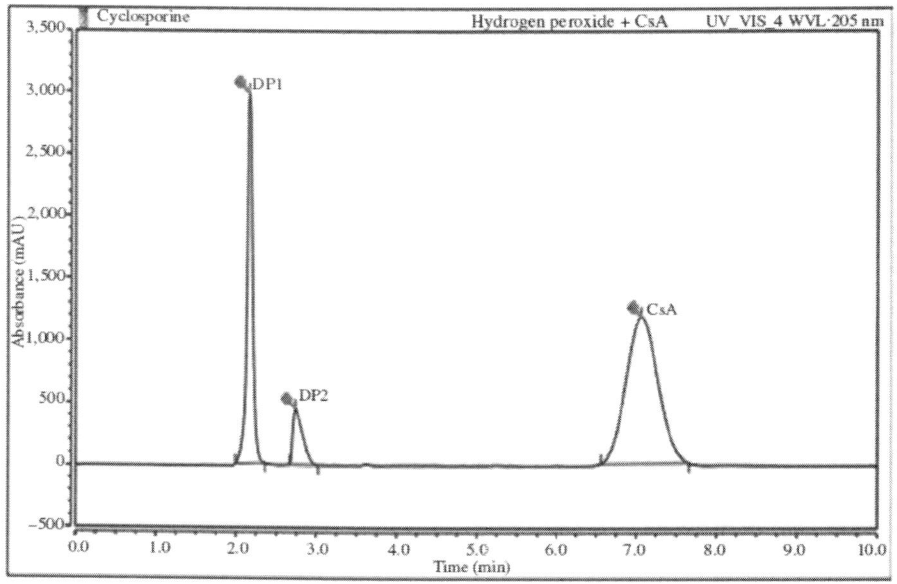

TABLE 3.4. ***Continued***

Untreated/Treated Samples	Inference	Peak Area (mAU·min) (Mean ± SD)	Chromatogram of CsA
Photolytic (sunlight)	No degradation	683.27 ± 1.85 (CsA)	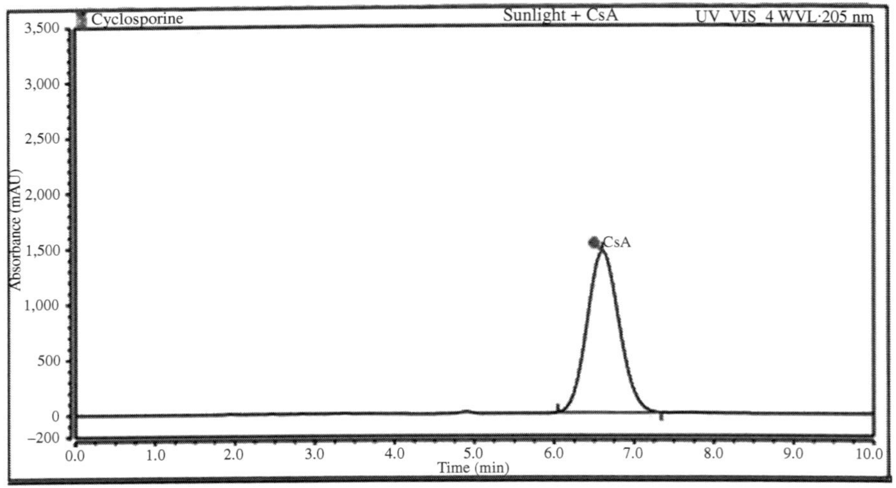

NA, not applicable.

time, samples were collected from all the three vials and diluted up to 1 ml with acetonitrile. The CsA contents in each sample were analyzed using the UHPLC system. Table 3.5 shows the peak area and UHPLC chromatograms obtained with the treated nanosized emulsions following subjection into different thermodegradation conditions along with the untreated/controlled CsA (0.05 and 0.1% w/w)-loaded nanosized emulsions.

The CsA in its two different concentration levels (0.05 and 0.1% w/w), when loaded into o/w nanosized emulsion, looks safely entrapped in oil phase, oil–water interface, and water phase of the emulsion following subjection into three different thermodegradation conditions. That is why no peaks to indicate the possible degradation of CsA were obtained in the UHPLC chromatograms of API-loaded emulsions after allowing them into different conditions. This indicates that the CsA structure-protecting ability of emulsion formulation over the CsA solution made from acetonitrile (Table 3.5) wherein the API was shown at two different peaks in UHPLC chromatograms signifying the occurrence of possible API degradation within the solution.

3.2.6. Particle Size and Interfacial Charge

Dynamic light scattering (DLS) or photon correlation spectroscopy (PCS) is used to analyze the fluctuations in the intensity of scattering by dispersed droplets/particles due to Brownian motion (Ruth et al. 1995). While measuring the hydrodynamic size of dispersed droplets/particles, which are undergoing Brownian motion, the DLS/PCS is generally limited to measuring the particle sizes in the range of submicron region. The Brownian motion of dispersed droplets/particles is basically measured by the scattering intensity fluctuations. Therefore, several experimentally determined controlled factors such as particle diameter (i.e., the diameter of a sphere that has the same translational diffusion coefficient as the particles of interest), dispersant viscosity, temperature, and the RI of the particulate material are likely to influence the final value of the particle size of dispersion. In addition, the DLS/PCS analysis relies on the Stokes–Einstein equation to correlate the velocity of the particles in Brownian motion to their hydrodynamic size. The viscosity of the dispersant is particularly important for DLS/PCS measurement as it is inversely proportional to the mobility of particles in Brownian motion and can be greatly affected by the presence of formulation excipients. The o/w nanosized emulsions' MPS, polydispersity index (PDI), and ZP values can be assessed by DLS/PCS. This instrument also measures PDI, which is a measure of the broadness of the size distribution derived from the cumulative analysis of DLS. Moreover, the PDI indicates the quality or homogeneity of the dispersion (Li et al. 2011).

Considering the importance of MPS value in the therapeutic applications of o/w nanosized emulsions, the primary and secondary factors influencing the MPS value are listed in Table 3.6.

TABLE 3.5. Peak Area and UHPLC Chromatograms Obtained with the Treated Cyclosporin A-Loaded Nanosized Emulsions Following Subjection into Different Thermodegradation Conditions Along with the Untreated/Controlled CsA (0.05 and 0.1% w/w)-Loaded Nanosized Emulsions

Treated/Untreated Conditions	Nanosized Emulsions Loaded With	Peak area (mAU·min) (Mean ± SD, $n = 3$)	UHPLC Chromatogram of CsA
Untreated condition	0.05% w/w CsA	27.715 ± 3.89	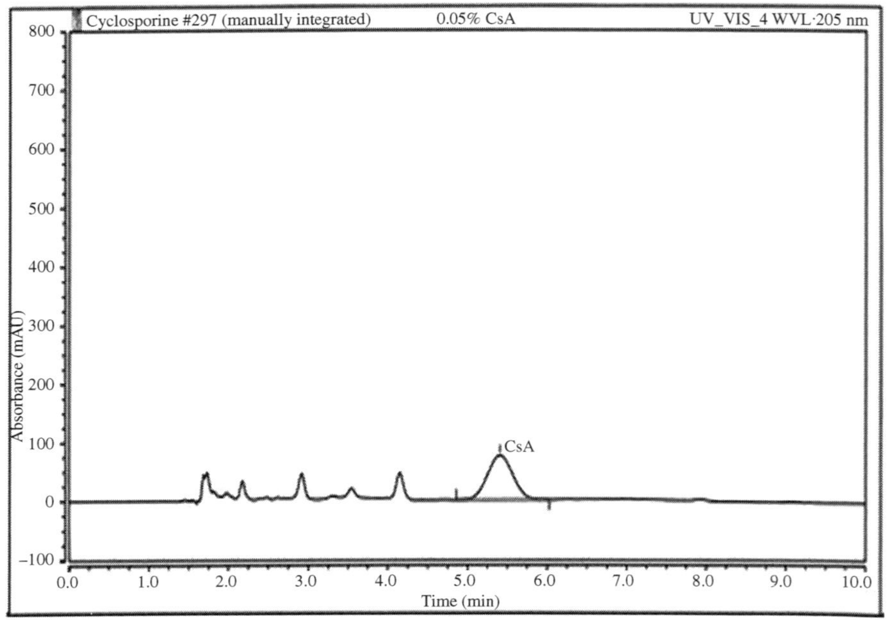

0.1% w/w CsA	52.611 ± 2.18	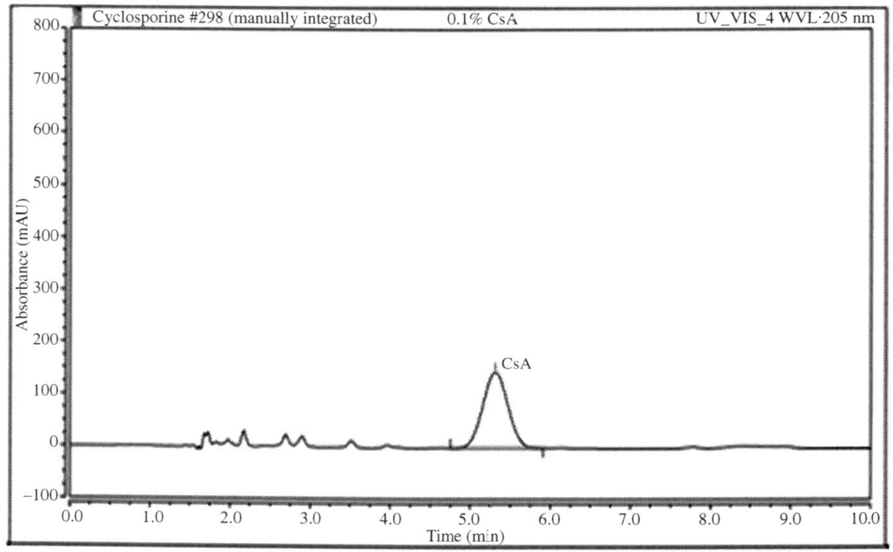

TABLE 3.5. *Continued*

Treated/Untreated Conditions	Nanosized Emulsions Loaded With	Peak area (mAU·min) (Mean ± SD, $n = 3$)	UHPLC Chromatogram of CsA
Autoclave condition	0.05% w/w CsA	25.575 ± 1.12	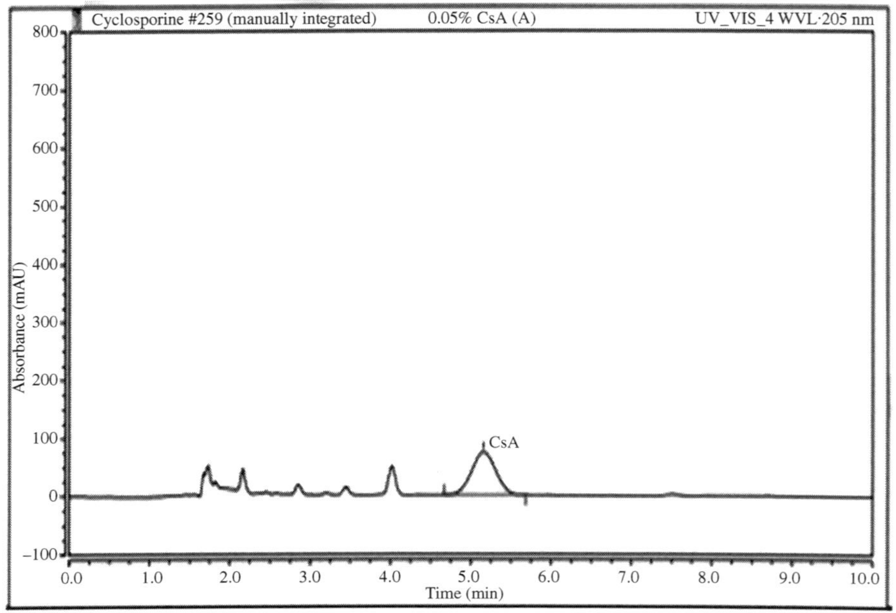

0.1% w/w CsA	51.003 ± 1.79	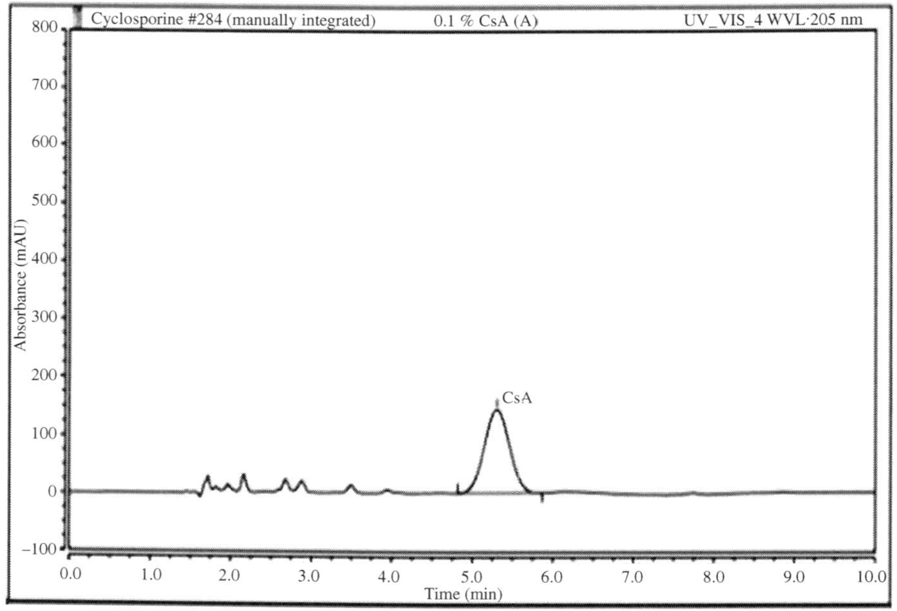

TABLE 3.5. ***Continued***

Treated/Untreated Conditions	Nanosized Emulsions Loaded With	Peak area (mAU·min) (Mean ± SD, $n = 3$)	UHPLC Chromatogram of CsA
Hot air oven condition	0.05% w/w CsA	26.249 ± 2.57	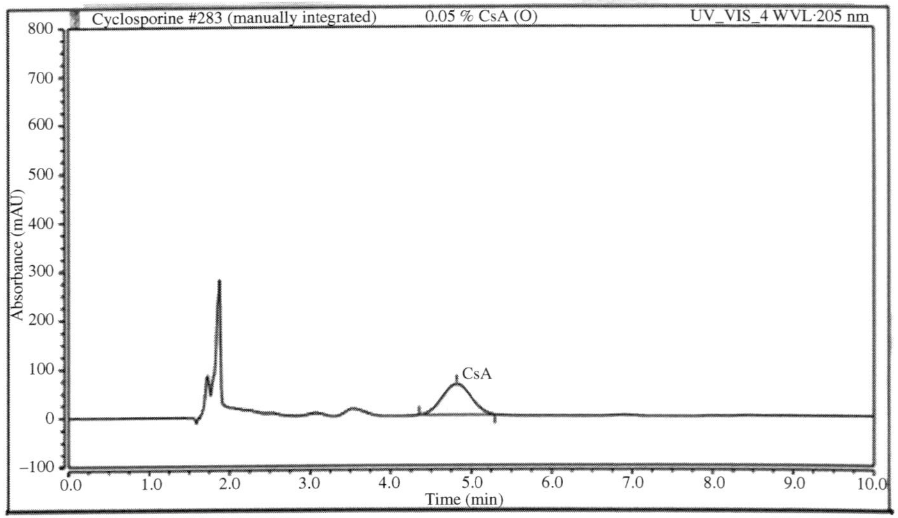

0.1% w/w CsA	51.003 ± 0.91	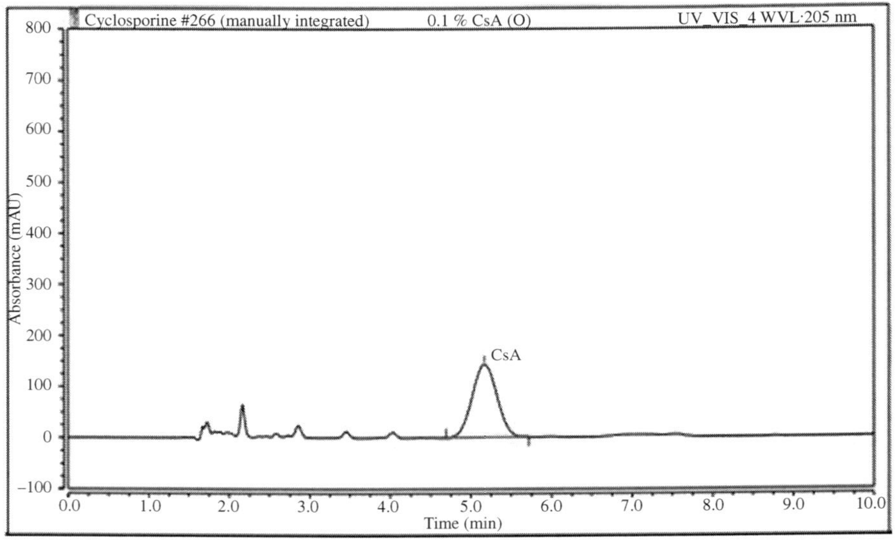

TABLE 3.5. ***Continued***

Treated/Untreated Conditions	Nanosized Emulsions Loaded With	Peak area (mAU·min) (Mean ± SD, $n = 3$)	UHPLC Chromatogram of CsA
Room temperature condition	0.05% w/w CsA	27.094 ± 4.16	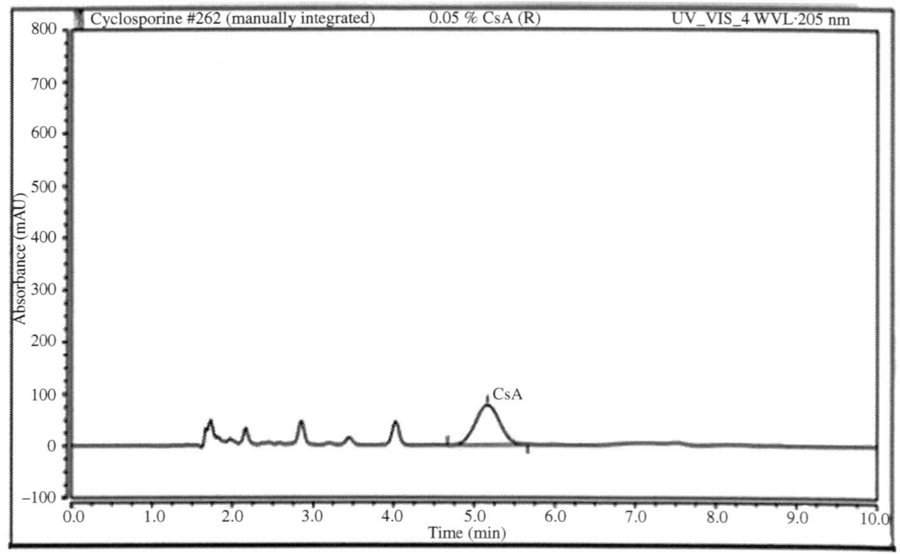

0.1% w/w CsA	52.498 ± 1.96	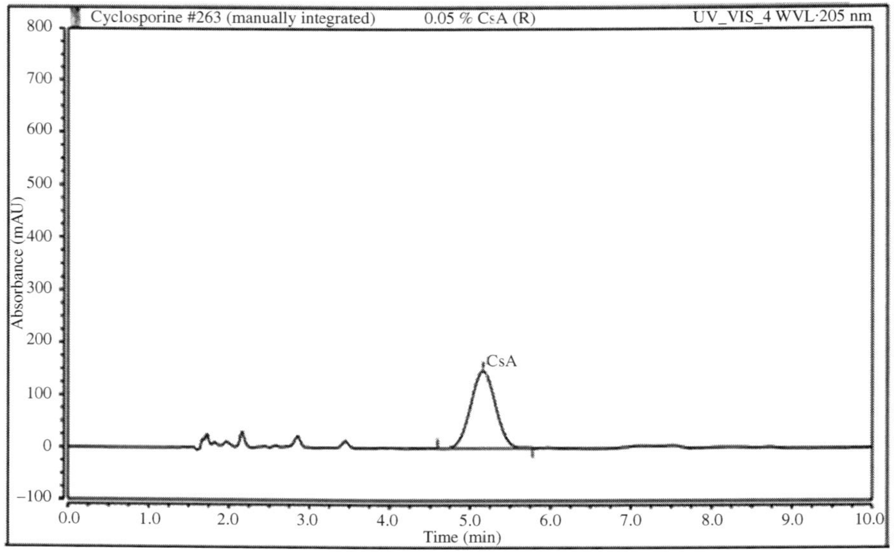

TABLE 3.6. Non-exhaustive List of Various Factors Influencing the Mean Particle Size Value of O/W Nanosized Emulsions

Primary Factors
Emulsifier molecules' concentration including charged/non-charged surfactant and steric hindrance providing emulsifier
Homogenization time
Homogenization speed
Single oil or combined oils concentration
Storage conditions (humidity, temperature, and time) of final emulsions
Secondary Factors
Homogenizer shaft position (30°, 60°, 90°, etc.) during emulsification
Oil–water phase volume ratio (though limited for o/w nanosized emulsion)
Pouring of oil phase into water phase or vice versa
Premixing time of oil and water phases or vice versa

3.2.6.1. Emulsion Sample Preparation to Determine the MPS Value Using DLS/PCS The MPS value of o/w nanosized emulsions can be measured at 25°C using Malvern zetasizer (model: Nano ZS, Malvern Instrument Ltd, Malvern, UK). The source of laser beam used in this instrument is He–Ne light source at 633-nm wavelength. This instrument possesses the sensitivity range of 0.3 nm^{-10} µm. Three samples of each formulation were tested under both undiluted and 10× diluted (ultrapure water, prefiltered through 0.02 µm syringe filter) conditions. Material RI was set to 1.48 with absorption set to 0.001. Dispersant viscosity and RI were set to that of water (i.e., 0.89 mPa·s and 1.330 at 25°C). Measurement was performed at 173°, 6 runs with 10 s run^{-1}. The *Z*-average size, PDI, D10, D50, and D90 (intensity weighted distribution) were obtained. The optimal measurement position was determined to be 3.0 mm using the "auto-seek" function for micro-cuvette (Malvern ZEN0040) and was kept constant for all subsequent measurements for all the formulations. The attenuation index for the undiluted and 10× dilution samples were fixed at 6 and 5, respectively, to compensate for scattering intensity variation due to the change of particle concentration (Petrochenko et al. 2018).

3.2.6.2. Emulsion Particle Size Measurement Using Laser Diffraction Technique Laser diffraction is another technique for measurement of particle size distributions in the micron range and may be less suited for characterization of nanoemulsions and micelles (with particle sizes under 20 nm). Laser diffraction technique is used to measure the droplet size distribution of emulsions. In fact, Malvern mastersizer works on the principle of laser light diffraction.

A monochromatic beam of laser light is transmitted through a dilute emulsion and the resulting diffraction pattern is measured using a series of light-sensitive detectors. The diffraction pattern is the result of scattering of the laser beam by the droplets in the emulsion being analyzed, and its precise form depends on the droplet characteristics (i.e., size, concentration, and complex RI). Information about droplet size distribution and concentration is determined by finding the values that give the best agreement between the measured diffraction pattern and that predicted by light-scattering theory. This theory assumes that each photon of light is only scattered by a single droplet, which means that the emulsion must be dilute (<0.1% droplets) to avoid multiple scattering. Most commercially available instruments can measure droplet diameters between 0.1 and 1,000 μm, although some have special features that enable them to analyze smaller droplets (McClements 2011).

The fundamental particle size distribution derived by this technique is volume based and is expressed in terms of the volume of equivalent spheres (DN%) and weighted mean of the volume distribution (mass mean diameter). Since the laser diffraction system is used for this analysis, a rough equivalent of particle polydispersity could be given by two factors/values namely, uniformity (how symmetrical the distribution is around the median point) and span (the width of the distribution). The instrument uses an approximation of Mie-scattering theory, which uses the RI of the dispersed phase and its absorption (Chime et al. 2014). The span value is defined by the following Eq. (3.9):

$$\text{Span} = \frac{(D90\% - D10\%)}{D50\%} \tag{3.9}$$

where DN% ($N = 10, 50, 90\%$) means that the volume percentage of particles with diameters up to DN% equals to N%. The smaller the span value the narrower the particle size distribution.

3.2.6.3. Measuring of ZP in Nanosized Emulsions The Greek letter ζ (zeta) is commonly used in mathematical equations while describing the electrokinetic potential value of colloidal dispersions and hence, the name zeta potential is awarded. The ZP is the potential at the slipping/shear plane of a colloid particle moving under electric field (Kaszuba et al. 2010). The earliest theory for calculating ZP from experimental data was developed by Marian Smoluchowski in 1903. The ZP reflects the potential difference between the electric double layer (EDL) of electrophoretically mobile particles and the layer of dispersant around them at the slipping plane. The ZP measures the magnitude of electrostatic attraction/ repulsion between particles and thus, it has become one of the fundamental parameters known to affect the stability of colloidal particles. When applied to colloidal dispersions such as o/w nanosized emulsions, the ZP value thus indicates the stability of the formulation, which generally means the resistance to change of the dispersion with time. If the ZP is reduced below a certain value, then the attractive forces exceed the repulsive forces, and the particles come together. This phenomenon is known as coalescence/flocculation.

From knowledge of the direction and rate of migration of charged particles, the sign and magnitude of the ZP in a colloidal system can be determined. The relevant equation, which yields the ZP, ζ, in volts, requires a knowledge of the velocity of migration, v, of the sol in centimeter per second in an electrophoresis tube of a definite length in centimeter, the viscosity of the medium, η, in poises (dynes s cm^{-2}), the dielectric constant of the medium, ε, and the potential gradient, E, in volts per centimeter. The term v/E is known as the mobility.

$$\zeta = \frac{v}{E} \times \frac{4\pi\eta}{\varepsilon} \qquad (3.10)$$

3.2.6.3.1. Understanding the EDL and slipping plane It is well known that when a charged particle is dispersed, an adsorbed double layer often referred to as EDL develops on its surface (Ruiz-Cabello et al. 2014). The inner layer consists predominantly of ions/molecules with opposite charge to that of the particle (Stern layer). Beyond Stern layer the electrostatic effects due to the surface charge on the particles decrease as per Debye's law, which states that with the distance of each Debye length the field decreases by a factor of $1/e$ (Chen et al. 2013).

Although mathematically this electrostatic effect extends till infinity, experimentally it is only present till few nm from particle surface. Due to the electrostatic field of the charged particles, a diffuse layer consisting of both same and opposite charged ions/molecules grow beyond the Stern layer which along with the Stern layer forms the EDL. When an electric field is applied to such dispersion, the charged particles move toward the opposite electrode (electrophoresis). Within this diffuse layer there is a hypothetical plane that acts as the interface between the moving particles and the layer of dispersant around it while electrophoresis. This plane is the characteristic slipping/shear plane and ZP is the potential at this particle–fluid interface. The potential on the particle surface itself is known as the Nernst potential (ψ_0) and cannot be measured (Vidal-Iglesias et al. 2012). Figure 3.4 shows pictorial explanation/representation for the behavior of charged particles in colloidal dispersion under electric current. Table 3.7 shows the different factors influencing the ZP value of o/w nanosized emulsions.

3.2.6.3.2. Factors influencing ZP

3.2.6.3.2.1. pH In general, the ZP value varies with pH and becomes more positive and negative in magnitude with acidic and basic pH, respectively (Usković et al. 2010). The pH of final autoclave-sterilized o/w nanosized emulsions is perhaps the most influential parameter in ZP measurements. The charge-producing excipients used in nanosized emulsions are the emulsifier molecules. Table 2.1 (Chapter 2) lists the various anionic, cationic, neutral, and

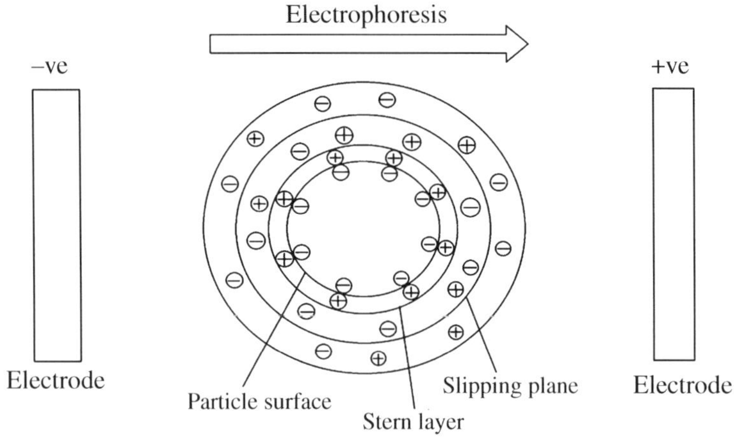

Figure 3.4. Pictorial representation to show the behavior of charged dispersed particles in colloidal dispersion upon application of electric current. [Modified from Bhattacharjee (2016).]

TABLE 3.7. List of Various Factors Influencing the Zeta Potential Value of O/W Nanosized Emulsions

pH
Ionic strength
Particle concentration

nonionic emulsifiers used for the emulsion formulation development. Sometimes, the emulsion-incorporated API molecules do contain/develop charges depending on the pH value of the final colloidal dispersion. Similarly, the anion or cation conferring emulsifier molecule does produce the negative or positive charge depending on the final colloidal dispersion pH value. More interestingly, the dispersed oil droplets of the commercial emulsions are stabilized by emulsifier combination rather than single emulsifier molecule. The multimolecular/layer emulsifier film formed over the dispersed oil droplets of the emulsion do contain final globular/net charges depending on the selected emulsifier molecules whose charge conferration ability is highly controlled/determined by pH value of the final emulsion colloidal dispersion. For example, the selected charge-containing API molecules and charge-conferring emulsifier molecules may interact during- and after-emulsification process depending on the variation of emulsion pH values. To expedite further this concept, the titration curve of ZP against different pH values developed by Kirkwood et al. (2015) is often very useful and helps to determine the isoelectric point, i.e., the pH where the ZP becomes zero. For aqueous dispersions—where H^+ and OH^- are major ionic constituents—the isoelectric point also denotes the point of zero charge (PZC) (Salis et al. 2011). Colloids lose stability and agglomerate/flocculate when the pH is close to the isoelectric point. The further discussion narrated below indicates an intermingling phenomenon comprising of emulsion stability, pH, and ZP values.

A very surprising result was noticed by seeing the ZP values of the macroemulsion with or without AZM. The macroemulsion without AZM did show the ZP value in the negative side (–2.60 ± 0.75 mV) while the AZM-loaded macroemulsion possessed the ZP value that was in the positive side (+6.52 ± 0.96 mV). It indicates that there was a reversal in ZP value from negative to positive side when 100 mg of AZM was incorporated into the macroemulsion. It should be noted that the macroemulsion without AZM was prepared based on the emulgator combination that consisted of 50 mg of chitosan and 1,500 mg of Tween 20. Naturally, the presence of chitosan in the emulgator combination should confer/provide a positive charge to the macroemulsion as the chitosan is a polycationic biopolymer having pK_a values of ~6.5, which can become protonated in weakly acidic conditions. It is this polycationic character that confers chitosan's antimicrobial properties, which favors interaction with negatively charged microbial cell walls and cytoplasmic membranes. This electrostatic interaction between protonated chitosan and microbial cell walls results in decreased osmotic stability, membrane disruption, and eventual leakage of intracellular elements. Since the macroemulsion without AZM showed a negative ZP value, it is clear that the formation of protonated chitosan is very low. But the question remains how the macroemulsion without AZM was able to be physically stable at room temperature. It should be remembered that there are two stabilizing forces acting inside the o/w emulsion system to disperse/stabilize the oil droplets in the water namely, cationization and steric hindrance. Indeed, Tween 20 is known for its steric hindrance effect and thus it basically stabilizes the o/w macroemulsion. By weight amount comparison, the ratio between chitosan and Tween 20 (as the current emulgator combination) is 1 : 30. So, naturally, the o/w macroemulsion without AZM was found to be stable because of the steric hindrance effect produced by the Tween 20, which, in turn, prevented oil droplet coalescence upon random collisions. This explanation seems to be a plausible reason why the o/w macroemulsion having the negative ZP value was able to maintain its physical stability at the room temperature condition. However, a long-term stability study is underway in our laboratory to see the shelf life of the o/w macroemulsion having chitosan and Tween 20 emulgator combination. When looking for the reason why the chitosan (and Tween 20-)-based macroemulsion exhibited a negative ZP value, the dispersed oil droplets are basically stabilized by a mono- or multi-molecular emulgator film comprising of intercalated two different emulgator molecules. In the current macroemulsion, such type of intercalation networking occurs between the two emulgator molecules by engulfing of smaller amount emulgator molecules (chitosan) with the high amount emulgator molecules (Tween 20). In other words, the direct exposure of chitosan to surrounding water medium is somewhat masked by the Tween 20, which, in turn, prevented the protonated chitosan form to occur and therefore no positive charge is conferred over the emulsified oil droplets of the macroemulsion. On the other hand, the AZM-loaded o/w macroemulsion was also found to be stable at room temperature condition and this emulsion possessed the positive ZP value of +6.52 ± 0.96 mV. The addition of a fixed amount (100 mg) of basic API like AZM into the macroemulsion should contribute a positive charge to the o/w interface of the emulsion by intercalating into the emulgator combination network. This is why the reversal in ZP value was noticed upon addition of AZM into the macroemulsion. A similar reversal in ZP value of o/w nanosized emulsion was noticed after inclusion of either AZM or CsA (Liu et al. 2009; Tamilvanan et al. 2014). Collectively, the stability of macroemulsion without AZM was primarily due to the steric hindrance effect of Tween 20, whereas for AZM-loaded macroemulsion, the physical stability of the macroemulsion was because of positive ZP value conferred by the AZM along with the Tween 20's steric hindrance effect (Shunmugaperumal and Kaur 2015).

3.2.6.3.2.2. IONIC STRENGTH With increasing ionic strength, the EDL becomes more compressed while the ZP decreases and vice versa. The valency of the ions is also important while measuring ZP. For ions with higher valency (e.g., Ca^{2+}, Al^{3+} having higher valency than monovalent Na^{+}, H^{+}, OH^{-}), the EDL becomes more compact and the ZP decreases in magnitude.

3.2.6.3.2.3. PARTICLE CONCENTRATION The relation between ZP and particle concentration is complex and usually determined by both surface adsorption and the effect of EDL. It is difficult to provide with a general guideline on the effect of concentration on ZP. However, it can be stated that overall in dilute conditions the surface adsorption phenomenon dominates and hence, the ZP increases with concentration. However, at higher concentration range the thickness of EDL dominates and then by increasing concentration an opposite effect, i.e., decrease in ZP with lesser stability of the dispersion is observed (Nägele 1989).

3.2.6.3.3. ZP and colloid stability One of the most popular uses of ZP data is to relate them with colloid stability. Guidelines classifying nanosized particle dispersions with ZP values of ±0–10, ±10–20, ±20–30, and >±30 mV as highly unstable, relatively stable, moderately stable, and highly stable, respectively, are common in API delivery literature (Patel and Agrawal 2011). Unfortunately, the reality is more complex than that. Although ZP does provide indications on colloid stability, it does not reflect the entire picture. As per the most widely accepted DLVO (named after inventors Derjaguin, Landau, Verwey, and Overbeek), theory colloid stability depends on the sum of van der Waals attractive forces and electrostatic repulsive forces due to the EDL (Missana and Adell 2000). While ZP provides information on the electrostatic repulsive forces, it does not provide any insight on the attractive van der Waals forces. Therefore, it is not uncommon to come across stable colloids with low ZP and vice versa. There is plenty of theory involved in understanding such attractive forces in nature like van der Waals, which falls beyond the scope of this discussion. One important point to note is that the van der Waals attractive force is dependent on the Hamaker constant (Leite et al. 2012), which indirectly corresponds to the difference between the RI of the particle and the dispersant. Therefore, if the Hamaker constant is low, the van der Waals attractive forces also become weak and then mild electrostatic repulsion reflected by low ZP (e.g., 10–15 mV) may be enough to ensure colloid stability. Materials like colloidal silica shows exceptional stability at very low ZP (Kim et al. 2014). It should be noted that steric interactions can also contribute to colloid stability. For example, some water-in-oil emulsions are highly stable despite having low ZP (Almeida et al. 2015). PEGylation is also known to facilitate stability of NPs while decreasing the ZP (Kouchakzadeh et al. 2010).

3.2.6.3.4. ZP and surface charge of nanoformulations Another widely popular use of ZP is to use it in assessing the surface charge of nanoformulations. The positive or negative dimensions of ZP are determined by identifying which electrode the particles are moving toward during electrophoresis. It should be noted that ZP never measures charge or charge density and rather deals with surface potential. Therefore, only the magnitude of ZP is important while the positive/negative finding associated with it is not robust and should not be related with surface charge or charge density or making comparisons between different nanoformulations. As indicated previously, a wide variety of factors (e.g., pH, which is relevant for nanoformulations) can change it from +ve to −ve and vice versa. Thus, the ZP value only provides with indicative evidence toward the nature of surface charge (positive/negative) assuming that the predominant ions in the EDL up to the slipping plane are similar (positive/negative) compared with the surface of the particle itself. Unfortunately, there are too many exceptions to such assumption. The practical way to confirm the nature (positive/negative) as well as to determine charge density on nanoformulations is to titrate it with known amounts of ions. A detailed description of such titration technique falls beyond the scope of this chapter although excellent reference literature is available (Clavier et al. 2015; Pillai et al. 2016). Charge on particle–dispersant interface (e.g., slipping plane) is a complicated and less understood phenomenon. Usually, almost all of the naturally occurring surfaces and molecules exhibit negative charge at physiological pH (e.g., cell membrane, proteins, lipids, mucus, …). On the contrary, cationic surfaces and molecules are often synthetic. It is also inappropriate to claim having "neutral" nanoformulations based on ZP value as there are never neutral nanoformulations in dispersion due to inevitable charge buildup on their surfaces. An interesting fact is that surface charge on nanoformulations can actually vary depending on the different phases within the colloid. As per Cohen's rule, if both solute and solvent are insulators, then the one with higher relative permittivity (ε_r) becomes positive at the interface. Hence, in room temperature silica ($\varepsilon_r = 3.9$) nanoformulations are negatively charged in water ($\varepsilon_r = 80$) but positively charged in benzene ($\varepsilon_r = 2.27$).

3.2.6.3.5. Physical constraints in measuring the ZP The ZP measurement also consists of many physical constraints as detailed below.

3.2.6.3.5.1. REFERENCE MATERIALS Unlike DLS, there are as such no reference or standard materials for ZP, which is inconvenient in practical terms. The National Institute of Standards and Technology (NIST) has suggested the use of Goethite (α-FeO(OH)), which upon preparation under specified conditions should deliver a ZP of +(32.5 ± 0.12) mV (Di Marco et al. 2007). However, the samples need to be made fresh every time and it may foul the cuvette. Slight variation in data based on instruments may also occur.

3.2.6.3.5.2. REUSING SAMPLES AFTER MEASURING ZP Electrophoresis may degrade some nanoformulations and hence, may render the samples unsuitable for reuse after measuring ZP. As a common guideline, it can be stated that if possible reusing samples for experiments after measuring ZP should be avoided. If that is not the case (e.g., due to small volume of sample), then adequate re-characterization of the particles including DLS should be done after measuring ZP to exclude any degradation of the particles under applied voltage.

3.2.6.3.5.3. USING BUFFERS WITH METALLIC IONS The electrodes in cells for ZP measurements are prone to react especially with metallic ions (e.g., Fe^{3+}) (Hedberg et al. 2013). Such reactions can destroy the electrode and compromise the quality of data. Therefore, regular checking of the electrode is advisable. In case the contact between the electrodes and the ions cannot be prevented in spite of deleterious reactions, then the diffusion barrier method (Tucker et al. 2015) can be used where the electrophoretic mobility of the particles can still be measured while preventing any contact between the electrode and the buffer. However, it requires additional expertise.

3.2.6.3.5.4. MEASURING ZP IN CELL CULTURE MEDIUM It can be challenging to measure ZP in cell culture medium. Enriched with plenty of ions, the cell culture mediums have very high conductivity and interfere with ZP measurements. Such high conductivity can generate enough heat under constant voltage, which may degrade the sample. Using higher concentrations of nanoformulations (5–10 mg ml^{-1}) under low voltage can be helpful to obtain a stable ZP reading. Unfortunately, it becomes much more complex in cell culture medium carrying fetal calf serum (FCS), which contains plenty of proteins.

The available protein molecules get adsorbed on the nanoformulation surfaces and influence both the dispersion and ZP (Satzer et al. 2016). The protein molecules also sometimes make small nanoagglomerates, which interfere with the readings and can generate additional aberrant peaks.

3.2.6.4. Current Thought About DLS/PCL and ZP Determination in Nanoformulations The fundamental principles of both DLS and ZP are rooted within the realms of physical colloid chemistry and it is essential to have strong grip over the core physical and mathematical principles in order to understand their applied aspects. The basics of DLS and ZP are taught already to the undergraduate students in physical chemistry in most universities. Unfortunately, similar training and structured grooming process for young researchers performing DLS and ZP are often missing in API delivery research groups. This gap in knowledge as well as lack of proper training need to be addressed.

The DLS/ZP measurements are based on light scattering and hence, only clear samples can be subjected to this technique. Additionally, this technique is not capable to handle concentrated samples. Just to exemplify, the Stokes–Einstein equation—which is the backbone of particle size measurements based on light scattering—is only mathematically feasible at infinitely dilute concentrations. In reality, usually 50–100 $\mu g\ ml^{-1}$ concentrations are used. Unfortunately, it hardly correlates with therapeutically relevant doses, which are much higher in concentration with particles frequently >200 nm present in it. As a result, the prepared nanoformulation samples are often neither clear nor dilute enough to be fit for ZP measurement. It needs to be emphasized that ZP have their own shares of limitations and its inability to handle high concentrations is a major weakness of this technique. Surface chemistry is important in measuring the DLS/ZP and any change in surfaces of the particles will alter the results. In therapeutically relevant samples with high concentration containing complex engineered nanoconstructs, plenty of parameters (e.g., viscosity, pH, dielectric constant, RI, …) change and hence, the dilute samples used for DLS/ZP measurements are never an adequate representation of the therapeutic formulations to be used *in vivo*. Hence, the results from DLS/ZP should not provide grounds for rushing nanoformulations for *in vivo* studies. The DLS/ZP are run mostly on aqueous dispersants with known ionic strength and pH under controlled laboratory environments that are hardly comparable with *in vivo* circumstances where the dispersion medium often is blood with a complex matrix. The DLS/ZP neither can operate in blood nor can predict the behavior of nanoformulations in blood. This technique was originally developed for protein dispersions and although they work fine with engineered nanoformulations within certain operational conditions, the scope of this technique in characterizing nanoformulations *in vivo* is limited. Unfortunately, at this stage, adequate analytical tools capable to handle complex biological matrices (e.g., blood) are missing and the focus of ongoing research work should try to address this issue urgently in order to facilitate translation. In the last few years, a wide range of nanoformulations have emerged although their translational impact overall has been disappointing with poor *in vitro–in vivo* correlation (IVIVC). There are challenges in characterization of nanoformulations under physiologically relevant conditions. DLS/ZP is the excellent tool to characterize nanoformulations at their initial stages of development. However, the scopes for this technique become increasingly limited in further phases of nanoformulation preparation or to provide sound data for *in vivo* correlation (Bhattacharjee 2016).

3.2.7. Microscopic Evaluations of Nanosized Emulsions

3.2.7.1. Transmission Electron Microscopy (TEM) The TEM operates on the same basic principles as the light microscope but uses electrons instead of light. Because the wavelength of electrons is much smaller than that of light, the optimal resolution attainable for TEM images is many orders of magnitude better than that from a light microscope. Thus, TEM can reveal the finest details of internal structure and in some cases as small as individual atoms. TEM has three essential systems: (1) an electron gun, which produces the electron beam, and the condenser system, which focuses the beam onto the object; (2) the image-producing system, consisting of the objective lens, movable specimen stage, and intermediate and projector lenses, which focus the electrons passing through the specimen to form a real, highly magnified image; and (3) the image-recording system, which converts the electron image into some form perceptible to the human eye. The image-recording system usually consists of a fluorescent screen for viewing and focusing the image and a digital camera for permanent records. In addition, a vacuum system, consisting of pumps and their associated gauges and valves, and power supplies are required (Fig. 3.5).

3.2.7.1.1. Particle size reduction between emulsion preparation steps assessed by TEM Pharmaceutical emulsions are colloidal API delivery system that ultimately uses the externally applied electrical, mechanical, and ultrasonic forces to break up the dispersed oil droplets into the desired size ranges particularly at the nanometer level. Due to the involvement of all of these forces, the MPS value of the dispersed oil droplets are gradually reduced to reach up to the nanometer size ranges. Even the produced multivariate particles by the application of the external forces are further subjected into equipment, which works on the principle of homogenization under highly pressurized condition in order to convert the bigger size oil droplets into uniform and small-sized particles. In a laboratory condition, the initial mixing of oil and water phases of the emulsion is usually performed with the help of mechanical stirrer not only for mixing the two phases but also to produce the dispersed oil droplets at macron/micron sizes. Figure 3.6 shows the TEM images of dispersed oil droplets taken at and during the initial mixing of castor oil with the water phase containing a suitable emulsifier. It can clearly be seen from Fig. 3.6 that

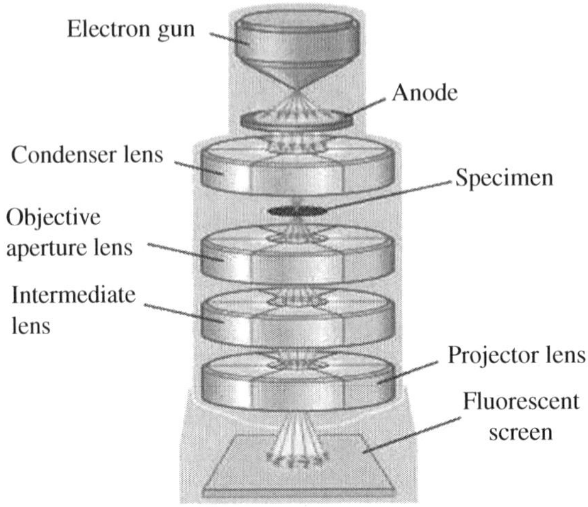

Figure 3.5. Essential parts of transmission electron microscope. [Adapted from https://www.britannica.com/technology/transmission-electron-microscope, accessed on November 28, 2019.] (See color insert.)

the size range of the oil droplets falls in macron/micron and therefore, the produced emulsions at this stage could be termed as coarse or macroemulsion.

The produced coarse or macroemulsion now undergoes the next stage of size reduction wherein the high-speed electrical homogenizer involves to destruct/destroy the macron-sized oil droplets into small sizes. These phenomenon results in the production of dispersed oil droplets of the emulsion to possess the mean particle diameter close or nearer to 1 μm level. Figure 3.6 shows the TEM images of dispersed oil droplets taken at and during the high-speed electrical homogenization stage. From Fig. 3.6, one can visualize the produced dispersed oil droplets having the micron-level size ranges.

The next stage of emulsion production method involves the utilization of either high-pressure homogenizer or probe and/or ultrasonicator. The function of homogenizer or sonicator is just to convert and to confine the multivariate particle sizes of the dispersed oil droplets into a more uniform-sized particles that too at the sub/under micron level preferably even below 100 μm. Figure 3.6 shows the TEM images of dispersed oil droplets taken at and during the high-pressure homogenization or probe and/or ultrasonication. Most of the intravenously (IV) administered o/w nanosized emulsions are prepared by following all these three stages with the involvement of atleast three different size-reduction machineries. It should be emphasized that the formation of emulsion (simply called emulsification) is occurring at all of these three different stages. However, the efficiency of the size-reduction machineries/equipment determines the mean particle diameter of dispersed oil droplets of the emulsion. Since the emulsification occurs at all of these three stages, the

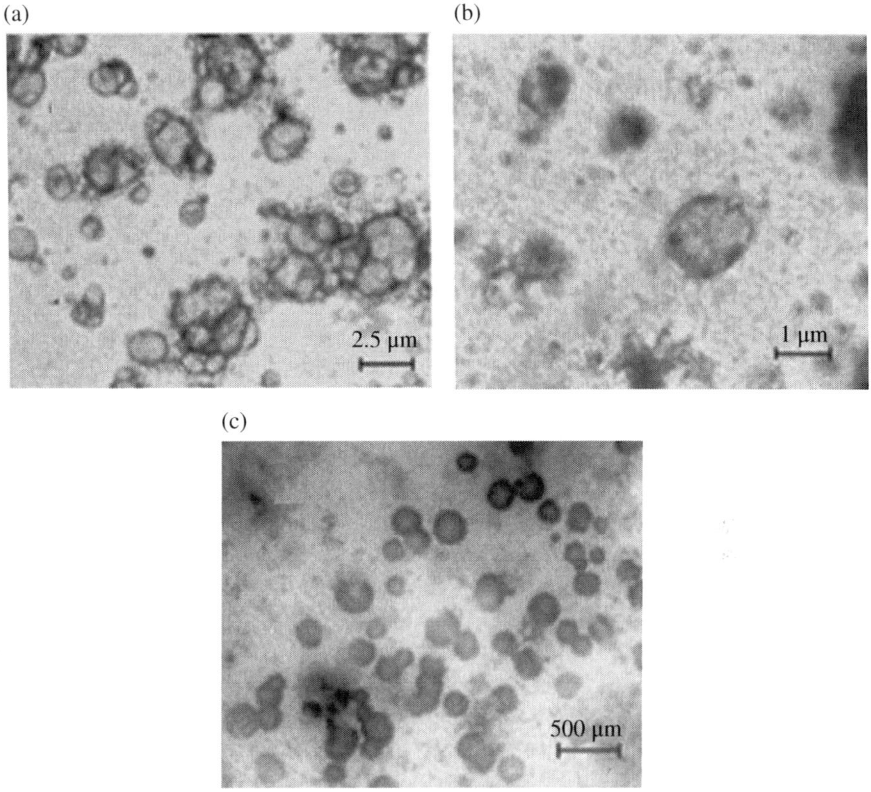

Figure 3.6. TEM images of emulsion samples taken at each one of the particle size reduction steps. Key: (a) Initial mixing of oil phase with water phase using magnetic stirrer to form crude emulsion. (b) Rapid homogenization of crude emulsion formed. (c) Probe sonication of fine emulsion. [Reproduced with permission from Tamilvanan and Kumar (2011).]

method used to produce the o/w nanosized emulsions can thus be named as "combined emulsification techniques." The importance of making the therapeutic o/w nanosized emulsions by the combined emulsification techniques with the involvement of multiple size-reduction machineries is realized when the emulsions are administered via or meant for IV route and topical nasal or ocular route. Because, these routes require the administration of sterilized o/w nanosized emulsions and interestingly, the stability of dispersed oil droplets of the emulsion solely depends on the MPS of the emulsion. The coarse- or macron-sized emulsions usually destabilize following sterilization using the gaseous (ethylene oxide), irradiation (gamma rays), and terminal (autoclaving) sterilization methods. Whatever the sterilization method chosen, the stress provided onto the emulsion is too strong to the dispersed oil droplets to resist/withstand against the applied stress and therefore, the emulsion allows/undergoes the breakage/separation of the constituent phases (oil and water) to occur. From this point of view, the stability study of the therapeutic emulsions should contain the protocol for emulsion stabilization against sterilization. The sterilization-induced destabilization of dispersed oil droplets of the emulsion is controlled/influenced and even judged by the mean particle diameter of the oil droplets that/which dispersed in the o/w emulsion. Overall, the mean particle diameter of dispersed oil droplets of the emulsion is based on the use of combined emulsification method utilizing the efficiency/efforts of multiple size-reduction machineries.

3.2.7.2. Atomic Force Microscopy (AFM) AFM is one of the most popular techniques for determining the nanoformulations' surface characteristics such as surface roughness due to its ability to quantitatively measure the lateral (x and y) and height (z) direction with nanoscale resolution. There are multiples mode available in AFM to perform surface roughness measurements, which differ by contact type. Basically used two modes include contact and gentler modes. The simplest mode is contact mode, where the tip is "dragged" across the surface at a constant cantilever deflection. The user defines the load at which the tip is "dragged" across the surface so that a heavier load can be selected for stiff, robust materials and a lighter load for softer materials. A feedback loop on the z piezo in the instrument then keeps the cantilever deflection constant throughout the image. The topography information is provided through the z piezo motion.

A gentler mode used to measure surface topography (and thus surface roughness) is in tapping mode. This is a dynamic mode where the tip is oscillated at a resonance frequency, and now the tip gently interacts with the surface at constant amplitude of oscillation. The user defines the amplitude of oscillation at which the tip images the surface so that larger amplitude can be selected for stiff, robust materials and smaller amplitude for softer materials. A feedback loop on the z piezo in the instrument then keeps the cantilever amplitude of oscillation constant throughout the image. The topography information is provided through this z piezo motion. With these two modes, practically any surface topography can be imaged from soft biological cells, to polymers, to stiffer semiconductors, and metals (Accessed from https://www.nanosurf.com/en/how-afm-works/afm-modes-overview/topography-and-surface-roughness-measurements, on November 28, 2019).

3.2.7.2.1. Working principle of AFM The AFM principle is based on the cantilever/tip assembly that interacts with the sample; this assembly is also commonly referred to as the probe. The AFM probe interacts with the substrate through a raster scanning motion. The up/down and side to side motion of the AFM tip as it scans along the surface is monitored through a laser beam reflected off the cantilever. This reflected laser beam is tracked by a position sensitive photodetector (PSPD) that picks up the vertical and lateral motion of the probe. The deflection sensitivity of these detectors has to be calibrated in terms of how many nanometers of motion correspond to a unit of voltage measured on the detector.

In order to achieve the AFM modes known as tapping modes, the probe is mounted into a holder with a shaker piezo. The shaker piezo provides the ability to oscillate the probe at a wide range of frequencies (typically 100 Hz–2 MHz). Tapping modes of operation can be divided into resonant modes (where operation is at or near the resonance frequency of the cantilever) and off-resonance modes (where operation is at a frequency usually far below the cantilever's resonance frequency) (Accessed from https://www.nanosurf.com/en/how-afm-works/afm-operating-principle, on November 28, 2019).

According to Sivagnanam et al. (2019), the AFM measurement of nanosized emulsions started with the preparation of diluted emulsion with water (1 : 1,000) and placed on cleaved mica surface and air dried at room temperature. The AFM experiment was carried out using a Brüker Multimode AFM nanoScope V microscope with a scan rate of 0.997 Hz and 512 × 512 pixels in 5 × 5 μm dimensions. Images were collected using the tapping mode. Figure 3.7 shows the 2D and 3D AFM images of the nanosized emulsions obtained at days 0 and 30 (Sivagnanam et al. 2019). This analysis confirms the particle size measurements of the nanosized emulsions. AFM showed us that the height sensor at day 0 was 47.6 nm, and at day 30 it was further increased to 59.2 nm. However, the AFM results recorded for size were smaller than the size measured by DLS. This may be because of the sample state during measurement of the AFM analysis. When using DLS, the size of the nanosized emulsions' hydrated layers surrounding the droplets may increase. However, when using AFM, the nanosized emulsions were completely dried before the measurement and therefore lack the hydrated layers; hence, a small size was observed, and this result is in agreement with earlier reports (Preetz et al. 2010).

3.2.8. Emulsion Stability Evaluation/Prediction Using TurbiScan®

The physical destabilization (creaming, sedimentation, flocculation, and coalescence) of concentrated/diluted colloidal dispersions such as emulsions and suspensions can be evaluated/predicted using TurbiScan® (Meunier 1994; Meunier and Mengual 1996). Furthermore, the shelf life of colloidal dispersions can also be monitored up to 50 times earlier than visual observation and thus, allowing the development time for new products to be significantly reduced. In addition, the stability measurements can be converted into a completely automated system using the robotized ageing station, called Turbiscan automated ageing station (Turbiscan ags).

The heart of the optical scanning analyser, Turbiscan®, is a detection head, which moves up and down along with a flat-bottom cylindrical glass cell

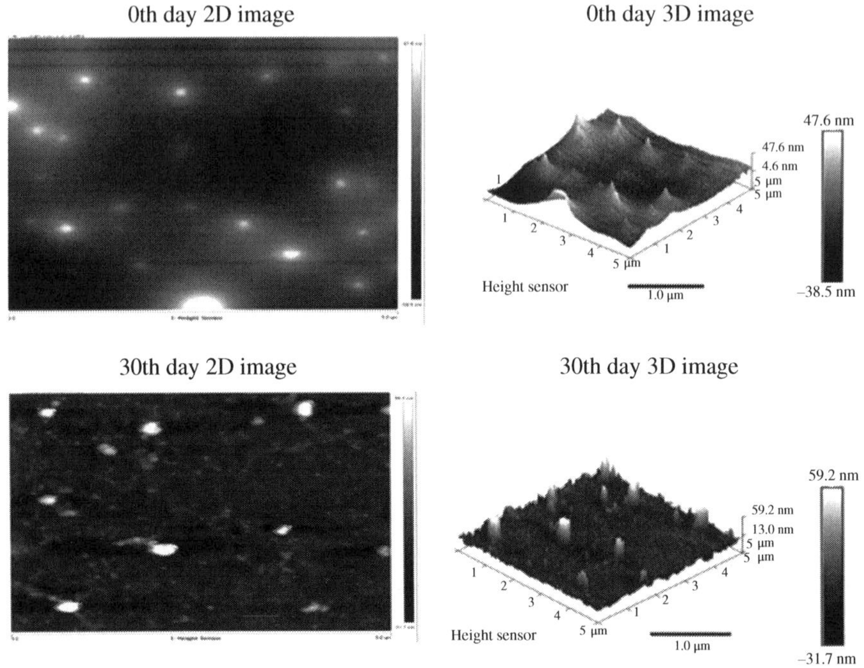

Figure 3.7. 2D and 3D AFM topography of seaweed oil nanosized emulsions. [Reproduced with permission from Sivagnanam et al. (2019).] (See color insert.)

(Fig. 3.8a). The detection head is composed of a pulsed near infrared light source ($\lambda = 880\,\text{nm}$) and two synchronous detectors. The transmission detector (at 180°) receives the light, which goes through the sample, while the backscattering detector (at 45°) receives the light scattered backward by the sample. The detection head scans the entire height of the sample, acquiring transmission and backscattering data every 40 µm. The Turbiscan LAB can be thermoregulated from 4 to 60°C and linked to a fully automated ageing station (Turbiscan ags) for long-term stability analyses. Increasing temperature is the ideal parameter to accelerate destabilization processes, while maintaining realistic testing conditions.

The Turbiscan® makes scans at various preprogrammed times and overlays the profiles on one graph in order to show the destabilization. Graphs are usually displayed in reference mode, whereby the first profile is subtracted to all other profiles, in order to enhance variations. A stable product has all the profiles overlaid on one curve (Fig. 3.8b), as an unstable formulation shows variations of the profiles (Fig. 3.8c). Backscattering and/or transmission fluxes are shown in ordinate and the height of the cell in abscissa (Fig. 3.8b and c). The first profile is displayed in pink, the last one in red (Formulaction Smart Scientific Analysis 2009).

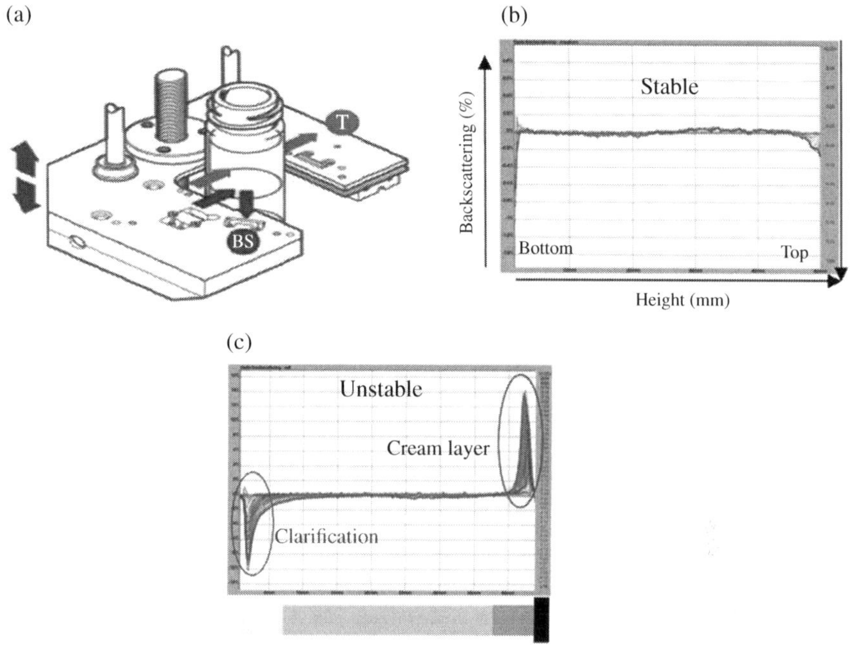

Figure 3.8. Turbiscan® assembly. Key: (a) Optical scanning analyser, Turbiscan®. (b) Superposition of scans with time for a stable sample. (c) Superposition of scans with time for an unstable sample (creaming). [Adapted from http://www.titanex.com.tw/doc/tecsupport/ANH-Turbiscan-application%20paper%20on%20%20pharmaceuticals.pdf, accessed on November 28, 2019.] (See color insert.)

3.2.8.1. Instability Detection Using Turbiscan® The measurement principle of the Turbiscan® range is based on multiple light scattering (MLS), where the photons are scattered many times by the particles/droplets of the dispersions before being detected by the backscattering detector. The intensity of the light backscattered by the sample depends on three parameters: the diameter of the particles, their volume fraction, and the relative RI between the dispersed and continuous phases. Therefore, any change due to a variation of the particle size (flocculation, coalescence) or a local variation of the volume fraction (migration phenomena: creaming, sedimentation) is detected by the optical device.

1. Particle size variation

   In Figure 3.9a, the variation of the backscattering level is shown as a function of the particle diameter for a fixed volume fraction of latex particles. The curve obtained is a bell-shaped curve, where the top is linked to the wavelength of the incident light (880 nm). For particles smaller than the incident light (left part of the curve), an increase of particle size is shown by an increase in backscattering. For particles bigger than the incident

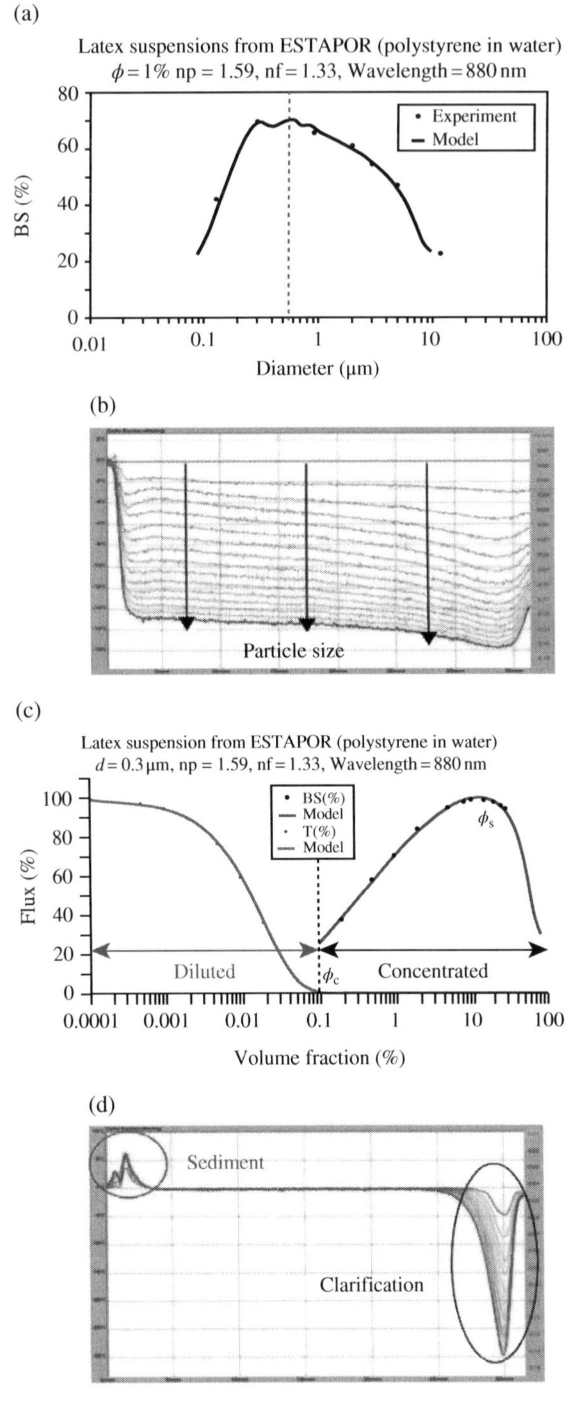

Figure 3.9. Assessment of colloidal dispersion using Turbiscan®. Key: (a) Backscattering level versus diameter for latex particles at 1%. (b) Typical profiles for flocculation phenomenon (initial size = 1 μm). (c) Transmission (red) and backscattering (blue) levels versus volume fraction for latex particles of 0.3 μm. (d) Typical backscattering profiles for a sedimentation phenomenon. (Adapted from http://www.titanex.com.tw/doc/tecsupport/ANH-Turbiscan-application%20paper%20on%20%20pharmaceuticals.pdf, accessed on November 28, 2019.) (See color insert.)

light (right part of the curve), an increase in size leads to a decrease in backscattering. On the Turbiscan® profiles, the particle size variations are displayed by a variation of the backscattering level over the total height of the sample (Fig. 3.9b).

2. Migration phenomena

   Migration phenomena (sedimentation or creaming) lead to local variation of the concentration of particles in the sample. In Fig. 3.9c, the variation of transmission and backscattering levels are shown as a function of the volume fraction for a fixed diameter of latex particles. If the concentration of particles is smaller than the critical concentration ($\varnothing_c$), the product can be considered as diluted and the transmission level decreases with an increase in concentration ($\varnothing$). When the concentration is sufficient ($\varnothing > \varnothing_c$), there is no transmission signal (opaque product) and the backscattering level increases with an increase of the volume fraction. At high or saturated concentration ($\varnothing_s$) of particles, i.e., ($\varnothing > \varnothing_s$), the backscattering level starts to decrease as the distance between particles is smaller than the wavelength of incident light. This phenomenon is called dependent diffusion and is mostly observed for small particles (<1 µm). On the Turbiscan® profiles, migration phenomena are displayed by local variations of the backscattering. In Fig. 3.9d, the backscattering level decreases at the top (right part of the graph), due to a decrease of the concentration of particles, hence a clarification, while it increases at the bottom due to the increase of particle concentration consecutive to the sediment formation. It is interesting to note that there is no variation in the middle of the sample, indicating no particle size variation.

3.2.9. Spreading Coefficient

The spreading coefficient or parameter is a measure of the tendency of a liquid phase (1) to spread or complete wetting on a second, liquid or solid phase (2). The spreading coefficient (S) is the difference between the work of adhesion ($W_{1,2}$) between the phases and the work of cohesion ($W_{1,1}$) of the phase under consideration (https://www.kruss-scientific.com/services/education-theory/glossary/spreading-coefficient/, accessed on November 15, 2019).

$$S = W_{1,2} - W_{1,1} \tag{3.11}$$

The spreading behavior of oil on water can be understood through the spreading coefficient, which is given below [Eq. (3.12)].

$$S_{ow} = \gamma_{a/w} - \gamma_{o/w} - \gamma_{o/a} \tag{3.12}$$

where $\gamma_{a/w}$, $\gamma_{o/w}$, and $\gamma_{o/a}$ are the air/water, oil/water, and oil/air interfacial tensions, respectively. The spreading coefficient predicts whether the oil phase will spontaneously spread over the surfaces of condensed water drops at the water/air interface, cloaking the water drops with a thin film of oil. A positive spreading coefficient indicates the oil will spread; a negative spreading coefficient predicts the oil will not spread (Anand et al. 2012, 2015).

Spreading coefficient can be employed as a parameter to predict the occurrence of oil spreading at the air–water interface (Harkins and Feldman 1922). An oil droplet in the vicinity of an air–water interface can adopt one of three conformations: a non-entered droplet, a lens, or a spread-oil layer (Hotrum et al. 2004). The balance of $\gamma_{a/w}$, $\gamma_{o/w}$, and $\gamma_{o/a}$ determines whether or not an oil droplet can enter or spread at the air–water interface. The parameters $\gamma_{a/w}$, $\gamma_{o/w}$, and $\gamma_{o/a}$ refer to the interfacial tension at the air–water (a/w), oil–water (o/w), and oil–air (o/w) interface, respectively.

Robinson and Woods (1948) defined the entering coefficient (E) as a parameter to predict whether an emulsion droplet will enter the air–water interface or not. This parameter is reflective of the decrease in free energy at the air–water interface when an emulsion droplet enters the interface.

$$E = \gamma_{a/w} + \gamma_{o/w} - \gamma_{o/a} \tag{3.13}$$

$$S = \gamma_{a/w} - \gamma_{o/w} - \gamma_{o/a} \tag{3.14}$$

All the interfacial tensions indicated in Eqs. (3.13) and (3.14) are referred to the initial state in the system, which means before the emulsion droplet reaches the air–water interface. It has been hypothesized that when $S > 0$, a spreading oil layer will form replacing the air–water interface with an oil–water interface. When $S < 0 < E$, the oil droplet will enter the air–water interface and forms a lens. When $E < 0$, the oil droplet will not enter the air–water interface at all. This hypothesis was questioned by a number of researchers, as it was not consistent with experimental findings (Koczo et al. 1992; Aveyard et al. 1993; Lobo and Wasan 1993; Hotrum et al. 2002). It has been hypothesized that in addition to considering the balance of $\gamma_{a/w}$, $\gamma_{o/w}$ and $\gamma_{o/a}$, the formation of a pseudoemulsion film needs to be taken into consideration in the case of an entering or spreading event. The pseudoemulsion film has been defined as a thin water film formed between an approaching emulsion droplet and the air–water interface, acting as a kinetic barrier and effectively inhibits oil entering/spreading (Wasan et al. 1994). Neither the spreading nor entering coefficients account for the presence of this pseudoemulsion film. Hence, entering or spreading of emulsion droplets might have not been observed even when E or S is positive due to the presence of this thin film. Hotrum et al. (2004) reported the entering of immersed emulsion droplets at the air–water interface only when the surface tension is greater than $57\,\mathrm{mN\,m^{-1}}$. However, according to the entering and spreading coefficient calculation, entering and spreading of droplets should be observed more often in their study. Therefore, they concluded that the emulsifiers such as proteins or low molecular weight surfactants that adsorb onto the air–water interface may contribute to the stability of the thin film through electrostatic and/or steric repulsion (Harkins and Feldman 1922; Hotrum et al. 2004). Therefore, in order to eliminate emulsion droplet entering and spreading at the air–water interface, besides understanding the classical entering and spreading theory, the complexity of the formulations need to be taken into consideration, and more techniques need to be developed to facilitate the monitoring entering/spreading of oil droplets at interfaces.

The presence of charge (either negative or positive) on the dispersed oil droplets of the nanosized emulsion influences the spreading coefficient value. For instance, the cationic emulsions were proved to possess better on the cornea and conjunctiva than conventional eye drops and anionic emulsions. This improved spreading coefficient leads to better ocular surface wettability. Optimal spreading of the cationic emulsion confers protective filmogenic properties and reduces tear washout. Figure 3.10 illustrates the behavior of the cationic emulsion that spreads over the eye very rapidly compared with other formulations. It has been well described that o/w emulsions enhance API absorption by facilitating corneal or conjunctival absorption or prolonging the contact with the eye, thus improving API delivery (Aiache et al. 1997).

In addition, the electroattractive interactions between the positively charged oil droplets of the cationic emulsion and the negatively charged ocular surface cell epithelia might also explain the 50% lower contact angle observed with cationic emulsions versus anionic (negatively charged) emulsions, and the higher spreading coefficient (Klang et al. 2000). A low contact angle, better spreading coefficient, and an increased residence time of the cationic emulsions may all contribute to the better API absorption of lipophilic APIs solubilized in cationic emulsions (Lallemand et al. 2012).

3.3. CHEMICAL AND BIOLOGICAL EVALUATIONS

3.3.1. Confirmation of API Encapsulation into Emulsion

The quantification of API entrapped into the API delivery system is prime importance to see the effectiveness/performance of formulation. Determining the drug (API) entrapment efficiency in percentage (DEE%) looks quite a straight forward way for single-unit solid dosage form such as tablet or capsules and multiple-unit solid dosage form like microcapsules, microspheres, or microparticles. Even the DEE determination in polymer-based nanoparticles (suspended in surrounding aqueous or nonaqueous medium) looks easier. All of these solid- and suspension-type dosage forms do not possess multiple phases wherein the API can localize. In other words, these solid- and suspension-type dosage forms consist of a structure that is basically rigid in nature and even the structure won't allow getting deformation at normal environmental conditions

Figure 3.10. Dynamic contact angle measurement and base width of an eye drop instilled on rabbit eyes. Photos taken at 0, 0.66, 1.33, and 3.32 s after instillation of hyaluronate hydrogel (Hylo-COMOD), anionic emulsion (Refresh Endura), and cationic emulsion (Cationorm). Contact angle and base width values confirm the optimal and faster spreading of cationic emulsions compared with anionic emulsions and hyaluronic acid-based product. [Reproduced from Lallemand et al. (2012).]

after their formation. So, basically the API of interest is simply accommodated into/adsorbed onto these solid- and suspension-type dosage forms. Thus, the DEE determination simply follows the equation comprising the ratio between loaded API amount and experimentally determined API amount with conversion of the obtained value into percentage.

Coming to the o/w nanosized emulsions, this type of colloid-like dispersion consists of multiple phases namely, oil phase, oil–water interface, and water phase. Even micelles-like structure is also frequently seen in the liquid-retentive o/w nanosized emulsions. The beauty of all these phases/structure is that they are able to accommodate the hydrophobic API molecules in considerable or significant amount. If the API of interest is hydrophilic in nature, then the water phase and oil–water interface are the structures for API incorporation. Reversibly, if the API of interest is hydrophobic in nature, then the oil phase, oil–water interface, and micelles are the structures for API incorporation. In this respect, the use of DEE% formula mentioned for solid- and suspension-type dosage forms in determining the DEE% for liquid-retentive o/w nanosized emulsions is just allowing the API amount determination in whole of emulsion without any discrimination of phases present in the emulsion system. Moreover, the o/w nanosized emulsions consist of dispersed oil droplets with very flexible structures and even the observed structures are often very fragile or interconvertible in nature depending on the employed storage conditions. Keeping all these points concerning the presence of multiple phases, fragility or interconvertibility of structures, and existence of micelles-like structure in the o/w nanosized emulsions into consideration, the DEE determination in all of these multiple phases might be of interest at least in early emulsion formulation development studies. But the availability of sensitive analytical instruments to detect/follow/monitor the interconvertibility of all these API-entrapped structures over the storage time periods of emulsion is one of the limiting factor in the emulsion science. Still, at early formulation development stage, the possibility of breaking the emulsion structure by centrifugal or mechanical force or destabilizing the emulsion by chemical or other physical way allows the partial or full separation of oil and water phases. Since the API solubility values in pure oil or oil combination with or without emulsifier molecules as well as the aqueous solubility can be obtained either experimentally or theoretically, the amounts of API entrapped in oil phase, oil–water interface, micelles-like structure, and water phases following emulsification can indirectly be computed for a particular emulsion formula. This type of indirect way of API's DEE% computation is discussed below with acetazolamide (ACZM) as a model API of interest entrapped into the o/w nanosized emulsions.

The DEE% of the emulsions was determined by measuring the concentration of ACZM in the aqueous layer obtained by ultracentrifugation (UC) (Wang et al. 2006). Centrifugation was carried out using a HITACHI UC apparatus, operated at 50,000 rpm (~162,000*g*) at 4°C for 2 h. Polyallomer tubes were used and their bottoms were pricked after centrifugation with a syringe needle to collect the aqueous phase. Concentrations of ACZM in both the aqueous layer and the whole emulsion were determined by high-performance liquid chromatography (HPLC). Using the following Eq. (3.15), the DEE% was calculated (Groves et al. 1985; Férézou et al. 1994):

$$\text{DEE\%} = \frac{\{(C_{\text{total}} \times V_{\text{total}}) - (C_{\text{water}} \times V_{\text{water}})\}}{C_{\text{total}} \times V_{\text{total}}} \times 100 \tag{3.15}$$

where C_{total} is total API concentration (50 mg), V_{total} is total volume of emulsion (50 ml), and C_{water} is API concentration in water phase.

The above-described UC method utilizes the mechanical cum centrifugal force to separate transiently the water phase from oil phase of the emulsion. Apart from UC method, the API amount present at the o/w interface can also be determined using ultrafiltration (UF) method. Zhang et al. (2007) performed the UF method using VIVASPIN 4 filters (molecular weight cutoff of ~10,000 Da; VIVASCIENCE Ltd. Co., Germany) at 810*g* for 30 min with all types of the developed emulsions. The amount of ACZM in the separated aqueous phase was measured by HPLC. The API amount (%) at the o/w interface of the emulsion was computed using the following Eq. (3.16):

$$\begin{aligned}&\text{API amount } (\%)\ \mathit{at}\ \text{o / w interface of the emulsion}\\&=\frac{(\text{API amount in whole emulsion} - \text{API amount in water phase})}{(\text{Initial API amount added during emulsion preparation})}\times 100.\end{aligned} \tag{3.16}$$

The DEE% of cationic emulsion was higher followed by neutral and anionic emulsions (Table 3.8). Similarly, the order for API amount (%) at the o/w interface of the emulsion was cationic > neutral > anionic (Tamilvanan and Kumar 2011).

3.3.2. *In Vitro* API Release from Nanosized Emulsions

Performing *in vitro* API release from colloidal API delivery carriers is not a trivial exercise. Because the two major experimental difficulties encountered are (1) the quick/rapid release of API from nanosized carriers and (2) appropriate separation of dissolved API from the carrier components (Levy and Benita 1990; Santos Magalhaes et al. 1991; Magenheim et al. 1993). Many

TABLE 3.8. Drug Entrapment Efficiency (DEE) % and Acetazolamide (ACZM) Amount (%) at Oil–Water Interface of ACZM-Loaded Anionic, Cationic, and Neutral-Charged Nanosized Emulsions Determined by Ultracentrifugation and Ultrafiltration Methods, Respectively

Formulations	DEE (% ± SD, $n = 3$)	ACZM Amount (%) at Oil–Water Interface of the Emulsion
ACZM-loaded anionic emulsion	69 ± 2	25
ACZM-loaded cationic emulsion	99 ± 1	48
ACZM-loaded neutral-charged emulsion	85[a] ± 2	39*

*Statistically significant when compared with cationic emulsion ($P < 0.05$).

different methods have been described, such as the dialysis sac diffusion technique (Levy and Benita 1990) and bulk equilibrium reverse dialysis (Santos Magalhaes et al. 1991). It is claimed that with these methods, the carrier is never fully diluted with the release solution, and hence transport through the membrane may be a rate-limiting step in the release process. Centrifugal UF has no such drawback but suffers from possible changes in API distribution profile as a result of the strong centrifugal force. An UF technique at low pressure has also been proposed (Magenheim et al. 1993).

In recent years, a membrane-free API release study is also proposed (Rahman et al. 2020) in addition to the UF technique. Using the UF technique, the results of *in vitro* celecoxib (CXB) release from both non-phospholipid-based cationic (MPS value: 238 ± 20 and ZP value: 40.2 ± 2.81) and phospholipid-based anionic (MPS value: 285 ± 23 and ZP value: −33.9 ± 3.18) nanosized emulsions are discussed below.

3.3.2.1. Ultrafiltration Technique as a Model to Study In Vitro API Release from Nanosized Emulsions In preliminary experiments, we have determined that following soaking in isopropyl myristate for 2 h, an artificial membrane, cellulose acetate, did not adsorb CXB, and therefore, it was decided to perform the *in vitro* API release by using the cellulose acetate membrane. Nevertheless, CXB-loaded anionic and cationic emulsions showed controlled release characteristics compared with the burst release observed with CXB solution (Fig. 3.11).

This result suggests the significance of emulsions over solution for topical application onto skin. There has been considerable interest in the *in vitro* release and *in vivo* fate of lipophilic API molecules within oil droplets stabilized by classical and phospholipid emulsifiers (Washington and Evans 1995; Klang and Benita 1998; Nishikawa et al. 1998). The kinetics of release is mostly governed by the API partition coefficient ($\log PC_{oct}$). Although APIs with $\log PC_{oct} > 9$ are generally retained in emulsion droplets and their *in vivo* disposition follows that of the droplet, APIs with $\log PC_{oct} < 9$ are rapidly released under sink conditions and their biofate is consequently independent of the droplets (Nishikawa et al. 1998). API release kinetics is therefore driven by API partitioning, and lack of sustained release properties is a characteristic of conventional emulsions (Washington and Evans 1995; Klang and Benita 1998; Nishikawa et al. 1998). For example, Washington and Evans (1995) also reported a rapid release (within 500 s) of a probe hydrophobic API (chlorpromazine) from lecithin (phospholipid)-stabilized submicron triglyceride emulsions, and this was attributed to the fact that commonly used emulsifiers such as lecithin and Pluronics® do not act as a strong interfacial transport barrier. CXB is a lipophilic compound ($\log P = 3.5$) (Shono et al. 2009) and it is expected to be more soluble in the inner oil phase of the emulsions. Therefore, API amount determined at the oil–water interface of the emulsion was found to be low (ranging from 5 to 8%) for both anionic and cationic emulsions

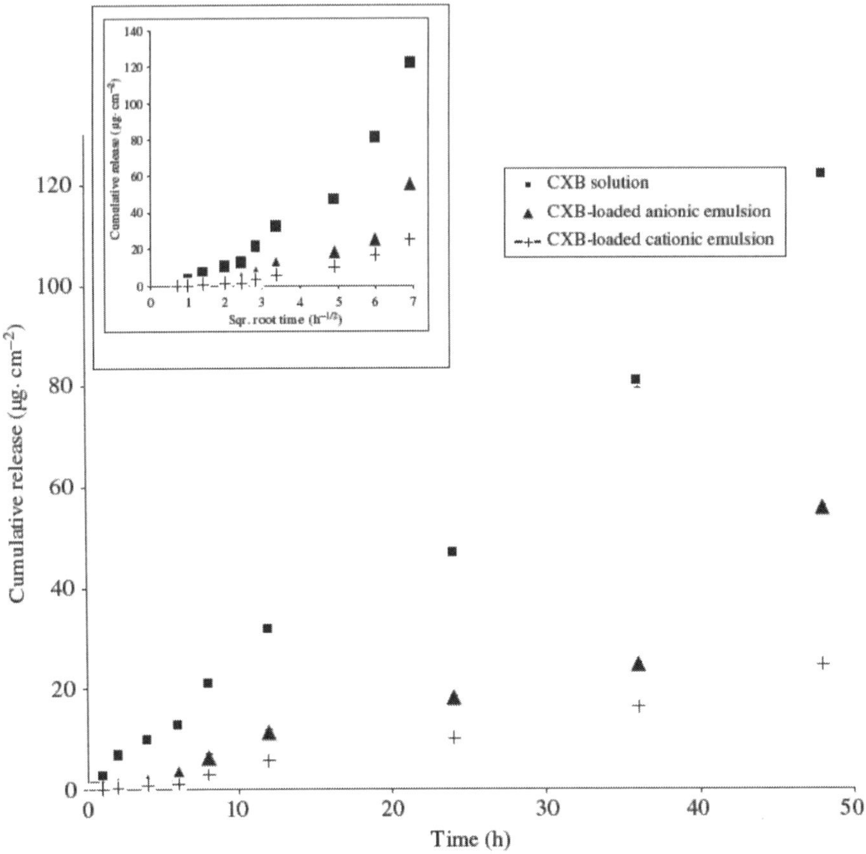

Figure 3.11. *In vitro* release of celecoxib (CXB) through cellulose acetate membrane for the solution and nanosized emulsions within 48 h (mean ± SE, $n = 3$). The inset graph shows linear fits of the release data from time 0 to 12 h against square root of time. The linear regression for the solution and anionic emulsion look in a biphasic model while for the cationic emulsion, it is linear. [Reproduced with permission from Tamilvanan and Baskar (2013).]

(Table 3.9). In the current investigation, non-phospholipid-based cationic emulsion, however, showed more sustained API release characteristics than the phospholipid-based anionic emulsion. A similar release profile was also observed by Tamilvanan and Kumar (2011) when studying the influence of ACZM loading on the *in vitro* performances of non-phospholipid-based cationic nanosized emulsion in comparison with phospholipid-based anionic and neutral-charged nanosized emulsions. The observed disparity in the release characteristics of phospholipid-based anionic and non-phospholipid-based cationic emulsions could be corroborated to the difference in the interfacial

TABLE 3.9. Drug Entrapment Efficiency (DEE) % and Celecoxib (CXB) Amount (%) at Oil–Water Interface of CXB-Loaded Anionic and Cationic Nanosized Emulsions

Formulations	Drug Entrapment Efficiency (%), (±SD, $n = 3$)	CXB Amount (%) at Oil–Water Interface of the Emulsion, (±SD, $n = 3$)
CXB-loaded anionic emulsion	98 ± 2	5 ± 3
CXB-loaded cationic emulsion	99 ± 1	8 ± 4

TABLE 3.10. *In Vitro* Release Characteristics of Celecoxib (CXB) Through Cellulose Acetate Membrane for Solution and Nanosized Emulsions

Formulations	Diffusional Flux J_{SS} (ng·cm^{-2}·h$^{-1/2}$) (±SD, $n = 3$)	R^2	K_P (cm·h$^{-1/2}$)	Cumulative release (μg·cm^{-2}, 48 h) (±SD, $n = 3$)
CXB solution	6,825.79 ± 920.86	0.9125	0.142 ± 0.04	122.00 ± 0.70
CXB-loaded anionic emulsion	2,513.15 ± 382.71	0.9857	0.073 ± 0.0007	55.80 ± 0.70
CXB-loaded cationic emulsion	1,925.67 ± 147.42	0.9982	0.018 ± 0.0009	24.79 ± 0.90

layer structure formed over the dispersed oil droplets of these two emulsions. The slower the API release from cationic emulsion, the lower is the flux at the steady state (J_{SS}) through the membrane. Therefore, the substantial difference in J_{SS} value (588 ng cm^{-2} h$^{-1/2}$) was also obtained with non-phospholipid-based cationic emulsion over phospholipid-based anionic emulsion (Table 3.10). The calculated permeability coefficient and enhancement ratio values for solution and emulsion formulations also followed the trend similar to that observed in the J_{SS} values of these formulations (Table 3.10). Interestingly, both emulsions and solution formulations demonstrated higher correlation coefficient values when the API release data were fitted with Higuchi model rather than with first-order model (Table 3.11). It is known that there was no release-retarding barrier effect due to the excipients used in manufacturing of both emulsion and solution formulations. Despite that both emulsions presented a low API release rate compared with the API release rate obtained with solution, between anionic and cationic emulsions, the latter one showed somewhat higher retardation in the API release rate (2.62 ± 0.08 vs. 4.90 ± 0.10) (Tamilvanan and Baskar 2013).

TABLE 3.11. Celecoxib (CXB) Release Rates Assessed by Two Different Model Equations

Formulations	Higuchi Model ($min^{-1/2}$) (±SD, n = 3)	First-Order Model (min^{-1}) (±SD, n = 3)
CXB solution	12.70 ± 0.07 (0.8138)	0.23 ± 0.05 (0.6947)
CXB-loaded anionic emulsion	4.90 ± 0.10 (0.9555)	0.18 ± 0.01 (0.7653)
CXB-loaded cationic emulsion	2.62 ± 0.08 (0.9826)	0.11 ± 0.05 (0.7823)

Note: Values in parentheses indicate correlation coefficient (R^2).

3.3.2.2. Membrane-Free Model to Study the In Vitro API Release from Nanosized Emulsions In the context of membrane-free API release studies, the *in vitro* release kinetics of topical ophthalmic emulsions loaded with two different concentrations (0.05 and 0.1% w/w) of CsA were carried out in simulated tear fluid (STF). The STF was prepared using 192.4 mg $NaHCO_3$, 111.0 mg KCl, 2.29 mg $CaCl_2$, 672.8 mg NaCl, 669.0 mg albumin, 2.5 mg glucose, and double-distilled water up to 100 ml (Moses 1981). To simulate the intimate mixing condition that occurred following the topical application of a single drop of emulsion with the tear fluid present in the eye, a 1 : 2.5 dilution ratio was arbitrarily selected to perform the *in vitro* API release study (Tamilvanan and Kumar 2011). So, 1,600 µl of CsA (0.05 and 0.1% w/w)-loaded emulsions were diluted or mixed up to 4 ml with STF in glass vials and the temperature was maintained throughout the study at 37°C using a water bath. Four hundred microliters of samples were withdrawn from the vials at post-mixing time period of 1, 16, 36, 56, and 86 min (corresponding to the equivalent time intervals of 5, 20, 40, 60, and 90 min including centrifugation time) and the samples were centrifuged at 4,000 rpm for 4 min. The clear supernatant portion from each of the centrifuged samples were taken and diluted up to 1 ml with acetonitrile before injecting them into the UHPLC system to determine CsA contents. Irrespective of the CsA loadings (0.05 and 0.1% w/w) into the emulsions, the DEE% determined ranged from 73.20 ± 0.13 to 74.42 ± 0.15.

In a previous report, it was shown that the availability of electrolytes under the physiological conditions existing in the lachrymal sac in the presence of tear can be monitored by performing the *in vitro* release study of API-loaded non-phospholipid-based cationic nanosized emulsions in STF with varying dilution ratios (from 1 : 5 to 1 : 40). In the current investigation, we have further reduced the dilution ratio from 1 : 5 to 1 : 2.5 to see the *in vitro* CsA release from non-phospholipid-based cationic nanosized emulsions. Figure 3.12a depicts the cumulative percent API release from non-phospholipid-based CsA (0.05 and 0.1% w/w)-loaded topical ophthalmic emulsions at 1 : 2.5 dilution in STF. Up to 20 min post-mixing time, the cumulative percent CsA release values observed for emulsions having both of the API loadings were confined from

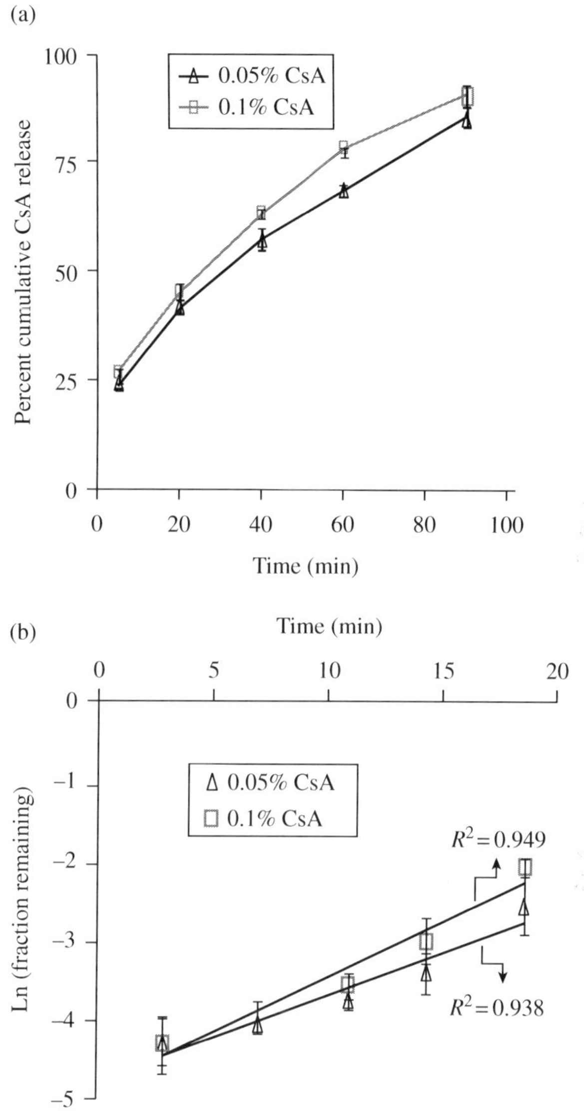

Figure 3.12. *In vitro* cyclosporin A (CsA) release behavior at 1 : 2.5 dilution in simulated tear fluid (STF) of CsA-loaded topical ophthalmic emulsions. Key: (a) Plot of time versus cumulative percent CsA release. (b) Ion-exchange equation release plot between Ln(1 − F) and $t^{0.65}$. [Reproduced with permission from Rahman et al. (2020).]

24.5 ± 1.34 to 45.75 ± 1.38. However, a significant difference (two-way ANOVA) in the cumulative percent CsA release values was noticed at 40 and 60 min post-mixing time wherein the observed cumulative percent CsA release value for 0.1% w/w API-loaded emulsion was higher than the observed cumulative percent CsA release value for 0.05% w/w API-loaded emulsion (63.88 ± 1.22 versus 57.63 ± 2.31 at 40 min post-mixing time and 78.96 ± 2.14 versus 69.13 ± 1.22 at 60 min post-mixing time). It indicates that higher the CsA loading into the non-phospholipid-based emulsion, the faster was the API release. This explanation was also further substantiated through the result of $t_{50\%}$ values wherein the 0.05% w/w CsA-loaded emulsion showed the $t_{50\%}$ value of 30 ± 0.5 min and the 0.1% w/w CsA-loaded emulsion yielded the $t_{50\%}$ value of 25 ± 0.7 min only. But, at 90 min post-mixing time, the cumulative percent CsA release values observed for emulsions having two different API loadings did not show significant difference (86.13 ± 2.36 % for 0.05% w/w API-loaded emulsion and 91.25 ± 2.51 % for 0.1% w/w API-loaded emulsion).

Although CsA is neutral in nature, the presence of anions such as carbonates and chlorides at 7.6 and 130 mM, respectively, in STF induced the CsA release from the emulsion possibly via an anion-exchange process or the rapid API partitioning out of the charged emulsifier (mono- or multi-layer) film. Instead of using the traditional Fick's diffusion law-based equations to study the API release kinetics, it is common in recent years to apply another mathematical kinetics model for describing the particle diffusion controlled release of API from the phospholipid-based emulsions (Hagigit et al. 2008). Equation (3.17) shows a mathematical kinetics model suggested assessing the API release kinetic profile from the emulsion. In the present study, the same model equation is employed for the non-phospholipid-based emulsions having two different CsA loadings.

$$-\mathrm{Ln}\left(1-F\right)-1.59\left(\frac{3}{r}\right)^{1.3}\times \mathrm{Di}^{0.65}\times t^{0.65} \tag{3.17}$$

By plotting Ln(1− F) versus $t^{0.65}$, the particle diffusion control can be assessed by simply seeing the linearity (r^2 value) in the plot. In general, either the simple diffusion of counterions or an actual chemical exchange reaction occurring at the fixed ionic groups was the rate-determining step for the so-called ion-exchange process. However, Eq. (3.17) is meant for studying the CsA release profile from the emulsions based on counterions' diffusion only. The ion-exchange equation release plots of CsA from typical non-phospholipid-based topical ophthalmic emulsions containing 0.05 and 0.1% w/w API in 1 : 2.5 dilution with STF is shown in Fig. 3.12b. It can be seen from Fig. 3.12b that the kinetic data do not match with counterions' diffusion since the r^2 value is not close to 1 (0.938 and 0.949, respectively, for emulsions having 0.05 and 0.1% w/w CsA loadings). Because, the anions exchange process from STF should not be as simple as possible as significant deviation might occur leading to put the ion-exchange process in a more complex situation where it is difficult to characterize using mathematical fitting equations. The nonmatching does not exclude potential therapeutic efficacy since it is highly likely that despite infinite dilutions of applied single drop of emulsion with tear fluid and the presence of anions, a majority of the entrapped CsA molecule, which remain mainly within the cationic oil droplets, would elicit the desired pharmacological actions once the CsA-loaded emulsion droplets reached the target ocular tissues following topical application into eye. It can be deduced that non-phospholipid-based topical ophthalmic emulsions should maintain initial physicochemical properties, entrapped CsA molecules, and release them under appropriate physiological conditions. Furthermore, the CsA release kinetic process will not be governed by a film diffusion process but rather by low oil degradation at the tissue of target.

Whether membrane-free method or UF technique is suitable for assessing the *in vitro* API release, it is up to the formulator to decide the therapeutic applicability of the developed o/w nanosized emulsions. If any formulator designs a new *in vitro* API release apparatus/equipment, then it will be of a welcome addition in nanosized emulsion characterization.

3.3.3. Techniques Used to Study the *In Vitro* API Leakage Before Release in the Targeted Site

Although o/w nanosized emulsions are very fragile in nature, the selected API molecules can safely be made to sit in the dispersed oil droplets and o/w interface of emulsion meaning that the API molecules might be in their original conformation before reaching the intended/targeted site inside the human body. This assumption could be checked via *in vitro* API leakage study using AZM as a model API incorporated into the o/w nanosized emulsions.

About 500 µl of AZM-loaded emulsion (equivalent to 1,000 µg of AZM) was taken in a vial, and 50 ml of pH 7.4 phosphate buffer was added into it. In another vial, 1 ml of AZM solution (1 mg ml^{-1}) in pH 7.4 phosphate buffer was taken and diluted with 50 ml buffer. These two vials were kept in (1) an autoclave at 121°C and 15 lb in.$^{-1}$ pressure for 15 min, (2) a hot air oven at 70°C for 30 min, and (3) at room temperature of 25°C for 30 min. The API leakage (in percentage) from both emulsion and solution at these three different storage conditions were calculated according to Eq. (3.18):

$$\text{API leakage\%} = \frac{\text{API amount leached}}{\text{Initial API amount added}} \times 100 \quad (3.18)$$

Figure 3.13 depicts the thermodegradation behavior of AZM-loaded macroemulsion in three different storage conditions. It should be added that keeping the 10-ml AZM solution (2 mg ml^{-1}) in pH 7.4 phosphate buffer at all the three studied storage conditions led to the formation of API precipitation at the bottom of the vials. In contrast, the macroemulsion allowed the AZM leakage from the inner oil phase or the oil/water interface of the emulsion that depended on the storage conditions used. The autoclave condition showed the

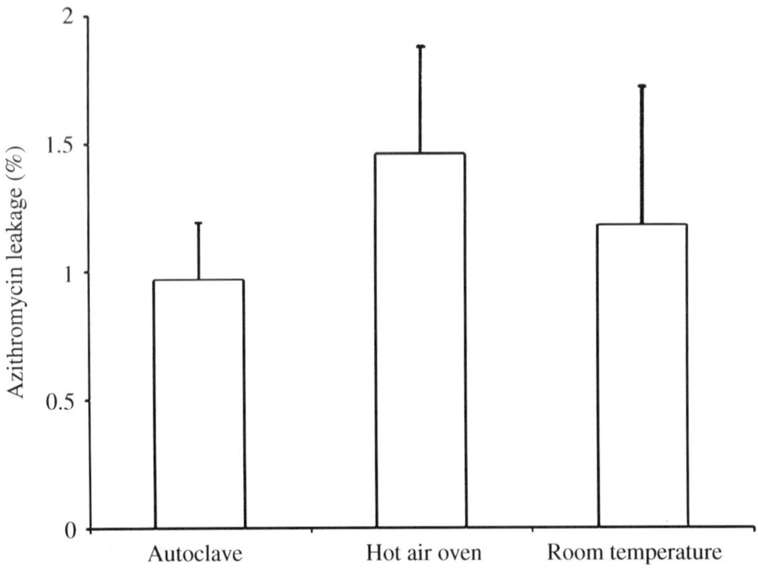

Figure 3.13. Azithromycin (AZM) leakage percentage observed following the storage of AZM-loaded oil-in-water macroemulsion at three different storage conditions. [Reproduced with permission from Shunmugaperumal and Kaur (2016).]

API leakage percentage value of 0.972 ± 0.02 followed by hot air oven with the API leakage percentage value of 1.46 ± 0.72, and only the API leakage percentage value of 1.18 ± 0.54 was noticed for room temperature.

Keeping the AZM molecule in phosphate buffer solution (pH 7.4) always led to the precipitation of the API molecules in all of the studied three different (stipulated) storage conditions. According to Bowman et al. (2009), the AZM molecule is prone to undergo hydrolytic degradation in aqueous-based solution that lead to the formation of its major degradant product, DES. It is our belief that the total precipitation of AZM molecule in pH 7.4 phosphate buffer should be corroborated with the complete API DP formation.

On the other hand, the AZM molecule after its incorporation into chitosan- and PEG-based o/w macroemulsion exhibited the API leakage percentage value of 0.972 ± 0.02 in the autoclave storage condition followed by 1.46% ± 0.72% API leakage in hot air oven and 1.18% ± 0.54% in room temperature conditions (Fig. 3.13). In our previous experiment, the AZM molecule was incorporated into PEG-based microparticles wherein about 13% API leakage was noticed after storing of the microparticles in autoclave conditions followed by 8.65% ± 0.72% API leakage in hot air oven and 4.15% ± 1.54% in room temperature conditions (Tamilvanan et al. 2014). Switching from microparticles to macroemulsion, the AZM molecule appears to be more tightly bound to the inner castor oil core or o/w interface of the macroemulsion. Therefore, the chitosan- and PEG-based o/w macroemulsion shows a lesser AZM leakage percentage value than the API leakage percentage value observed with PEG-based microparticles at all of the studied three different storage conditions. Furthermore, the observed low API leakage percentage values following even in the autoclave storage condition of chitosan- and PEG-based o/w macroemulsion (when compared with the precipitation of API in phosphate buffer solution of pH 7.4) could provide an impetus to work further with the macroemulsion (Shunmugaperumal and Kaur 2016).

3.3.4. Transcorneal Permeation Study

For laboratory study, the transcorneal permeation experiments were performed with a modified diffusion chamber. The cell, made of acrylic plastic, consisted of a donor and a receiving compartment (volumes 2.0 and 5.0 ml, respectively) (Camber 1985). No significant adsorption of the tested formulations to the diffusion chamber was observed over the 2 h period of the permeability experiments. The receptor solution consisted of STF. Before use, the receptor solution was aerated with the mixture of 95% O_2 and 5% CO_2 to maintain oxygenation of cornea. Following humanely killing of the goats at the local slaughter house, three to four eyes were enucleated and placed over a separate plastic bag containing the aerated fresh receptor medium to preserve the eyes during transportation/before dissection. Sterile surgical procedures were used and for avoiding cross-contamination, excision of cornea along with 2–4 mm ring of surrounding scleral tissue was done from a single eye at a time. The excised corneas were then preserved separately in the medium. One cornea with 2–4 mm ring of sclera was mounted by sandwiching surrounding scleral tissue between clamped donor and receptor compartments of a perfusion apparatus in such a way that its epithelial surface faced the donor compartment. The corneal area available for diffusion was 1.0 cm^2. A 5 ml aliquot of the receptor solution was added to the endothelial side, whereas 1.0 ml (1,000 μg) of either the anionic, cationic, and neutral-charged nanosized emulsions or the ACZM solution (1% w/v or 10,000 μg) was added to the epithelial side. The temperature in the diffusion chamber was maintained at 37 ± 0.5°C by a thermostatic water bath. Sample aliquots from the receptor chamber were withdrawn at 15, 30, 60, 75, 90, 100, and 120 min and immediately replaced by previously aerated fresh receptor medium. Samples were filtered through a 0.45 μm microporous membrane and the filtrate was kept at 4°C until analyzed by HPLC. According to previous studies, it could be corroborated that at the end of the study, the final concentration of ACZM was at least 20-fold below the maximal solubility in the acceptor medium.

The apparent corneal permeability coefficient (P_{app}) of these formulations was determined according to Eq. (3.19) (Schoenwald and Huang 1983):

$$P_{app} = \frac{\Delta Q}{\Delta t \times C_0 \times A \times 60\left(\text{cm s}^{-1}\right)} \quad (3.19)$$

where C_0 is the initial concentration of ACZM in the donor compartment, 60 represents the conversion of minutes to seconds, and A is the area of the cornea. For the calculation of the apparent permeation coefficient in the present study, A is the surface area of goat cornea, $1.0\,\text{cm}^2$. $\Delta Q/\Delta t$ is the steady-state rate of API permeation across the intact cornea, as obtained from the slope of the straight line relating corneal permeability (API amount, Q) to time (t) plot (Q vs. t). The lag time was also determined from this graph by extrapolating the linear portion of the x-axis.

The flow rate of the steady state (J_{ss}) can be calculated by Eq. (3.20).

$$J_{SS} = C_0 \times P_{app} \quad (3.20)$$

Figure 3.14 shows the *in vitro* transcorneal permeation profiles of the ACZM solution, ACZM-loaded anionic, cationic, and neutral-charged emulsions in

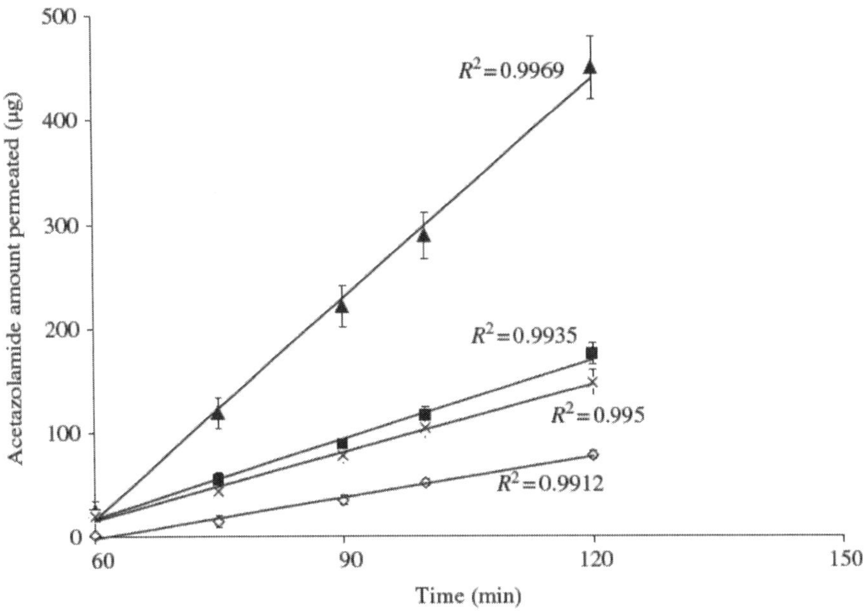

Figure 3.14. Acetazolamide (ACZM) amount (microgram) permeated through goat cornea versus time (minute). Key: ◊, ACZM solution; ■, ACZM-loaded anionic emulsion; ▲, ACZM-loaded cationic emulsion; ×, ACZM-loaded neutral-charged emulsion. [Reproduced with permission from Tamilvanan and Kumar (2011).]

goat cornea. All the formulations showed linear relationships in the time versus API amount permeated plot as the correlation (R^2) values were in the range from 0.9912 to 0.9969. Permeation data of ACZM solution and ACZM-loaded emulsion formulations are shown in Table 3.12. All of the tested formulations had similar lag time of 60 min to exert their permeation effect through goat cornea epithelial cells. Cationic emulsion possessed the % permeation value of 45.00 ± 4.3 followed by anionic and neutral-charged emulsions (14.80–17.50%) and ACZM solution (7.80 ± 0.86%). The steady-state flux (J_{ss}) value was also higher for cationic emulsion (2.22 ± 0.1 μg min^{-1}). Again the ACZM solution exhibited the lowest J_{ss} value (0.25 ± 0.3 μg min^{-1}), whereas the neutral and anionic emulsions possessed 1.10 ± 0.2 and 0.96 ± 0.1 μg min^{-1}, respectively. As expected, the cationic emulsion showed a higher apparent P_{app} value of 13.32 ± 0.92 × 10^{-6} cm s^{-1}, whereas the anionic and neutral-charged emulsions demonstrated, respectively, the P_{app} values of 5.64 ± 0.14 × 10^{-6} cm s^{-1} and 4.87 ± 0.52 × 10^{-6} cm s^{-1}. The ACZM solution illustrated the lowest P_{app} values of 2.52 ± 0.42 × 10^{-6} cm s^{-1}. ACZM is a poorly water- and lipid-soluble API. Its poor lipid solubility limits its transit through the corneal epithelium and endothelium, whereas poor aqueous solubility prevents the transit through the hydrophilic stroma (Barar et al. 2008). The analysis of data from Table 3.12 indicates that the ACZM-loaded cationic emulsion was the most efficient in API permeation according to % of permeation, steady-state flux, and P_{app} values. The anionic and neutral-charged emulsions showed lower permeation than the cationic emulsion. On the other hand, the ACZM from its solution was permeated in a much lesser extent compared with the emulsions. It is important to highlight that the ACZM concentration in its solution is 10-fold higher than in the case of the currently developed ACZM-loaded anionic, cationic, and neutral-charged

TABLE 3.12. Comparison of Acetazolamide (ACZM) Formulations in Terms of % Permeation, Steady-State Flux, and Apparent Permeability Coefficient P_{app} from *In Vitro* Permeation Studies in Simulated Tear Fluid Using Goat Cornea

Formulations	% Permeation (±SD, n = 3)	Lag Time (min)	Steady-State Flux (μg min^{-1}) (±SD, n = 3)	Apparent Permeability Coefficient, P_{app} (cm s^{-1}) (×10^{-6}) (±SD, n = 3)
ACZM solution	7.80 ± 0.86	60	0.25 ± 0.3	2.52 ± 0.42
ACZM-loaded anionic emulsion	17.50 ± 1.4	60	1.10 ± 0.2	5.64 ± 0.14
ACZM-loaded cationic emulsion	45.00 ± 4.3	60	2.22 ± 0.1	13.32 ± 0.92
ACZM-loaded neutral-charged emulsion	14.80 ± 1.3	60	0.96 ± 0.1	4.87 ± 0.52

nanosized emulsions. The observed noticeable differences in permeation efficiency between solution and emulsion formulations can be attributed to the fact that the ACZM was entrapped at the oil–water interface of the emulsion or simply dissolved in a vesicular-type oil droplet protected by mono- or multi-layer emulsifier films. Moreover, a hydrophilic solution form of the API is not at all favorable for the permeation phenomenon to occur through goat cornea probably due to the lesser contact time between the API and cornea. A remarkable fact from the results of the present study is that the cationic emulsion provided higher ACZM permeation coefficient value than anionic and neutral-charged emulsions. There are evidences that colloidal delivery systems can facilitate the penetration of APIs into ocular surface tissues through an endocytic mechanism (Calvo et al. 1996a). As previously suggested, the endocytic effect is probably more pronounced with the phospholipid-based cationic emulsion (Abdulrazik et al. 2001). In experiments of keratocyte cell-culture exposed to either anionic or cationic blank emulsion (phospholipid-based), only the positively charged oil droplets were internalized in the keratocytic cells (unpublished data). In spite of differences in blink frequency, ocular surface permeability, and aqueous humor dynamics between goat and man, a mean ACZM% permeation of 45.00 ± 4.3 as observed with non-phospholipid-based cationic emulsion, in this present study, was of further clinical interest. A high and fast penetration achievement after *in vitro* transcorneal permeability through goat cornea might be an indication that the topical administration of ACZM-loaded cationic emulsion to patients with glaucomatous condition should provide adequate levels of ACZM in aqueous humor resulting in a decreased production of aqueous humor and hence a lowering of intraocular pressure (IOP) (Tamilvanan and Kumar 2011).

3.3.5. *Ex Vivo* Diffusion Studies Using Animal Tissues

Experimental conditions and the apparatus for skin studies were equivalent to the above-described method for *in vitro* release studies using a synthetic membrane. Male Wistar rats (weight, 180–220 g; age, 6–7 weeks) were sacrificed with prolonged ether anesthesia, and the abdominal skin of each rat was excised. Previously reported method for removing and further processing of stratum corneum (SC) samples was followed (Shakeel et al. 2008). In brief, hairs on the skin of animals were removed with electrical clipper, subcutaneous tissues were surgically removed, and dermis side was wiped with isopropyl alcohol to remove residual adhering fat. The skin was then washed with distilled water, wrapped in aluminum foil, and stored in a deep freezer at −20°C till further use. The skin was treated with 1 M sodium bromide solution in distilled water for 4 h. The epidermis from full-thickness skin was separated using cotton swab, which was moistened with water. Epidermal sheet was cleaned by washing with distilled water and dried under vacuum and examined for cuts or holes if any. SC samples were prepared by floating freshly prepared epidermis membrane on 0.1% trypsin solution for 12h. Then, SC sheets were cleaned by washing with distilled water. In each run of skin retention experiments, 100 µl of the CXB solution or CXB-loaded emulsions was placed on the skin surface with uniform spreading and then covered with parafilm. After exposure times of 6, 12, and 24h, skin samples were removed, and the surface was rinsed three times with ethanol/water and MilliQ water, respectively, to eliminate the residual formulation. The excess water was absorbed with tissue, and the skin was then cut into 3×3 mm pieces. CXB was extracted with acetonitrile using sonication for 15 min and then quantified using HPLC. The API concentration was also analyzed in the receptor medium to determine the permeation of CXB through rat skin after 2, 4, 6, 8, and 12h treatment with control solution and emulsions.

Rat skin uptake of CXB at 6, 12, and 24h after topical application of solution, phospholipid-based anionic emulsion, and non-phospholipid-based cationic emulsion is depicted in Fig. 3.15. The CXB solution showed limited skin uptake; less than 0.5% of the topically applied CXB was recovered in the whole skin after 24h. Significantly higher levels of CXB

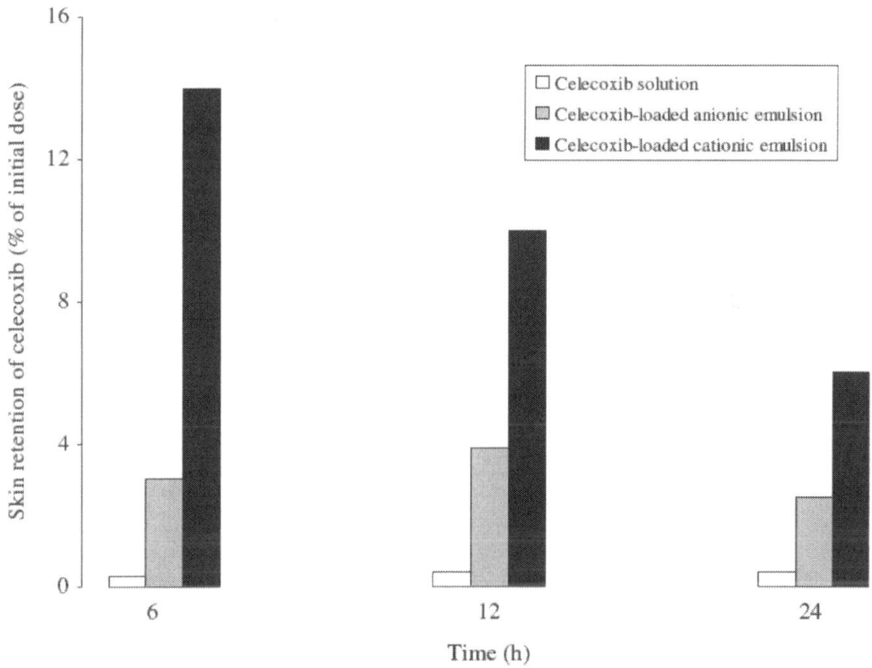

Figure 3.15. Skin retention of celecoxib (CXB) from solution and anionic and cationic nanosized emulsions. Statistical differences between the anionic and cationic nanosized emulsions were determined using Wilcoxon rank-sum test and indicated by an asterisk (mean ± SE, $n = 3, P \leq 0.05$). [Reproduced with permission from Tamilvanan and Baskar (2013).]

($p \leq 0.05$) were detected in full-thickness rat skin from non-phospholipid-based cationic emulsion in comparison with the phospholipid-based anionic emulsion. In addition, the kinetics of skin uptake and transport of CXB were both altered in the presence of a cationic charge over the oil droplets of the emulsion. When cationic charge was conferred to the oil droplets, the maximum skin uptake of CXB (14.0 ± 1.0, 10.0 ± 0.9, and 6.0 ± 0.4% of the initial dose, respectively) was observed at 6, 12, and 24 h after topical application. However, only 2.5 ± 0.5 to 3.9 ± 0.4% of the initial dose was detected in full-thickness rat skin (over the period of 24 h) from phospholipid-based anionic emulsion followed by the elimination of the API from the skin, which was evidenced by the appearance of CXB in the receptor phase after 24 h. The maximum permeation of CXB through full-thickness rat skin, i.e., the concentration detected in the receptor medium after 24 h, was 1.45 and 0.42% of the topically applied dose for anionic and cationic emulsions, respectively.

The results of the skin retention study showed that the skin retention of CXB was also dependent on the initial emulsifier type (Fig. 3.15). The order for skin retention of CXB is cationic emulsion > anionic emulsion > solution. More importantly, the skin retention of CXB was three to four times higher for cationic emulsion compared with anionic emulsion. Because the isoelectric point of the skin is between 3.5 and 4.8, the skin is negatively charged under physiologic conditions (Wilkerson 1935; Higaki et al. 2003). It seems that an electrostatic attraction occurred between the cationic dispersed oil droplets of the emulsion and the anionic skin membranes. Furthermore, the higher skin uptake of CXB in the presence of chitosan as observed in this study can also be related to the perturbation in the lipid structure of the SC. Because Takeuchi et al. (1992) demonstrated that the presence of chitosan might alter the frequency alterations in the CH-asymmetric stretching band near 2,920 cm^{-1} in the SC as analyzed with Fourier transform infrared/attenuated total reflection (Tamilvanan and Baskar 2013).

3.3.6. Techniques Used to Study the *In Vitro/In Vivo* Integrity of the Droplets Before and After It Arrives on the Targeted Site

The movement of IV-administered o/w nanosized emulsions can be monitored in a noninvasive manner as per the report provided by Bouchaala et al. (2016). These authors used the Forster resonance energy transfer (FRET) for monitoring the movement of lipid nanocarrriers (NCs) encapsulated with two dyes lipophilic cyanine 5.5 (Cy5.5LP) and 7.5 (Cy7.5LP). A 1% of Cy5.5LP (with respect to Labrafac WL® used to make lipid nanocarriers) acts as energy donor and a 1% of Cy7.5LP operates as energy acceptor.

The NCs were diluted 10,000 times from the original formulation and incubated in water and 100% of fetal bovine serum (FBS). High dilution was needed to avoid saturation of serum by lipids of NCs. The donor in the NCs was excited at 670 nm. The semiquantitative parameter of FRET efficiency was calculated according to the following equation (Preus and Wilhelmsson 2012):

$$E = \frac{A}{(A + D)} \tag{3.21}$$

where A and D are the maximum of fluorescence intensity of the acceptor and donor, respectively.

Due to an excellent FRET couple with ideal spectral properties for *in vivo* whole-animal imaging, Cy5.5 and 7.5 LP dyes were selected for encapsulation into NCs. Indeed, Cy 5.5 LP derivatives are among the most commonly used NIR imaging agents with convenient excitation of 650–700 nm and emission in the NIR region (>700 nm). Moreover, Cy 7.5 LP absorption spectrum overlaps perfectly with the emission spectrum of Cy 5.5 LP while its emission in the NIR region at 840 nm is perfectly separated from the emission of Cy 5.5 LP, making this couple perfectly tailored for ratiometric FRET imaging. To achieve both strong brightness and highly efficient FRET inside NCs, Cy5.5 and 7.5 LP should be encapsulated at very high concentrations up to 1 wt% in oil, as shown for other FRET pair (Klymchenko et al. 2012). Then, nano-emulsions (lipid NCs) were generated by the spontaneous emulsification of dye-loaded oil and nonionic surfactant (Cremophor® ELP), giving rise to fluorescent PEGylated droplets sizing around 90–100 nm. To generate FRET NCs, these authors prepared NCs encapsulating increasing amount of Cy5.5LP and Cy7.5LP. As expected, the emission spectra recorded after excitation of the energy donor showed strong dependence on the concentration of the FRET pair. Thus, at 0.1% of dyes, the emission spectrum was very close to that of Cy5.5LP alone, while at higher concentrations, a long wavelength emission, corresponding to the FRET acceptor, appeared and became dominant at ≥0.5% of the dyes. It should be noted that the emission maximum of Cy7.5LP shifted gradually to the red with increase in the dye content. Similar concentration-dependent red shifts were previously observed for Cy3 dye in lipid droplets (Kilin et al. 2014); they could be related to some aggregation of the dyes, homo-FRET, as well as to small changes in the oil core properties at higher dye concentrations. Remarkably, the total (donor + acceptor) fluorescence quantum yield (QY) of NCs encapsulating FRET pair at 1% loading of each dye was 11%. Although this value was lower than the QY of NCs with 1% of Cy5.5LP alone (27%), it remained relatively high to ensure high brightness to these NCs, important for *in vivo* imaging. For comparison, QY of NIR dye indocyanine green measured in phosphate-buffered saline (PBS) is 2.4% (Russin et al. 2010). Here, these authors achieved much higher QY already for the FRET system. Moreover, because NCs minimize potential interactions of loaded dyes with tissues *in vivo*, they allow the use of higher dye concentrations, which is essential for achieving optimal imaging contrast for NIR imaging.

The FRET methodology revealed that the integrity of NCs in the blood circulation of healthy mice is preserved at 93% at 6h of post-administration, while it drops to 66% in the liver (half-life is 8.2h). Moreover, these NCs show fast and efficient accumulation in tumors, where they enter in nearly intact form (77% integrity at 2h) before losing their integrity to 40% at 6h (half-life is 4.4h). Thus, the ratiometric FRET imaging *in vivo* can be used to evaluate quantitatively NC integrity in small animals. Furthermore, the FRET methodology also demonstrated that nanosized emulsion droplets are remarkably stable nano-objects that remain nearly intact in the blood circulation and release their content mainly after entering the target site (Bouchaala et al. 2016).

3.3.7. *In Vitro* Mono-Layer Cell-Culture/Transfection Efficiency Studies to Substantiate API Targeting Concept

Due to their subcellular sizes, the nanosized particle delivery systems can be absorbed directly by cells (Chai et al. 2016). The cellular uptake of nanosized emulsions may be through different pathways, such as macropinocytosis, clathrin-mediated endocytosis, caveolae-mediated endocytosis, and clathrin- and caveolae-independent endocytosis (Conner and Schmid 2003). It was reported that siRNA-loaded lipid nanoparticles enter cells by both constitutive and inducible pathways in a cell type-specific manner using clathrin-mediated endocytosis as well as macropinocytosis (Gilleron et al. 2013). The mechanism of uptake of lipid-based nanoparticles has also been reported to vary appreciably due to differences in emulsifiers, interface characteristics, and particle size (Chithrani and Chan 2007).

In order to characterize endocytosis of o/w nanosized emulsions, Nile red, a fluorescent lipid-dissoluble dye, was used as a probe. Numerous studies (Jiang et al. 2008; Zheng et al. 2014; Yao et al. 2015) have indicated that increased cellular uptake and transport of encapsulated bioactive compound are linked to decreased droplet size; however, other factors including emulsifiers, carrier oil, interfacial characters, or loading amount represent significant confounding effects (Ribeiro et al. 2006; Harush-Frenkel et al. 2008; Luo et al. 2013).

The Nile red incorporated in the corn oil showed that the lipid-based nanoemulsions were clearly internalized into Caco-2 cells (Roger et al. 2009). The nanosized emulsions in the cells labeled by Nile red are shown to be surrounded by β-actin labeled with green fluorescing Alexa Fluor 488. Three different nanosized emulsions with different droplet sizes (556, 265, and 170nm) were prepared and used to evaluate the effects of droplet size on the uptake of lipid-based emulsions. There was a positive correlation between reduced droplet size and increased cellular uptake. The decreases of droplet size increase the adhesive contacts between cellular membrane and droplet, thereby significantly decreasing the energy requirements for deforming the membrane around the droplets, which led to the increase of cellular uptake (Herant et al. 2005; Yi et al. 2011). Furthermore, the decreases of deformability of the droplets due to the decrease of droplet size may have also resulted in the increase in cellular internalization (Yi et al. 2011; Anselmo et al. 2015).

3.3.7.1. *Effects of Four Inhibitors on the Cellular Uptake of Lipid-Based Emulsions* All three various droplet size emulsions (170, 265, and 556 nm) were used for intake mechanism study. Inhibitors (5-(*N*-ethyl-*N*-isopropyl) amiloride (EIPA), phenylarsine oxide (PAO), nystatin, and sodium azide) of four different endocytosis routes were used to determine the mechanism of uptake of lipid-based three different droplet size emulsions.

Liu et al. (2011) showed that sodium azide decreased emulsion droplets intake by about 50, 40, and 22%, for 170, 265, and 556 nm emulsions, respectively, indicating that at least some intake is by an endocytosis mechanism.

EIPA have been reported to be an effective inhibitor of macropinocytosis (Ivanov et al. 2004). The cellular uptake of all three lipid-based emulsion droplets decreased approximately 70, 28, and 65% for 170, 265, and 556 nm emulsions, respectively, with EIPA, indicating macropinocytosis was involved in the internalization and played the most important role for the largest droplet size emulsions.

PAO is an inhibitor of clathrin-mediated endocytosis through reacting with vicinal dithiol-containing molecules (He et al. 2013). The internalization of emulsions was significantly inhibited by about 25, 55, and 30% by PAO for 170, 265, and 556 nm emulsions, respectively, suggesting clathrin-mediated endocytosis may also play a vital role in the uptake process.

Nystatin is used to inhibit cholesterol-dependent uptake by caveolin- and lipid-raft-mediated endocytosis (Zhao et al. 2012). Nystatin had highest inhibition effects on both 170 and 265 nm emulsions, indicating caveolae/lipid raft-dependent endocytosis may be the most important manner in the internalization of emulsions with relative smaller droplet diameters. Compared with the other two endocytosis routes, clathrin-mediated pathway was relatively less important for 170 and 556 nm emulsions.

3.3.7.2. *Effects of Temperature on the Cellular Uptake of Nanosized Emulsions* Endocytosis is affected by temperature due to effects on membrane fluidity. The cellular uptake of Nile red-labeled nanosized emulsions was determined at two different temperatures within 4 h (Fan et al. 2017). Lipid-based nanosized emulsions were obviously internalized into Caco-2 cells at both temperatures observed by the fluorescence intensity of Caco-2 cells with CLSM. At 37°C, fluorescence intensity was significantly higher than at 4°C. The quantitative analysis also showed that the cellular uptake of nanosized emulsions increased gradually with incubation time and the internalization efficiency at 37°C was greater than at 4°C. The uptake was 3.7% after 30 min and 9.6% after 2 h incubation at 37°C. After 4 h incubation, the uptake efficiencies were 10.3 and 8.4% for 37 and 4°C, respectively ($p < 0.05$). The lower uptake amount at 4°C was attributed to the low enzyme metabolic activities and poor membrane fluidity.

The results showed that the cellular uptake of nanosized emulsion is time- and energy-dependent, consistent with the study of Luo et al. (2013) who showed that the uptake mechanism of sodium caseinate-encapsulated zein nanoparticles by Caco-2 cell was by an energy-dependent endocytosis.

3.4. SAFETY AND EFFICACY ASSESSMENTS

Efficacy and safety are the two exclusively and mutually running two concepts of prime importance in the development of pharmaceutical products (Talbot 2008). Regulatory bodies require the efficacy and safety data in a pooled form right from pre-formulation studies, prototype development, product optimization and subsequent characterizations, different phases of clinical trial, and information gathered from patient population after API product's usage following the API product positioning and pharmacovigilance. Moreover, very dedicated and sincerity are required to collect the "signals" arised from pharmacovigilance and post-marketing surveillance studies of already marketed pharmaceutical products (Yao et al. 2013). There are few safety assessment tests in non- or pre-clinical stage of development of o/w nanosized emulsions intended to administer either ophthalmic or parenteral routes.

3.4.1. Non- or Pre-clinical Safety Assessments for Ophthalmic Emulsions

Table 3.13 displays different parameters investigated so far to assess the safety aspects of ophthalmic emulsions with respect to both regulatory toxicity studies and safety screening. The ophthalmic emulsions do contain tonicity modifier like glycerol (2.25–2.5% w/w) and an osmometer was used to determine tonicity of the emulsion containing 2.5% w/w glycerol ($283.9 \pm 1.73\,\text{mosmol}\,\text{kg}^{-1}$) in comparison with the tonicity value of normal saline ($285–290\,\text{mosmol}\,\text{kg}^{-1}$). The closeness in the isotonicity values of glycerol-containing emulsions and normal saline indicates the prepared o/w ophthalmic emulsions were indeed iso-osmotic with ocular/tear fluids. Furthermore, the use of sorbitol or xylitol as suitable substances to isotonize the ophthalmic emulsions has also been recommended (Jumaa and Müller 1999). For ophthalmic emulsions, the glycerol is the widely used substance to adjust the tonicity of the vehicle with respect to tear fluid (Klang et al. 2000).

3.4.1.1. Draize Rabbit Eye Test Eye irritation is the production of changes in the eye following the application of a test substance to the anterior surface of the eye of rabbits, which are followed for reversibility for 21 days after application (OECD Test Guideline 405, *in vivo*) (OECD 2017), also known as the Draize rabbit eye test.

Following instillation into eye, the ocular tolerance of administered o/w ophthalmic nanosized emulsions with the eye tissues is one of the efficacy and

TABLE 3.13. Different Parameters Investigated So Far to Assess the Safety Aspects of Ophthalmic Emulsions with Respect to Both Regulatory Toxicity Studies and Safety Screening

Regulatory Toxicity Studies	Safety Screening
In vitro evaluation of the cytotoxic potential by indirect contact	Biomicroscopic evaluation of the lipid film
Delayed-type hypersensitivity evaluation in the Guinea pig	Break-up-time (BUT)
Ocular irritation test in the rabbit (short term: 72h) following a single application	Corneal surface imaging by scanning electron microscopy
Determination of the physical compatibility of nanoemulsion with contact lenses	Draize test
28-Day ocular tolerance in the rabbit	Demonstration in a repeated acute rabbit toxicity model
Evaluation of the potential to induce delayed contact hypersensitivity (local lymph node assay)	*In vivo* toxicity evaluation in the rabbit
Evaluation of the corneal sensitivity following repeated applications in albino rabbits	Isotonicity of the emulsion (hemolysis test)
6-Month ocular toxicity in the dog and rabbit	Ocular safety evaluation of newly developed *in vitro* corneal wound healing model and in an acute *in vivo* rabbit model
Phototoxicity and photoallergic potential evaluation following topical applications in the Guinea pig	Ophthalmic parameters like Rose Bengal staining
	Schirmer tear test
	Superficial punctuate keratitis
	Symptoms of ocular discomfort

safety assessments for topically applied ocular formulations. *Ex vivo*, *in vitro*, and *in vivo* evaluations should be done on both blank (API-free) and API-loaded ophthalmic emulsions. Irritation of ocular tissues may arise due to one of the components of ophthalmic emulsions, usually from a high concentration of oil or emulsifiers. In recent years, the oil concentration is kept at 5% w/w or less to prevent possible blurred vision after topical application of emulsions into eye. If the emulsifier molecules impart a potential irritation effect on the ocular surface tissues, a reflex lachrymation can be seen followed by inflammatory symptoms (edema) on the precorneal area of the eye, such as conjunctival congestion, swelling, corneal opacification, etc. Ocular irritation assessments are important in establishing the safety profile of topically applied ophthalmic emulsions. Despite interspecies differences in blink frequency, ocular surface permeability and aqueous humor dynamics between animals and humans (Maurice and Mishima 1984), the rabbits are used as the standard animal model for assessing the ocular irritancy. Rabbits test procedures are based on the methods originally developed by Dr. John Draize and coworkers at the U.S. Food and Drug Administration in the 1940s. The Draize rabbit test involves introduction of 0.1 ml of a test substance into the lower conjunctival cul-de-sac of albino rabbit eyes (Draize et al. 1944).

The lids are then gently held together for about one second in order to prevent loss of the material. The other eye, which remains untreated, serves as a control. Observations of various criteria (i.e., corneal opacity and area of corneal involvement, conjunctival hyperemia, chemosis, ocular discharges, and iris abnormalities) are taken at predefined time points after administration (Wilhelmus 2001). Responses of the cornea, conjunctiva, and iris are graded at 1, 24, and 48 h (and longer if needed) after exposure to the test materials. A scoring system is used to assign numerical grades for ocular tissues that are subjected to ophthalmic emulsion exposer. However, concerns have been expressed over Draize rabbit eye testing, from both the standpoint of animal use as well as the accuracy of the Draize procedure for prediction of human eye response. Objections and limitations to the Draize rabbit's eye test and the necessity to develop nonanimal tests in assessing the ocular irritancy of a topically applied API were raised and discussed in book chapters (Hutak and Jacaruso 1996; Spielmann 1997).

Modified eye irritation tests have been proposed, including the low-volume eye test (LVET). In the LVET procedure (Griffith et al. 1980), 0.01 ml or mg equivalent of test material (ophthalmic emulsion) is placed directly on the central cornea to obtain a better model of accidental human eye exposures (Freeberg et al. 1986). In this study, groups of eight human volunteers and eight albino rabbits, under controlled laboratory conditions, were exposed in one eye without subsequent rinsing to the same concentrations and volumes of four prototype consumer products: fabric softener, shampoo, hand soap, and laundry detergent. Two irritation scales were employed with both human and rabbit eyes: the Draize scale by a technician and a medical scale used with slit lamp examination by an ophthalmologist. Eyes were examined by both graders before and after dosing at specified intervals until recovery. Mean and maximum irritation scores (IS) are presented for each grading time, method, and exposure, as are the mean hours to recovery (clearing) for each exposure. For surfactant-based consumer products, the LVET more accurately predicts human exposures and rates of recovery than does the Draize procedure. Additionally, the LVET meets criteria suggested by the interagency regulatory alternatives group (IRAG) for alternative tests to the Draize (Bruner and Kohrman 1993).

From 1992 to 1994, a worldwide international validation study on nine alternatives to the Draize eye test was conducted in 37 laboratories on 60 coded chemicals, sponsored by the European Union (EU) and the British Home Office (Balls et al. 1995). The principal goal of the study was to establish whether or not one or more of the nine nonanimal tests (Table 3.14) could be used to replace the Draize eye test for all severely irritating materials or for severely irritating materials belonging to specific classes and to replace the animal test completely for chemicals with or without regard to chemical class. However, with the possible exception of predicting the irritancy of surfactants, none of the nine tests met any of the four performance goals.

A potential-induced toxicity study was conducted using a novel cationic emulsion vehicle prepared from a combination of emulsifiers consisting of phospholipids (Lipoid E 80), poloxamer 188 (Pluronics F68), and stearylamine (SA) in rabbit and rat animal models (Klang et al. 1994). Despite the presence of a cationic primary amine SA, which may be suspected of being irritant in the pure form, in the emulsifier combination, the hourly instillation of SA-based cationic emulsion vehicle into rabbit eye was well tolerated without any evidence of any toxic or inflammatory response to the ocular surface during 5 days of the study (40 single drop instillations between 8 a.m. and 4 p.m. each day). Following 0.2, 0.4, and 0.6 ml single bolus injections of the same emulsion vehicle, representing a huge single administered dose of 30 ml kg^{-1}, no animal deaths were noted over a period of 30 days apparently indicating the absence of marked acute toxicity. Furthermore, the same SA-based cationic emulsion vehicle did not cause acute neurotoxicity in rats when a continuous IV infusion (3.3 ml) for 2 h at a rate of 27.4 μl min^{-1} was administered through jugular vein. Another study suggests that the long-term sub-chronic toxicity examination in rabbit eye (healthy) following thrice-daily one single drop topical instillation of the SA-based emulsion elicited an almost similar nonirritating effect to eye tissues in comparison with the thrice-daily one single drop topical instillation of the normal saline-treated control rabbit eyes (data yet to be published). Similar experimental protocols were also followed in evaluating the ophthalmic tolerance of an amphotericin B ophthalmic emulsion, which has been

TABLE 3.14. Nonanimal Test Alternatives to Draize Eye Test for Classification and Labeling of Chemicals as Shown in EU/Home Office International Validation Study

Organotypical Tests	Physiochemical and Cellular Test
Bovine corneal opacity	EYTEX test (*in vitro* International, Irvine, USA)
Chicken enucleated eye test	Fluorescein leakage test
Hen's Egg Test-Chorioallantoic Membrane (HET-CAM) Test	Neutral red uptake cytotoxicity test
Isolated rabbit eye test	Hemolysis test
Permeability test	Silicon microphysiometer test

Adapted from Balls et al. (1995).

found to be better tolerated than the instillation of commercial Fungizone® eye drops in rabbits (Calvo et al. 1996b). With a view to finding irritation-free ocular colloidal API carriers, Calvo et al. (1996b) prepared an anionic ophthalmic emulsion based on Miglyol 840 oil and lecithin. This formulation was tested on rabbit for acute ocular tolerance following topical instillation (30 μl) every 30 min for 6 h. A microscopic evaluation using a slit lamp, at 3, 6.5, and 24 h after the first instillation, was carried out to study possible damage in the conjunctiva, cornea, and iris and to determine the irritant indices according to a scale of predetermined scores. Calvo et al. (1996b) also showed the ocular lesion index (OLI) values of 1.2 ± 1.1, 5.6 ± 1.7, and 0, respectively, at 3, 6.5, and 24 h after the first instillation of the anionic ophthalmic emulsion into rabbit eye. Nevertheless, the OLI values observed at all time points are less than 5% of the limit value of acceptability ($OLI_{max} = 110$). This means that the ophthalmic emulsion, prepared with Miglyol 840 oil and lecithin, is well tolerated and can be useful for topical ophthalmic therapy. The results of all of these safety evaluation studies support the statement of Rieger (1990) that ophthalmic emulsion most closely resemble nearer to natural tears.

An immortalized human corneal epithelial cell line has also been demonstrated to be an interesting tool for predicting topical irritation or toxicity of applied formulations (Saarinen-Savolainen et al. 1998). This cell culture technique obviously reduces the time and expense that the formulator requires for animal experiments in order to assess the formulation irritation effect. Moreover, a mitochondrial assay based on MTT formation is also a very useful tool to address the formulation toxicity or irritation. Furthermore, a human cell-based *in vitro* method, termed the tissue equivalent assay, is a valuable tool to screen for the ocular irritancy potential of various formulations including ophthalmic emulsion to mucosal tissues such as cornea and conjunctiva (Osborne et al. 1995). These authors developed methods for topical application (an exposure that mimics *in vivo* testing) and wash-off of test substances on the epithelial surface of human skin-derived epithelial-fibroblast cocultures. These cultures contain noncornified stratified squamous epithelium, providing a three-dimensional *in vitro* model that resembles noncornified mucosal epithelium, such as cornea and conjunctiva. The hypothesis tested and confirmed in this work was that the rate of cytotoxicity induced by topical application of test substances to the stratified epithelial cell cultures would correlate with ocular irritancy.

Apart from these described general parameters concerning the ocular safety assessment of blank ophthalmic emulsion, the concentration of API obtained *in vivo* in aqueous humour, cornea, conjunctiva, vitreous, sclera, or retina can also be utilized to partially evaluate the safety assessment of the API-loaded ophthalmic emulsion. The IOP as a function of time is also often considered for antiglaucoma agent. Muchtar et al. (1997) conducted an *ex vivo* permeation study with a new corneal diffusion model and the parameters such as miotic effects (Sznitowska et al. 1999), alkali burn model, and re-epithelization healing process (Klang et al. 1999) have been proposed depending on the intended pharmacological action of incorporated APIs. Further experimental works have focused on ophthalmologic parameters such as Rose Bengal staining, tear breakup-time (TBUT), biomicroscopic evaluation of the lipid film, superficial punctate keratis, Schirmer tear test, and ocular surface disease impact (Rieger 1990; Selek et al. 2000; Stevenson et al. 2000).

3.4.1.2. Schirmer Testing The Schirmer test is a simple test that was first described in 1903 and it is still commonly performed in the office to assess aqueous tear production. There are three variations of this test, but the most popular is the Schirmer I test, which measures both basal and reflex tear production. In Schirmer I test, a strip of filter paper is placed on the lower eyelid margin without anesthesia, after 5 min, the strip is removed, and the amount of wetting is measured in millimeters (Jeng 2013). Although this test is used frequently in the office, it has been found to lack accuracy and reproducibility: the same person's test results taken at the same time each day for several days can fluctuate widely, and the mean Schirmer I test results for normal individuals have been reported to range from 8.1 to 33.1 mm (Savini et al. 2008). As such, many ophthalmologists do not even use this test anymore, but for those who do, in general, any value below 10 mm is considered abnormal. Many other ophthalmologists consider this test as a reasonable diagnostic tool only for severe dry eyes, where there is moderate reproducibility, with many practitioners only considering values of less than 5 mm to be significant.

3.4.1.3. Tear Break-Up-Time The TBUT is defined as the time interval between a complete blink and the first appearance of a dry spot in the tear film after fluorescein administration (Lemp 1973). It is believed that this represents an unstable tear film, whereby the mucous layer may rupture, allowing the aqueous to come in contact with exposed epithelium (Sharma and Ruckenstein 1985), but the exact mechanism is poorly understood. Like the Schirmer test, the TBUT test has been criticized as being unreliable and not reproducible. Many factors may lead to its non-reproducibility, including the volume of fluorescein administered, as well as the presence of preservatives, such as benzalkonium chloride, which may shorten TBUT. In normal eyes, TBUT values range from 3 to 132 s, with an average of 27 s (Norn 1969). TBUT less than 10 s suggests an abnormal tear film, with values of 5–10 s considered marginal, and less than 5 s indicative of dry eye (Pflugfelder et al. 1998). The sensitivity and specificity of TBUT is 75 and 60%, respectively. Given the lack of TBUT reproducibility, numerous noninvasive techniques have been reported. These techniques avoid instillation of fluorescein and there is no contact between the measuring instrument and the eye or eyelids. It is also important that the methodology does not substantially alter ocular environment such as ocular surface temperature from illumination systems (Szczesna and Iskander 2010), enabling tear film assessment in its unaltered condition. Noninvasive testing is considered more precise, repeatable, and therefore preferable to invasive testing in tear film stability assessment. Generally, in normal population, TBUT and noninvasive tear breakup time (NIBUT) are poorly correlated, 4 ± 12 s, with NIBUT being longer (Tonge et al. 1991; Nichols et al. 2002). The range in normal population was 4–214 s, median 4–19 s (Madden et al. 1994; Willcox et al. 2017). However, in patients with dry eye, NIBUT and TBUT values are almost the same (Nichols et al. 2002; Lan et al. 2014). The cutoff values for positive finding can be as low as 2.7 s for automated algorithms and up to 10 s for subjective observation techniques (Jones et al. 2017).

The NIBUT test is usually performed by sophisticated and expensive instruments, Tearscope, keratometers, or complicated computerized systems, topographic analysis systems including videokeratoscopy, ocular surface thermography, or lateral shearing interferometry (Pauk et al. 2019).

3.4.1.3.1. Case study of nanosized emulsions with respect to Schirmer and TBUT tests CsA is a topically applicable anti-inflammatory agent for use in dry eye syndrome. Restasis®, which is the first introduced 0.05% topical CsA, has been shown to be effective in patients with DE including SS. However, Restasis® is plagued by issues of long-term instability, which leads to separation of the contents of the eye drop over time and irritation due to the contents of the API; thus, new options are needed to overcome the lipophilic nature and poor aqueous solubility of CsA (Sheppard et al. 2011; Di Tommaso et al. 2012). On the contrary, a novel CsA nanoemulsion (Cyporin N®) demonstrates a transparent appearance and higher bioavailability (Talegaonkar et al. 2008), so it is expected to be able to overcome the problems of Restasis®.

In a recent clinical study conducted by Kang et al. (2019), the efficacy and safety evaluation was performed using Schirmer and TBUT tests for 0.05% CsA-loaded nanosized emulsions [Cyporin N® (Taejoon Pharm, Seoul, Korea)] and the obtained results were compared with a conventional emulsion [Restasis® (Allergan Inc., Dublin, Ireland)] in primary Sjögren's syndrome (SS) dry eyes (DE).

The standardized Schirmer strips (Eagle Vision, Memphis, TN, USA) were placed in the lateral one third of the lower eyelid of patients, without anesthesia. After 5 min, the length of the portion of each strip that was wetted was measured. For assessment of TBUT, instillation of fluorescein into one drop (15–30 ml) of saline from a conventional fluorescein strip (Haag-Streit AG, Köniz, Switzerland) was performed; each solution was applied according to the respective manufacturer's instructions, with excess fluid shaken off prior to application. The fluorescein was applied to the superotemporal bulbar conjunctiva with participants instructed to look inferonasally, and the TBUT was measured using a slit lamp under cobalt blue light. The average value of three TBUT measurements was used for analysis (Hwang et al. 2014).

Schirmer I value did not significantly change from baseline to any time point in either group (Table 3.15). TBUT was improved at 12 weeks from

TABLE 3.15. Schirmer I and Tear Break-Up-Time (TBUT) Test Values Obtained at 4- and 12-Weeks Postocular Topical Instillation of Cyporin N® and Restasis® in Clinical Study Using Real Patient Eyes

Test Name	Time Period	Cyporin N®		Restasis®		Between Group
		Mean ± SD	*P* Value	Mean ± SD	*P* Value	*P* Value
Schirmer I test	Baseline	3.56 ± 3.865		3.5 ± 2.324		0.941
	4 Weeks	3.72 ± 5.824		3.97 ± 3.308		0.823
	12 Weeks	3.36 ± 4.987		3 ± 2.255		0.693
	Δ 4 Weeks to baseline	0.17 ± 4.754	0.835	0.47 ± 2.261	0.219	0.729
	Δ 12 Weeks to baseline	−0.19 ± 4.99	0.817	−0.5 ± 2.624	0.261	0.746
Tear break-up-time test	Baseline	3.33 ± 1.927		2.72 ± 1.279		0.117
	4 Weeks	3.92 ± 2.761		2.92 ± 1.381		0.057
	12 Weeks	4.75 ± 2.359		3.08 ± 1.574		0.001*
	Δ 4 Weeks to baseline	0.58 ± 2.634	0.192	0.19 ± 1.305	0.378	0.431
	Δ 12 Weeks to baseline	1.42 ± 2.771	0.004*	0.36 ± 1.93	0.269	0.065

Δ = post treatment baseline; SD = standard deviation.
*Indicates statistical significance.

baseline in the Cyporin N® group (from 3.33 ± 1.927 to 4.75 ± 2.359; $p = 0.004$), but there was no change of TBUT in the Restasis® group throughout the 12 weeks. TBUT was longer in the Cyporin N® group than in the Restasis® group at 12 weeks, with statistical significance ($p = 0.001$).

Authors evaluated other dry eye parameters; the Cyporin N group showed a significant increase of TBUT with a difference at 12 weeks compared with the Restasis group in primary SS DE. An improvement of TBUT in the Cyporin N group was noted later than the improvements in corneal and conjunctival staining. This trend was also seen in the results of Kim et al. (2017) who evaluated a different CsA nanoemulsion eye drop, Clacier®, which is similar to Cyporin N except for the components of the surfactant and cosurfactant used to prepare the nanoemulsion. This finding implies that tear film stability is slowly and progressively improved more so than ocular surface staining score. Authors thought that the restoration of corneal epithelium and glycocalyx on microvilli could result in attachment of mucous/aqueous layer to the glycocalyx, resulting in tear film stability. Therefore, TBUT shows improvement later than the corneal surface staining score. Interestingly, Schirmer I value did not improve in either group. Because the Schirmer I test reflects both baseline and reflex tearing from the lacrimal glands (Sall et al. 2000; Zhou and Wei 2014;

Kim et al. 2017), authors thought that primary SS DE with the associated deterioration of lacrimal gland function did not show any improvement after 12 weeks of 0.05% CsA treatment. Some previous studies have reported similar results of the Schirmer I test. A dose-ranging, randomized trial in patients with moderate-to-severe DE including SS reported that the results of the Schirmer I test did not improve significantly after 12 weeks with 0.05, 0.1, 0.2, or 0.4% topical CsA. A long-term study of 0.05% topical CsA in severe DE including SS reported that Schirmer I test results did not improve at 6 months, but the improvement was noted at 10 years (Straub et al. 2016). This suggests that the lacrimal gland can be damaged irreversibly in SS; therefore, the Schirmer I value did not improve after treatment or 12 weeks of treatment may not be long enough for functional recovery of a damaged lacrimal gland. Taken together with the results of corneal and conjunctival staining scores and TBUT, authors speculate that the use of a CsA nanoemulsion eye drop may enhance the bioavailability of the API compared with the original formulation (Kang et al. 2019).

In conclusion, this study demonstrated that the novel CsA nanoemulsion Cyporin N® improved ocular symptoms and signs after 12 weeks of treatment in primary SS DE. Both Cyporin N® and Restasis® reduced ocular surface staining scores with similar degrees of subjective symptom change, but Cyporin N® produced faster improvement than Restasis® due to its enhanced bioavailability.

3.4.1.4. In Vitro Cytotoxicity Assay Cytotoxicity of the o/w nanosized emulsions was determined using the 3-(4,5-dimethylthiazol-2-yl)-2,5 diphenyltetrazolium bromide (MTT) cell viability assay. MTT is a yellow compound that when reduced by functioning mitochondria, produces purple formazan crystals that can be measured spectrophotometrically. The quantity of formazan is presumably directly proportional to the number of viable cells and their metabolic activity (Ke et al. 2011). According to ISO 10993-5 (International Organization for Standardization 2009) recommendations, if viability is reduced to <70% of the control, the treatment tested has a cytotoxic potential.

3.4.1.4.1. Case study Saxena et al. (2017) performed the MTT assay to see the cytotoxicity of the α-tocopherol-loaded o/w nanosized emulsions prepared based on coconut oil, triton X-100, and water. Triton X-100, a nonionic surfactant, is widely used for protein extraction from bacterial system and for 18 cell permeabilization during florescence imaging. Koley and Bard (2010) reported that concentration of triton X-100 $\leq$ 0.15 mM is nontoxic to cell (no change in cell viability) and does not produce any change in cell permeability. The overall concentration of triton X-100 in the prepared nanosized emulsion added in the assay media was very less (0.026 mM) than toxic level as reported previously (Koley and Bard 2010). The cell line used by these authors for cytotoxicity assay was L929 (mouse fibroblast, NCCS India) cell line. Cells were maintained in Dulbecco's modified Eagle's medium (DMEM, Invitrogen) supplemented with 10% FBS (Invitrogen) and 1% antibiotic (Pen Strep, Invitrogen) at 37°C, 5% CO_2, and 85% humidity in a CO_2 incubator.

Briefly, 1 × 104 cells per well per 100 µl of DMEM were added to 96 well plate and left for incubation for 24 h. Post incubation, spent culture media was replaced with fresh complete media containing pre-dissolved unloaded and loaded (α-tocopherol) nanosized emulsions at 10 and 20 µg ml^{-1} concentrations. Cells were incubated for 1, 2, and 3 days to determine cell viability. After each specified time interval, media was replaced with fresh media containing 20 µl of MTT solution (5 mg ml^{-1}, PBS pH 7.4) and incubated for 4 h at 37°C. Later, optical density (OD) was measured at 570 nm (Infinite 200 Pro, Tecan) after incubating in DMSO (150 µl) for 10 min. Experiment was performed in triplicate to determine standard deviation.

The MTT assay was performed taking blank media (without any nanosized emulsion) as a control for three consecutive days and was compared with samples as shown in Table 3.16. Results revealed no significant change in cell viability and was comparable at both 10 and 20 µg ml^{-1} concentrations of unloaded and loaded nanoemulsions. Hence, this delivery carrier can be potentially used as a vehicle for various cargoes like biological molecules, APIs, and macromolecules with reasonably good biocompatibility.

3.4.1.5. Hen's Egg Test-Chorioallantoic Membrane (HET-CAM) Test Method The Hen's Egg Test-Chorioallantoic Membrane (HET-CAM) test procedure is based on the one suggested by Luepke in 1985. The HET-CAM test was used to determine the potential ocular irritancy of the topically applied API delivery system. The test analyzes the appearance of irritative reactions on the chorioallantoic membrane (CAM) of fertilized chicken eggs, as a response to the exposition of the membrane to the test sample. Because, the CAM of egg is a highly vascularized structure present inside it. Its exposure allows the direct observation of blood vessels. The effects are usually measured by the onset of hemorrhage, coagulation, and vessel lysis. These assessments are considered individually and then combined to derive a score, which is used

TABLE 3.16. The Cytotoxicity Assay of the α-Tocopherol-Loaded and Unloaded O/W Nanosized Emulsions

Formulations	Cell Viability (% ± SD, n = 3)		
	Day 1	Day 2	Day 3
Control (blank media)	100 ± 2	100 ± 3	100 ± 1
α-Tocopherol-loaded emulsions (10 µg ml^{-1})	91 ± 1	88 ± 1	97 ± 2
α-Tocopherol-loaded emulsions (20 µg ml^{-1})	93 ± 2	87 ± 2	98 ± 1
Blank emulsion (10 µg ml^{-1})	92 ± 4	94 ± 1	95 ± 3
Blank emulsion (20 µg ml^{-1})	95 ± 2	93 ± 1	98 ± 1

to classify the irritancy level of the test substance. The observation of reactions such as coagulation (intra- and extra-vascular protein denaturation), hemorrhage (bleeding from the vessels), and vascular lysis (blood vessel disintegration) are used as evaluation parameters over a period of 300 s. The time for the appearance of each of the noted endpoints should be monitored and recorded, in seconds. The collection of additional information and data may be useful in further analyses and conducting retrospective studies. Tamilvanan et al. (2013) studied the possible ocular irritancy of AZM-loaded emulsions using a modified HET-CAM test reported by Luepke (1985) and further modified by Balls et al. (1995) and Gilhotra et al. (2011). In brief, fresh (not older than 7 days) White Leghorn chicken eggs weighing between 50 and 60 g were purchased from local poultry farm. The eggs were candled to discard the nonviable or defective ones. Extreme care has been taken to avoid unnecessary tilting, knocking, shaking, and all other mechanical irritation of the eggs during the experiment. Three eggs were allocated for each one of the three test samples (anionic, cationic, and neutral emulsions), negative control (0.9% sodium chloride solution), and positive control [0.1 N sodium hydroxide solution or 1% sodium dodecyl sulphate (SDS) solution]. These eggs were incubated in humidified incubator at a temperature of $37 \pm 0.5°C$ for 3 days. The trays containing eggs were rotated manually in a gentle manner after every 12 h. On day 3, egg albumin (3 ml) was removed by using aseptic techniques from the pointed end of the egg. The hole was sealed by 70% alcohol-sterilized parafilm (American Can Company, NJ, USA) with the help of a heated spatula. The eggs were kept in the equatorial position for the development of CAM away from the shell. The eggs were again candled on the fifth day of incubation and every day, thereafter, nonviable embryos were removed. On the tenth day, a window (2×2 cm) was made on the equator of the eggs through which the formulations (0.3 ml) were instilled. An IS analysis method was followed to assess the possible ocular irritancy of the test samples in comparison with the negative and positive controls. Below mentioned Eq. (3.22) is used to generate the IS value for each tested samples (ICCVAM Test Method Evaluation Report: Appendix G 2006).

$$IS = \left(\frac{301 - \text{Hemorrhage time}}{300} \times 5\right) + \left(\frac{301 - \text{Lysis time}}{300} \times 7\right) + \left(\frac{301 - \text{Coagulation time}}{300} \times 9\right) \tag{3.22}$$

where Hemorrhage time = observed start (in seconds) of hemorrhage (bleeding from the vessels) reactions on CAM, Lysis time = observed start (in seconds) of vessel lysis (blood vessel disintegration) on CAM, and Coagulation time = observed start (in seconds) of coagulation (intra- and extra-vascular protein denaturation) formation on CAM. The IS values attained by each samples were recorded according to the scoring schemes as shown in Table 3.17 and the scores obtained by each of them are given in Table 3.18.

According to the scoring scheme as shown in Table 3.17, the scores obtained by the tested three emulsions along with negative and positive controls are presented in Table 3.18. The positive controls obtained the IS value of 1 immediately after 30 s and reached the highest IS value of 3 at 300 s. Whereas the non-phospholipid-based cationic emulsion acquired the IS value of 1 only at 300 s, the phospholipid-based anionic- and neutral-charged emulsions attained the IS value of 1 at 150 and 120 s, respectively. No IS value was seen for negative control.

In the HET-CAM test, 0.9% sodium chloride solution acts as a negative control for inducing a nonirritating response, while 0.1 N sodium hydroxide solution or 1% SDS solution entertains as positive control to induce an

TABLE 3.17. Scoring Chart for Modified Hen's Egg Chorioallantoic Membrane (HET-CAM) Test

Effect	Scores	Inference
No visible hemorrhage	0	Nonirritant
Just visible membrane discoloration	1	Mild irritant
Structures are covered partially due to membrane discoloration or hemorrhage	2	Moderately irritant
Structures are covered totally due to membrane discoloration or hemorrhage	3	Severely irritant

According to ICCVAM Test Method Evaluation Report: Appendix G (2006).

TABLE 3.18. Scores Obtained by Various Test Formulations in Modified Hen's Egg Chorioallantoic Membrane (HET-CAM) Test

Test Formulation		Scores in Different Time (s)							
		30	60	90	120	150	180	240	300
Non-phospholipid-based cationic emulsion		0	0	0	0	0	0	0	1.0
Phospholipid-based anionic emulsion		0	0	0	0	1.0	1.0	1.0	1.0
Neutral-charged emulsion		0	0	0	1.0	1.0	1.0	1.0	1.0
Negative control (0.9% sodium chloride solution)		0	0	0	0	0	0	0	0
Positive control	0.1 N sodium hydroxide solution	18.0	18.0	19.0	19.0	19.0	19.0	19.0	19.0
	1% sodium dodecyl sulfate (SDS) solution	9.0	9.0	9.0	10.0	10.0	10.0	10.0	10.0

irritating response. It should be mentioned from historical control studies (ICCVAM Test Method Evaluation Report: Appendix G 2006) that the negative control should produce the IS value of 0.0. Similarly, using 1% SDS and 0.1 N sodium hydroxide as positive controls should show the IS values ranged between 10 and 19, respectively. Furthermore, the test substance will be categorized under the severe irritancy classification when the IS value is greater than nine. From Table 3.18, it is clear that the cationic emulsions scored the IS value of 1 only at the 300 s of the post-start experiment time period. Both anionic- and neutral-charged emulsions attained the IS value of 1 much earlier than the cationic emulsion (150 and 120 s of the post-start experiment time periods, respectively). It appears, therefore, that the non-phospholipid-based cationic emulsion should confer less irritation following ocular administration into eye.

3.4.2. Non- or Pre-clinical Safety Assessments for Parenteral Emulsions

3.4.2.1. In Vitro and In Vivo Myotoxicity Studies By utilizing AZM-incorporated non-phospholipid-based cationic, phospholipid-based anionic- and neutral-charged nanosized emulsions (Tamilvanan et al. 2013), the myotoxicity effects of the formulations were studied by *in vitro* and *in vivo* methods as reported by Al-Suwayeh et al. (1996) and Rungseevijitprapa et al. (2008). The corresponding blank emulsions were also subjected to myotoxicity tests.

3.4.2.1.1. In vitro myotoxicity study The acute myotoxicity was evaluated *in vitro* using an isolated rodent skeletal muscle model and measuring the cumulative efflux of the cytosolic enzyme creatine kinase (CK) from the isolated rat extensor digitorum longus (EDL) muscle over a 120-min period. In brief, the rodents were administered an anesthetic dose of sodium pentobarbital and sacrificed via cervical dislocation. Approximately 150 mg of EDL muscle was isolated from male Sprague-Dawley rats. The negative (isotonic sodium chloride solution, 0.9%) and positive [phenytoin (50 mg ml^{-1} in normal saline)] controls, AZM solution, or emulsions with (100 μg) or without AZM (15 μl) were injected into EDL muscles using a 100 μl Hamilton syringe equipped with a needle guard to control the depth and angle of injection. The injected muscles were placed into a Teflon-coated plastic basket and immersed in 9 ml of a carbogenated (95% O_2/5% CO_2) balanced salt solution (BSS). The solutions were drained and fresh BSS was added at 30-min intervals. The drained solutions at 30, 60, 90, and 120 min were then analyzed for CK using a commercially available spectrophotometric kinetic assay (Sigma Chemical No. 47-20 UV). Myotoxicity was calculated from the cumulative sum of the CK values (U l^{-1}) over a 120-min period.

Figure 3.16 depicts a linear CK release profile from the EDL muscles treated with the various solutions or emulsions for a 120 min-period. Muscles injected with isotonic normal saline solution were used as negative control to rule out muscle damage caused by needle puncture. The injection of phenytoin solution

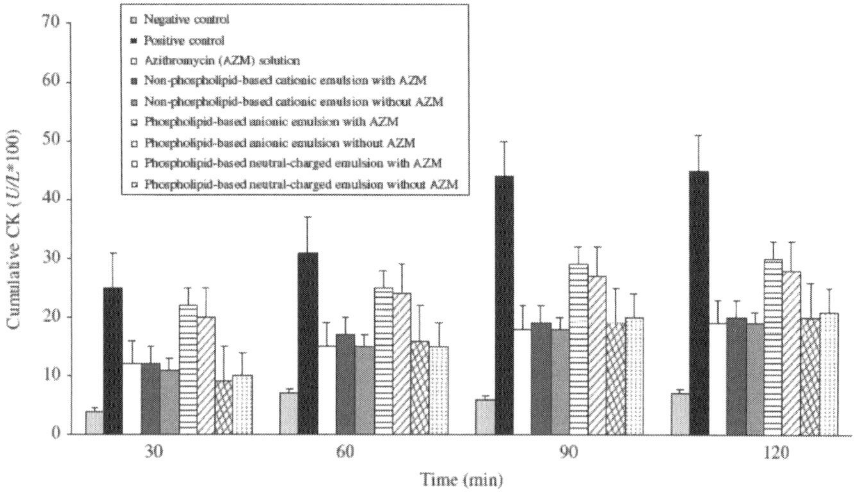

Figure 3.16. Cumulative creatine kinase (CK) release from 150 mg of the isolated rat extensor digitorum longus (EDL) muscle of male Sprague-Dawley rats after the treatment over a 120-min period with the various formulations containing equivalent amount of (100 μg) azithromycin (AZM) in comparison with the positive control (phenytoin) and the negative control (0.9% NaCl). [Adapted from Tamilvanan et al. (2013).] (See color insert.)

(positive control) was approximately 5–8 times more myotoxic than the negative control. For example, the CK value observed for negative control at 30 min was only 4 whereas for positive control, it was 25 and hence, suggesting the potency to use the EDL muscle model to distinguish between the myotoxicity of various emulsions and AZM solution by measuring the CK efflux. The cumulative release amount of the intracellular CK caused by all of the tested AZM-loaded emulsions and the AZM solution was always less than that of the positive control indicating their possible lesser myotoxicity potential at the injection site. It is also noteworthy to mention here that the CK release from muscles treated with the emulsions or the AZM solution was significantly different from that of the negative control (ANOVA; $p < 0.05$). The CK values shown by the API solution and the non-phospholipid-based cationic and the phospholipid-based neutral-charged emulsions were almost similar (9–25 U l^{-1} × 100) for the 30 and 120 min incubations. However, the observed CK values of phospholipid-based anionic emulsion (20–35 U l^{-1} × 100) was significantly higher (ANOVA; $p < 0.05$) than the obtained CK values of other two emulsions (9–25 U l^{-1} × 100), and significantly lower (ANOVA; $p < 0.05$) than the CK values observed for positive control (25–60 U l^{-1} × 100). Furthermore, irrespective of the studied cumulative CK release time of 30, 60, 90, and 120 min, the emulsions prepared with AZM did show the CK values almost similar to the CK values that were observed with their respective charge emulsions prepared without AZM. This indicates that the presence of AZM in the emulsions did not cause any additional myotoxicity.

The *in vitro* myotoxicity studies indicate that neither the tested emulsions with or without AZM nor the AZM solution possessed CK values that were higher than the CK values of the positive control (Fig. 3.16). However, among the tested three emulsions having different charges and prepared with or without AZM, the phospholipid-based anionic emulsion produced the CK values that were significantly higher than the CK values of other two emulsions. The EDL muscles present at the injection site turned pale immediately after the parenteral administration of positive control that indicates as the most myotoxic one. None of the tested emulsions or solution following their injection showed this color change in the EDL muscle. It can therefore be deduced that both emulsions and solution could be suitable for parenteral administration. But considering the higher CK values produced by the phospholipid-based anionic emulsion in comparison with the non-phospholipid-based cationic and phospholipid-based neutral-charged emulsions, the potential of anionic emulsion to be useful for parenteral application should be considered only after performing other safety/toxicity confirmation tests.

3.4.2.1.2. In vivo myotoxicity study An *in vivo* myotoxicity study was conducted using Sprague-Dawley male rats (3-months-old, 225–250 g, 3 rats per study per formulation). Briefly, rats were catheterized and allowed to recover for 3 days prior to the study to allow CK levels to stabilize at baseline. Following intramuscular injection (0.3 ml) of the various solutions or emulsions in the right thigh muscle, blood samples (0.5 ml) were collected via the carotid artery at 0, 30, 60, 120, and 240 min. The blood samples were centrifuged immediately and plasma were stored at −20°C for analysis of CK level, while blood cells were reconstituted in heparinized (40 U ml^{-1}) normal saline solution (0.25 ml) and reinjected into the rat following the next sample to maintain blood volume.

Figure 3.17 shows the mean plasma CK values ($n = 3$) obtained after the parenteral administration of various solutions or emulsions (0.3 ml) in the right thigh muscle of male Sprague-Dawley rats. At 2-h postinjection time period, the positive control showed high plasma CK value (810 U l^{-1}) followed by AZM solution (400 U l^{-1})/AZM-loaded anionic emulsion (400 U l^{-1}), AZM-loaded neutral/cationic-charged emulsions (220 and 200 U l^{-1}, respectively), and then by the negative control (120 U l^{-1}). The same rank order was observed for 4, 6, 8, 10, and 12-h postinjection time periods. All of the peak plasma CK levels occurred at 4-h postinjection time period with an exception of negative control, which reduced from 120 to 85 U l^{-1}. From the 6-h postinjection time period onward, the mean plasma CK values appeared to reduce precipitously for all of the tested emulsions and solutions. Although the emulsions with AZM or without AZM showed marked differences in the obtained mean plasma CK values at all of the studied postinjection time periods, they were not statistically significant (ANOVA).

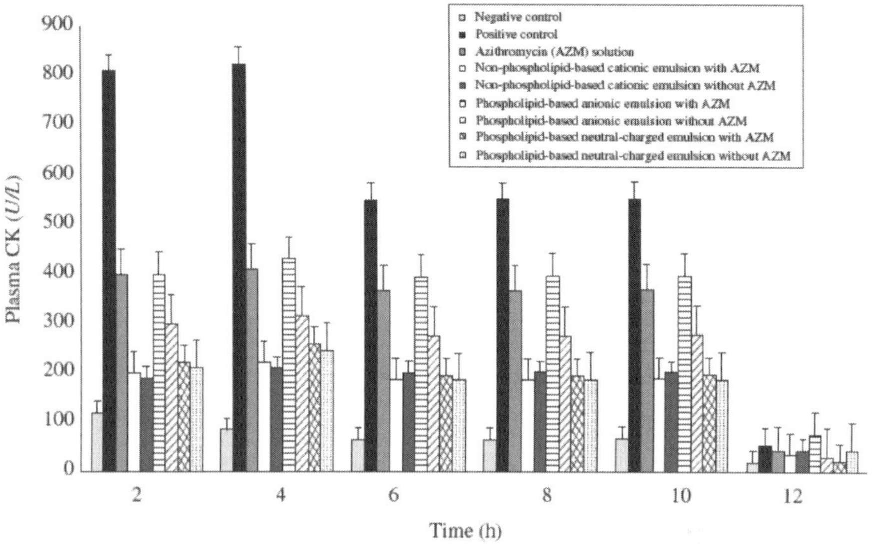

Figure 3.17. Mean plasma creatine kinase (CK) levels versus time following the intramuscular injection (0.3 ml) in the right thigh muscle of male Sprague-Dawley rats with various formulations containing equivalent amount of (100 μg) azithromycin (AZM), the positive control (phenytoin), and the negative control (0.9% NaCl). (See color insert.)

Although the *in vitro* myotoxicity can be used to determine the acute effect of formulations on muscle tissue, this method cannot be utilized to investigate long-acting or repeated injection formulations for their potential to cause tissue damage at the injection site. Therefore, in some circumstances, it becomes necessary to employ the *in vivo* release of cytosolic enzymes as a marker of tissue damage caused by parenteral products. The major advantage is that the nervous, vascular, and immune systems associated with skeletal muscle remain intact and can contribute to parenteral product-induced tissue toxicity (Brazeau and Fung 1989). From Fig. 3.17, it is clear that at 4-h postinjection time period, all of the tested solutions and emulsions attained highest mean plasma CK values with the exception of the negative control, which reduced from 120 to 85 U l^{-1}. Here too, both of the phospholipid-based emulsions showed a difference in the CK levels. While phospholipid-based anionic emulsion attained the CK levels higher than the phospholipid-based neutral-charged emulsion at all of the studied postinjection time periods. The difference in the composition of these two phospholipid-based anionic- and neutral-charged emulsions is the presence of 125 mg of oleic acid in the 50 ml of anionic emulsion, which is absent in the neutral-charged emulsion. The effect of oleic acid on the EDL muscle damage and the intracellular CK release are currently undergoing investigations in our laboratory. In spite of that, the *in vivo* toxicity results correlated very well with the *in vitro* model, thus providing their values for rapid screening of pharmaceutical formulations meant for parenteral administration.

3.4.2.2. In Vitro Hemolysis Test Tamilvanan et al. (2013) have prepared AZM-loaded non-phospholipid-based cationic, phospholipid-based anionic- and neutral-charged nanosized emulsions, and the hemolysis effects of these emulsions were assessed by *in vitro*, red blood cell (RBC) lysis test. Freshly collected blood from a healthy sheep was mixed with anticoagulant solution (74.8 mM sodium citrate/38.07 mM citric acid/124.3 mM glucose 1 : 1.19 v/v) and centrifuged at 650 × *g* for 10 min. The supernatant was discarded and the erythrocytes were resuspended in PBS (5 mM phosphate, 150 mM NaCl, pH = 7.4). Next, the erythrocytes were washed three times with an isotonic buffer (PBS), and the upper phase with a buffy coat containing precipitated debris and serum proteins was carefully removed at each wash step. After the last washing, the packed cells were suspended in a buffer to a hematocrit of 50%. All the erythrocyte suspensions used in the experiments were prepared daily. To determine the hemolytic effect, 100 µl of each emulsions containing 1,000 µg of AZM was diluted with 10 ml of 2.25% sorbitol and one ml from this diluted emulsion (containing 100 µg of AZM) was added to 20 µl of erythrocyte suspension (50% hematocrit) and adjusted to a 4-ml volume with PBS. The samples were stirred and incubated for 30 min at 37°C. Debris and intact erythrocytes were removed by centrifugation at 650 × *g* for 10 min. The hemoglobin released into the supernatant was detected spectrophotometrically at 540 nm against a corresponding blank sample (Jaromin et al. 2006).

The hemolytic effect, measured as the % of hemolysis (H), was determined according to the following Eq. (3.23):

$$\text{Haemolysis}(\%) = \frac{\text{Abs}_{\text{sample}} - \text{Abs}_{0\%\text{hemolysis}}}{\text{Abs}_{100\%\text{hemolysis}} - \text{Abs}_{0\%\text{hemolysis}}} \times 100 \qquad (3.23)$$

where $\text{Abs}_{\text{sample}}$ = absorbance of sample, $\text{Abs}_{0\%\ \text{hemolysis}}$ = absorbance of negative control, and $\text{Abs}_{100\%\ \text{hemolysis}}$ = absorbance of positive control.

At the same concentration of 100 µg of AZM, the free API-induced 100% hemolysis and the various tested emulsions showed RBCs' destruction only in the range of 7–16% depending on the emulsifier combination and presence or absence of API into them (Fig. 3.18). Particularly, the phospholipid-based anionic emulsion exhibited 15 and 16% hemolysis when loaded with AZM and without AZM, respectively. Those % hemolysis values shown by the anionic emulsion loaded with or without AZM (15 and 16, respectively) was significantly higher than the % hemolysis values obtained by non-phospholipid-based cationic emulsion with or without AZM (7 and 8, respectively) (ANOVA, $p < 0.05$). Although the % hemolysis values found for plain and AZM-loaded phospholipid-based neutral-charged emulsion (10 and 9, respectively) were

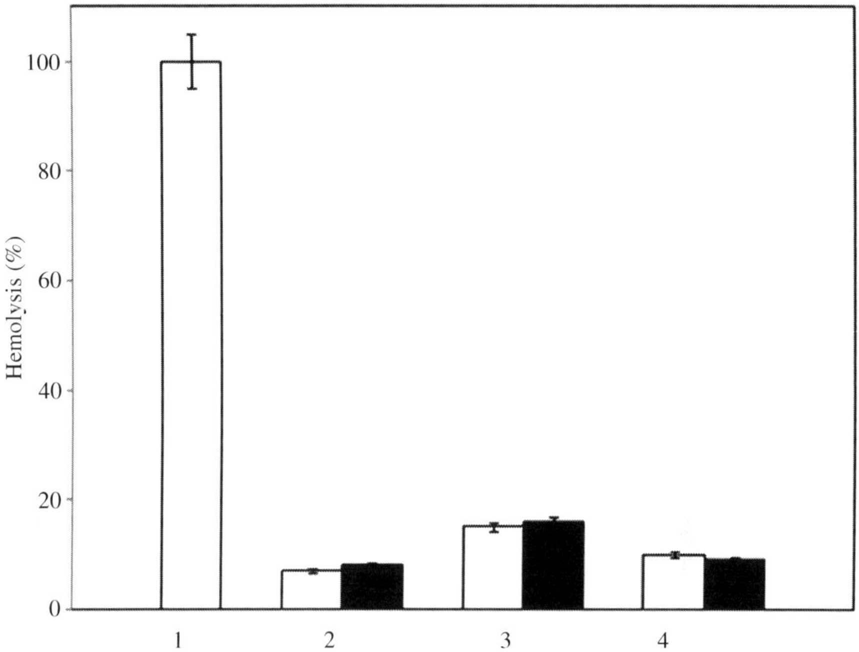

Figure 3.18. Hemolysis of sheep erythrocytes induced by free AZM (number 1 in *x*-axis) and by various emulsion formulations with (numbers 2–4 in *x*-axis) (open bars) or without AZM (numbers 2–4 in *x*-axis) (darken bars). AZM was present at a concentration of 100 μg in all tested formulation. The number 2 indicates non-phospholipid-based cationic emulsion, whereas the numbers 3 and 4 represent phospholipid-based anionic and neutral-charged emulsions, respectively. Each point represents the mean ± SD ($n = 3$).

low when compared with the values obtained by the anionic emulsion loaded with or without AZM (15 and 16, respectively), no statistically significant difference was noticed (ANOVA).

As shown in Fig. 3.18, the encapsulation of AZM into the emulsions resulted in an almost complete loss of its hemolytic action, and only a very low-level lytic activity (~7–16%) against sheep RBCs was observed. This indicates that the encapsulation of AZM neutralized its lytic activity and resulted in the protection of cells against its deleterious effect. However, the non-phospholipid-based cationic- and phospholipid-based neutral-charged emulsions showed % hemolysis values that were relatively lower than the % hemolysis value shown by the phospholipid-based anionic emulsion (Fig. 3.18). As demonstrated by Nagasaka and Ishii (2001), the % of hemolysis in this type of experiment is related to the phospholipid content, chitosan content, or poloxamer content as the emulsifying agents in the emulsions tested. Therefore, this assay should also be carried out with the same emulsions, but prepared without AZM. The % hemolysis values obtained with the emulsions without AZM were almost

TABLE 3.19. Physicochemical Characteristics of Azithromycin (AZM)-Loaded Anionic, Cationic, and Neutral-Charged Nanosized Emulsions

Formulations	Mean Droplet Diameter (nm ± SD, $n = 3$)	Zeta Potential (mV ± SD, $n = 3$)	AZM Entrapment Efficiency (%)	AZM Amount (%) at Oil–Water Interface of the Emulsion
AZM-loaded anionic emulsion	295 (±21)	−34.2* (±1.23)	87* (±2)	13* (±0.3)
AZM-loaded cationic emulsion	200 (±29)	+42.6 (±1.45)	99 (±1)	0.8 (±0.4)
AZM-loaded neutral charged emulsion	434 (±13)	No charge	90* (±1)	10* (±0.4)

Figures in parentheses indicate standard deviation ($n = 3$).
*Statistically significant when compared with cationic emulsion ($p < 0.05$).

identical (only 1% difference) with that of the % hemolysis values attained by the emulsions with AZM (Fig. 3.18) after a 30-min incubation at 37°C. AZM, like other hemolytic hydrophobic/lipophilic compounds, is probably incorporated into the lipophilic core of oil droplets or into the o/w interface of the non-phospholipid-based cationic and phospholipid-based anionic- and neutral-charged emulsions. This hypothesis is further corroborated with the results of API entrapment efficiency and API at the o/w interface of the emulsion as depicted in Table 3.19. It is also possible that API solubilization occurs within the emulsifier molecular structures like the micelles or liposomes present in the aqueous phase (Sznitowska et al. 2002). Consequently, AZM has limited direct contact with the tested blood cells, and is not easily partitioned into them, resulting in the loss of its hemolytic activity. Such a mechanism has been suggested for other lytic agents (Jumaa and Müller 2000).

3.5. CONCLUSION

The characterization and safety assessment tests conducted for o/w nanosized emulsions at least in laboratory/pre- or non-clinical scale were arranged serially or systematically. This serial or systematic arrangement helps the formulator of o/w nanosized emulsions to plan the research and development work at an industrial level. Moreover, further clinical safety assessment tests will be of interest for regulatory purpose and the safety assessment tests are also changing time-to-time to ensure the administration of o/w nanosized emulsions into the human body. The formulators are requested to follow-up with the recent regulatory guidelines issued by the major pharmaceutical product's regulatory authorities of the world.

REFERENCES

Abdulrazik, M., Tamilvanan, S., Khoury, K. et al. (2001), Ocular delivery of cyclosporin A. Part II. Effect of submicron emulsion's surface charge on ocular distribution of topical cyclosporin A, *STP Pharm Sci.*, 11 (6), 427–432.

Aiache, J.M., El Meski, S., Beyssac, E., et al. (1997), The formulation of drug for ocular administration, *J. Biomater. Appl.*, 11 (3), 329–348. doi:10.1177/088532829701100305.

Akoh, C., and Min, D. (2002), *Food Lipids: Chemistry, Nutrition, and Biochemistry*, 2nd ed., Marcel Dekker, New York.

Almeida, T.C.A., Larentis, A.L., and Ferraz, H.C. (2015), Evaluation of the stability of concentrated emulsions for lemon beverages using sequential experimental designs, *PLoS One*, 10 (3), e0118690. doi:10.1371/journal.pone.0118690

Al-Suwayeh, S.A., Tebbett, I.R., Wielbo, D. et al. (1996), In vitro-in vivo myotoxicity of intramuscular liposomal formulations, *Pharm. Res.*, 13 (9), 1384–1388. doi:10.1023/A:1016082218942

Anand, S., Paxson, A.T., Dhiman, R. et al. (2012), Enhanced condensation on lubricant-impregnated nanotextured surfaces, *ACS Nano*, 6 (11), 10122–10129. doi:10.1021/nn303867y

Anand, S., Rykaczewski, K., Subramanyam, S.B. et al. (2015), How droplets nucleate and grow on liquids and liquid impregnated surfaces, *Soft Matter*, 11 (1), 69–80. doi:10.1039/c4sm01424c

Anselmo, A.C., Zhang, M., Kumar, S. et al. (2015), Elasticity of nanoparticles influences their blood circulation, phagocytosis, endocytosis, and targeting, *ACS Nano*, 9 (3), 3169–3177. doi:10.1021/acsnano.5b00147

Aveyard, R., Binks, B.P., Fletcher, P.D.I. et al. (1993), Entry and spreading of alkane drops at the air/surfactant solution interface in relation to foam and soap film stability, *J. Chem. Soc. Faraday Trans.*, 89 (24), 4313–4321. doi:10.1039/FT9938904313

Baboota, S., Shakeel, F., and Ahuja, A. (2007), Design development and evaluation of novel nanoemulsion formulations for transdermal potential of Celecoxib, *Acta Pharm.*, 57 (3), 315–332. doi:10.2478/v10007-007-0025-5

Balls, M., Botham, P.A., Bruner, L.H. et al. (1995), The EU/Home Office validation study on alternatives to the Draize eye irritation test for the classification and labelling of chemicals, *Toxicol. In Vitro*, 9 (6), 871–929. doi:10.1016/0887-2333(95)00092-5

Barar, J., Javadzadeh, A.R., and Omidi, Y. (2008), Ocular novel drug delivery: impacts of membranes and barriers, *Expert Opin. Drug Deliv.*, 5 (5), 567–581. doi:10.1517/17425247.5.5.567.

Bhattacharjee, S. (2016), DLS and zeta potential—what they are and what they are not? *J. Control. Release*, 235, 337–351. doi:10.1016/j.jconrel.2016.06.017

Bouchaala, R., Mercier, L., Andreiuk, B. et al. (2016), Integrity of lipid nanocarriers in bloodstream and tumor quantified by near-infrared ratiometric FRET imaging in living mice, *J. Control. Release*, 236, 57–67. doi:10.1016/j.jconrel.2016.06.027

Bowman, L.M., Si, E., Pang, J. et al. (2009), Development of a topical polymeric mucoadhesive ocular delivery system for azithromycin, *J. Ocul. Pharmacol. Ther.*, 25 (2), 133–139. doi:10.1089/jop.2008.0066

Brazeau, G.A., and Fung, H.L. (1989), Use of an in vitro model for the assessment of muscle damage from intramuscular injections: in vitro-in vivo correlation and predictability with mixed solvent systems, *Pharm. Res.*, 6 (9), 766–771. doi:10.1023/A:1015971429942

Bruner, L.H. and Kohrman, K.A. (1993), The low volume eye test as a refinement to the Draize eye irritation test, in: Niemi, S.M. and Wilson, J.E., Ed., *Refinement and Reduction in Animal Testing*, Scientists Centre for Animal Welfare, Bethesda, pp. 81–93.

Calvo, P., Alonso, M.J., Vila-Jato, J.L. et al. (1996a), Improved ocular bioavailability of indomethacin by novel ocular drug carriers, *J. Pharm. Pharmacol.*, 48 (11), 1147–1152. doi:10.1111/j.2042-7158.1996.tb03911.x

Calvo, P., Vila-Jato, J.L., and Alonso, M.J. (1996b), Comparative in vitro evaluation of several colloidal systems, nanoparticles, nanocapsules, and nanoemulsions, as ocular drug carriers, *J. Pharm. Sci.*, 85 (5), 530–536. doi:10.1021/js950474

Camber, O. (1985), An *in vitro* model for determination of drug permeability through the cornea, *Acta Pharm. Suec.*, 22 (6), 335–342. PMID:3832784

Chai, G.H., Xu, Y., Chen, S.Q. et al. (2016), Transport mechanisms of solid lipid nanoparticles across Caco-2 cell monolayers and their related cytotoxicology, *ACS Appl. Mater. Interfaces*, 8 (9), 5929–5940. doi:10.1021/acsami.6b00821

Chen, Z., Wei, Z., Chen, Y. et al. (2013), Anisotropic Debye model for the thermal boundary conductance, *Phys. Rev. B*, 87 (12), 125426. doi:10.1103/PhysRevB.87.125426

Chime, S.A., Kenechukwu, F.C., and Attama, A.A. (2014), *Nanoemulsions* – advances in formulation, characterization and applications in drug delivery, Application of nanotechnology in drug delivery, Ali Demir Sezer, IntechOpen. doi:10.5772/58673, Available from: https://www.intechopen.com/books/application-of-nanotechnology-in-drug-delivery/nanoemulsions-advances-in-formulation-characterization-and-applications-in-drug-delivery

Chithrani, B.D. and Chan, W.C.W. (2007), Elucidating the mechanism of cellular uptake and removal of protein-coated gold nanoparticles of different sizes and shapes, *Nano Lett.*, 7 (6), 1542–1550. doi:10.1021/nl070363y

Clavier, M., Seijo, F., and Carnal, S. (2015), Stoll, surface charging behavior of nanoparticles by considering site distribution and density, dielectric constant and pH changes—a Monte Carlo approach, *Phys. Chem. Chem. Phys.*, 17, 4346–4353. doi:10.1039/C4CP04733H

Conner, S.D. and Schmid, S.L. (2003), Regulated portals of entry into the cell, *Nature*, 422 (6927), 37–44. doi:10.1038/nature01451

Czogalla, A. (2009), Oral cyclosporine A—the current picture of its liposomal and other delivery systems, *Cell. Mol. Biol. Lett.*, 14 (1), 139–152. doi:10.2478/s11658-008-0041-6

Deshmukh, G.S. and Sharma, A.K. (2016), Stability-indicating validated HPLC method for assay of cyclosporine-A in bulk drug and ophthalmic formulations, *Pharm. Biol. Eval.*, 3 (6), 547–553.

Di Marco, M., Sadun, C., Port, M. et al. (2007), Physicochemical characterization of ultrasmall superparamagnetic iron oxide particles (USPIO) for biomedical application as MRI contrast agents, *Int. J. Nanomedicine*, 2 (4), 609. PMID:18203428

Di Tommaso, C., Valamanesh, F, Miller, F. et al. (2012), A novel cyclosporin A aqueous formulation for dry eye treatment: in vitro and in vivo evaluation, *Invest. Ophthalmol. Vis. Sci.*, 53 (4), 2292–2299. doi:10.1167/iovs.11-8829

Dickinson, E. (1992), *Introduction to Food Colloids*, Oxford University Press, Oxford. doi:10.1002/food.19920360540

Draize, J.H., Woodard, G., and Calvery, H.O. (1944), Methods for the study of irritation and toxicity of substances applied topically to the skin and mucous membranes, *J. Pharmacol. Exp. Ther.*, 82, 377–390. doi:10.1016/S0928-0987(99)00059-7

Fan, Y., Zhang, Y., Yokoyama, W. et al. (2017), Endocytosis of corn oil-caseinate emulsions in vitro:impacts of droplet sizes, *Nanomaterials*, 7, 349. doi:10.3390/nano7110349

Férézou, J., Nguyen, T.L., Leray, C. et al. (1994), Lipid composition and structure of commercial parenteral emulsions, *Biochim. Biophys. Acta*, 1213 (2), 149–158. doi:10.1016/0005-2760(94)90021-3

Formulaction Smart Scientific Analysis (2009), Stability of pharmaceutical products. Application paper 1–5, http://www.formulaction.com, Accessed on November 15, 2019.

Freeberg, F.E., Nixon, G.A., Reer, P.J. et al. (1986), Human and rabbit eye responses to chemical insult, *Fundam. Appl. Toxicol.*, 7 (4), 626–634. doi:10.1016/0272-0590(86)90112-0

Gilhotra, R.M., Nagpal, K., and Mishra, D.N. (2011), Azithromycin novel drug delivery system for ocular application, *Int. J. Pharm. Investig.*, 1 (1), 22. doi:10.4103/2230-973X.76725

Gilleron, J., Querbes, W., Zeigerer, A. et al. (2013), Image-based analysis of lipid nanoparticle-mediated siRNA delivery, intracellular trafficking and endosomal escape, *Nat. Biotechnol.*, 31, 638–646. doi:10.1038/nbt.2612

Griffith, J.F., Nixon, G.A., Bruce, R.D. et al. (1980), Dose-response studies with chemical irritants in the albino rabbit eye as a basis for selecting optimum testing conditions for predicting hazard to the human eye, *Toxicol. Appl. Pharmacol.*, 55, 501–513. doi:10.1016/0041-008X(80)90052-6

Groves, M.J., Wineberg, M., and Brain, A.P.R. (1985), The presence of liposomal material in phosphatide stabilized emulsions, *J. Dispers. Sci. Technol.*, 2, 237–247. doi:10.1080/01932698508943947

Hagigit, T., Nassar, T., Behar-Cohen, F. et al. (2008), The influence of cationic lipid type on in-vitro release kinetic profiles of antisense oligonucleotide from cationic nanoemulsions, *Eur. J. Pharm. Biopharm.*, 70, 248–259. doi:10.1016/j.ejpb.2008.03.005

Hansrani, P.K., Davis, S.S., and Groves, M.J. (1983), The preparation and properties of sterile intravenous emulsions, *J. Parenter. Sci. Technol.*, 37 (4), 145–150. PMID:6620048

Harika, K., Debnath, S., and Babu, M.N. (2015), Formulation and evaluation of nanoemulsion of amphotericin B, *Int. J. Novel Trends Pharm. Sci.*, 5, 114–122.

Harkins, W.D. and Feldman, A. (1922), Films: the spreading of liquids and the spreading coefficient, *J. Am. Chem. Soc.*, 44 (12), 2665–2685. doi:10.1021/ja01433a001

Harush-Frenkel, O., Rozentur, E., Benita, S. et al. (2008), Surface charge of nanoparticles determines their endocytic and transcytotic pathway in polarized MDCK cells, *Biomacromolecules*, 9, 435–443. doi:10.1021/bm700535p

He, B., Lin, P., Jia, Z. et al. (2013), The transport mechanisms of polymer nanoparticles in Caco-2 epithelial cells, *Biomaterials*, 34, 6082–6098. doi:10.1016/j.biomaterials.2013.04.053

Hedberg, Y., Wang, X., Hedberg, J. et al. (2013), Surface-protein interactions on different stainless steel grades: effects of protein adsorption, surface changes and metal release, *J. Mater. Sci. Mater. Med.*, 24, 1015–1033. doi:10.1007/s10856-013-4859-8

Herant, M., Heinrich, V., and Dembo, M. (2005), Mechanics of neutrophil phagocytosis: behavior of the cortical tension, *J. Cell Sci.*, 118, 1789–1797. doi:10.1242/jcs.02275

Higaki, K., Amnuaikit, C., and Kimura, T. (2003), Strategies for overcoming the stratum corneum: chemical and physical approaches, *Am. J. Adv. Drug Deliv.*, 1, 187–214. doi:10.2165/00137696-200301030-00004

Hotrum, N.E., Stuart, M.A.C., van Vliet, T. et al. (2004), Spreading of partially crystallized oil droplets on an air/water interface, *Colloids Surf. A Physicochem. Eng. Aspects*, 240, 83–92. doi:10.1016/j.colsurfa.2004.03.015

Hotrum, N.E., van Vliet, T., Stuart, M.A.C. et al. (2002), Monitoring entering and spreading of emulsion droplets at an expanding air/water interface: a novel technique, *J. Colloid Interface Sci.*, 247 (1), 125–131. doi:10.1006/jcis.2001.8124

Hutak, C.M. and Jacaruso, R.B. (1996), Evaluation of primary ocular irritation: alternatives to the Draize test, in: Reddy IK, Ed., Ocular Therapeutics and Drug Delivery, a Multi-disciplinary Approach, Technomic Publishing Company Inc., Pennsylvania, pp. 489–525.

Hwang, J., Chung, S.H., Jeon, S. et al. (2014), Comparison of clinical efficacies of autologous serum eye drops in patients with primary and secondary Sjogren syndrome, *Cornea*, 33 (7), 663–667. doi:10.1097/ICO.0000000000000147

ICCVAM Test Method Evaluation Report: Appendix G, November 2006, Available in http://iccvam.niehs.nih.gov/docs/ocutox_docs/OTeval/appG.pdf, Accessed on January 10, 2020.

ISO 10993-5:2009. Biological evaluation of medical devices — Part 5: Tests for in vitro cytotoxicity, ICS>11>11.100>11.100.20, 2009, https://www.iso.org/standard/36406.html, accessed on December 19, 2019

Ivanov, A.I., Nusrat, A., and Parkos, C.A. (2004), Endocytosis of epithelial apical junctional proteins by a clathrin-mediated pathway into a unique storage compartment, *Mol. Biol. Cell*, 15, 176–188. doi:10.1091/mbc.E03-05-0319

Jaromin, A., Żarnowski, R., and Kozubek, A. (2006), Emulsions of oil from *Adenanthera pavonina* L. seeds and their protective effect, *Cell. Mol. Biol. Lett.*, 11 (3), 438. doi:10.2478/s11658-006-0037-z

Jeng, B.H. (2013), Diagnostic techniques in ocular surface disease, in: Holland, E. J., Mannis, M. J., Lee, W. B., Eds., *Ocular Surface Disease: Cornea, Conjunctiva and Tear Film*, Elsevier, pp. 47–54. doi:10.1016/b978-1-4557-2876-3.00007-9

Jiang, W., KimBetty, Y.S., Rutka, J.T. et al. (2008), Nanoparticle-mediated cellular response is size-dependent, *Nat. Nanotechnol.*, 3, 145–150. doi:10.1038/nnano.2008.30

Jones, L., Downie, L.E., Korb, D. et al. (2017), TFOS DEWS II management and therapy report, *Ocul. Surf.*, 15 (3), 575–628. doi:10.1016/j.jtos.2017.05.006

Jumaa, M. and Müller, B.W. (1999), In vitro investigation of the effect of various isotonic substances in parenteral emulsions on human erythrocytes, *Eur. J. Pharm. Sci.*, 9, 207–212. doi:10.1016/S0928-0987(99)00059-7

Jumaa, M. and Müller, B.W. (2000), Lipid emulsions as a novel system to reduce the hemolytic activity of lytic agents: mechanism of the protective effect, *Eur. J. Pharm. Sci.*, 9 (3), 285–290. doi:10.1016/S0928-0987(99)00071-8

Kang, M.J., Kim, Y.H., Chou, M. et al. (2019), Evaluation of the efficacy and safety of a novel 0.05% cyclosporin a topical nanoemulsion in primary Sjögren's syndrome dry eye, *Ocul. Immunol. Inflamm.*, 28 (3), 1–9 doi:10.1080/09273948.2019.1587470

Kaszuba, M., Corbett, J., Watson, F. M. et al. (2010), High-concentration zeta potential measurements using light-scattering techniques, *Philos. Trans. R. Soc. A Math. Phys. Eng. Sci.*, 368 (1927), 4439–4451. doi:10.1098/rsta.2010.0175

Ke, N., Wang, X., Xu, X. et al. (2011). The xCELLigence system for real-time and label-free monitoring of cell viability, *Methods Mol. Biol.*, 740, 33–43. doi:10.1007/978-1-61779-108-6_6

Kilin, V.N., Anton, H., Anton, N. et al. (2014), Counterion-enhanced cyanine dye loading into lipid nano-droplets for single-particle tracking in zebrafish, *Biomaterials*, 35 (18), 4950–4957. doi:10.1016/j.biomaterials.2014.02.053

Kim, H.S., Kim, T.I., Kim, J.H. et al. (2017), Evaluation of clinical efficacy and safety of a novel cyclosporin A nanoemulsion in the treatment of dry eye syndrome, *J. Ocul. Pharmacol. Ther.*, 33 (7), 530–538. doi:10.1089/ jop.2016.0164

Kim, K.M., Kim, H.M., Lee, W.J. et al. (2014), Surface treatment of silica nanoparticles for stable and charge-controlled colloidal silica, *Int. J. Nanomedicine*, 9 (Suppl 2), 29–40. doi:10.2147/IJN.S57922

Kirkwood, J., Hargreaves, D., O'Keefe, S. et al. (2015), Using isoelectric point to determine the pH for initial protein crystallization trials, *Bioinformatics*, 31 (9), 1444–1451. doi:10.1093/bioinformatics/btv011

Klang, S., Abdulrazik, M., and Benita, S. (2000), Influence of emulsion droplet surface charge on indomethacin ocular tissue distribution, *Pharm. Dev. Technol.*, 5 (4), 521–532. doi:10.1081/PDT-100102035

Klang, S. and Benita, S. (1998), Design and evaluation of submicron emulsions as colloidal drug carriers for intravenous administration, in: Benita, S., Ed., *Submicron Emulsions in Drug Targeting and Delivery*, Harwood Academic Publishers, London, pp. 119–152

Klang, S., Frucht-Pery, J., Hoffman, A. et al. (1994), Physicochemical characterization and acute toxicity evaluation of a positively-charged submicron emulsion vehicle, *J. Pharm. Pharmacol.*, 46 (12), 986–993. doi:10.1111/j.2042-7158.1994.tb03254.x

Klang, S.H., Siganos, C.S., Benita, S. et al. (1999), Evaluation of a positively-charged submicron emulsion of piroxicam in the rabbit corneum healing process following alkali burn, *J. Control. Release*, 57, 19–27. doi:10.1016/S0168-3659(98)00107-2

Klymchenko, A.S., Roger, E., Anton, N. et al. (2012), Highly lipophilic fluorescent dyes in nano-emulsions: towards bright non-leaking nano-droplets, *RSC Adv.*, 2, 11876–11886. doi:10.1039/C2RA21544F

Koczo, K., Lobo, L.A., and Wasan, D.T. (1992), Effect of oil on foam stability: aqueous foams stabilized by emulsions, *J. Colloid Interface Sci.*, 150 (2), 492–506. doi:10.1016/0021-9797(92)90218-B

Koley, D. and Bard, A.J. (2010), Triton X-100 concentration effects on membrane permeability of a single HeLa cell by scanning electrochemical microscopy (SECM), *Proc. Natl. Acad. Sci.*, 107 (39), 16783–16787. doi:10.1073/pnas.1011614107

Kouchakzadeh, H., Shojaosadati, S.A., Maghsoudi, A. et al. (2010), Optimization of PEGylation conditions for BSA nanoparticles using response surface methodology, *AAPS PharmSciTech.*, 11, 1206–1211. doi:10.1208/s12249-010-9487-8

Krishna, G., Wood, G.C., and Sheth, B.B. (1998), Improving emulsification efficacy of lecithin by formulation design I: effect of adding a secondary surfactant, *PDA J. Pharm. Sci. Technol.*, 52, 331–336. PMID:10050132

Kumar, S. (2014), Role of nano-emulsion in pharmaceutical sciences—a review, *Asian J. Res. Pharm. Sci. Biotechnol.*, 2, 1–15.

Lallemand, F., Daull, P., Benita, S. et al. (2012), Successfully improving ocular drug delivery using the cationic nanoemulsion, novasorb, *J. Drug Deliv.*, 2012, 1–16. doi:10.1155/2012/604204

Lallemand, F., Felt-Baeyens, O., Besseghir, K. et al. (2003), Cyclosporine A delivery to the eye: a pharmaceutical challenge, *Eur. J. Pharm. Biopharm.*, 56 (3), 307–318. doi:10.1016/S0939-6411(03)00138-3

Lan, W., Lin, L., Yang, X. et al. (2014), Automatic noninvasive Tear Breakup Time (TBUT) and conventional fluorescent TBUT, *Optom. Vis. Sci.*, 9, 1412. doi:10.1097/OPX.0000000000000418

Leite, F.L., Bueno, C.C., Da Róz, A.L. et al. (2012), Theoretical models for surface forces and adhesion and their measurement using atomic force microscopy, *Int. J. Mol. Sci.*, 13 (10), 12773–12856. doi:10.3390/ijms131012773

Lemp, M.A. (1973), Breakup of the tear film, *Int. Ophthalmol. Clin.*, 13, 97–102.

Levy, M.Y. and Benita, S. (1990), Drug release from submicronized o/w emulsion: a new in vitro kinetic evaluation model, *Int. J. Pharm.*, 66 (1–3), 29–37. doi:10.1016/0378-5173(90)90381-D

Li, X., Anton, N., Ta, T.M.C. et al. (2011), Microencapsulation of nano-emulsions:novel trojan particles for bioactive lipid molecule delivery. *Int. J. Nanomedicine*, 6, 1313–1325. doi:10.2147%2FIJN.S20353

Liu, C., Zhen, X., Wang, X. et al. (2011), Cellular entry fashion of hollow milk protein spheres, *Soft Matter*, 7 (24), 11526–11534. doi:10.1039/C1SM06388J

Liu, Y., Lin, X., and Tang, X. (2009), Lipid emulsions as a potential delivery system for ocular use of azithromycin, *Drug Dev. Ind. Pharm.*, 35 (7), 887–896. doi:10.1080/03639040802680271

Lobo, L. and Wasan, D.T. (1993), Mechanisms of aqueous foam stability in the presence of emulsified non-aqueous-phase liquids: structure and stability of the pseudoemulsion film, *Langmuir*, 9 (7), 1668–1677. doi:10.1021/la00031a012

Luepke, N. (1985), Hen's egg chorioallantoic membrane test for irritation potential, *Food Chem. Toxicol.*, 23 (2), 287–291. doi:10.1016/0278-6915(85)90030-4

Luo, Y., Teng, Z., Wang, T.T.Y. et al. (2013), Cellular uptake and transport of zein nanoparticles: effects of sodium caseinate, *J. Agric. Food Chem.*, 61 (31), 7621–7629. doi:10.1021/jf402198r

Madden, R.K., Paugh, J.R., and Wang, C. (1994), Comparative study of two non-invasive tear film stability techniques, *Curr. Eye Res.*, 13, 263–269. doi:10.3109/02713689408995787.

Magenheim, B., Levy, M.Y., and Benita, S. (1993), A new in vitro technique for the evaluation of drug release profile from colloidal carriers-ultrafiltration technique at low pressure, *Int. J. Pharm.*, 94 (1–3), 115–123. doi:10.1016/0378-5173(93)90015-8.

Martin, A.N., Banker, G.S., and Chun, A.H. (1964), Rheology, in: Bean, H.S., Ed., *Advances in Pharmaceutical Sciences. Ch. 1*, Academy Press, London.

Maurice, D.M. and Mishima, S. (1984), Ocular pharmacokinetics, in: *Pharmacology of the Eye*, Springer, Berlin, pp. 19–116. doi:10.1007/978-3-642-69222-2_2

McClements, D.J. (2011), Edible nanoemulsions: fabrication, properties, and functional performance, *Soft Matter*, 7 (6), 2297–2316. doi:10.1039/C0SM00549E

Meunier, G. (1994), Spectra Anal., 179, 53–58.

Meunier, G. and Mengual, O. (1996), 4th World surfactants congress, CESIO, 4, 301–313.

Missana, T. and Adell, A. (2000), On the applicability of DLVO theory to the prediction of clay colloids stability, *J. Colloid Interface Sci.*, 230 (1), 150–156. doi:10.1006/jcis.2000.7003

Moses, R.A. (1981), *ADLER'S Physiology of the Eye Clinical Application*, The C.V. Mosby Company, St. Louis, pp. 16–23.

Muchtar, S., Abdulrazik, M., Frucht-Pery, J. et al. (1997), Ex vivo permeation study of indomethacin from a submicron emulsion through albino rabbit cornea, *J. Control. Release*, 44 (1), 55–64. doi:10.1016/S0168-3659(96)01503-9

Nagasaka, Y. and Ishii, F. (2001), Interaction between erythrocytes from three different animals and emulsions prepared with various lecithins and oils, *Colloids Surf. B Biointerfaces*, 22 (2), 141–147. doi:10.1016/S0927-7765(01)00148-5

Nägele, E.W. (1989), The transient zeta potential of hydrating cement, *Chem. Eng. Sci.*, 44, 1637–1645. doi:10.1016/0009-2509(89)80006-5

Nichols, J.J., Nichols, K.K., Puent, B. et al. (2002), Evaluation of tear film interference patterns and measures of tear break-up time, *Optom. Vis. Sci.*, 79, 363–369. doi:10.1097/00006324-200206000-00009

Nishikawa, M., Takakura, Y., and Hasida, M. (1998), Biofate of fat emulsions, in: Benita S, Ed., *Submicron Emulsions in Drug Targeting and Delivery*, Harwood Academic Publishers, London, pp. 99–118.

Norn, M.S. (1969), Dead, degenerate and living cells in conjunctival fluid and mucous thread, *Acta Ophthalmol.*, 47, 1102–1115. doi:10.1111/j.1755-3768.1969.tb02509.x

OECD, (2017), *Test No. 405: Acute Eye Irritation/Corrosion*, OECD Guidelines for the Testing of Chemicals, Section 4, OECD Publishing, Paris, https://doi.org/10.1787/9789264185333-en

Osborne, R., Perkins, M.A., and Roberts, D.A. (1995), Development and intralaboratory evaluation of an in vitro human cell-based test to aid ocular irritancy assessments, *Fundam. Appl. Toxicol.*, 28 (1), 139–153. doi:10.1023/A:1011956327987

Patel, V.R. and Agrawal, Y.K. (2011), Nanosuspension: an approach to enhance solubility of drugs, *J. Adv. Pharm. Technol. Res.*, 2 (2), 81. doi:10.4103%2F2231-4040.82950

Pathak, M. (2007), Nanoemulsions and their stability for enhancing functional properties of food ingredients, In: Oprea, A. E., and Grumezescu, A. M., Eds., *Nanotechnology Applications in Food*, Academic Press, pp. 87–106.

Pauk, S.V., Petriček, I., Jukić, T. et al. (2019), Noninvasive tear film break-up time assessment using handheld lipid layer examination instrument, *Acta Clin. Croat.*, 58 (1), 63–71. doi:10.20471/acc.2019.58.01.09

Petrochenko, P.E., Pavurala, N., Wu, Y. et al. (2018), Analytical considerations for measuring the globule size distribution of cyclosporine ophthalmic emulsions, *Int. J. Pharm.*, 550, 229–239. doi:10.1016/j.ijpharm.2018.08.030.

Petrucci, R.H., Herring, G.F., Madura, J.D. et al. (2011), *General Chemistry: Principles and Modern Applications*, 10th ed., Pearson Canada, New Jersey.

Pflugfelder, S.C., Tseng, S.C., Sanabria, O. et al. (1998), Evaluation of subjective assessments and objective diagnostic tests for diagnosing tear-film disorders known to cause ocular irritation, *Cornea*, 17, 38–56. doi:10.1097/00003226-199801000-00007

Pillai, P. P., Kowalczyk, B., Pudlo, W.J. et al. (2016), Electrostatic titrations reveal surface compositions of mixed, on-nanoparticle monolayers comprising positively and negatively charged ligands, *J. Phys. Chem. C*, 120 (7), 4139–4144. doi:10.1021/acs.jpcc.5b12599

Preetz, C., Hauser, A., Hause, G. et al. (2010), Application of atomic force microscopy and ultrasonic resonator technology on nanoscale: distinction of nanoemulsions from nanocapsules, *Eur. J. Pharm. Biopharm.*, 39 (1–3), 141–151. doi:10.1016/j.ejps.2009.11.009.

Preus, S. and Wilhelmsson, L.M. (2012), Advances in quantitative FRET-based methods for studying nucleic acids, *Chembiochem.*, 13 (14), 1990–2001. doi:10.1002/cbic.201200400

Rabinovich-Guilatt, L., Couvreur, P., Lambert, G. et al. (2004), Extensive surface studies help to analyse zeta potential data: the case of cationic emulsions, *Chem. Phys. Lipids*, 131, 1–13. doi:10.1016/j.chemphyslip.2004.04.003

Rahman, S.N.R., Pawde, D.M., Katari, O. et al. (2020), Systematic optimization, in vitro drug release, and preliminary nonclinical toxicity assessment of nonphospholipid-based topical ophthalmic emulsions containing 0.05 or 0.1% w/w cyclosporin A for dry-eye syndrome management, *AAPS Pharm. Sci. Tech.*, 21, 36 doi:10.1208/s12249-019-1587-5

Ribeiro, H.S., Guerrero, J.M.M., Briviba, K. et al. (2006), Cellular uptake of carotenoid-loaded oil-in-water emulsions in colon carcinoma cells in vitro, *J. Agric. Food Chem.*, 54, 9366–9369. doi:10.1021/jf062409z

Rieger, G. (1990), Lipid-containing eye drops: a step closer to natural tears, *Ophthalmologica*, 201 (4), 206–212. doi:10.1159/000310154

Robinson, J.V. and Woods, W.W. (1948), A method of selecting foam inhibitors, *J. Ind. Chem. Soc.*, 67 (9), 361–365. doi:10.1002/jctb.5000670908

Roger, E., Lagarce, F., Garcion, E. et al. (2009), Lipid nanocarriers improve paclitaxel transport throughout human intestinal epithelial cells by using vesicle-mediated transcytosis, *J. Control. Release*, 140, 174–181. doi:10.1016/j.jconrel.2009.08.010

Ruiz-Cabello, F.J.M., Trefalt, G., Maroni, P. et al. (2014), Electric double-layer potentials and surface regulation properties measured by colloidal-probe atomic force microscopy, *Phys. Rev. E*, 90 (1), 012301. doi:10.1103/PhysRevE.90.012301

Rungseevijitprapa, W., Brazeau, G.A., Simkins, J.W. et al. (2008), Myotoxicity studies of o/w-in situ forming microparticle systems, *Eur. J. Pharm. Biopharm.*, 69 (1), 126–133. doi:10.1016/j.ejpb.2007.10.009

Russin, T.J., Altinoğlu, E.İ., Adair, J.H. et al. (2010), Measuring the fluorescent quantum efficiency of indocyanine green encapsulated in nanocomposite particulates, *J. Phys. Condens. Matter*, 22 (33), 334217.

Ruth, H.S., Attwood, D., Ktistis, G. et al. (1995), Phase studies and particle size analysis of oil-in-water phospholipid microemulsions, *Int. J. Pharm.*, 116 (2), 253–261. doi:10.1016/0378-5173(94)00316-W

Saarinen-Savolainen, P., Järvinen, T., Araki-Sasaki, K. et al. (1998), Evaluation of cytotoxicity of various ophthalmic drugs, eye drop excipients and cyclodextrins in an immortalized human corneal epithelial cell line, *Pharm. Res.*, 15 (8), 1275–1280. doi:10.1023/A:1011956327987

Salis, A., Boström, M., Medda, L. et al. (2011), Measurements and theoretical interpretation of points of zero charge/potential of BSA protein, *Langmuir*, 27 (18), 11597–11604. doi:10.1021/la2024605

Sall, K., Stevenson, O.D., Mundorf, T.K. et al. (2000), Two multi- center, randomized studies of the efficacy and safety of cyclosporine ophthalmic emulsion in moderate to severe dry eye disease. CsA phase 3 study group, *Ophthalmology*, 107, 631–639.

Santos Magalhaes, N.S., Cave, G., Seiller, M. et al. (1991), The stabilityand in vitro release kinetics of a clofibride emulsion, *Int. J. Pharm.*, 76 (3), 225–237. doi:10.1016/0378-5173(91)90275-S

Satzer, P., Svec, F., Sekot, G. et al. (2016), Protein adsorption onto nanoparticles induces conformational changes: particle size dependency, kinetics, and mechanisms, *Eng. Life Sci.*, 16 (3), 238–246. doi:10.1002/elsc.201500059

Savini, G., Prabhawasat, P., Kojima, T. et al. (2008), The challenge of dry eye diagnosis, *Clin. Ophthalmol. (Auckland, NZ)*, 2 (1), 31 doi:10.2147/opth.s1496

Saxena, V., Hasan, A., Sharma, S. et al. (2017), Edible oil nanoemulsion: an organic nanoantibiotic as a potential biomolecule delivery vehicle, *Int. J. Polym. Mater. Polym. Biomater.*, 67 (7), 410–419. doi:10.1080/00914037.2017.1332625

Schoenwald, R.D. and Huang, H.S. (1983), Corneal penetration behavior of beta-blocking agents I: physiochemical factors, *J. Pharm. Sci.*, 72, 1266–1272.

Selek, H., Ünlu, N., and Orhan, M. (2000), Evaluation of retinoic acid ophthalmic emulsion in dry eye, *Eur. J. Ophthalmol.*, 10 (2), 121–127. doi:10.1177%2F112067210001000205

Shakeel, F., Baboota, S., Ahuja, A. et al. (2008), Celecoxib nanoemulsion: skin permeation mechanism and bioavailability assessment, *J. Drug Target.*, 16 (10), 733–740. doi:10.1080/10611860802473402

Sharma, A. and Ruckenstein, E. (1985), Mechanism of tear film rupture and its implications for contact lens tolerance, *Am. J. Optom. Physiol. Opt.*, 62 (4), 246–253. doi:10.1097/00006324-198504000-00003

Sheppard, J.D., Scoper, S.V., and Samudre, S. (2011), Topical loteprednol pretreatment reduces cyclosporine stinging in chronic dry eye disease, *J. Ocul. Pharmacol. Ther.*, 27 (1), 23–27. doi:10.1089/jop.2010.0085

Shono, Y., Jantratid, E., Janssen, N. et al. (2009), Prediction of food effects on the absorption of celecoxib based on biorelevant dissolution testing coupled with physiologically based pharmacokinetic modeling, *Eur. J. Pharm. Biopharm.*, 73 (1), 107–114. doi:10.1016/j.ejpb.2009.05.009

Shunmugaperumal, T. and Kaur, V. (2015), In vitro anti-inflammatory and antimicrobial activities of azithromycin after loaded in chitosan- and tween 20-based oil-in-water macroemulsion for acne management, *AAPS Pharm. Sci. Tech.*, 17 (3), 700–709. doi:10.1208/s12249-015-0401-2

Shunmugaperumal, T. and Kaur, V. (2016), In vitro anti-inflammatory and antimicrobial activities of azithromycin after loaded in chitosan- and tween 20-based oil-in-water macroemulsion for acne management, *AAPS PharmSciTech.*, 17 (3), 700–709. doi:10.1208/s12249-015-0401-2

Silla, E., Arnau, A., and Tuñón, I. (2019), Fundamental principles governing solvents use, in: Wypych, G., Ed., *Handbook of Solvents*, ChemTec Publishing, Ontario, Vol. 1, 3rd ed., pp. 11–77. doi:10.1016/b978-1-927885-38-3.50004-8

Sivagnanam, S.P., Shanmugapriya, K., Rudolf Nobbs Gereniu, C. et al. (2019), Ultrasound-mediated fucoxanthin rich oil nanoemulsions stabilized by κ-carrageenan: process optimization, bio-accessibility and cytotoxicity, *Ultrason. Sonochem.*, 55, 105–116. doi:10.1016/j.ultsonch.2019.03.014

Spielmann, H. (1997), Ocular irritation, in: Castell, J.V., Gŏmez-Lechŏn, M.J., Eds., *In Vitro Methods in Pharmaceutical Research*, Academic Press, London, pp. 265–287.

Stevenson, D., Tauber, J., and Reis, B.L. (2000), Efficacy and safety of cyclosporin A ophthalmic emulsion in the treatment of moderate-to-severe dry eye disease: a dose-ranging, randomized trial. The Cyclosporin A Phase 2 Study Group, *Ophthalmology*, 107 (5), 967–974. doi:10.1016/S0161-6420(00)00035-X

Straub, M., Bron, A.M., Muselier-Mathieu, A. et al. (2016), Long-term outcome after topical ciclosporin in severe dry eye disease with a 10-year follow-up, *Br. J. Ophthalmol.*, 100 (11), 1547–1550. doi:10.1136/ bjophthalmol-2015-306930

Szczesna, D.H. and Iskander, D.R. (2010), Lateral shearing interferometry for analysis of tear film surface kinetics, *Optom. Vis. Sci.*, 87, 513–517. doi:10.1097/OPX.0b013e3181e17279

Sznitowska, M., Dabrowska, E.A., and Janicki, S. (2002), Solubilizing potential of submicron emulsions and aqueous dispersions of lecithin, *Int. J. Pharm.*, 246 (1–2), 203–206. doi:10.1016/S0378-5173(02)00395-2

Sznitowska, M., Zurowska-Pryczkowska, K., Janicki, S. et al. (1999), Miotic effect and irritation potential of pilocarpine prodrug incorporated into a submicron emulsion vehicle, *Int. J. Pharm.*, 184 (1), 115–120. doi:10.1016/S0378-5173(99)00106-4

Takeuchi, Y., Yasukawa, H., Yamaoka, Y. et al. (1992), Destabilization of whole skin lipid bio-liposomes induced by skin penetration enhancers and FT-IR/ATR (Fourier transform infrared/attenuated total reflection) analysis of stratum corneum lipids, *Chem. Pharm. Bull.*, 40, 484–487. doi:10.1248/cpb.40.484

Talbot, G.H. (2008), Efficacy as an important facet of "safety" in clinical trials: how can we do our best for our patients? *Clin. Infect. Dis.*, 47 (Suppl. 3), S180–S185. doi:10.1086/591401

Talegaonkar, S., Azeem, A., Ahmad, F. et al. (2008), Microemulsions: a novel approach to enhanced drug delivery, *Recent Pat. Drug Deliv. Formul.*, 2 (3), 238–257. doi:10.2174/187221108786241679

Tamilvanan, S. and Baskar, R. (2013), Effect of non-phospholipid-based cationic and phospholipid-based anionic nanosized emulsions on skin retention and anti-inflammatory activity of celecoxib, *Pharm. Dev. Technol.*, 18 (4), 761–771. doi:10.3109/10837450.2011.586038

Tamilvanan, S., Khanum, R., Senthilkumar, S.R. et al. (2013), Studies on ocular and parenteral application potentials of Azithromycin-loaded anionic, cationic and neutral-charged emulsions, *Curr. Drug Deliv.*, 10, 572–586. doi:10.2174/1567201811310990036

Tamilvanan, S. and Kumar, B.A. (2011), Influence of acetazolamide loading on the (*in vitro*) performances of non-phospholipid-based cationic nanosized emulsion in comparison with phospholipid-based anionic and neutral-charged nanosized emulsions, *Drug Dev. Ind. Pharm.*, 37 (9), 1003–1015. doi:10.3109/03639045.2011.555407

Tamilvanan, S., Kumar, B.A., Senthilkumar, S.R. et al. (2010), Stability assessment of injectable castor oil-based nano-sized emulsion containing cationic droplets stabilized by poloxamer–chitosan emulsifier films, *AAPS PharmSciTech.*, 11, 904–909. doi:10.1208/s12249-010-9455-3

Tamilvanan, S., Kumar, V., Sharma, D. et al. (2014), In vitro evaluation of polyethylene glycol based microparticles containing azithromycin, *Drug Deliv. Transl. Res.*, 4 (2), 139–148. doi:10.1007/s13346-013-0187-2

Tonge, S.R., Hunsaker, J., and Holly, F.J. (1991), Non-invasive assessment of tear film break-up time in a group of normal subjects-implications for contact lens wear, *J. Br. Contact Lens Assoc.*, 14, 201–205. doi:10.1016/0141-7037(91)80022-E

Tucker, I.M., Corbett, J.C.W., Fatkin, J. et al. (2015), Laser Doppler electrophoresis applied to colloids and surfaces, *Curr. Opin. Colloid Interface Sci.*, 20 (4), 215–226. doi:10.1016/j.cocis.2015.07.001

Uskoković, V., Castiglione, Z., Cubas, P. et al. (2010), Zeta-potential and particle size analysis of human amelogenins, *J. Dent. Res.*, 89 (2), 149–153. doi:10.1177%2F0022034509354455

Vidal-Iglesias, F.J., Solla-Gullón, J., Rodes, A. et al. (2012), Understanding the Nernst Equation and other electrochemical concepts: an easy experimental approach for students, *J. Chem. Educ.*, 89 (7), 936–939. doi:10.1021/ed2007179

Wang, L.X., He, H.B., Tang, X. et al. (2006), A less irritant norcantharidin lipid microspheres: formulation and drug distribution, *Int. J. Pharm.*, 323 (1–2), 161–167. doi:10.1016/j.ijpharm.2006.05.060

Wasan, D.T., Koczo, K., and Nikolov, A.D. (1994), Mechanisms of aqueous foam stability and antifoaming action with and without oil-a thin film approach, *Adv. Chemother.*, 242, 47–111. doi:10.1021/ba-1994-0242.ch002

Washington, C. and Davis, S.S. (1987), Ageing effects in parentral fat emulsions: the role of fatty acids, *Int. J. Pharm.*, 39 (1–2), 33–37. doi:10.1016/0378-5173(87)90195-5

Washington, C. and Evans, K. (1995), Release rate measurements of model hydrophobic solutes from submicron triglyceride emulsions, *J. Control. Release*, 33 (3), 383–390. doi:10.1016/0168-3659(94)00110-G

Wilhelmus, K.R. (2001), The Draize eye test, *Surv. Ophthalmol.*, 45 (6), 493–515. doi:10.1016/s0039-6257(01)00211-9

Wilkerson, V.A. (1935), The chemistry of human epidermis II. The isoelectric points of the stratum corneum, hair, and nails as determined by electrophoresis, *J. Biol. Chem.*, 112 (1), 329–335.

Willcox, M.D.P., Argüeso, P., Georgiev, G.A. et al. (2017), TFOS DEWS II tear film report, *Ocul. Surf.*, 15 (3), 366–403. doi:10.1016/j.jtos.2017.03.006

Yao, B., Zhu, L., Jiang, Q. et al. (2013), Safety monitoring in clinical trials, *Pharmaceutics*, 5 (1), 94–106. doi:10.3390/pharmaceutics5010094

Yao, M., He, L., McClements, D. J. et al. (2015), Uptake of gold nanoparticles by intestinal epithelial cells: impact of particle size on their absorption, accumulation, and toxicity, *J. Agric. Food Chem.*, 63 (36), 8044–8049. doi:10.1021/acs.jafc.5b03242

Yi, X., Shi, X., and Gao, H. (2011), Cellular uptake of elastic nanoparticles, *Phys. Rev. Lett.*, 107 (9), 098101. doi:10.1103/PhysRevLett.107.098101

Zhang, H.Y., Tang, X., Li, H.Y. et al. (2007), A lipid microsphere vehicle for vinorelbine: stability, safety and pharmacokinetics, *Int. J. Pharm.*, 348, 70–79. doi:10.1016/j.ijpharm.2007.07.013

Zhao, S., Dai, W., He, B. et al. (2012), Monitoring the transport of polymeric micelles across MDCK cell monolayer and exploring related mechanisms, *J. Control. Release*, 158 (3), 413–423. doi:10.1016/j.jconrel.2011.12.018

Zheng, J., Li, Y., Song, M. et al. (2014), Improving intracellular uptake of 5-demethyltangeretin by food grade nanoemulsions, *Food Res. Int.*, 62, 98–103. doi:10.1016/j.foodres.2014.02.013

Zhou, X.Q. and Wei, R.L. (2014), Topical cyclosporine A in the treatment of dry eye: a systematic review and meta-analysis, *Cornea*, 33 (7), 760–767. doi:10.1097/ICO.0000000000000123

CHAPTER 4

MANUFACTURING AND POSITIONING (GENERATIONS) OF OIL-IN-WATER NANOSIZED EMULSIONS

OLY KATARI* AND TAMILVANAN SHUNMUGAPERUMAL

Department of Pharmaceutics, National Institute of Pharmaceutical Education and Research (NIPER), Guwahati, Assam, India

*The funding provided by Department of Pharmaceuticals, Ministry of Chemicals and Fertilizers, Govt. of India to the author Oly Katari is greatly acknowledged.

Oil-in-Water Nanosized Emulsions for Drug Delivery and Targeting,
First Edition. Tamilvanan Shunmugaperumal.
© 2021 John Wiley & Sons, Inc. Published 2021 by John Wiley & Sons, Inc.

EXPANSION OF ABBREVIATIONS

apos	apolipoproteins
API	active pharmaceutical ingredient
AS	atherosclerosis
BAK	benzalkonium chloride
BRCA	BReast CAncer
CK	cytokeratins
CKC	cetalkonium chloride
CO	cholesteryl oleate
CT	computed tomography
CVD	cardiovascular disease
CWS	cell wall skeleton
DFS	disease-free survival
DHA	docosahexaenoic acid
DOPE	1,2-dioleoyl-*sn*-glycero-3-phosphoethanolamine
DOTAP	1,2-dioleoyl-*sn*-glycero-3-trimethylammonium-propane
DPPC	dipalmitoyl phosphatidylcholine
ED	endothelial dysfunction
EGFR	epidermal growth factor receptor
EPA	eicosapentaenoic acid
EPR	enhanced permeability and retention
ER	estrogen
EYP	egg yolk phosphatides
FCA or CFA	Freund's Complete Adjuvant
FDG	fluorodeoxyglucose
FIA or IFA	Freund's Incomplete Adjuvant
FMT	fluorescence molecular tomography
GRAS	generally regarded as safe
HCO	hydrogenated castor oil
HER2+	human epidermal growth factor receptor 2-positive
ICI	imperial chemical industries
IGDD	image-guided drug delivery
IHC	immunohistochemistry
LCT	long-chain triglyceride
LPL	lipoprotein lipase
MCT	medium-chain triglyceride
MPL	monophosphoryl-lipid A
MPS	mononuclear phagocyte system
MRI	magnetic resonance imaging
MUFA	monounsaturated fatty acids
NIR	near infrared
O/W	oil-in-water

OS	overall survival
PEG	polyethylene glycol
PEG-PE	phosphatidylethanolamine derivative with polyethylene glycol
PET	positron emission tomography
PgR	progesterone receptor
POE	polyoxyethylene
PUFA	polyunsaturated fatty acids
QDs	quantum dots
RAS	Ribi Adjuvant System
RECIST	response evaluation criteria in solid tumors
RES	reticuloendothelial system
SAF	Syntex Adjuvant Formulation
SFA	saturated fatty acids
SL	structured lipids
SLM	structured lipids having medium-chain (C_8–C_{10}) fatty acids
SLS	structured lipids having short-chain (C_4) fatty acids
SM	sphingomyelin
SO	soybean oil
TDM	trehalose dicorynomycolate
TG	triglyceride
TNBC	triple negative breast cancer
TO	triolein
TPN	total parenteral nutrition
VLDL	very low density lipoproteins

4.1. INTRODUCTION

Initially, the oil-in-water (o/w) emulsions developed consist of phospholipid component as one of the emulsifier molecule to stabilize the emulsions as well as to accommodate the nutrient substances like mineral salts, glucose molecule, and vitamins. The purpose of these initially developed emulsions confined to the aim of maintaining/supplying energy to the bedridden patients. Therefore, these emulsions are grouped under parenteral nutrition category (i.e., total parenteral nutrition, TPN) and thus termed as first generation. As per the advice/direction from medical doctors, the health-care providers like nurses or clinical pharmacist tried to inject the active pharmaceutical ingredient (API) molecules such as anesthetics, antibiotics, anti-inflammatory, etc., into the infusion set of TPN so that the API molecules will run into the patient body along with nutrient materials contained in the TPN. This type of extemporaneous addition of API molecules into the infusion set of TPN did not show any incompatibility issues between few of the API molecules and the TPN components (mineral salts, glucose molecule, and vitamins). On the other hand, the direct injection of API molecules into the infusion set of TPN was usually done by nurses or clinical pharmacist on those days. These two health-care professionals sometimes have noticed the formation of crystals/sediment formation in the TPN dispersion immediately after or in due course of time of the API molecule injection into it. The crystal/sediment formation became a serious regulatory issue and even prompted to raise the therapeutic efficacy of the infused API molecules together with possible side/toxic effect of crystals/sediment formed due to the incompatibility between TPN components and API molecules. Keeping these points in mind, the extemporaneous addition of API molecules into the TPN did not regard as a standard method to prepare API-loaded emulsions. In fact, the next generations (second, third, and fourth) of emulsions involve the API-incorporation not to the free-formed emulsions (such as TPN) but dispersing/dissolving separately the API molecules into the oil phase of the emulsions. Thus, it becomes mandatory to identify a suitable oil or oil combinations to disperse/dissolve the API molecules and then go for emulsion preparation steps (as shown in Chapter 2). This type of practical/systematic identification of oil or oil combinations before the emulsion preparation is termed as "emulsion *de novo* method". In this respect, the o/w nanosized emulsions could be considered as API-cargo carrying API delivery system. The blank emulsions (produced without API molecules) thus produced were proven to possess many appealing properties such as biocompatible, biodegradable, physically stable, and relatively easy to produce on a large scale using emulsion *de novo* technology (Fukushima et al. 2000). Nevertheless, their subcellular and submicron size makes the emulsion particles to penetrate deep into tissues through fine capillaries and even cross the fenestration present in the epithelial lining in liver after parenteral administration into human body. This allows efficient delivery of therapeutic agents to target sites in the body. In addition to consider as delivery carrier for lipophilic and hydrophobic APIs, the nanosized emulsions can also be viewed nowadays as adjuvant to enhance the potency of vaccines (influenza, polio, etc.). It should be emphasized that many liquid-retentive and semisolid-natured topical emulsion formulations are available in the market and they consisted of API or non-API molecules. Many of these products did not fall under nanosized emulsion category but they do contain either dispersed oil or water droplets in their respective continuous medium (water or oil). Table 4.1 shows few selected list of topical emulsion formulations useful especially for managing skin infections/inflammations and pain.

The sterile emulsions (both o/w and w/o types) used as adjuvants are also shown in Table 4.1. An adjuvant is an ingredient added to a vaccine that helps create a stronger immune response to vaccination. Four different adjuvant emulsions of o/w types and two different adjuvant emulsions of w/o types are useful to enhance the vaccine potency. All these four o/w adjuvant emulsions can conveniently be classified into cationic and nonionic types depending on the emulsifier combinations used to form mono or multi emulsifier film over

TABLE 4.1. Non-Exhaustive or Selected List of Marketed Medical and Nonmedical Emulsions

Parenteral Fat Emulsions (O/W Type) for Nutrition	
Product	Producer
Abbolipid/Liposyn	Abbott, Illinois, USA
Cleviprex	Chiesi, Parma, Italy
ClinOleic	Clintec/Baxter Healthcare, Illinois, USA
Clinolipid	Baxter SA, Lessines, Belgium
Cinvanti	Heron, California, USA
Intralipid	Pharmacia & Upjohn, Michigan, USA
Intralipos	Green Cross, Geneva, Switzerland
Ivelip/Salvilipid	Clintec/Baxter Healthcare, Illinois, USA
Kabimix	Pharmacia & Upjohn, Michigan, USA
Lipofundin N/Endolipide	B. Braun, Melsungen, Germany
Lipofundin MCT/LCT/	B. Braun, Melsungen, Germany
Lipoplus (Lipidem)	B. Braun, Melsungen, Germany
Lipovenos	Fresenius Kabi, Bad Homburg, Germany
Medialipide/Vasolipid	B. Braun, Melsungen, Germany
Medianut	B. Braun, Melsungen, Germany
Nutrilipid	B. Braun, Melsungen, Germany
Omegaven	Fresenius Kabi, Bad Homburg, Germany
Perikabiven	Fresenius Kabi, Bad Homburg, Germany
SMOFLipid	Fresenius Kabi, Bad Homburg, Germany
Structolipid	Fresenius Kabi, Bad Homburg, Germany
Trivè 1000	Baxter SA, Illinois, USA

Therapeutic Emulsions (O/W Type) With or Without Active Pharmaceutical Ingredient (API)			
Product	API	Producer	Application
Cationorm	API-free	Santen Pharmaceutical, Osaka, Japan	Ocular topical use
Catioprost	Latanoprost	Santen Pharmaceutical, Osaka, Japan	Ocular topical use
Clacier	Cyclosporin A	Huons Co. Ltd., Seongnam, Korea	Ocular topical use
Cyclokat	Cyclosporin A	Santen Pharmaceutical, Osaka, Japan	Ocular topical use
Cyporin N	Cyclosporin A	Taejoon Pharmaceutical Co. Ltd., Seoul, Korea	Ocular topical use
Diazepam-Lipuro	Diazepam	Braun Melsungen, Melsungen, Germany	Intravenous
DiprivanPropofol	AstraZeneca	Cambridge, UK	Intravenous
Emustil	API-free	SIFI S.p.A, Catania, Italy	Ocular topical use

TABLE 4.1. *Continued*

Parenteral Fat Emulsions (O/W Type) for Nutrition			
Product	API	Producer	Application
Etomidat-Lipuro	Etomidate	Braun Melsungen, Melsungen, Germany	Intravenous
Gengraf	Cyclosporin A	Abbott, Illinois, USA	Oral
Lipotalon (Limethason)	Dexamethasone Palmitate	Merckle, Blaubeuren, Germany	Intraarticular
Neoral	Cyclosporin A	Novartis, Basel, Switzerland	Oral
Norvir	Ritonavir	Abbott, Illinois, USA	Oral
Refresh Endura	API-free	Allergan, Dublin, Ireland	Ocular topical use
Restasis	Cyclosporin A	Allergan, Dublin, Ireland	Ocular topical use
Sandimmune	Cyclosporin A	Novartis, Basel, Switzerland	Oral
Stesolid	Diazepam	Dumex, Paris, France	Intravenous
Vekacia	Cyclosporin	Santen Pharmaceutical, Osaka, Japan	Ocular topical use
Xalatan	Latanoprost	Pfizer, New York, USA	Ocular topical use

Perfluorocarbon Emulsions (Fluorocarbon-in-Water Emulsions)		
Product	Producer	Application
Fluosol DA	Green Cross, Geneva, Switzerland	Blood supplement or O_2 carrier
Imagent	Alliance Pharmaceutical Corp, San Diego, California, USA	Contrast agent to image heart
Oxyfluor	Hemagen/pfc, Missouri, USA	O_2 carrier
Oxygent	Alliance Pharmaceutical Corp, San Diego, California, USA	Blood supplement or O_2 carrier
Perftoran	Institute of Theoretical and Experimental Biophysics, Pushchino, Russia	O_2 carrier

Selected Topical Formulations Based on O/W or W/O Emulsion	
Product	Producer
Daivonex cream and ointment	Laboratoires Leo, Vernouillet, France
Diclomax Emulgel	Torrent Pharmaceuticals, Ahmedabad, India
EMLA cream	Astra, Södertälje, Sweden
Levorag	THD SpA, Correggio RE, Italy
Miconaz-H-emulgel	Medical union Pharmaceuticals, Cairo, Egypt
Physiogel	Stiefel Laboratorium GmbH, Offenbach, Germany
Voltaren emulgel	Ciba-Geigy, Basel, Switzerland
Voltarol 1.16% emulgel	Novartis, Basel, Switzerland

the dispersed oil droplets of the emulsions. The Fluad (MF59) is the only cationic emulsion used for adjuvant purpose (Ott et al. 2002). Fluad is a standard-dose, three-component (trivalent) inactivated flu vaccine, manufactured by using an egg-based process and it is formulated with the adjuvant MF59. The oil used for MF59 is squalene, which is a naturally occurring triterpene hydrocarbon with a simple composition ($C_{30}H_{50}$), but a complex structure, which is found in many plants and animals, including humans. The final composition of MF59 emulsion consists of squalene oil 4.3%, Tween 80 0.5%, span 85 0.5%, a cationic lipid 1,2-dioleoyl-*sn*-glycero-3-trimethylammonium-propane (DOTAP) and water. The Fluad (MF59) have shown an interaction of cationic emulsion droplets with DNA and the formed DNA-adsorbed emulsion had a higher antibody response in mice *in vivo* while maintaining the cellular responses equivalent to that seen with naked DNA at the same doses.

The nonionic adjuvant emulsion category consists of Ribi Adjuvant System (RAS) resulting from US patent literatures (Ribi 1984; Ribi et al. 1984; Myers and Truchot 1990) and Syntex Adjuvant Formulation (SAF). The RAS can be classified into two types depending on the animal species used. Those RAS meant for use in mice termed as Monophosphoryl-lipid A + Trehalose dicorynomycolate emulsion (MPL + TDM emulsion) and another for use in rabbits, goats, and larger animals called as Monophosphoryl-lipid A + Trehalose dicorynomycolate + Cell wall skeleton emulsion (MPL + TDM + CWS emulsion). Strikingly, the MPL + TDM and MPL + TDM + CWS emulsions are also prepared using 2% squalene along with Tween 80 and water. These adjuvants are derived from bacterial and mycobacterial cell wall components that have been prepared to reduce the undesirable side effects of toxicity and allergenicity, but still provide potent stimulus to the immune system. The SAF consists of a preformed o/w emulsion stabilized by Tween 80 and Pluronic® L121 (Allison and Byars 1986).

The two different w/o adjuvant emulsions are named after Jules T. Freund, that is, Freund's adjuvant. It is a solution of antigen emulsified in mineral oil and used as an immunopotentiator (booster). The complete form, Freund's Complete Adjuvant (CFA or FCA), is composed of inactivated and dried mycobacteria (usually *Mycobacterium tuberculosis*), whereas the incomplete form (FIA or IFA) lacks the mycobacterial components (hence just the water in oil emulsion). In applications, the CFA or FCA is usually only necessary for the initial immunization, while FIA or IFA is the adjuvant of choice for subsequent immunizations. The Freund's adjuvants have the disadvantage of difficulty to mix with immunogen and can cause tissue at the injection site. When mixed with an antigen or immunogen, Freund's adjuvants help to deposit or sequester the injected material thereby helping to increase antibody response. However, the Freund's adjuvants do not confer immunogenicity to non-immunogenic haptens. Specifically, the CFA is meant for research purpose only and not for use in humans.

Keeping in mind the potentiality of o/w nanosized emulsions, the purpose of this chapter is to classify the emulsions according to the generations and to focus a short overview on preparation of the new (second, third, and fourth) generation emulsions followed by a description about an unique property of the third-generation emulsion and examples of selected excipients used for emulsion preparation. Given a specific interest especially on parenteral route and then on ocular topical route, the o/w nanosized emulsions for both routes share a common platform on strict criteria concerning the maximum globule size and requirement of sterility in the final emulsions. Nasal and topical routes are also covered based on the published research works with nanosized emulsions. It should be clearly emphasized that the actual chapter focused only on preformed nanosized emulsions (having size distribution ranging between 50 and 1,000 nm with a mean droplet size of about 250 nm), which should not be confused with self-microemulsifying API delivery systems or preformed microemulsions that are transparent thermodynamically stable dosage forms.

4.2. GENERATIONS OF O/W NANOSIZED EMULSIONS

In this chapter, the o/w nanosized emulsions are positioned in year-wise with respect to four generations (see Flowchart 4.1) based on their development to make ultimately a better colloidal carrier for a target site within the organs/parts of the body and eye, thus allowing site-specific API delivery and/or enhanced API absorption.

4.2.1. First-Generation Nanosized Emulsions

The desire of all human beings is to be healthy with quality life style. According to WHO definition, the health is "Health is a state of complete physical, mental and social well-being and not merely the absence of disease or infirmity" (https://www.who.int/about/who-we-are/constitution, accessed on December 16, 2019). However, this definition do not mention how to preserve health. When you go back to age-old literatures documented in

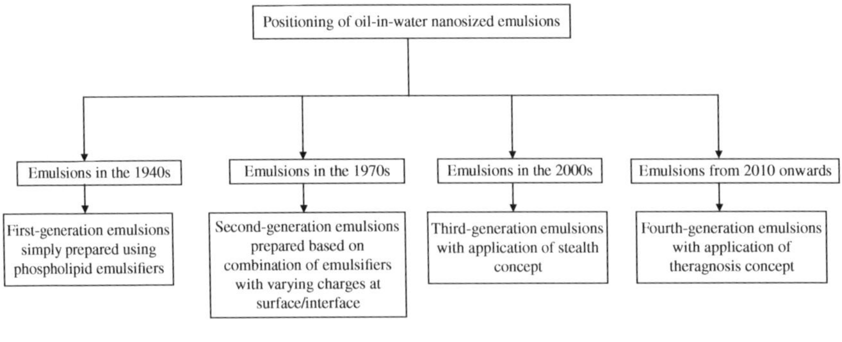

Flowchart 4.1. Year-wise positioning of oil-in-water nanosized emulsions.

Indian scriptures in 5,000 BCE, the health of a human being could be maintained and even preserved by the intake of balanced/correct nutrition (Chandra 1985). The nutritional depletion in the human body occurs due to either changes in the quality and amount of dietary fat intake or abnormalities in lipid metabolism. This results in immunosuppression and therefore host defense impairment, favoring increased infection and finally the mortality rates of humans.

If the dietary fats depletion happened especially in malnourished or hypercatabolic patient, then it is compensated through intravenous feeding of a solution containing amino acids, glucose, electrolytes, and vitamins or intravenous infusion of nanosized emulsions having all of these essential nutrients. As per Shils (1998), the o/w type emulsion consists of triglyceride (TG) droplets enveloped with a stabilizing superficial layer of phospholipids. However, in order to make autoclave-sterilizable and reasonably stable emulsions for parenteral use, many other excipients (other than TG) such as phosphatidylcholine, glycerol, and α-tocopherol in variable amounts need to be incorporated into the emulsions. Interestingly, the o/w nanosized emulsions often show myriad complex inner structures when envisioned under microscopic techniques. The complex inner structures of the o/w nanosized emulsions include oil droplets covered by an emulsifier monolayer, oil droplets covered by emulsifier oligolayers, double-emulsion droplets, and possibly small unilamellar vesicles. Very recently, a delicate/sessile structure was found depending on the oil combination used to generate oil phase of the emulsions. The delicate/sessile structure found in the dispersed oil phase of the emulsion is termed as "bi-compartmental or Janus or hand-bag." A detailed discussion about this new structure and its importance to make dual API delivery using the o/w nanosized emulsions is given in Chapter 7. Coming back to the first-generation emulsions, all of the commercially available nanosized emulsions used as intravenous high-calorie nutrient fluids have particle size normally around 160–400 nm in diameter and typically, their surfaces are normally negatively charged. However, larger droplets can also be detected in commercially available emulsions (Koster et al. 1996). The larger-sized droplets found in the final o/w nanosized emulsions could be corroborated with the amount of oils used and length of fatty acid chains in TG. For instance, Lutz et al. (1989) reported that the mean diameter of particles in the 20% emulsions is larger than in the 10% emulsions.

Importantly, the TG used to make the nanosized emulsions may be presented structurally in long or medium chains, respectively, named as LCT and MCT. The LCT comprises of both saturated and unsaturated fatty acid chains with 14, 16, 18, 20, and 22 carbon atoms. The number of double bonds present defines the fatty acids in LCT as saturated, mono, or polyunsaturated. If the first double bond is on carbon number 3, 6, or 9 from the methyl end of the carbon chain, then the fatty acid is *n*–3, *n*–6, or *n*–9, respectively. Purified soybean or safflower oil contains LCT with a high proportion of *n*–6 polyunsaturated fatty acids (PUFA), whereas olive oil has LCT with *n*–9 monounsaturated fatty acids (MUFA). Fish oil includes LCT with 20 or more carbon atoms where the first double bond is located between the third and the fourth carbons from the methyl terminal of the fatty acid chain (omega-3 or *n*–3). Coming to the MCT, it is derived from coconut oil and possesses saturated fatty acids (SFA) with chains containing carbon atoms at 6, 8, 10, or 12 positions.

The selected fatty acids of interest to make the parenteral lipid emulsions (first generation) along with their common names, shorthand nomenclature, and sources are listed in Table 4.2. The fatty acids are also referred to by a shorthand nomenclature that denotes the number of C in the chain, the number of double bonds, and the position of the first double bond relative to the methyl-C (*n*; also termed ω; Table 4.2).

The simplest *n*–6 fatty acid is linoleic acid (18:2*n*–6) and the simplest *n*–3 fatty acid is α-linolenic acid (18:3*n*–3). Although mammalian cells cannot synthesize linoleic and α-linolenic acids, they can metabolize them by further desaturation and elongation. Linoleic acid can be converted to γ-linolenic (18:3*n*–6), then to dihomo-γ-linolenic acid (20:3*n*–6), and then to arachidonic acid (20:4*n*–6). Using the same series of enzymes, α-linolenic acid is converted to eicosapentaenoic acid (EPA) (20:5*n*–3). A complex pathway for further conversion of EPA to docosahexaenoic acid (DHA) (22:6*n*–3) exists (Sprecher 2002).

The lipid typically used in parenteral nutrition is soybean oil, in which linoleic acid comprises about 50% of the fatty acids present. Commercially available soybean oil-based nanosized emulsions include: Intralipid® (Fresenius Kabi, Bad Homberg, Germany); Lipovenoes® (Fresenius Kabi); Lipofundin® (B Braun, Melsugen, Germany); Ivelip® (Baxter Healthcare, Maurepas, France). A study in patients following major gastrointestinal surgery has identified that the amount of *n*–6 PUFA (i.e., linoleic acid) infused is one of two predictors of the length of hospital stay (increased by 1.6 days/100 g *n*–6 PUFA

TABLE 4.2. Selected List of Fatty Acids Used in Parenteral Lipid Emulsions (First Generation) Along with Their Common Names, Shorthand Nomenclature, and Sources

Common Name	Shorthand Nomenclature	Typical Source(s)
Caprylic acid	8:0	Coconut oil
Capric acid	10.0	Coconut oil
Myristic acid	14.0	Coconut oil
Palmitic acid	16.0	Olive oil, soybean oil, fish oil
Oleic acid	18:1*n*–9	Olive oil, soybean oil
Linoleic acid	18:2*n*–6	Soybean oil
α-Linolenic acid	18:3*n*–3	Soybean oil
Eicosapentaenoic acid (EPA)	20:5*n*–3	Fish oil
Docosahexaenoic acid (DHA)	22:6*n*–3	Fish oil

infused), the other being the delay in the onset of initiating nutritional support (Koch and Heller 2005). Furthermore, the concern about potential harm, based mainly on the notion that n–6 PUFA might be "proinflammatory, immunosuppressive and pro-coagulatory," has led to the development of alternative first-generation nanosized emulsions for parenteral applications. Two alternative philosophies to reducing the amount of linoleic acid have been adopted. The first philosophy has been to simply dilute soybean oil with another oil that is fairly inert. Examples of this strategy include the use of so-called MCT and the use of olive oil.

The second philosophy has been to partially replace soybean oil with another oil that is believed to exert benefits in its own right. An example of this strategy is the use of fish oil. Soybean oil is often referred to as "LCT," but this nomenclature is an incorrect use of this term since the lipids found in olive oil, fish oil, and other oils not used in parenteral nutrition also contain LCT. The following FDA-approved nanosized emulsions are available as alternatives to pure soybean oil emulsions: Lipofundin MCT/LCT® (B Braun), a 50:50 (v/v) mixture of MCT (in the form of coconut oil) and soybean oil; Lipovenoes MCT® (Fresenius Kabi), a 50:50 (v/v) mixture of MCT (in the form of coconut oil) and soybean oil; Structolipid® (Fresenius Kabi), produced by interesterification of a 50:50 (v/v) mixture of MCT (in the form of coconut oil) and soybean oil; ClinOleic® (Baxter Healthcare), an 80:20 (v/v) mixture of olive and soybean oils; Lipoplus® (also known as Lipidem®; B Braun), a 50:40:10 (by vol.) mixture of coconut, soybean, and fish oils; SMOFLipid® (Fresenius Kabi), a 30:30:25:15 (by vol.) mixture of coconut, soybean, olive, and fish oils. In addition, the product Omegaven® (Fresenius Kabi), which is 100% fish oil, is available for use as a supplement to be diluted with another lipid emulsion of choice. In Europe, nanosized emulsion containing LCT/MCT enriched with fish oil became available for research. Fish oil consists of two major omega-3 fatty acids, namely, EPA and DHA. Both EPA and DHA have been shown to be active in a number of biological processes, including retinal and brain development, immune function, blood clotting, and prevention of cardiac arrhythmias (Horrocks and Yoe 1999). Thus, omega-3 fatty acids-laden nanosized emulsions made from fish oil are likely to be increasingly used not only for nutrition support but also for modification of a number of biological and pathological processes.

Both LCT and MCT either alone or MCT in combination with LCT have known for their long-term commercial acceptability in parenteral emulsions and even these two are found in several FDA-approved first-generation emulsion products. In European academic institutions, the o/w nanosized emulsions containing LCT/MCT enriched with fish oil became available for research purpose. With LCT/MCT combinations in a specific ratio, the o/w nanosized emulsions appear to provide a more readily metabolizable source of energy (Rubin et al. 1991). But, the LCT-based emulsions were used in clinical practice for over 50 years. Keeping the point of API's solubility enhancement in the oil phase and oil–water interface of the emulsions, the MCT is reported to be

100 times more soluble in water than LCT and thus to have an escalated solubilizing capability of APIs. Moreover, the mean particle diameter value obtained with LCT-based emulsions is greater than the mean particle diameter value observed with MCT-based emulsions (Lutz et al. 1989).

The rationale for using newer first-generation nanosized emulsions, prepared from olive oil and fish oil, as parenteral nutrition and the clinical trials performed using these nanosized emulsions in adult patients post-surgery (mainly gastrointestinal) or critically ill adults was reviewed recently by Calder (2009).

4.2.2. Second-Generation Nanosized Emulsions

The second-generation emulsions are grouped together with a view of the capability of the emulsion structures that include easy and substantial associability of hydrophobic APIs as cargo, improved stabilization of entrapped APIs, sustained API release, and even site-specific API delivery. Fittingly the anionic and cationic nanosized emulsions can be administered by almost all available routes including topical percutaneous (Amselem and Friedman 1998) and ocular (Tamilvanan and Benita 2004), parenteral, oral, nasal (Tirucherai et al. 2002), and even aerosolization to the lungs (Mizushima et al. 1983). In spite of differences in routes of administration, the selected examples of commercially available emulsion-based formulations utilized for medical and nonmedical applications purposes are shown in Table 4.1. Typical o/w nanosized emulsion is basically comprising about 0.5–50% of a first component of an oil or oil mixture, about 0.1–10% of a second component of an emulsifier, about 0.05–5% of a nonionic surfactant, and an aqueous component with the mean droplet size being in the submicron range, that is, below about 500 nm and preferably between about 100 and 300 nm.

Apart from serving as excellent vehicles for API encapsulation, the o/w nanosized emulsions also mitigate the toxicities associated with surfactant and ethoxylated castor oil (Cremophor® EL)-based formulations. Cremophor® EL is associated with nephrotoxicity, hypotension, and bronchospasms and can produce anaphylactic reaction. For example, propofol injectable solution was prepared in Cremophor® EL by Imperial Chemical Industries (ICI) was gone into clinical usage but later withdrawn due to toxicity of Cremophor® EL. Propofol was reformulated in o/w nanosized emulsion using generally regarded as safe (GRAS)-grade excipients such as soybean oil, glycerol, egg lecithin, and disodium edetate and launched with the trade name Diprivan® by ICI (now Astra Zeneca) (Cockshott 1985). Diprivan® is used in intensive care medicine as a short-acting, intravenous sedative and known to have low toxicity, controlled sedation effect, rapid onset, and quick recovery despite prolonged usage (Cockshott 1985). Emulsion prepared using the edible lipidic substances (soybean and safflower oils) have been extensively used in clinical practice to administer nutritive substances parenterally (e.g., Intralipid®). These edible lipids serve as a wealthy resource of essential fatty acids such as omega-3 and omega-6 PUFA, non-glucose-based calories, and vitamins E and K. Similarly, a Cremophor® EL-free formulation of paclitaxel was prepared using TOCOSOL nanoemulsion to overcome anaphylactic reactions associated with Cremophor® formulation. Paclitaxel-loaded TOCOSOL formulation was approved by FDA in 2003 for the treatment of non-superficial urothelial cancer. In another example, a Cremophor® EL/ethanol-based cyclosporine (Sandimmune® Injection) was reformulated into nanoemulsion using soybean oil and egg lecithin composition. These studies indicated that a change in the vehicle may reduce the acute nephrotoxic side effects associated with cyclosporine in the Cremophor® EL formulation (Tibell et al. 1993). Alprostadil palmitate, amphotericin B, dexamethasone, flurbiprofen axetil, and vitamins A, D, E, and K are some other examples of therapeutically relevant compounds that have been formulated in nanoemulsions for clinical applications.

4.2.2.1. Second-Generation Nanosized Emulsions Versus Opsonization

The lipid-induced enhancement in oral bioavailability of many hydrophobic APIs is a well-known documented fact when these APIs are incorporated into oil phase and oil–water interface of the emulsions (Craig et al. 2000; Hauss 2002). However, direct intravascularly or locally administered conventional first- and second-generation emulsion particles could be cleared off/taken up quickly by the circulating monocytes to clear them by the reticuloendothelial cells (through organs such as the liver, spleen, and bone marrow) (Sakaeda and Hirano 1998). Regardless of the residence time of o/w nanosized emulsions within the vascular system, much of an injected dose is taken up via endocytosis by the cells of reticuloendothelial system (RES) to end up in the lysosomal apparatus. The o/w nanosized emulsions can be considered as being artificial chylomicrons and enter the fat metabolism pathway through the adsorption of apolipoproteins (apos) and the subsequent action of lipoprotein lipase (LPL). It appeared that uptake of conventional o/w nanosized emulsions by the RES cells is lysosomotropic resulting in the localization of emulsion droplets inside the lysosomes where they are degraded by local enzymes.

Furthermore, the extent of clearance is enhanced by the adsorption of opsonic plasma proteins onto the dispersed oil droplet surfaces of emulsion. This type of opsonic plasma proteins-directed/enhanced emulsion particle clearance by macrophage is simply termed as opsonization. Another way by which the dispersed oil droplet surfaces of emulsion would also be cleared from the blood circulation/stream by macrophages is without the involvement of opsonic plasma proteins. This type of emulsion particle clearance by the macrophage without opsonic plasma proteins' involvement occurs when the oil phase should liberate from the emulsion through any one of the destabilization processes mentioned in Chapter 1. Both of these processes occur inside the blood compartment immediately after administering the emulsion intravascularly or locally. It is well known that the inner oil core of the o/w emulsion is lipophilic in nature but the whole emulsion envelop is indeed hydrophilic. That is why the exposure of the hydrophobic part to the aqueous medium will therefore destabilize the emulsion. Moreover, Sasaki et al. (1996) have assumed that when the castor oil-based emulsions interact with the tears in the eye, the electrolytes in the tears elicit a physical emulsion instability resulting in some release of the oil. The electrolytes present in blood or cellular fluids could also cause a similar type of emulsion destabilization process resulting in separation of the oil and water phases from the parenterally administered emulsions. It is thus reasonable to say that the resultant oily hydrophobic particles of the emulsions would also be taken up by macrophages independent of an opsonization process.

An opsonization process is the adsorption of protein entities capable of interacting with specific plasma membrane receptors on monocytes and various subsets of tissue macrophages (see Flowchart 4.2 for list), thus promoting particle recognition by these cells. Classical examples of opsonic molecules include various subclasses of immunoglobulins (Harnisch and Müller 1998, 2000), complement proteins like C1q and generated C3 fragments (C3b, iC3b) (Szebeni 2001), apos (Harnisch and Müller 1998, 2000), von Willebrand factor, thrombospondin, fibronectin (Price et al. 2001), and mannose-binding protein. Reports that the coemulsification or incubation of nanosized emulsion with gelatin prior to intravenous injection enhanced its rate of clearance from the blood by fixed RES cells of the liver, lung, spleen, and bone marrow have led to the development of an artificial nanosized emulsion for RES functionality tests (Illum et al. 1989). The mechanisms employed are interactions with circulating plasma opsonin proteins, thereby exploiting a receptor-mediated process involving fibronectin. It should also be emphasized that the interaction of particles with blood protein may have effects beyond opsonization. These may include interference with the blood-clotting cascade, a process that may lead to fibrin formation, and anaphylaxis because of complement activation.

When given by other parenteral routes, for example, intraperitoneally, subcutaneously, or intramuscularly, the majority of emulsion droplets enter the lymphatic system and eventually to the blood circulation where particles behave as if given intravenously. Liver, spleen, and bone marrow uptake is

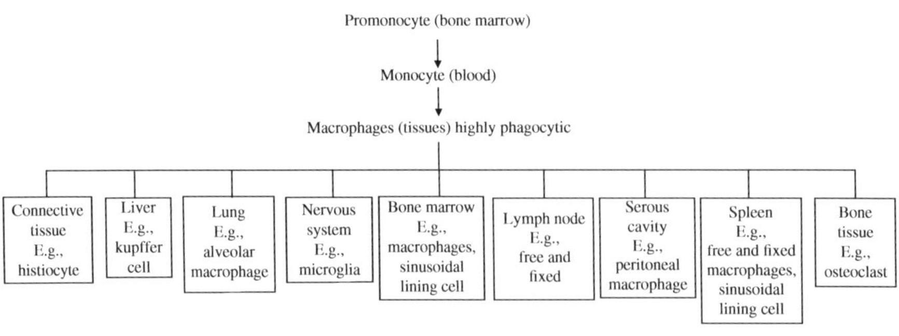

Flowchart 4.2. Monocytes and various subsets of tissue macrophages.

significantly lower. Indeed, relative to the emulsion droplet size, lymph nodes take up a much greater (over 100 fold) proportion than any other RES tissue. Lymphatic transport was predominantly associated with chylomicron-based transport (Porter et al. 1996).

4.2.2.2. Second-Generation Nanosized Emulsions and Long Circulation Concept There is an increasing interest in developing injectable emulsions that is not cleared quickly from the circulation when emulsions are designed to reach non-RES tissues in the vascular system, extravascular sites of action, or to act as circulating API reservoirs. It has been thought that the initial clearance rate by the RES can be affected by the presence of large numbers of nanosized emulsion particles occupying available RES receptors or exhausting opsonizing factors (Saba 1970). Overloading or saturating the RES by single large doses or repeated administration may lead to subsequent remain of injected nanosized emulsion particles in the circulation and could be used to affect distribution patterns. Davis et al. (1992) confirmed that the infusion of nanosized emulsions into the rabbit can cause slight temporary impairment of the RES as determined by subsequent administration of a radiolabeled colloid probe. An infused nanosized emulsion can cause RES blockade by one or both of the two mechanisms. Immediately after infusion, some of the nanosized emulsion particle may be recognized as foreign and are then cleared by the RES (largely the Kupffer cells of the liver). In addition, or alternatively, the Kupffer cells of the liver may become overloaded by nanosized emulsion remnants that will result from normal metabolism of the nanosized emulsion by tissue lipases (Davis et al. 1992). The study by Ueda et al. (2001) is supporting in a way the RES overloading or saturation concept. They prepared nanosized emulsions consisting of soybean oil and egg yolk phosphatides and studied the influence of the volume of intravenously injected nanosized emulsions on the pharmacokinetic parameters of the dispersed oil particles and incorporated cargo molecules like menatetrenone (vitamin K_2). At $3.0\,ml\,kg^{-1}$ injection volume, which equates to 180 ml for a person weighing 60 kg, these nanosized emulsions showed a prolonged plasma half-life in rats following intravenous administration. In contrast, at $0.1\,ml\,kg^{-1}$ injection volume, which equates to 6 ml for a 60 kg person, these nanosized emulsions were shown to disappear from the circulation soon after intravenous administration to rats.

4.2.2.3. Feasible Approaches for Making Long-Circulating Second-Generation Nanosized Emulsions To augment substantially the nanosized emulsion half-life in blood circulation, the two different approaches that are being investigated most actively so far on second-generation nanosized emulsion involve the use of either structured lipids (SL) as oil core in final nanosized emulsion or emulsion droplet surface modification using a coemulsifier with highly hydrophilic chains like polyoxyethylene (POE) and amphipathic polyethylene glycol (PEG) derivatives.

Many earlier approaches to make the long-circulating emulsion concept include changes in oil phase of the emulsion such as MCT versus LCT (Deckelbaum et al. 1990), use of SL having medium-chain (C_8–C_{10}) fatty acids (SLM) and short-chain (C_4) fatty acids (SLS) (Hedeman et al. 1996), addition of sphingomyelin (SM) (Redgrave et al. 1992, 1993; Takino et al. 1994; Arimoto et al. 1998) and cholesterol (Handa et al. 1994) to the emulsion, and use of hydrogenated castor oil (HCO) with atleast 20 oxyethylene units (HCO20) (Lin et al. 1992; Sakaeda et al. 1994; Kurihara et al. 1996a, b; Yamaguchi et al. 1995; Ueda et al. 2003).

Deckelbaum et al. (1990) explored how enzyme affinity and enzyme activity (LPL and hepatic lipase) regulate hydrolysis of phospholipid-stabilized emulsions of MCT versus LCT. It was shown that MCT-based nanosized emulsion is more readily hydrolyzed by LPL and hepatic lipase than LCT-based nanosized emulsion because of greater MCT solubility and mobility at the oil–water interface. The major factors involved are the affinity of the lipases for the interface and accessibility of individual substrate molecules to the lipases. In mixtures of LCT- and MCT-based nanosized emulsions, a higher affinity for the LCT-containing particle results in partitioning of the lipases away from the MCT-based nanosized emulsion with consequently diminished MCT hydrolysis. Hedeman et al. (1996) prepared nanosized emulsion using SL and explored its potential to prolong the *in vivo* circulation time. Takino et al. (1994) developed nanosized emulsion for lipophilic APIs with the potential for prolonged circulation in the blood or hepatic targeting. A coating with SM in the surface of the oil droplets resulted in avoidance of the RES (Takino et al. 1994). In an another study, Takino et al. (1993) demonstrated further that ^{14}C cholesteryl oleate (CO) administered with oil droplets of a conventional nanosized emulsion was rapidly taken up from the circulation by the RES cells, while those containing SM survived in the circulation for a considerably longer period. Similarly, it has further been confirmed that the presence of SM in nanosized emulsion delays the removal of emulsion particles from rat plasma (Arimoto et al. 1998). The results of Kurihara et al. (1996a) indicated that HCO-60 emulsions, when compared with conventional lecithin-stabilized emulsions, are more stable to LPL and show low uptake by RES organs, long circulations in the plasma, and high distribution in tumors. Lin et al. (1992) confirmed that HCO-60 is a good emulsifier for the preparation of nanosized emulsion with better stability and prolonged and selective delivery properties. Thus, these sterically stabilized nanosized emulsions could show potential as effective carriers for highly lipophilic antitumor agents to enhance the API delivery in tumors. This was confirmed by Sakaeda et al. (1994) who found that the rate of selective delivery of Sudan II to liver, lungs, and spleen could be suppressed by using HCO-60-based nanosized emulsion. Conversely, the use of saturated MCT in nanosized emulsion was the most effective way to increase blood concentration of Sudan II, resulting in higher distribution to liver, lungs, spleen, and brain (Sakaeda and Hirano 1995). Furthermore, an o/w-type nanosized emulsion containing HCO-60 was shown to be superior in the selective distribution of adriamycin–HCl to the liver and in decreasing concentration in heart and kidney (Yamaguchi et al. 1995). Again, Ueda et al. (2003) reported the effect of using a series of HCOs having different oxyethylene numbers such as HCO10, HCO20, HCO30, HCO60, and HCO100 on the pharmacokinetics of menatetrenone (vitamin K_2) incorporated in soybean oil (SO)-based nanosized emulsion in rats. Plasma half-life of menatetrenone after administration as the nanosized emulsion prepared by HCO with 10 oxyethylene units (SO/HCO10) was similar to that after the administration as SO/egg yolk phosphatides (SO/EYP), but was shorter than that as the nanosized emulsions prepared by HCOs with >20 oxyethylene units (SO/HCO20, SO/HCO30, SO/HCO60 and SO/HCO100). These findings clearly demonstrate that 20 oxyethylene units in HCOs are minimum requirement for the prolongation of the plasma circulation time of the incorporated API in SO/HCO-based nanosized emulsions. The above-described studies suggest the involvement of oil or SL in the enhancement of systemic circulation of the nanosized emulsion.

The long-circulating emulsion concept utilizes the process wherein the emulsion particle clearance takes place by macrophages with the involvement of opsonic plasma proteins. In order to achieve the diversion of opsonic plasma protein adsorption onto the intravenously administered o/w nanosized emulsion, an established formulation approach by which the emulsion droplet surfaces could be altered might, however, be of more realistic and even further useful for a wide array of API targeting purposes. Steric barrier or enhanced hydrophilicity effect exerted by POE chain having surfactants when added as coemulsifier into the phospholipid-stabilized first-generation emulsions allows, to some extent, the passive/inverse API targeting to the lung, kidneys, and areas of inflammation (Lee et al. 1995; Liu and Liu 1995). Addition of POE-based surfactants into the otherwise amphipathic phospholipid-stabilized emulsion is particularly effective against plasma protein adsorption onto emulsion surfaces because of the hydrophilicity and unique solution properties of POE-based surfactants, including minimal interfacial free energy with water, high aqueous solubility, high mobility, and large exclusion volume (Lee et al. 1995). In addition, colloidal particles presenting hydrophilic surfaces with a low contact angle will be almost ignored by phagocytic cells (Davis and Hansrani 1985), although emulsion particles are not supposed to be recognized as foreign by the body to some extent. Examples of POE chain containing surfactants employed so far in emulsion are Brij, poloxamer 188 (commercially named as Pluronic® F 68 or Lutrol® F 68), span 80, and Tween 80. The effectiveness of these polymeric surfactant molecules to intercalate at the oil–water interface with strong bounding to the phospholipid molecules could also be checked/judged through an *in vitro* monolayer experiment (Levy et al. 1991).

In general, the modification of particulate carriers using amphipathic PEG-containing molecules results in a prolongation of their blood circulation time (Harris et al. 2001; Bhadra et al. 2002). A phosphatidylethanolamine derivative with polyethylene glycol (PEG-PE) is widely used to increase the plasma retention of particulate carriers such as liposomes (Klibanov et al. 1990; Allen et al. 1991; Woodle and Lasic 1992), polystyrene microspheres (Dunn et al. 1994), and nanospheres (Gref et al. 1994). Therefore, similar to POE, the PEG-PE is also incorporated as a coemulsifier into emulsion (termed as pegylated emulsion) to augment its circulation half-life time (Wheeler et al. 1994). Liu and Liu (1995) studied the biodistribution of emulsion particles coated with phosphatidylethanolamine derivatives with three different molecular weight PEG (mw 1,000, 2,000, and 5,000). Among them, the PEG-2000 was able to prolong the circulation time of emulsion probably due to the increased hydrophilicity of the droplet surface and/or the formation of a steric barrier. However, Tirosh et al. (1998) assumed, while characterizing the PEG-2000-grafted liposome by differential scanning calorimetry, densitometry, and ultrasound velocity, and absorption measurements, that the steric stabilization is much more than increasing hydrophilicity. In addition, PEG-containing compounds also decrease the lipolysis of emulsion particles (Kurihara et al. 1996b) and prevent the uptake by the mononuclear phagocytes (Papahadjopoulos et al. 1991).

A dipalmitoyl phosphatidylcholine (DPPC)-stabilized emulsion was prepared by Lundberg et al. (1996) and the effect of addition of PEG-PE, polysorbate 80, or Pluronic F-68 on the metabolism of DPPC-stabilized emulsion was studied. Two different radioactive markers were used to investigate the fate of emulsion particles following injection into tail vein of female BALB/c inbred mice. While ^{14}C-triolein (TO) is susceptible to the action of LPL, ^{3}H-CO ether is not. Hence the removal of ^{14}C-TO represents the triglyceride metabolism, whereas the other one is a core marker to represent whole particle removal by RES organ uptake. The emulsions with DPPC as sole emulsifier were rapidly cleared from the blood with only 10–11% of CO or TO left in circulation after 1 h. However, addition of PEG-PE gave a prolonged clearance rate especially during the first 3 h. A further addition of cosurfactant polysorbate 80 or Pluronics F-68 resulted in a marked extension of the circulation life time during the first 6 h. The notable effects of polysorbate 80 and Pluronic F-68 can apparently be attributed mainly to the decrease in droplet size, although an additional influence due to the increased hydrophilicity may not be ruled out.

The *in vivo* disposition of emulsions administered as nutrients (surface-unmodified first-generation emulsions), as well as administered as API carriers (surface-modified second-generation emulsions), would depend on the particle properties such as the size (Connelly and Kuksis 1981; Allen and Everest 1983; Senior et al. 1985; Liu et al. 1991), zeta-potential (see details in the following third-generation emulsions section and again under medical applications Chapter 6), and compositions of phospholipids and oil phase (see details in the above paragraphs), which may vary among different products and the batches of each products. The size of particulate carriers like liposomes is known to influence both the phagocytic uptake by the mononuclear phagocyte system

(MPS) (Allen and Everest 1983; Senior et al. 1985; Liu et al. 1991) and the binding of apos to emulsions (Connelly and Kuksis 1981). Furthermore, the particle size is a major determinant of the transfer to extravascular spaces from the blood compartment. The capillaries of the vascular system can be classified into three categories: the continuous, the fenestrated, and the discontinuous (sinusoidal) ones (Bundgaard 1980). Particulate carriers including nanosized emulsions are considered to pass through capillaries and reach extravascular cells only in organs having discontinuous capillaries such as liver, spleen, and bone marrow. In such tissues, the extravasation of particles should be regulated by their size since the largest pores in the capillary endothelium is reported to be about 100 nm (Wisse 1970). In addition, tumor capillaries have unique characteristics in their structures and functions in comparison with normal tissues such as muscle (Jain 1987; Takakura and Hashida 1995), which results in the enhanced distribution of particulate carriers to tumor tissues (Gabizon et al. 1990; Wu et al. 1993; Yuan et al. 1995). The distribution of emulsions within a tumor tissue was also regulated by the size of particulate carriers (Nomura et al. 1995). Obviously, because of the submicron size range (175–400 nm in diameter) of the emulsions, the longer the time they circulate, the greater their chance of reaching respective targets. More specifically, growing solid tumors as well as regions of infection and inflammation have capillaries with increased permeability as a result of the disease process [e.g., tumor angiogenesis (Jain 1987)]. Pore diameters in these capillaries can range from 100 to 800 nm. Thus, API-containing emulsion particles are small enough to extravasate from the blood into the tumor interstitial space through these pores (Yuan et al. 1994). Normal tissues, by and large, contain capillaries with tight junctions that are impermeable to emulsions and other particles of this diameter. This differential accumulation of emulsion-laden API in tumor tissues relative to normal cells is the basis for the increased tumor specificity for the emulsion-laden API relative to free (non-emulsion) API. In addition, tumors lack lymphatic drainage and therefore there is low clearance of the extravasated emulsion from tumors. Thus, long-circulating lipid carriers, such as POE/PEG-coated nanosized emulsions, tend to accumulate in tumors as a result of increased microvascular permeability and defective lymphatic drainage, a process also referred to as the enhanced permeability and retention (EPR) effect (Gabizon et al. 2004). Figure 4.1 displays the diagrammatic representation of EPR effect. Table 4.3 lists various formulation factors affecting the metabolism as lipoproteins, the recognition by the MPS, and the elimination from the blood circulation of both second- and third-generation nanosized emulsions after parenteral administration.

4.2.2.4. Long-Circulating Second-Generation Nanosized Emulsions Conjugated with Antibodies In order to bring the colloidal carrier more closer to otherwise inaccessible pathological target tissues, homing devices such as antibodies, cell recognizing proteins, etc., are usually linked somehow

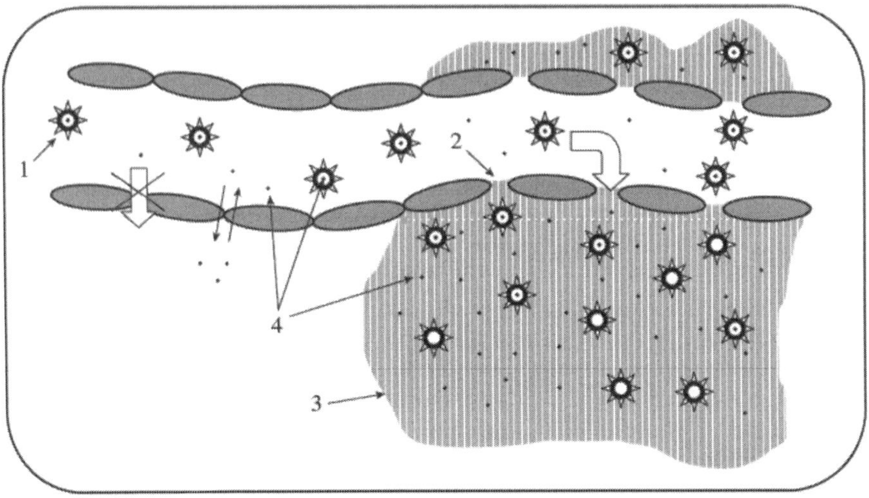

Figure 4.1. Hypothetical representation of enhanced permeability and retention (EPR) effect of multifunctional o/w nanosized emulsions. Key: Long-circulating API carriers (1) penetrate through the leaky pathological vasculature (2) into the tumor interstitium (3) and degrade there, releasing a free API (4) and creating its high local concentration.

onto the particle surfaces. Antibodies are proteins able to specifically recognize an antigen (e.g., a pathogen or a tumor agent). It has been observed over the last two decades that progression of cancer is often accompanied by the overexpression of one or several proteins, called tumor antigens. The use of monoclonal antibodies for the treatment of cancer has been suggested as a means of targeting cancer cells while sparing normal cells (Farah et al. 1998). To avoid interference of PEG chains of PEG-PE-containing nanosized emulsion with antibody localization on the surface of the colloidal carrier, the coupling of antibodies to the terminal ends of the PEG chains has attracted much attention in recent years. The methods used to link the homing devices such as antibodies onto microspheres (another API delivery system) consist of simple, direct adsorption to covalent bonds. However, the PEG-PE emulsifier-containing nanosized emulsion can usually be linked only through a conjugation process (through a covalent link) with antibodies. Such an anchorage of antibodies achieved by the conjugation process at the distal ends of PEG chain emulsifiers orienting from oil–water interface of nanosized emulsion may provide the "active" targeting of biological APIs to life-threatening diseases including various forms of malignancy (Song et al. 1996).

Decker et al. (1995) defined two fundamental properties dominating the delivery of APIs from nanosized emulsions: first, the concentration of the compound in the lipid phase of the emulsion is directly proportional to the concentration of the compound in the cell at equilibrium, and second, the rate of transfer is directly proportional to the concentration of particles in contact

TABLE 4.3. Influence of Dependent and Independent Variables of Both Second- and Third-Generation Nanosized Emulsions After Parenteral Administration on the *In Vivo* Particle Clearance Processes Like Metabolism as Lipoproteins, the Recognition by the Mononuclear Phagocyte System (MPS), and the Elimination from the Blood Circulation

In Vivo Process of	Nature of *In Vivo* Process	Independent Variables											Dependent Variable
		Cationic Lipid	Non-phospholipid Emulsifier					Oil Phase		Phospholipid Emulsifier			Particle Size
Metabolism as lipoproteins	Poor	SA/OA	Poloxamers	HCO-60	PEG-PE	Polysorbates	Solutol	LCT	—	DPPC	DSPC	SM	Large
	Extensive	—	—	—	—	—	—	MCT	—	EYPC	—	—	Small
Recognition by MPS	Poor	SA/OA	Poloxamers	HCO-60	PEG-PE	Polysorbates	Solutol	—	—	DPPC	SM	—	Small
	Extensive	—	—	—	—	—	—	—	—	DSPC	—	—	Large
Elimination from the blood	Slow	SA/OA	Poloxamers	HCO-60	PEG-PE	Polysorbates	Solutol	LCT	SLS	DPPC	SM	—	Small
	Rapid	—	—	—	—	—	—	MCT	SLM	EYPC	DSPC	—	Large

DPPC, dipalmitoylphosphatidylcholine; DSPC, distearoylphosphatidylcholine; SM, sphingomyelin; EYPC, egg yolk phosphatidylcholine; HCO-60, polyoxyethylene-(60)-hydrogenated castor oil; PEG-PE, phosphotidylethanolamine derivative with polyethylene glycol; SA, stearylamine; OA, oleylamine; LCT, long-chain triglycerides; MCT, medium-chain triglycerides; SLS, structured lipid with short-chain fatty acids, C_8-C_{10}; SLM, structured lipid with medium-chain fatty acids, C_4.

with cells. Moreover, the transfer is consistent with direct partitioning from the lipid phase of the emulsion to cells and occurs by the direct collision of emulsion particles with cells.

On the other hand, essential requirements of this “active” targeting approach include identification of recognition features (receptors) on the surface of the target, and the corresponding molecules (ligands) that can recognize the surface. Indeed, emulsions with appropriate ligands anchored on their surface must be able to access the target, bind to its receptors, and, if needed, enter it. In the case of cells, intracellular entry will possibly be carried out by receptor-mediated endocytosis or through an adsorptive pinocytosis to a minor extent. Access to the target is expected to occur easily in the intravascular space, as, for instance, is the case of populations of circulating cells, or where cells (e.g., certain malignant cells) are separated from the blood or other biological fluids by leaky capillaries. Obviously, because of the submicron size range (175–400 nm in diameter) of the nanosized emulsion, the more they circulate, the greater their chance of reaching respective targets. Thus, antibody-associated nanosized emulsions are more likely to offer a means to introduce APIs into the desired intracellular sites. Conjugation of an anti-B-cell lymphoma monoclonal antibody, LL2, to long-circulating API-carrier nanosized emulsions has been reported by Lundberg et al. (1999).

Furthermore, in order to bring the colloidal carrier more closer to otherwise inaccessible pathological target tissues, homing devices/ligands such as antibodies, cell recognizing proteins, etc., are usually linked somehow onto the particle surfaces. Various methods have been employed to couple ligands to the surface of the nanosized lipidic and polymeric carriers with reactive groups. These can be divided into covalent and non-covalent couplings. Non-covalent binding by simple physical association of targeting ligands to the nanocarrier surface has the advantage of eliminating the use of rigorous, destructive reaction agents. Common covalent coupling methods involve formation of a disulfide bond, cross-linking between two primary amines, reaction between a carboxylic acid group and primary amine, reaction between maleimide and thiol, reaction between hydrazide and aldehyde, and reaction between a primary amine and free aldehyde (Nobs et al. 2004). For antibody-conjugated second-generation anionic emulsions, the reaction of carboxyl derivative of coemulsifier molecule with free amine groups of the antibody, and disulfide bond formation between coemulsifier derivative and reduced antibody were the two reported conjugation techniques so far (Song et al. 1996; Lundberg et al. 1999, 2004). However, by the formation of thio-ether bond between the free maleimide reactive group already localized at the o/w interface of the emulsion oil droplets and a reduced monoclonal antibody, the antibody-tethered cationic emulsion was developed for active targeting to tumor cells (Goldstein et al. 2005). It should be added that the cationic emulsion investigated for the tumor targeting purpose belongs to the third-generation emulsion category and it will also be described in the following section of this chapter.

Benita and colleagues have prepared anti-ferritin mAb (AMB8LK) conjugated cationic nanoemulsions (Goldstein et al. 2005). Neither the physicochemical properties of the prepared immunoemulsions (average droplet size of 120 nm with a zeta potential of +50 mV) nor the immunoreactivity of the mAb was altered during the emulsion preparation. In the subsequent study from the same research group, trastuzumab conjugated cationic nanoemulsion was prepared and the emulsion was found to be attached significantly to the breast cancer cell line SKBR-3 overexpressing the HER2 receptor (Goldstein et al. 2007). Here too, the emulsion stability at room temperature conditions and other physicochemical properties such as size, zeta potential, pH, and API content were followed up during all the period of the animal experimentation and no marked changes were noticed.

For the treatment of highly advanced and refractory cancers occurring at breast and ovarian regions of the body, Cremophor EL-based paclitaxel i.v. formulation (Taxol) was initially developed. However, the presence of Cremophor EL (as a solubility enhancer) leads to the problems of safety and tolerance. Therefore, reformulation strategy could be applied in order to improve the safety and efficacy of paclitaxel-based anticancer therapy. Since, paclitaxel is a hydrophobic compound and has poor aqueous solubility, Lundberg et al. (2003) incorporated the API into the dispersed oil phase of an emulsion. Unfortunately, the paclitaxel separated from the emulsion system after i.v. injection and mixed with the blood serum because the API possesses $\log P$ value of 4.7. Due to this API partitioning effect from emulsion to blood serum following the i.v. injection of API-loaded emulsion, there was no difference in pharmacokinetic parameters of paclitaxel-loaded emulsion and the commercial paclitaxel formulation (Lee et al. 2005). If an API needs to remain inside the dispersed oil droplets of the emulsion following i.v. administration, then the $\log P$ value of the selected API should be >7 (Takino et al. 1994). If not so, upon i.v. administration, the emulsion formulation will dilute infinitely and then release the API quickly in the blood serum. This results in losing the advantage of long blood circulation and possibility of organ-passive targeting. Therefore, it becomes necessary to enhance the lipophilicity of the paclitaxel molecules by esterification reaction with a fatty acid as shown by Lundberg et al. (2003). These authors incorporated the esterified paclitaxel molecules (paclitaxel oleate) into a negatively charged (anionic) emulsion. By following the similar strategy, Goldstein et al. 2007 have produced paclitaxel palmitate (calculated $\log P$ value of 9) and entrapped the synthesized esterified paclitaxel into the positively charged (cationic) emulsion before antibody conjugation. Furthermore, the complement system activation following the i.v. administration of anti-HER2 immunoemulsions loaded with paclitaxel palmitate was also studied in order to show that the antiangiogenic mAb will diminish the resistance of the cancer cells and should enhance the cell uptake of paclitaxel palmitate (Goldstein et al. 2007).

4.2.2.5. RES-Related Disease Management by Second-Generation Nanosized Emulsions Apart from non-RES-related disease treatment through target-specific ligand conjugation, the second-generation emulsions may also be useful for RES-related disease treatment. Certain lipoprotein or polysaccharide moiety inclusion into the emulsions would help to achieve this concept. In general, uptake of small colloidal API carriers by the phagocytotic mononuclear cells of RES in the liver can be exploited to improve the treatment of parasitic, fungal, viral, and bacterial diseases such as, for example, leishmaniasis, AIDS, and hepatitis B. The approach to use emulsions as an API carrier against microbial diseases is superior to free antimicrobial agents both in terms of distribution to the relevant intracellular sites and in treating disseminated disease states effectively. As already discussed, conventional emulsion particles are capable of localizing in liver and spleen, where many pathogenic microorganisms reside.

Rensen et al. (1995) demonstrated the active/selective liver targeting of an antiviral prodrug (nucleoside analogue and iododeoxyuridine) incorporated in an emulsion complexed with ligands like recombinant apo E using Wistar rat as animal model because its apo E-receptor system is comparable to that of humans (Mahley 1988). Whereas the parent API did not show any affinity for emulsion due to hydrophilic property, derivatization with hydrophobic anchors allowed not only incorporating at least 130 prodrug molecules per emulsion particle but also without imparting any effect on the emulsion structure and apo E association to emulsion droplets. The authors did not describe on where do the 130 prodrug molecules reside in the emulsion and what is the emulsion/medium partition coefficient of the prodrug. The prodrug molecules might have reasonably higher solubility in the oil or o/w interface of the emulsion possibly due to a high partition coefficient value. Plausibly, this high partition value for prodrug molecules will determine the kinetic parameter k_{off} (desorption rate of an emulsion component from the assembly) as suggested by Barenholz and Cohen (1995) for liposomal technology. Furthermore, without being bound by theory, the apo E component helps to disguise the emulsion particles so that the body does not immediately recognize it as foreign, but may allow the body to perceive it as native chylomicrons or very low density lipoproteins (VLDL). The small size and the approximately spherical shape allow the emulsion particles to exhibit similar physicochemical properties to native chylomicrons or VLDL (hydrolyzed by LPL) whereas the incorporated prodrug remained associated with the emulsion remnant particles following injection into the blood circulation of the rat (Rensen et al. 1995). Because the carrier particles are not recognized as foreign, the systemic circulation of the API increases, thus increasing the likelihood of API delivery to the target tissues (up to 700 nM API concentration in liver parenchymal cells). Additionally, the clearance rate of the API decreases, thereby reducing the likelihood of toxic effects of the API on clearance tissues since accumulation of the API in other part of the clearance tissues is reduced. Thus, specific organs may be targeted by using nanosized emulsion particles as described above, due to target cells comprising high levels of specific receptors, for example, but not limited to apo E receptors.

To address this issue, the saccharide moieties of glycolipids and glycoproteins on the cell surface are considered to play an important role in various intercellular recognition processes. For instance, Iwamoto et al. (1991) investigated the influence of coating of the oil droplets in emulsion with cell-specific cholesterol bearing polysaccharide, such as mannan, amylopectin, or pullulan on the target ability of those formulations. They have observed a higher accumulation of mannan-coated emulsion in the lung in guinea pigs. Thus, selective API targeting through emulsion-bearing ligands would not only lead to an improved API effectiveness and a reduction in adverse reactions but also offers possibility of applying highly potent APIs. Hence, the composition of emulsion plays an important role concerning intercellular cell recognition processes and indeed, cell recognizability is also being improved by incorporation of apoproteins or galactoproteins onto the emulsion particles to enhance their specificity (Grolier et al. 1992).

Overall, although second-generation emulsions are usually used as a means of administering aqueous insoluble APIs by dissolution of the APIs within the oil phase of the emulsion, employing surface modification/pegylation by the attachment of targeting ligands (apo E, polysaccharide, and antibody) onto the droplet surface of emulsions may be useful in both passive and active API targeting purposes. Thus, receptor-mediated API targeting using ligands-attached emulsions seems to hold a promising future to the achievement of cell-specific delivery of multiple classes of therapeutic cargoes and this approach will certainly make a major contribution in treating many life-threatening diseases with a minimum of systemic side effects.

4.2.3. Third-Generation Nanosized Emulsions

To increase cellular uptake, cationization strategy is one of the change that progressively occurred in last decades particularly on the surfaces of nonviral, colloidal carrier systems such as liposomes, nano- and micro-particulates, and nanocapsules (Barratt 2003). For making the surface of these lipidic and polymeric carrier systems a cationic property, some cationic lipids/polymers are usually added into these systems during/after preparation. But, adding alone the cationic substances in phospholipid-stabilized first-generation emulsion does not help to obtain a physically stabilized emulsion on a prolonged storage period. However, using different cationic lipids as emulsifier and additional helper lipids as coemulsifier, for example, DOTAP, 1,2-dioleoyl-*sn*-glycero-3-phosphoethanolamine (DOPE), and 1-palmitoyl-2-oleoyl-*sn*-glycero-3-phosphoethanolamine-*N*-[poly(ethylene glycol)$_{2000}$] (PEG$_{2000}$PE), reports are available to prepare emulsions with positive charges on their droplets' surface (Kim et al. 2001a, 2001b). Alternatively or on the other hand, after the inclusion of cation-forming substances like lipids (stearyl or oleyl chain having primary amines) (Elbaz et al. 1993; Rabinovich-Guilatt et al. 2004), polymers (chitosan) (Calvo et al. 1997; Jumaa and Müller 1999), and surfactants (cetyltrimethylammonium bromide) (Samama et al. 1987) during the preparation of second-generation emulsion allows the formation of stabilized system with positive charges over on it. Further, the positive charge caused by stearylamine was also confirmed by a selective adsorption of thiocyanate. Its adsorption was correlated with increasing stearylamine concentration (Elbaz et al. 1993). So, nanosized emulsion consisting of complex emulsifiers, that is, phospholipid-polyoxyethylene surfactant-cationized primary amine or polymer combination can conveniently be termed as third-generation emulsions.

4.2.3.1. Third-Generation Nanosized Emulsions to Deliver Genetic Materials (Genes) The extemporaneous addition of the solid API or API previously solubilized in another solvent or oil to the preformed first- and second-generation emulsions is not a favored approach technology-wise as it might compromise the integrity of the emulsion. However, since therapeutic DNA and single-stranded oligos or siRNA are water soluble due to their polyanionic character, the aqueous solution of these compounds need to be added directly to the preformed third-generation emulsion in order to interact electrostatically with the cationic emulsion droplets and thus associate/link superficially at the oil–water interface of the emulsion (Teixeira et al. 1999; Choi et al. 2002). During *in vivo* condition when administered via parenteral, nasal, and ocular routes, the release of the DNA and oligos from the associated emulsion droplet surfaces should therefore initially be dependent solely on the affinity between the physiological anions of the biological fluid and cationic surface of the emulsion droplets. For instance, the mono- and di-valent anions containing biological fluid available in the parenteral route is plasma and in the ocular topical route is tear fluid, aqueous humor, and vitreous. Moreover, these biofluids contain multitude of macromolecules and nucleases. There is a possibility that endogenous negatively charged biofluid's components could dissociate the DNA and oligos from cationic emulsion. It is noteworthy to conduct, during the preformulation development stages, an *in vitro* release study for therapeutic DNA and oligos-containing emulsion in these biological fluids and this type of study could be considered as an indicator for the strength of the interaction that occurred between DNA or oligo and the emulsion particles. However, it is interesting to see what could happen when the third-generation emulsions are applied to *in vitro* cell culture models in the presence of serum. The serum stability of emulsion/DNA complex was reported (Yi et al. 2000; Kim et al. 2003). Further interesting investigations using third-generation emulsions in gene delivery purposes are briefly summarized in a review article (Tamilvanan 2004).

4.2.3.2. Third-Generation Nanosized Emulsions and Their Distinctive/ Typical Feature Extracting some beneficial property from the excipients added in the formulations is an attractive way to enhance API targeting efficacy of colloidal carriers like nanospheres and liposomes. Prolongation in plasma residence time through an escaping tendency from RES uptake and penetration of therapeutic agents into cell surface possibly via endocytotic mechanism are the two main facts that are needed for parenterally administered colloidal carrier systems. Conferring positive-charge (cationization) onto the surface of colloidal carrier systems will enhance the endocytosis of therapeutic agents into cell. Similarly, making hydrophilic surface (PEGylation) onto the surface of colloidal carriers will allow the particles to escape from RES uptake and thus prolong the longevity of the injected particles in the blood plasma. Therefore, the longer is the time the API-loaded particles circulate in the blood plasma, the better will be the chance of particles to reach ultimately the target site in the human body. When talking about nanospheres and liposomes, the cationization and pegylation were performed as a separate entity. That is, both nanospheres and liposomes did not possess these two facts together at a single time because of their structural conformation. However, the third-generation nanosized emulsions reported so far have differed significantly in a way that they do hold the combination of cationic and pegylatic surface properties onto them (Floyd 1999).

It has been reported in ocular pharmacokinetic study of cyclosporin A incorporated in deoxycholic acid-based anionic and stearylamine-based cationic emulsions in rabbit that when compared to anionic emulsion, the cationic emulsion showed a significant API reservoir effect of more than 8h in corneal and conjunctival tissues of the rabbit eye following topical application (Abdulrazik et al. 2001). Since cornea and conjunctiva are anionic nature at physiological pH (Rojanasakul and Robinson 1989), the cationic emulsion would interact with these tissues electrostatically to implicate the observed cyclosporin A reservoir effect. This hypothesis is supported, in principle, by an *ex vivo* study which showed that cationic emulsion carrier exhibited better wettability properties on rabbit cornea than either saline or anionic emulsion carrier (Klang et al. 2000).

Studies (Wretlind 1981; Davis 1982) have shown that small changes in physical properties of emulsions can influence the elimination rate of these formulations from the blood. Indeed, an organ distribution study of stearylamine-based cationic or deoxycholic acid-based anionic nanosized emulsions and Intralipid®, a well-known commercial anionic emulsion, containing ^{14}C–CO was carried out following injection into the tail vein of male BALB/c mice (20–26g) at a volume of 5 ml kg^{-1} (Klang et al. 1998; Yang and Benita 2000). Since cholesteryl oleate (CO) is one of the most lipophilic compounds used in biopharmacy and is not prone to degradation in the body (which remains within particles even after lipolysis of emulsion), its *in vivo* behavior can be regarded as reflecting that of the injected nanosized emulsion in the early phase (Takino et al. 1994, 1998).

Following intravenous administration of the various emulsions having ^{14}C–CO to BALB/c mice, the ^{14}C–CO was found to accumulate in organs such as lung and liver. Furthermore, it was observed that the concentration of ^{14}C–CO in the lung decreased, but was again elevated over time for both the developed cationic and anionic emulsion formulations, with a concomitant decrease in the concentration of the radiolabeled compound in the liver. However, within the various emulsion distribution patterns observed in liver, a lower ^{14}C–CO concentration was observed for stearylamine-based cationic emulsion when compared to Intralipid®, while for deoxycholic acid-based anionic emulsion the observed concentration of ^{14}C–CO was relatively very low when compared to cationic emulsion and Intralipid®. In addition, in comparison to both of the anionic emulsions, the stearylamine-based cationic emulsion elucidated a much longer retention time of ^{14}C–CO in the plasma, indicating clearly a long-circulating half-life for cationic emulsion in the blood. Thus, the cationic nanosized emulsion can be considered a stealth® long-circulating emulsion.

The above two studies described clearly the distinctive/typical characteristics of third-generation emulsions in enhancing ocular API bioavailability and on the other hand, the same emulsion has the property to circulate longer time in blood following parenteral administration. Excess positive charge at the oil–water interface in conjunction with the projection of highly hydrophilic POE chain (due to the presence of poloxamer 188) toward aqueous phase of the o/w-type nanosized emulsion is the main reason behind the emulsion to attain its unique property, which is absent in first- and second-generation emulsions. Still, however a better understanding of the structure of the third-generation emulsions in terms of forces involved in its formation and stabilization must ultimately be obtained in an effort to provide a clearly understood physical basis for uniqueness in its biological efficacy following parenteral and ocular administrations.

It should be added that the use of stearylamine in intravenous administrated emulsion might be problematic. Stearylamine is a single-chain amphiphile having relatively high critical micellar concentration although its concentration used in the studied emulsion is much higher than the critical micellar concentration. Therefore, due to the dilution in plasma as well as plasma lipoproteins and blood cells, there is a high probability that the emulsion will lose its stearylamine almost instantaneously. To substantiate indirectly this issue, Klang et al. (1994) showed the lack of potential-induced toxicity of stearylamine-based cationic emulsion in animal models *in vivo* and Korner et al. (1994) investigated the surface properties of mixed phospholipid-stearylamine monolayers and their interaction with a nonionic surfactant (poloxamer) *in vitro*. Despite the presence of the stearylamine, which may be suspected of being irritant in the pure form, in the emulsifier combination, the hourly instillation of stearylamine-based cationic emulsion vehicle into rabbit eye was well tolerated without any evidence of any toxic or inflammatory response to the ocular surface during the 5 days of the study (40 single-drop instillation between 8 am and 4 pm in each days) (Klang et al. 1994). Following 0.2, 0.4, and 0.6 ml single bolus injections of the same emulsion vehicle, representing a huge single-administered dose of 30 ml kg^{-1}, no animal deaths was noted over a period of 30 days apparently indicating the absence of marked acute toxicity (Klang et al. 1994). Furthermore, the same stearylamine-based cationic emulsion vehicle did not cause acute neurotoxicity in rats when a continuous intravenous infusion (3.3 ml) for 2 h at a rate of 27.4 μl min^{-1} was administered through jugular vein (Klang et al. 1994). A long-term sub-chronic toxicity examination in rabbit eye (healthy) following thrice-daily one single-drop topical instillation of the stearylamine-based emulsion elicited an almost similar nonirritating effect to eye tissues in comparison to the thrice-daily one single-drop topical instillation of the normal saline-treated control rabbit eyes (Unpublished data). Thus, overall results clearly indicated that the stearylamine was strongly bound at a molecular level to the mixed interfacial film formed by Lipoid E 80 and poloxamer 188 at the oil–water interface system (Korner et al. 1994). Such an intercalation between the emulsifiers is responsible for emulsified oil droplet stability and in fact, prevented the stearylamine from leaking and exerting any intrinsic possible local or systemic adverse effects in model animals. However, stearylamine is a primary amine and is very reactive toward other excipients and APIs. Moreover, stearylamine is not described in any pharmacopoeias and therefore it was not a reasonable choice for pharmaceutical products development. Oleylamine is another cationic lipid that has been used to manufacture ophthalmic emulsions (Rabinovich-Guilatt et al. 2004), but this lipid also has stability concerns due to its primary amine function and the presence of an unsaturated site in the aliphatic chain.

Repurposing of already regulatorily approved excipients of other purpose are now utilized as cation charge-conferring molecules in designing the o/w cationic nanosized emulsions. Quaternary ammoniums usually used as preservatives have surfactant properties and the potential to give a cationic charge to the nanosized emulsions. These agents include cetrimide, benzalkonium chloride (BAK), benzethonium chloride, benzododecinium bromide, and cetylpyridinium. As preservatives, these products protect against infectious contaminants by electrostatically binding to the negatively charged surface of bacteria and mycoplasma and disrupting their cell membranes. The disadvantage of quaternary ammoniums is that their effect on cell membranes is not limited only to microorganisms but they are also capable of injuring epithelial cells lining the ocular surface by the same mechanism of action. It was consequently not obvious to foresee these molecules as cationic agents, therefore, quaternary ammoniums were not initially considered for use in emulsions. In 2002, Sznitowska revealed findings that the preservative efficacy of this class of surfactants was diminished or neutralized in the presence of emulsions (Sznitowska et al. 2002). Part of the quaternary ammonium is bound to the emulsion, resulting in the presence of less free surfactant molecules in the aqueous phase to exert their antimicrobial action, and, consequently, their toxic effect on the ocular surface epithelia. This physicochemical property to make a new type of cationic nanovector using BAK and cetalkonium chloride (CKC) as cationic agents. CKC is a highly lipophilic ($\log P = 9.5$) component of BAK.

It is hence mostly included in the oily phase providing a higher zeta potential on surface of the oil droplets while leaving relatively no free molecules to induce ocular surface toxicity. BAK (and CKC as a component of BAK) has been routinely used as a preservative in other marketed eye drop solutions (e.g., BAK is used in Xalatan, Pfizer Canada ULC) and is accepted as compliant with regulatory requirements for ophthalmic products. These excipients used in lower concentrations as cationic agents in emulsions have been demonstrated to be safe for the eye. More importantly, the use of BAK and CKC as cationic surfactants only in emulsions are now protected by several granted and pending European and US patents [e.g., EP1655021 (Bague et al. 2008), EP1809237 (Bague et al. 2007a), EP1809238 (Philips et al. 2008), and EP1827373 (Bague et al. 2007b), which are granted].

4.2.4. Fourth-Generation Nanosized Emulsions

In the modern health-care system, the progression of any disease or infection usually starts with initial diagnosis, getting prescription, buying medicines, and intaking the medicine into human body. With the lone exception of intaking the medicines into human body, all of the preceding steps require the involvement/spending of money from the patient pocket. If any one of the preceding steps will combine with another step, then the possibility of cutting-down the medical expenses becomes reality. For instance, diagnosis and pharmaceutical formulations containing the medicines could at least be combined together. Recently, convergence of diagnosis, API delivery, and real-time noninvasive therapeutic monitoring took place (within the big pharma companies) that leads to creation of the word "image-guided drug delivery (IGDD) or theragnosis" for the implementation of these three distinct pursuits simultaneously (Accardo et al. 2013).

4.2.4.1. Imaging Basically the imaging may be of two types: anatomical/physiological imaging and biomedical imaging. Whereas the anatomical/physiological imaging helps to visualize the internal structures of various organs of human body, the biomedical imaging plays a vital role to explore the essential information regarding management of several diseases. Specifically, for management of cancer, the imaging techniques help to derive estimation of morphological characters, structure, and metabolic and functional information about the site to be examined. The purpose of the detection techniques are to make the therapy effective and to reduce the mortality rate by grabbing the exact cause along with origin of the disease like cancer (Fass 2008). The steps involved in cancer management includes: prediction, screening, staging,

TABLE 4.4. Steps Involved for Cancer Management Along with the Research Reported in Each Steps

S. No.	Stages of Cancer Management	References
1	Prediction	De Torres et al. (2007)
2	Screening	Lehman et al. (2007), Paajanen (2006), Sarkeala et al. (2008)
3	Biopsy guidance for detection	Nelson et al. (2007)
4	Staging	Kent et al. (2004), Brink et al. (2004), Shim et al. (2005)
5	Prognosis	Lee et al. (2004)
6	Therapy planning	Ferme et al. (2005), Ciernik et al. (2003)
7	Therapy guidance	Ashamalla et al. (2005)
8	Therapy response	Neves and Brindle (2006), Stroobants et al. (2003), Aboagye et al. (1998), Brindle (2008)
9	Recurrence	Keidar et al. (2004)
10	Palliation	Belfiore et al. (2004), Tam and Ahra (2007)

prognosis, therapy planning, and response. Table 4.4 shows the various steps involved for cancer management along with the research reported in each steps. The importance of diagnostic approaches in API development is highlighted by the growing global cancer diagnostics market, which is expected to reach an estimated value of USD 168.6 billion by the year 2020 [Transparency Market Research. Cancer diagnostics market (tumor biomarker tests, imaging, endoscopy and biopsy)-global industry analysis, size, share, growth, trends and forecast, 2014–2020. Available from: http://www.transparencymarketresearch.com/cancer-diagnostics-market.html, accessed on July 20, 2015].

The currently available noninvasive biomedical technologies for macroscopically visualizing the tumors comprise of both electromagnetic and optical-based imaging techniques such as X-ray computed tomography (CT) scans, magnetic resonance imaging (MRI) scans, positron emission tomography (PET) scans, single photon emission CT, ultrasound scans, and optical imaging. It is unfortunate to specify that all the above-said imaging techniques have limitations like sensing of the unusual growth of tissue in a nomothetic manner, that is, microscopic. In order to achieve information at the molecular level, many biomarkers have also been associated with the imaging systems. These biological agents can selectively target the tumors beside amplification of imaging signals; it also helps to identify the stage and aggressiveness of the tumor (Mishra and Verma 2010). Understanding of the imaging and response requires spatial mapping at the level of gene expression, which makes the techniques not only complicated but it also leads to extraction of extraordinary information regarding stages of the disease (Smith et al. 2003). Many of the advanced imaging techniques are based upon the interaction of the electromagnetic waves, ultrasonic waves, microwaves, X-rays, and other radiation either with the body tissues or fluid-surrounding cells (Table 4.5). Sometimes, the hybrid of the techniques are also been employed to collect all the desired information about cell proliferation, gene expression, angiogenesis, and vascular infunctionalities. The selection of techniques for imaging depends upon the sensitivity, specificity, targeting, and availability of the imaging agent (Fass 2008).

As the use of diagnostic imaging techniques became widespread in clinical trials, a set of standardized imaging assessment criteria from the World Health Organization were established (Miller et al. 1981). In the year 2000, a modified set of criteria called the Response Evaluation Criteria In Solid Tumors

TABLE 4.5. Imaging Techniques for Diagnosis of Cancer

Techniques	Principle	Applications	References
Ultrasound elastography	Based upon less elastic cancer tissues	Diagnosis of breast cancer, prostate cancer, and liver fibrosis	Hui et al. (2007), Lerner et al. (1990), Miyanaga et al. (2006), Pallwein et al. (2007), Tsutsumi et al. (2007)
Endoscopic ultrasound elastography	Electromagnetic radiation	Imaging of lymph nodes, pancreatic masses, adrenal and submucosal tumors	Saftoiu and Vilman (2006)
Near infrared spectroscopy	Non-ionizing electromagnetic radiation	Breast imaging	Poplack et al. (2004)
Electrical impedance spectroscopy and tomography	Non-ionizing electromagnetic radiation	Breast imaging	Tromberg et al. (2000)
Microwave imaging spectroscopy	Non-ionizing electromagnetic radiation	Breast imaging	Pogue et al. (2001)
Photoacoustic and thermoacoustic imaging	Non-ionizing electromagnetic radiation	Breast imaging	Franceschini et al. (1997), Grosenick et al. (1999)
Clinical optical imaging	Fluorescence and bioluminescence-based imaging	Cancer imaging	He et al. (2007)
Magnetic resonance imaging (MRI)	Magnetic radiations (superior to mammography and ultrasound)	Cancer detection, breast cancer	Ross et al. (1982), Bhattacharyya et al. (2008)

TABLE 4.5. *Continued*

Techniques	Principle	Applications	References
Contrast-enhanced MRI	Gadolinium as contrasting agent (darkening of image in highly perfused tissues indicative of tumor)	Evaluation of microcalcifications before biopsy and evaluation of neo-angiogenesis	Takayoshi et al. (2007), Folkman (1992)
Diffusion weighted imaging (DWI)	Image contrast due to difference in the movement of water molecules within the tissues	Detection of cytotoxic oedema in stroke, Identification of tumor and metastases. Tumor evaluation in abdomen and pelvis and characterization of benign and malignant breast lesions	Le Bihan et al. (1985)
DWI MRI	Image contrast due to difference in the movement of water molecules within the tissues and magnetic radiations	Detection and evaluation of the pancreatic, bladder carcinomas, prostate cancer, lung carcinomas, and brain tumor	Liapi et al. (2008), Matsuki et al. (2007), Matoba et al. (2007), Provenzale et al. (2006)
Proton magnetic resonance spectroscopy	MR spectroscopic imaging	Localization of tumors and guiding biopsies in breast, brain and prostate, ductal carcinomas	Kurhanewicz et al. (2000), Yeung et al. (2002)
High intensity focused ultrasound (HIFU)	Acoustic wave-induced hyperthermia	Treatment of localized cancer	Medel et al. (2012)
Positron-emitting radioisotopes	Radio-isotope imaging	Used for imaging of bones, proteins, and tracing cancer	Langstrom et al. (2007)

(RECIST 1.0) was introduced as part of collaborative efforts between the European Organization for Research and Treatment of Cancer, the National Cancer Institute in the USA, and the National Cancer Institute of Canada Clinical Trials Group (Therasse et al. 2000).

RECIST refined an objective set of criteria that defined when tumor lesions in cancer patients improve (partial or complete response), remain unchanged (stable disease), or worsen (progressive disease) during treatment. Since its introduction, RECIST has been updated (RECIST 1.1) to introduce standards for the assessment of lymph nodes, redefine "measurable" lesions and assessment of disease progression, as well as establish recommendations for standardized image acquisition (Eisenhauer et al. 2009).

Today, a large number of oncology clinical trials employ RECIST to objectively assess cancer treatment response in solid tumors. Advances in imaging technologies and our understanding of disease have resulted in additional consortia guidelines for standardizing diagnostic imaging in oncology clinical trials. Most notably, the Cheson criteria (1999, 2007, and 2014) have established guidelines for the use of diagnostic imaging using CT, MRI, and fluorodeoxyglucose (FDG)-PET as well as clinical findings for the assessment of lymphoma patients (Cheson et al. 2006).

In addition, the RANO criteria have been established for gliomas, and a number of other criteria have been introduced to specifically assess hepatocellular carcinoma, acute myeloid leukemia, prostate cancer, and the effects of immunotherapies on tumor responses (Cheson et al. 2003; Scher et al. 2008; Wolchok et al. 2009; Lencioni and Llovet 2010; Van den Bent et al. 2011).

As these criteria have evolved, it has become clearer that conventional anatomical imaging techniques, although very useful, have not supplied the entire objective assessments needed to make accurate early phase go/no go decisions based on clinical trial results by US-FDA. Initiatives by the Radiological Society of North America, including the Quantitative Imaging Biomarker Alliance, to advance volumetric assessments of tumor lesions continue to gain momentum, and researchers are showing increased interest in developing tools for the evaluation of metrics derived from CT and MRI studies. As an example, techniques such as dual-energy CT and spectral CT imaging are being used to better differentiate and characterize certain cancers. These types of image analysis in conjunction with efforts to assess the relationship of CT and MRI to the molecular biology of various tumors are helping to shape the new fields of radiomics and radiogenomics. Although these approaches hold great potential for oncology clinical trials, it is likely to be several years or more before they can be implemented in a clinical environment.

On the other hand, nanometer-sized semiconductor particles such as quantum dots (QDs) with unique photochemical and photophysical properties are of considerable interest in many research areas, because QDs can be covalently linked to peptides, antibodies, nucleic acids, or small molecule ligands as fluorescent probes (Bruchez et al. 1998; Chan and Nie 1998; Bailey et al. 2004; Medintz et al. 2005; Michalet et al. 2005; Iga et al. 2007). Moreover, fluorescent QDs have higher levels of brightness and photostability compared with conventional organic fluorophores, and are well suited for optical encoding and multiplexing applications due to their broad excitation profiles and narrow/symmetric emission spectra. In addition, the new generations of QDs have far-reaching potential for the study of intracellular processes at the single molecule level, high-resolution cellular imaging, long-term *in vivo* observation of cell trafficking, tumor targeting, and diagnostics (Alivisatos 1996, 2004). Despite all the above-said advantages, the direct use of QDs for API delivery remains questionable due to their potential long-term toxicity (Iga et al. 2007). However, if the QD core could be encapsulated with other API delivery vehicles like lipid- and polymer-based nanoparticle systems, then its inherent toxicity should significantly be reduced.

4.2.4.2. Inculcation of Theragnostic Concept Using Multifunctional O/W Nanosized Emulsions To make a complete structure for nanomedicine-based theragnostic approaches, the selection of appropriate API delivery carrier is very important. In this regard, the potential of long-circulating and multifunctional o/w nanosized emulsions on the atherosclerosis (AS) and cancer theragnostic procedures is being discussed one by one in this chapter.

The o/w nanosized emulsions possess an ideal prototype platform to resolve the challenge of IGDD or theragnosis owing to their complex self-assembled nature. Even, the emulsion components are capable of attracting multiple responsive functionalities. Figure 4.2 shows the schematic structure of the assembly of the multifunctional nanosized emulsions. In the scheme, the possibilities of developing magnetic nanocarrier (3) and contrast nanocarrier for imaging purposes (5), cell-penetrating nanocarrier for intracellular API delivery purposes (6), and DNA-carrying nanocarrier such as lipoplex for correcting genetically determined diseases (7) are not investigated yet using nanosized emulsions. But these four innovative avenues, if successful, should revolutionize the medical field due to their selective improvement in imaging and therapeutic efficacies. The IGDD or theragnosis concept lies in the Scheme 5 of multifunctional nanosized emulsions.

4.2.4.3. IGDD or Theragnostic Concept in Breast Cancer Treatment Cancer is primarily characterized by the uncontrolled proliferation of cells and their ability to metastasize. Despite significant advancements in the fight against cancer, it remains a challenging medical problem, particularly in the lungs, breast, liver, prostate, pancreas, and brain. To date, systemic chemotherapy is a common approach to the chronic management of cancer in patients. However, systemic toxicity is a major drawback, limiting the utility and effectiveness of chemotherapeutics. Recent research efforts in the development of API delivery systems have concentrated on targeted delivery and controlled release of the API or other agents in the tumor in order to increase the therapeutic ratio. Oncologic IGDD is a therapeutic method where tumor localization and API delivery are guided and monitored through noninvasive imaging. The goal in IGDD is to optimize local delivery of the therapeutic pharmaceutical to the

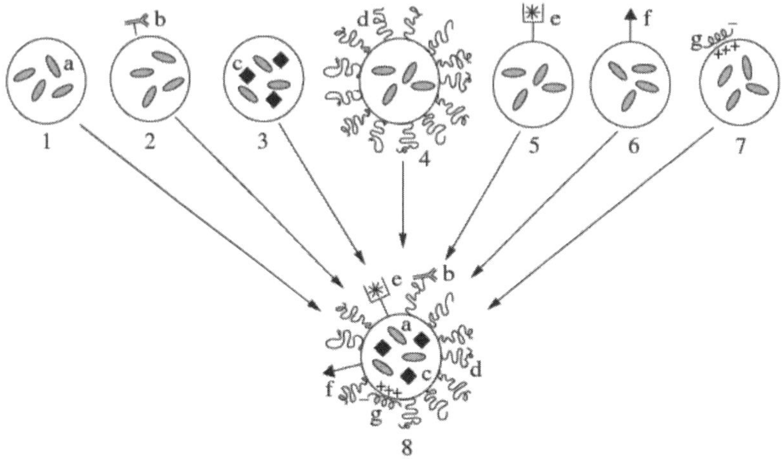

Figure 4.2. Schematic structure of the assembly of the multifunctional nanosized emulsions. Key: (1) Traditional "plain" nanocarrier (a: API loaded into the carrier); (2) targeted nanocarrier or immunocarrier (b: specific targeting ligands, usually monoclonal antibody, attached to the carrier surface); (3) magnetic nanocarrier (c: magnetic particles loaded into the carrier together with the API and allowing for the carrier sensitivity toward the external magnetic field and its use as a contrast agent for magnetic resonance imaging); (4) long-circulating nanocarrier [d: surface-attached protecting polymer (usually PEG) allowing for prolonged circulation of the nanocarrier in the blood]; (5) contrast nanocarrier for imaging purposes (e: heavy metal atom—^{111}In, ^{99m}Tc, Gd, Mn-loaded onto the nanocarrier via the carrier-incorporated chelating moiety for gamma- or MR imaging application); (6) cell-penetrating nanocarrier (f: cell-penetrating peptide, CPP, attached to the carrier surface and allowing for the carrier-enhanced uptake by the cells); (7) DNA-carrying nanocarrier such as lipoplex (g: DNA complexed by the carrier via the carrier surface positive charge); (8) hypothetical multifunctional pharmaceutical nanocarrier combining the properties of the carriers # 1–7.

target tissue and provide microanatomical and functional imaging feedback on the therapeutic process(es), including during treatment and monitoring. This research proposal shows a full implementation of IGDD for APIs that can be imaged or identified in the body as they enter the blood stream, are localized at the target, and are then released or otherwise activated to provide focal treatment.

Breast cancer classification is in constant evolution as advances in DNA and RNA microarrays as well as immunohistochemistry (IHC)-based staining that allow the researchers to define the molecular heterogeneity of different disease subtypes and to guide the selection of appropriate treatment (Minami et al. 2011). Triple negative breast cancer (TNBC) is a highly diverse group of cancers that is generally considered as an aggressive disease associated with both poor disease-free survival (DFS) and overall survival (OS) rates, higher risk of early recurrence, and lack of prognostic markers or therapeutic targets in contrast with hormonal receptor-positive and human epidermal growth factor receptor 2-positive (HER2+) breast cancers. Although TNBC accounts for only 7–20% of breast cancer, it causes a disproportionate number of deaths, particularly among young, black, and Hispanic women, and those with *BRCA1* and *BRCA2* (BReast CAncer *genes* 1 and 2) mutations (Brower 2009). Moreover, recent gene expression microarray studies have revealed at least six subtypes based on common molecular features, most frequently basal-like and claudin-low (Perou 2010; Yue et al. 2016). The basal-like group is composed almost entirely of the so-called "triple negative" cancers, characterized by the lack of any estrogen (ER) and progesterone receptor (PgR) immunoreactivity and of HER2/neu overexpression. Positive markers of this group of tumors are basal-cell cytokeratins including cytokeratins (CK) 5/6, 14, and 17, which are normally found in the basal layer of stratified epithelia. Whether identified by the expression of basal IHC markers or by a basal-like RNA expression profile, these tumors represent about 15% of all breast cancers and are characterized by an adverse clinical course, with an increased likelihood of disease recurrence and death. There is currently no specific targeted treatment for patients with triple-negative breast cancers, due to the lack of data on which to base treatment selection. But some TNBC patients, for example, luminal androgen receptor or molecular apocrine cancers, may have better prognosis than the remaining majority (Bidard et al. 2007). Therefore, the selection of gene expression subtypes for a single cancer is challenging in the clinics (Chen et al. 2012). The most widely accepted method in clinical practice is through identification of significant IHC surrogates for basal-like breast cancer, for example, epidermal growth factor receptor (EGFR, also known as HER1) and basal cytokeratins (CK5/6). However, prognostic stratification relying on individual biomarkers is prone to lack of standardization in patient risk management due to the heterogeneity in the staining and the absence of defined cutoffs of IHC-based surrogates. Viale et al. (2009) and Zhang et al. (2014) reported that EGFR immunoreactivity correlated significantly with worse prognosis in their TNBC patients. Thike et al. (2010) reported that basal cytokeratins had significant prognostic values in their cohort of patients. In addition, proliferation biomarker Ki-67 (a protein antigen and also known as MKI67) is also reported to be significantly associated with a high histologic grade and poor survival in TNBC patients (Keam et al. 2011). High Ki-67 is associated with a higher histologic grade, larger tumor size, presence of axillary lymph node metastasis, and worse outcome. Since TNBCs typically exhibit high tumor grade and high proliferation rate, the expression of Ki-67 is usually high in most TNBCs. In instances if the expression of these biomarkers is not in concert with each other, then it is challenging to determine appropriate treatment based on an individual biomarker. Taken together, all these studies indicate that designing a targeted API delivery system is one of the most viable options to manage TNBC.

In fact, the development of tumor-homing multifunctional nanosized emulsion for breast cancer theragnosis will allow a new IGDD to monitor online the tumor location, tumor targeting levels, intratumoral localization, and API release kinetics prior and during radio- and/or chemotherapeutic treatment. Furthermore, if an API delivery carrier possesses both a diagnostic/imaging agent and a API for targeting to a particular organ of the human body, then such delivery carrier could eliminate the unnecessary treatment of patients for whom therapy is not appropriate, resulting in significant API cost savings for the healthcare system. In simple words, there is a hybrid carrier system that combines nanotechnology and molecular imaging together for simultaneously facilitating the cancer theragnostics.

4.2.4.4. IGDD or Theragnostic Concept in AS It is a chronic and degenerative disease of the large artery walls. It is a leading cause of mortality and morbidity (Libby et al. 2011) and is the single-most important cause of cardiovascular disease (CVD), a predominant health problem worldwide (Murray and Lopez 1997). AS is characterized by dysfunctioning of endothelial layer of cells, which activates the inflammatory mediators followed by cell proliferation and plaque development. As the time passes, the plaque tends to get expanded, then ruptures which in turn causes thrombosis and infarction of tissues and organs (Gimbrone 1995; Lamberts et al. 1997; Libby et al. 1997; Markowitz 1997; Alexander and Dzau 2000; Baralle and Baralle 2000). The molecular pathways for disease progression depict many causes behind its development. Oxidative stress, infections and inflammation, apoptosis, effect of sex hormones, intake of toxins or mutagens, and DNA instability are likely to have a profound effect on development and aggravation in the state of the disease. Clinical manifestations of AS include myocardial infarction, heart failure, stroke, and peripheral artery disease, all of which resulting in irreversible organ damage (Werner and Nickenig 2006). The incidence rate of atherosclerotic disease has been on the rise due to dietary habits that originated in the West, insufficient exercise brought about by persistent motorization, increased mental stress from social downturn, and the rapid progression of aging in society (Fuster et al. 2011). As the disease AS has a long asymptomatic phase and the first manifestation of the disease may be sudden cardiac death, it is essential to find effective strategies to prevent it (Napoli et al. 2006). Current guidelines for the prevention of atherosclerotic diseases focus on treatment of established cardiovascular risk factors to attenuate the subsequent endothelial dysfunction (ED) and damage (Mensah et al. 2007). Because ED is an early event in AS and plays a pivotal role in the atherogenesis process (Bonetti et al. 2003).

As outlined in the introduction of fourth-generation nanosized emulsions, the imaging part of the theragnostic nanomedicine/approach may conventionally be classified into two types. The first type of imaging includes *ex vivo* assessment of diseased tissues postmortem and noninvasive imaging primarily of the structural and anatomic features of atherosclerotic plaques. This preliminary and preclinical imaging helps the formulator to identify the best nanoscale formulation platform for investigating and applying the theragnostic principle. For instance, the utility of near infrared (NIR) protease-sensing, ProSense® 750 fluorescent preclinical imaging agent in combination with FMT® (Fluorescence Molecular Tomography) 2,500 quantitative preclinical imaging system for the noninvasive quantitative measurement of atherosclerotic disease biology and its response to therapy in apo E-deficient mice *in vivo* was studied by Peterson (2012). Significant progression of disease was detected in apo E-deficient animal even after ingestion of high cholesterol diet over 20–30 weeks. However, treatment by using atorvastatin has shown an anticipated modest decrease in plasma cholesterol level. This study indicates the progression of AS in conjunction with high fatty food intake might be monitored by FMT and NIR in apo E-deficient mice model of AS.

The second-type imaging explores how to use dual-modal imaging in a single nanoscale delivery system for diagnosing the life-threatening diseases. For instance, macrophages have become widely recognized as a key target for AS imaging since they contribute significantly to the progression of AS. Dual-modal imaging contrast agents with unique X-ray CT and optical imaging capabilities have great potential in disease diagnosis because of complementary combination of the high spatial resolution of CT with the high sensitivity of optical imaging. In order to combine different imaging modalities like MRI, X-ray, CT, and optical imaging into one single nanomaterial, Ding et al. (2013) have developed a multifunctional and dual-modal nanoemulsion platform to investigate atherosclerotic plaques. These authors have developed a kind of QD-iodinated oil nanoemulsion having 80 nm droplet size as a CT/fluorescence dual-modal contrast agent. They have selected hydrophobic QDs and embedded them in iodinated oil phase, which subsequently dispersed in water phase to form the oil-in-water nanoemulsion. The PEGylated lipids were used as one of the mixed emulsifier molecule in order to improve the circulation time of nanoemulsion. The longer the circulation time of the nanoscale delivery system allows more of the nanoemulsion droplet to accommodate in macrophages of the atherosclerotic plaques for providing a better contrast effect. In this way, the developed nanoemulsion-based QD encapsulated hybrid formulation could make an efficient optical/CT contrast enhancement for cell imaging. Further studies have also showed that the nanoemulsion has great potential of targeting and visualizing atherosclerotic plaques with uptake *in vivo* by macrophages. Therefore, this nanoemulsion holds a great potential for future applications.

When analyzing these two different research works, the work by Peterson (2012) clearly indicated that the developed delivery system did contain an anti-atherosclerotic API molecule in the formulation while the work by Ding et al. (2013) did not incorporate any anti-atherosclerotic API molecule in their hybrid nanoscale formulation. However, both of these research works have contributed much on the future direction of managing the AS disease progression, monitoring, and treating of the same by theragnostic principle.

When compared to the polymer-based nanoparticle platform, which is considered to have a more rigid structural conformation, the more flexible structural components of lipid-based nanoscale formulation like o/w nanosized emulsions should be used for the development of theragnostic nanoemulsion to manage atherosclerotic plaques. It should be mentioned that Ding et al. (2013) have already made a strong foundation and in fact, this foundation should give an impetus to work with nanoemulsion containing the good combination of contrasting agents. One more step forward is just including the anti-atherosclerotic agent into the nanoemulsion system. Although currently many research laboratories are already working on this direction, fruitful results are yet to come for the effective management of atherosclerotic plaques.

4.3. PREPARATION METHODS FOR API-FREE/LOADED O/W NANOSIZED EMULSIONS

For getting better idea of how one would formulate the nanosized emulsion delivery systems suitable for parenteral, ocular, percutaneous, and nasal uses, there are more detailed description on methods of nanosized emulsion preparation (Benita and Levy 1993). A hot-stage high-pressure homogenization technique or combined emulsification technique (*de novo* production) is frequently employed in order to prepare nanosized emulsions with desired stability even after subjection into autoclave sterilization. Therefore, the various steps involved in this technique in making blank anionic and cationic emulsions were arranged in the following order:

1. Weigh the oil and water-soluble ingredients in separate beakers.
2. Heat the both oil and water phases separately to 70°C.
3. Add the oil phase into the water phase and continue the heating up to 80°C with constant stirring to form a coarse emulsion.
4. High shear mixing to make a fine emulsion.
5. Cooling the fine emulsion formed in ice bath.
6. Homogenize the fine emulsion.
7. Cooling the homogenized emulsion in ice bath.
8. Filter the emulsion using a 0.5 μm size membrane filter.
9. Adjust the emulsion to 7 using 0.1 N hydrochloric acid or 0.1 N sodium hydroxide solution.
10. Passing of nitrogen gas into the vials containing the emulsion.
11. Sterilization of the emulsion using an autoclave.

The traditional droplet size-reducing steps involved during preparation include constant mild stirring using a magnetic stirrer when initially mixing of

TABLE 4.6. Typical Formula to Make Oil-in-Water (O/W) Anionic and Cationic Nanosized Emulsions

Oil Phase	Water Phase
Natural/semisynthetic oils	Poloxamer 188
Phospholipid mixture	Glycerol
Stearylamine/oleylamine[a]	Double distilled water
Deoxycholic acid/oleic acid[b]	
α-Tocopherol (Vit. E)	

a and b are necessary ingredients for cationic and anionic emulsions, respectively.

oil and water phases, rapid Polytron mixing at a high speed, and the final droplet-size homogenization using a two-stage homogenizer valve assembly. The initial heating is vital for the effective solubilization of the respective oil and water phase's components in their corresponding phases. Mixing of two phases with constant mild stirring and subsequent raising of temperature to 85°C are needed to form an initial coarse emulsion and to localize the surfactant molecules for a better adsorption at the oil–water interface, respectively. A typical formula to make anionic and cationic nanosized emulsions is given in Table 4.6.

There are three different approaches to incorporate lipophilic APIs into the oil phase or at the o/w interface of the nanosized emulsions, namely, extemporaneous API addition, *de novo* emulsion preparation, and interfacial incorporation approach, which includes recently developed SolEmul® technology (Tamilvanan and Benita 2004). In principle, the lipophilic API molecules should however be incorporated by a *de novo* process. Thus, the API is initially solubilized or dispersed together with an emulsifier in suitable single oil or oil mixtures by means of slight heating. The water phase containing the osmotic agent with or without an additional emulsifier is also heated and mixed with the oil phase by means of high-speed mixers. Further homogenization takes place to obtain the needed small droplet size range of the emulsion. A terminal sterilization by filtration, steam, or autoclave then follows. The emulsion thus formed contains most of the API molecules within its oil phase. This is a generally accepted and standard method to prepare lipophilic API-loaded nanosized emulsions for parenteral, ocular, percutaneous, and nasal uses, as illustrated in Figure 4.3. This process is normally carried out under aseptic conditions and nitrogen atmosphere to prevent both contamination and potential oxidation of sensitive excipients.

4.4. CONCLUSION

Among the so far developed generations of emulsions, the third-generation nanosized emulsions are commercially exploited to administer via ocular, oral, and parenteral routes for managing/treating different infections. Some of the other functionalities like diagnostic agent-carrier function together with

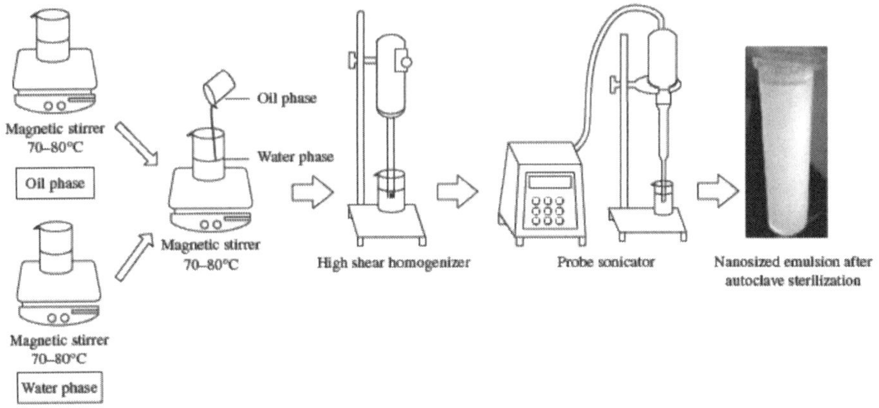

Figure 4.3. Oil-in-water nanosized emulsion preparation by *de novo* method. (See color insert.)

contrasting functionality, and even target-specific imaging functionality receive much attention in different multicentric studies in recent years. On the other hand, theranostic approach by the parenterally administered o/w nanosized emulsions especially mainly for AS and cancer is an interesting avenue for the interdisciplinary research at laboratories across the world. In the similar fashion, personalized approach is also progressing in the direction for detecting the atherosclerotic plaque at coronary arteries by using fourth-generation multifunctional nanosized emulsions. Considering the o/w nanosized emulsions as basic nutrient, lipid API carrier, stealth carrier, and cationic carriers were the research of previous decades. But, extending the utility of nanosized emulsions as multifunctional carrier property such as the IGDD is—and will be—the research for potential application to detect, manage, and monitor the life-threatening diseases like AS, cancer, etc.

REFERENCES

Abdulrazik, M., Tamilvanan, S., Khoury, K. et al. (2001), Ocular delivery of cyclosporin A II. Effect of submicron emulsion's surface charge on ocular distribution of topical cyclosporin A, *STP Pharm. Sci.*, 11, 427–432.

Aboagye, E.O., Bhujwalla, Z.M., Shungu, D.C. et al. (1998), Detection of tumour response to chemotherapy by ^{1}H nuclear magnetic resonance spectroscopy: effect of 5-fluorouracil on lactate levels in radiation-induced fibrosarcoma I tumours, *Cancer Res.*, 58 (5), 1063–1067.

Accardo, A., Tesauro, D., and Morelli, G. (2013), Peptide-based targeting strategies for simultaneous imaging and therapy with nanovectors, *Polym. J.*, 45, 481–493. doi:10.1038/pj.2012.215

Alexander, R.W. and Dzau, V.J. (2000), Vascular biology: the past 50 years, *Circulation*, 102 (20), 112–116. doi:10.1161/01.cir.102.suppl_4.iv-112

Alivisatos, A.P. (1996), Perspectives on the physical chemistry of semiconductor nanocrystals, *J. Phys. Chem.*, 100, 13226–13239. doi:10.1021/jp9535506

Alivisatos, A.P. (2004), The use of nanocrystals in biological detection, *Nat. Biotechnol.*, 72, 47–52. doi:10.1038/nbt927

Allen, T.M. and Everest, J.M. (1983), Effect of liposome size and drug release properties on pharmacokinetics of encapsulated drug in rats, *J. Pharmacol. Exp. Therap.*, 226 (2), 539–544.

Allen, T.M., Hansen, C., Martin, F. et al. (1991), Liposomes containing synthetic lipid derivatives of poly(ethylene glycol) show prolonged circulation half-lives in vivo, *Biochim. Biophys. Acta*, 1066, 29–36. doi:10.1016/0005-2736(91)90246-5

Allison, A.C. and Byars, N.E. (1986), An adjuvant formulation that selectively elicits the formation of antibodies of protective isotypes and of cell-mediated immunity, *J. Immunol. Methods*, 95, 157–168. doi:10.1016/0022-1759(86)90402-3

Amselem, S. and Friedman, D. (1998), Submicron emulsions as drug carriers for topical administration, in: Benita, S., Ed., *Submicron Emulsions in Drug Targeting and Delivery*, Harwood Academic Publishers, Amsterdam, The Netherlands, pp. 153–173.

Arimoto, I., Matsumoto, C., Tanaka, M. et al. (1998), Surface composition regulates clearance from plasma and triolein lipolysis of lipid emulsions, *Lipids*, 33, 773–779. doi:10.1007/s11745-998-0269-8

Ashamalla, H., Rafla, S., Parikh, K. et al. (2005), The contribution of integrated PET/CT to the evolving definition of treatment volumes in radiation treatment planning in lung cancer, *Int. J. Radiat. Oncol.*, 63, 1016–1023. doi:10.1016/j.ijrobp.2005.04.021

Bague, S., Philips, B., Garrigue, J.S. et al. (2008), Oil-in-water type emulsion with low concentration of cationic agent and positive zeta potential, EP1655021, Novagali Pharma, Evry, France.

Bague, S., Philips, B., Rabinovich-Guilatt, L. et al. (2007a), Ophthalmic oil-in-water type emulsion with stable positive zeta potential, US patent 20070248645, EP1809237, Novagali Pharma, Evry, France.

Bague, S., Philips, B., Rabinovich-Guilatt, L. et al. (2007b), Ophthalmic emulsions containing prostaglandins, EP1827373, Novagali Pharma, Evry, France.

Bailey, R.E., Smith, A.M., and Nie, S.M. (2004), Quantum dots in biology and medicine, *Physica E*, 25, 1–12. doi:10.1016/j.physe.2004.07.013

Baralle, M. and Baralle, F.E. (2000), Genetics and molecular biology, *Curr. Opin. Lipidol.*, 11, 653–656.

Barenholz, Y. and Cohen, R. (1995), Rational design of amphiphile-based drug carriers and sterically stabilized carriers, *J. Liposome Res.*, 5, 905–932. doi:10.3109/08982109509012690

Barratt, G. (2003), Colloidal drug carriers: achievements and perspectives, *Cell. Mol. Life Sci.*, 60, 21–37. doi:10.1007/s000180300002

Belfiore, G., Tedeschi, E., Moggio, G. et al. (2004), CT-guided radiofrequency palliative ablation of unresectable lung cancer: clinical and imaging findings at one year follow-up, *Radiology*, 233, 513.

Benita, S. and Levy, M.Y. (1993), Submicron emulsions as colloidal drug carriers for intravenous administration: comprehensive physicochemical characterization, *J. Pharm. Sci.*, 82, 1069–1079. doi:10.1002/jps.2600821102

Bhadra, D., Bhadra, S., Jain, P. et al. (2002), Pegnology: a review of PEG-ylated systems, *Pharmazie*, 57, 5–29.

Bhattacharyya, M., Ryan, D., Carpenter, R. et al. (2008), Using MRI to plan breast conserving surgery following neoadjuvant chemotherapy for early breast cancer, *Br. J. Cancer*, 98, 289–293. doi:10.1038/sj.bjc.6604171

Bidard, F.C., Conforti, R., Boulet, T. et al. (2007), Does triple-negative phenotype accurately identify basal-like tumour? An immunohistochemical analysis based on 143 "triple-negative" breast cancers, *Ann. Oncol.*, 18 (7), 1285–1286. doi:10.1093/annonc/mdm360

Bonetti, P.O., Lerman, L.O., and Lerman, A. (2003), Endothelial dysfunction: a marker of atherosclerotic risk, *Arterioscler. Thromb. Vasc. Biol.*, 23, 168–175. doi:10.1161/01.atv.0000051384.43104.fc

Brindle, K. (2008), New approaches for imaging tumour responses to treatment, *Nat. Rev. Cancer*, 8, 94–107.

Brink, I., Schumacher, T., Mix, M. et al. (2004), Impact of [^{18}F] FDG-PET on the primary staging of small-cell lung cancer, *Eur. J. Nucl. Med. Mol. Imaging*, 31, 1614–1620. doi:10.1007/s00259-004-1606-x

Brower, V. (2009), Search for new treatments intensifies for triple-negative breast cancer, *J. Natl. Cancer Inst.*, 101 (22), 1536–1537. doi:10.1093/jnci/djp422

Bruchez, M., Moronne, M., Gin, P. et al. (1998), Semiconductor nanocrystals as fluorescent biological labels, *Science*, 281, 2013–2016. doi:10.1126/science.281.5385.2013

Bundgaard, M. (1980), Transport pathways in capillaries-In search of pores, *Annu. Rev. Physiol.*, 42, 325–336. doi:10.1146/annurev.ph.42.030180.001545

Calder, P.C. (2009), Hot topics in parenteral nutrition. Rationale for using new lipid emulsions in parenteral nutrition and a review of the trials performed in adults, *Proc. Nutr. Soc.*, 68, 252–260. doi:10.1017/S0029665109001268

Calvo, P., Remuñá-López, C., Vila-Jato, J.L. et al. (1997), Development of positively charged colloidal drug carriers: chitosan-coated polyester nanocapsules and submicro-emulsions, *Colloid Polym. Sci.*, 275, 46–53. doi:10.1007/s003960050050

Chan, W.C.W. and Nie, S.M. (1998), Quantum dot bioconjugates for ultrasensitive nonisotopic detection, *Science*, 281, 2016–2018. doi:10.1126/science.281.5385.2016

Chandra, R.K. (1985), Grace A. Goldsmith Award lecture. Trace element regulation of immunity and infection, *J. Am. Coll. Nutr.*, 4 (1), 5–16. doi:10.1080/07315724.1985.10720062

Chen, X., Li, J., Gray, W.H. et al. (2012), TNBCtype: a subtyping tool for triple negative breast cancer, *Cancer Inform.*, 11, 147–156. doi:10.4137/CIN.S9983

Cheson, B.D., Bennett, J.M., Kopecky, K.J. et al. (2003), Revised recommendations of the international working group for diagnosis, standardization of response criteria, treatment outcomes, and reporting standards for therapeutic trials in acute myeloid leukemia, *J. Clin. Oncol.*, 21, 4642–4649. doi:10.1200/JCO.2003.04.036

Cheson, B.D., Greenberg, P.L., Bennett, J.M. et al. (2006), Clinical application and proposal for modification of the International Working Group (IWG) response criteria in myelodysplasia, *Blood*, 108, 419–425. doi:10.1182/blood-2005-10-4149

Choi, B.Y., Chung, J.W., Park, J.H. et al. (2002), Gene delivery to the rat liver using cationic lipid emulsion/DNA complex: comparison between intra-arterial, intraportal and intravenous administration, *Korean J. Radiol.*, 3, 194–198. doi:10.3348%2Fkjr.2002.3.3.194

Ciernik, I.F., Dizendorf, E., Baumert, B.G. et al. (2003), Radiation treatment planning with an integrated positron emission and computer tomography (PET/CT): a feasibility study, *Int. J. Radiat. Oncol. Biol. Phys.*, 57(3), 853–863. doi:10.1016/s0360-3016(03)00346-8

Cockshott, I.D. (1985), Propofol ("Diprivan") pharmacokinetics and metabolism-an overview, *Postgrad. Med. J.*, 61 (Suppl. 3), 45–50.

Connelly, P.W. and Kuksis, A. (1981), Effect of core composition and particle size of lipid emulsions on apolipoprotein transfer of plasma lipoproteins in vivo, *Biochim. Biophys. Acta*, 666, 80–89. doi:10.1016/0005-2760(81)90093-x

Craig, D.Q.M., Patel, M., and Ashford, M. (2000), Administration of emulsions to the gastrointestinal tract, in: Nielloud, F. and Marti-Mestres, G., Eds., *Pharmaceutical Emulsions and Suspensions*, Marcel Dekker, New York, pp. 323–360.

Davis, S.S. (1982), Emulsions systems for the delivery of drugs by the parenteral route, in: Bundgaard, H., Bagger Hansen, A., and Kofod, H., Eds., *Optimization of Drug Delivery*, Munksgaard, Copenhagen, pp. 333–346.

Davis, S.S. and Hansrani, P. (1985), The influence of emulsifying agents on the phagocytosis of lipid emulsions by macrophages, *Int. J. Pharm.*, 23, 69–77. doi:10.1016/0378-5173(85)90223-6

Davis, S. S., Illum, L., Washington, C. et al. (1992), Studies on the interaction of charge-reversed emulsions with the reticuloendothelial system, *Int. J. Pharm.*, 82, 99–105. doi:10.1016/0378-5173(92)90078-G

Deckelbaum, R.J., Hamilton, J.A., Moser, A. et al. (1990), Medium-chain versus long-chain triacylglycerol emulsion hydrolysis by lipoprotein lipase and hepatic lipase: implications for the mechanisms of lipase action, *Biochemistry*, 29, 1136–1142. doi:10.1021/bi00457a006

Decker, D.E., Vroegop, S.M., Goodman, T.G. et al. (1995), Kinetics and thermodynamics of emulsion delivery of lipophilic antioxidants to cells in culture, *Chem. Phys. Lipids*, 76, 7–25. doi:10.1016/0009-3084(94)02429-9

de Torres, J.P., Bastarrika, G., Wisnivesky, J.P. et al. (2007), Assessing the relationship between lung cancer risk and emphysema detected on low-dose CT of the chest, *Chest*, 132, 1932–1938. doi:10.1378/chest.07-1490

Ding, J., Wang, Y., Ma, M. et al. (2013), CT/fluorescence dual-modal nanoemulsion platform for investigating atherosclerotic plaques, *Biomaterials*, 34 (1), 209–216. doi:10.1016/j.biomaterials.2012.09.025

Dunn, S.E., Brindley, A., Davis, S.S. et al. (1994), Polystyrene-poly (ethylene glycol) (PS-PEG2000) particles as model systems for site specific drug delivery. 2. The effect of PEG surface density on the in vitro cell interaction and in vivo biodistribution, *Pharm. Res.*, 11 (7), 1016–1022. doi:10.1023/a:1018939521589

Eisenhauer, E.A., Therasse, P., Bogaerts, J. et al. (2009), New response evaluation criteria in solid tumours: revised RECIST guideline (version 1.1), *Eur. J. Cancer*, 45, 228–247. doi:10.1016/j.ejca.2008.10.026

Elbaz, E., Zeevi, A., Klang, S. et al. (1993), Positively charged submicron emulsion—a new type of colloidal drug carrier, *Int. J. Pharm.*, 96 (1–3), R1–R6. doi:10.1016/0378-5173(93)90237-A

Farah, R. A., Clinchy, B., Herrera, L. et al. (1998), The development of monoclonal antibodies for the therapy of cancer, *Crit. Rev. Eukaryot. Gene Exp.*, 8 (3–4), 321–356. doi:10.1615/critreveukargeneexpr.v8.i3-4.50

Fass, L. (2008), Imaging and cancer: a review, *Mol. Oncol.*, 2, 115–152. doi:10.1016/j.molonc.2008.04.001

Ferme, C., Vanel, D., Ribrag, V. et al. (2005), Role of imaging to choose treatment, *Cancer Imaging*, 5, 113–119. doi:10.1102/1470-7330.2005.0032

Floyd, A.G. (1999), Top ten considerations in the development of parenteral emulsions, *Pharm. Sci. Technol. Today*, 4 (2), 134–143. doi:10.1016/s1461-5347(99)00141-8

Folkman, J. (1992), The role of angiogenesis in tumor growth, *Semin. Cancer Biol.*, 3 (2), 65–71.

Franceschini, M.A., Moesta, K.T., Fantini, S. et al. (1997), Frequency-domain techniques enhance optical mammography: initial clinical results, *Proc. Natl Acad. Sci. USA*, 94 (12), 6468–6473. doi:10.1073%2Fpnas.94.12.6468

Fukushima, S., Kishimoto, S., Takeuchi, Y. et al. (2000), Preparation and evaluation of o/w type emulsions containing antitumor prostaglandin, *Adv. Drug Deliv. Rev.*, 45 (1), 65–75. doi:10.1016/s0169-409x(00)00101-0

Fuster, V., Kelly, B.B., Vedanthan, R. et al. (2011), Promoting global cardiovascular health: moving forward, *Circulation*, 123 (15), 1671–1678. doi:10.1161/CIRCULATIONAHA.110.009522

Gabizon, A., Price, D.C., Huberty, J. et al. (1990), Effect of liposome composition and other factors on the targeting of liposomes to experimental tumors: biodistribution and imaging studies, *Cancer Res.*, 50, 6371–6378.

Gabizon, A., Shmeeda, H., Horowitz, A.T. et al. (2004), Tumor cell targeting of liposome-entrapped drugs with phospholipid-anchored folic acid-PEG conjugates, *Adv. Drug Deliv. Rev.*, 56 (8), 1177–1192. doi:10.1016/j.addr.2004.01.011

Gimbrone, M.A. (1995), Vascular endothelium: an integrator of pathophysiologic stimuli in atherosclerosis, *Am. J. Cardiol.*, 75, 67B–70B. doi:10.1016/0002-9149(95)80016-l

Goldstein, D., Gofrit, O., Nyska, A. et al. (2007), Anti-HER2 cationic immunoemulsion as a potential targeted drug delivery system for the treatment of prostate cancer, *Cancer Res.*, 67 (1), 269–275. doi:10.1158/0008-5472.CAN-06-2731

Goldstein, D., Nassar, T., Lambert, G. et al. (2005), The design and evaluation of a novel targeted drug delivery system using cationic emulsion-antibody conjugates, *J. Control. Release*, 108 (2–3), 418–432. doi:10.1016/j.jconrel.2005.08.021

Gref, R., Minamitake, Y., Peracchia, M.T. et al. (1994), Biodegradable long-circulating polymeric nanospheres, *Science*, 263 (5153), 1600–1603. doi:10.1126/science.8128245

Grolier, P., Azais-Braesco, V., Zelmire, L. et al. (1992), Incorporation of carotenoids in aqueous systems: uptake by cultured rat hepatocytes, *Biochim. Biophys. Acta*, 1111, 135–138. doi:10.1016/0005-2736(92)90283-R

Grosenick, D., Wabnitz, H., Rinneberg, H.H. et al. (1999), Development of a time-domain optical mammograph and first in vivo appliations, *Appl. Optics*, 38 (13), 2927–2943. doi:10.1364/ao.38.002927

Handa, T., Eguchi, Y., and Miyajima, K. (1994), Effects of cholesterol and cholesteryl oleate on lipolysis and liver uptake of triglyceride/phosphatidylcholine emulsions in rats, *Pharm. Res.*, 11 (9), 1283–1287. doi:10.1023/A:1018990327466

Harnisch, S. and Müller, R.H. (1998), Plasma protein adsorption patterns on emulsions for parenteral administration: establishment of a protocol for two-dimensional polyacrylamide electrophoresis, *Electrophoresis*, 19 (2), 349–354. doi:10.1002/elps.1150190233

Harnisch, S. and Müller, R.H. (2000), Adsorption kinetics of plasma proteins on oil-in-water emulsions for parenteral nutrition, *Eur. J. Pharm. Biopharm.*, 49 (1), 41–46. doi:10.1016/S0939-6411(99)00064-8

Harris, J.M., Martin, N.E., and Modi, M. (2001), Pegylation: a novel process for modifying pharmacokinetics, *Clin. Pharmacokinet.*, 40 (7), 539–551. doi:10.2165/00003088-200140070-00005

Hauss, D.J. (2002), Lipid-based systems for oral drug delivery: enhancing the bioavailability of poorly water soluble drugs, *Am. Pharm. Rev.*, 5, 22–36.

He, W., Wang, H.F., Hartmann, L.C. et al. (2007), In vivo quantitation of rare circulating tumor cells by multiphoton intravital flow cytometry, *Proc. Natl. Acad. Sci. USA*, 104, 11760–11765. doi:10.1073/pnas.0703875104

Hedeman, H., Brøndsted, H., Müllertz, A. et al. (1996), Fat emulsions based on structured lipids (1,3-specific triglycerides): an investigation of the in vivo fate, *Pharm. Res.*, 13 (5), 725–728. doi:10.1023/a:1016095415849

Horrocks, L.A. and Yoe, Y.K. (1999), Health benefits of docosahexaenoic acid (DHA), *Pharmacol. Res.*, 40 (3), 211–225. doi:10.1006/phrs.1999.0495

Hui, Z., Bing, O., Bao, M. et al. (2007), Comparison of ultrasound elastography, mammography, and sonography in the diagnosis of solid breast lesions, *J. Nucl. Med.*, 26 (6), 807–815. doi:10.7863/jum.2007.26.6.807

Iga, A.M., Robertson, J.H.P., Winslet, M.C. et al. (2007), Clinical potential of quantum dots, *J. Biomed. Biotechnol.*, 2007 (10), 76087. doi:10.1155/2007/76087

Illum, L., West, P., Washington, C. et al. (1989), The effect of stabilising agents on the organ distribution of lipid emulsions, *Int. J. Pharm.*, 54, 41–49. doi:10.1016/0378-5173(89)90163-4

Iwamoto, K., Kato, T., Kawahara, M. et al. (1991), Polysaccharide-coated oil droplets in oil-in-water emulsions as targetable carriers for lipophilic drugs, *J. Pharm. Sci.*, 80 (3), 219–224. doi:10.1002/jps.2600800306

Jain, R.K. (1987), Transport of molecules across tumor vasculature, *Cancer Metastasis Rev.*, 6 (4), 559–593. doi:10.1007/BF00047468

Jumaa, M. and Müller, B.W. (1999), Physicochemical properties of chitosan-lipid emulsions and their stability during the autoclaving process, *Int. J. Pharm.*, 183 (2), 175–184. doi:10.1016/s0378-5173(99)00086-1

Keam, B., Im, S.-A., Lee, K.-H. et al. (2011), Ki-67 can be used for further classification of triple negative breast cancer into two subtypes with different response and prognosis, *Breast Cancer Res.*, 13 (2), R22. doi:10.1186/bcr2834.

Keidar, Z., Haim, N., Guralnik, L. et al. (2004), PET/CT using FDG in suspected lung cancer recurrence: diagnostic value and impact on patient management, *J. Nucl. Med.*, 45 (10), 1640–1646.

Kent, M.S., Port, J.L., and Altorki, N.K. (2004), Current state of imaging for lung cancer staging, *Thorac. Surg. Clin.*, 14, 1–13. doi:10.1016/S1547-4127(04)00031-3

Kim, T.W., Chung, H., Kwon, I.C. et al. (2001a), Optimization of lipid composition in cationic emulsion as in vitro and in vivo transfection agents, *Pharm. Res.*, 18, 54–60. doi:10.1023/A:1011074610100

Kim, Y.J., Kim, T.W., Chung, H. et al. (2001b), Counterion effects on transfection activity of cationic lipid emulsion, *Biotechnol. Bioprocess Eng.*, 6 (4), 279–283. doi:10.1007/BF02931990

Kim, Y.J., Kim, T.W., Chung, H. et al. (2003), The effects of serum on the stability and the transfection activity of the cationic lipid emulsion with various oils, *Int. J. Pharm.*, 252 (1–2), 241–252. doi:10.1016/S0378-5173(02)00676-2

Klang, S., Abdulrazik, M., and Benita, S. (2000), Influence of emulsion droplet surface charge on indomethacin ocular tissue distribution. *Pharm. Dev. Technol.*, 5, 521–532. doi:10.1081/PDT-100102035

Klang, S.H, Frucht-Pery, J., Hoffman, A. et al. (1994), Physicochemical characterization and acute toxicity evaluation of a positively-charged submicron emulsion vehicle, *J. Pharm. Pharmacol.*, 46, 986–993. doi:10.1111/j.2042-7158.1994.tb03254.x

Klang, S.H., Parnas, M., and Benita, S. (1998), Emulsions as drug carriers-possibilities, limitations, and future perspectives, in: Muller, R.H., Benita, S., and Bohm, H.L.,

Eds., *Emulsions and Nanosuspensions for the Formulation of Poorly Soluble Drugs*, Medpharm, Stuttgart, pp. 31–65.

Klibanov, A.L., Maruyama, K., Torchilin, V.P. et al. (1990), Amphipathic polyethyleneglycols effectively prolong the circulation time of liposomes, *FEBS Lett.*, 268, 235–237. doi:10.1016/0014-5793(90)81016-h

Koch, T. and Heller, A.R. (2005), Auswirkungen einer parenteralen ernahrung mit n-3-fettsauren auf das therapieergebnis- eine multizentrische analyse bei 661 patienten (Effects of parenteral nutrition with n-3-fatty acids on the result of therapy- a multicentre analysis with 661 patients), *Akt Ernahrungs*, 30, 15–22.

Korner, D., Benita, S., Albrecht, G. et al. (1994), Surface properties of mixed phospholipid–stearylamine monolayers and their interaction with a non-ionic surfactant (poloxamer), *Colloids Surf. B Biointerfaces*, 3 (1–2), 101–109. doi:10.1016/0927-7765(93)01111-4

Koster, V.S., Kuks, P.F.M., Langer, R. et al. (1996), Particle size in parenteral fat emulsions, what are the true limitations? *Int. J. Pharm.*, 134, (1–2), 235–238. doi:10.1016/0378-5173(95)04409-4

Kurhanewicz, J., Vigneron, D.B., and Nelson, S.J. (2000), Three dimensional magnetic resonance spectroscopic imaging of brain and prostate cancer, *Neoplasia*, 2 (1–2), 166–189. doi:10.1038/sj.neo.7900081

Kurihara, A., Shibayama, Y., Mizota, A. et al. (1996a), Pharmacokinetics of highly lipophilic antitumor agent palmitoyl rhizoxin incorporated in lipid emulsions in rats, *Biol. Pharm. Bull.*, 19 (2), 252–258. doi:10.1248/bpb.19.252

Kurihara, A., Shibayama, Y., Yasuno, A. et al. (1996b), Lipid emulsions of palmitoylrhizoxin: effects of particle size on blood dispositions of emulsion lipid and incorporated compound in rats, *Biopharm. Drug Dispos.*, 17, 343–353. doi:10.1002/(SICI)1099-081X(199605)17:4%3C343::AID-BDD958%3E3.0.CO;2-7

Lamberts, S.W., Van den Beld, A.W., and Van der Lely, A.J. (1997), The endocrinology of aging, *Science*, 278 (5337), 419–424. doi:10.1126/science.278.5337.419

Langstrom, B., Itsenko, O., and Rahman, O. (2007), [^{11}C] Carbon monoxide, a versatile and useful precursor in labelling chemistry for PET-ligand development, *J. Labelled Comp. Radiopharm.*, 50 (9), 794–810. doi:10.1002/jlcr.1446

Le Bihan, D., Breton, E., and Syrota, A. (1985), Imagerie de diffusion par resonance magnetique nucleaire, *C. R. Acad. Sci.*, 301, 1109–1112.

Lee, I.H., Park, Y.T., Ron, K. et al. (2005), Stable paclitaxel formulations in oily contrast medium, *J. Control. Release*, 102 (2), 415–425. doi:10.1016/j.jconrel.2004.10.023

Lee, J.H., Lee, H.B., and Andrade, J.D. (1995), Blood compatibility of polyethylene oxide surfaces, *Prog. Polym. Sci.*, 20 (6), 1043–1079. 10.1016/0079-6700(95)00011-4

Lee, K.S., Jeong, Y.J., Han, J. et al. (2004), T1 non-small cell lung cancer: imaging and histopathologic findings and their prognostic implications, *Radiographics*, 24, 1617–1636. doi:10.1148/rg.246045018

Lehman, C.D., Issacs, C., Schnall, M.D. et al. (2007), Cancer yield of mammography, MR and US in high-risk women: prospective multi-institution breast cancer screening study, *Radiology*, 244 (2), 381–388. doi:10.1148/radiol.2442060461

Lencioni, R. and Llovet, J.M. (2010), Modified RECIST (mRECIST) assessment for hepatocellular carcinoma. *Semin. Liver Dis.*, 30, 52–60. doi:10.1055/s-0030-1247132

Lerner, R.M., Huang, S.R., and Parker, K.J. (1990), "Sonoelasticity" images derived from ultrasound signals in mechanically vibrated tissues, *Ultrasound Med. Biol.*, 16 (3), 231–239. doi:10.1016/0301-5629(90)90002-T

Levy, M.Y., Benita, S., and Baszkin, A. (1991), Interactions of a non-ionic surfactant with mixed phospholipid-oleic acid monolayers. Studies under dynamic conditions, *Colloids Surf.*, 59, 225–241. doi:10.1016/0166-6622(91)80249-N

Liapi, E., Geschwind, J.F., Vossen, J.A. et al. (2008), Functional MRI evaluation of tumor response in patients with neuroendocrine hepatic metastases treated with transcatheter arterial chemoembolization. *Am. J. Roentgenol.*, 190 (1), 67–73. doi:10.2214/ajr.07.2550

Libby, P., Ridker, P.M., and Hansson, G.K. (2011), Progress and challenges in translating the biology of atherosclerosis, *Nature*, 473 (7347), 317–325. doi:10.1038/nature10146

Libby, P., Sukhova, G., Lee, R.T. et al. (1997), Molecular biology of atherosclerosis, *Int. J. Cardiol.*, 62 (2), 23–29. doi:10.1016/S0167-5273(97)00238-6

Lin, S.Y., Wu, W.H., and Lui, W.Y. (1992), In vitro release, pharmacokinetic and tissue distribution studies of doxorubicin hydrochloride (Adriamycin HCl) encapsulated in lipiodolized w/o emulsions and w/o/w multiple emulsions, *Pharmazie*, 47 (6), 439–443.

Liu, F. and Liu, D. (1995), Long-circulating emulsions (oil-in-water) as carriers for lipophilic drugs, *Pharm. Res.*, 12 (7), 1060–1064. doi:10.1023/a:1016274801930

Liu, D., Mori, A., and Huang, L. (1991), Large liposomes containing ganglioside GM1 accumulate effectively in spleen, *Biochim. Biophys. Acta*, 1066 (2), 159–165. doi:10.1016/0005-2736(91)90182-8

Lundberg, B.B., Griffiths, G., and Hansen, H.J. (1999), Conjugation of an anti-B-cell lymphoma monoclonal antibody, LL2, to long circulating drug-carrier lipid emulsions, *J. Pharm. Pharmacol.*, 51 (10), 1099–1105. doi:10.1211/0022357991776787

Lundberg, B.B., Griffiths, G., and Hansen, H.J. (2004), Cellular association and cytotoxicity of anti-CD74-targeted lipid drug-carriers in B lymphoma cells, *J. Control. Release*, 94, 155–161. doi:10.1016/j.jconrel.2003.09.016

Lundberg, B.B., Mortimer, B.-C., and Redgrave, T.G. (1996), Submicron lipid emulsions containing amphipathic polyethylene glycol for use as drug-carriers with prolonged circulation time, *Int. J. Pharm.*, 134 (1–2), 119–127. doi:10.1016/0378-5173(96)04456-0

Lundberg, B.B., Risovic, V., Ramaswamy, M. et al. (2003), A lipophilic paclitaxel derivative incorporated in a lipid emulsion for parenteral administration, *J. Control. Release*, 86, 93–100. doi:10.1016/s0168-3659(02)00323-1

Lutz, O., Meraihi, Z., Mura, J.L. et al. (1989), Fat emulsion particle size: influence on the clearance rate and the tissue lipolytic activity, *Am. J. Clin. Nutr.*, 50 (6), 1370–1381. doi:10.1093/ajcn/50.6.1370

Mahley, R.W. (1988), Apolipoprotein E: cholesterol transport protein with expanding role in cell biology, *Science*, 240 (4852), 622–630. doi:10.1126/science.3283935

Markowitz, S.D. (1997), Atherosclerosis: just another cancer? *J. Clin. Investig.*, 100 (9), 2143–2145. doi:10.1172%2FJCI119749

Matoba, M., Tonami, H., Kondou, T. et al. (2007), Lung carcinoma: diffusion-weighted MR imaging—preliminary evaluation with apparent diffusion coefficient, *Radiology*, 243 (2), 570–577. doi:10.1148/radiol.2432060131

Matsuki, M., Inada, Y., and Nakai, G. (2007), Diffusion-weighed MR imaging of pancreatic carcinoma, *Abdom. Imaging*, 32 (4), 481–483. doi:10.1007/s00261-007-9192-6

Medel, R., Monteith, S.J., Elias, W.J. et al. (2012), Magnetic resonance image-guided focused ultrasound surgery: part 2—a review of current and future applications, *Neurosurgery*, 71 (4), 755–763. doi:10.1227/NEU.0b013e3182672ac9

Medintz, I.L., Uyeda, H.T., Goldman, E.R. et al. (2005), Quantum dot bioconjugates for imaging, labeling and sensing, *Nat. Mater.*, 4, 435–446. doi:10.1038/nmat1390

Mensah, G.A., Ryan, U.S., Hooper, W.C. et al. (2007), Vascular endothelium summary statement II: cardiovascular disease prevention and control, *Vascul. Pharmacol.*, 46 (5), 318–320. doi:10.1016/j.vph.2006.10.019

Michalet, X., Pinaud, F.F., Bentolila, L.A. et al. (2005), Quantum dots for live cells, in vivo imaging, and diagnostics, *Science*, 307 (5709), 538–544. doi:10.1126/science.1104274

Miller, A.B., Hoogstraten, B., Staquet, M. et al. (1981), Reporting results of cancer treatment, *Cancer*, 47, 207–214. doi:10.1002/1097-0142(19810101)47:1%3C207::aid-cncr2820470134%3E3.0.co;2-6

Minami, C.A., Chung, D.U., and Chang, H.R. (2011), Management options in triple-negative breast cancer, *Breast Cancer*, 5, 175–199. doi:10.4137%2FBCBCR.S6562

Mishra, A. and Verma, M. (2010), Cancer biomarkers: are we ready for the prime time, *Cancer*, 2, 190–208. doi:10.3390/cancers2010190

Miyanaga, N., Akaza, H., Yamakawa, M. et al. (2006), Tissue elasticity imaging for diagnosis of prostate cancer: a preliminary report, *Int. J. Urol.*, 13 (12), 1514–1518. doi:10.1111/j.1442-2042.2006.01612.x

Mizushima, Y., Hoshi, K., Aihara, H. et al. (1983), Inhibition of bronchoconstriction by aerosol of a lipid containing prostaglandin E1, *J. Pharm. Pharmacol.*, 35 (6), 397. doi:10.1111/j.2042-7158.1983.tb02968.x

Murray, C.J. and Lopez, A.D. (1997), Global mortality, disability, and the contribution of risk factors: Global Burden of Disease Study, *Lancet*, 349 (9063), 1436–1442. doi:10.1016/S0140-6736(96)07495-8

Myers, K.R. and Truchot, A.T. (1990), Modified lipopolysaccharides and process of preparation, US Patent No. 4,912,094, March 27, 1990.

Napoli, C., Lerman, L.O, de Nigris F. et al. (2006), Rethinking primary prevention of atherosclerosis-related diseases, *Circulation*, 114, 2517–2527. doi:10.1161/CIRCULATIONAHA.105.570358

Nelson, E.D., Slotoroff, C.B., Gomella, L.G. et al. (2007), Targeted biopsy of the prostate: the impact of color Doppler imaging and elastography on prostate cancer detection and Gleason score, *J. Urol.*, 70 (6), 1136–1140. doi:10.1016/j.urology.2007.07.067

Neves, A.A. and Brindle, K.M. (2006), Assessing responses to cancer therapy using molecular imaging, *Biochim. Biophys. Acta*, 1766 (2), 242–261. doi:10.1016/j.bbcan.2006.10.002

Nobs, L., Buchegger, F., Gurny, R. et al. (2004), Current methods for attaching targeting ligands to liposomes and nanoparticles, *J. Pharm. Sci.*, 93, 1980–1992. doi:10.1002/jps.20098

Nomura, T., Yamashita, F., Takakura, Y. et al. (1995), Effect of cationic liposomes on intratumoral behavior of gene medicines after local injection, *Proc. Int. Sym. Control Release Bioactive Mater.*, 22, 420–421.

Ott, G., Singh, M., Kazzaz, J. et al. (2002), A cationic sub-micron emulsion (MF59/DOTAP) is an effective delivery system for DNA vaccines, *J. Control. Release*, 79 (1–3), 1–5. doi:10.1016/s0168-3659(01)00545-4

Paajanen, H. (2006), Increasing use of mammography improves the outcome of breast cancer in Finland, *Breast J.*, 12 (1), 88–90. doi:10.1111/j.1075-122X.2006.00197.x

Pallwein, L., Mitterberger, M., Struve, P. et al. (2007), Real-time elastography for detecting prostate cancer: preliminary experience, *BJU Int.*, 100 (1), 42–46. doi:10.1111/j.1464-410X.2007.06851.x

Papahadjopoulos, D., Allen, T.M., Gabizon, A. et al. (1991), Sterically stabilized liposomes: improvements in pharmacokinetics and antitumor therapeutic efficacy, *Proc. Natl. Acad. Sci. USA*, 88 (24), 11460–11464. doi:10.1073%2Fpnas.88.24.11460

Perou, C.M. (2010), Molecular stratification of triple-negative breast cancers, *Oncologist*, 15 (Suppl. 5), 39–48. doi:10.1634/theoncologist.2010-S5-39

Peterson, J.D. (2012), Non-invasive quantitative in vivo imaging of atherosclerosis disease progression and treatment response in apo E deficient mice using fluorescent molecular tomography and NIR fluorescent preclinical imaging agents, *Perkin Elmer Leaflet*, 127, 1–7.

Philips, B., Bague, S., Rabinovich-Guilatt, L. et al. (2008), Ophthalmic emulsions containing an immunosuppressive agent, EP 1809238, Novagali Pharma, Evry, France.

Pogue, B.W., Poplack, S.P., McBride, T.O. et al. (2001), Quantitative hemoglobin tomography with diffuse near infrared spectroscopy: pilot results in the breast, *Radiology*, 218, 261–266. doi:10.1148/radiology.218.1.r01ja51261

Poplack, S.P., Paulsen, K.D., Hartov, A. et al. (2004), Electromagnetic breast imaging: average tissue property values in women with negative clinical findings, *Radiology*, 231 (2), 571–580. doi:10.1148/radiol.2312030606

Porter, C.J.H., Charman, S.A., and Charman, W.N. (1996), Lymphatic transport of halofantrine in the triple-cannulated anesthetized rat model: effect of lipid vehicle dispersion, *J. Pharm. Sci.*, 85, 351–356. doi:10.1021/js950221g

Price, M.E., Cornelius, R.M., and Brash, J.L. (2001), Protein adsorption to polyethylene glycol modified liposomes from fibrinogen solution and from plasma, *Biochim. Biophys. Acta*, 1512, 191–205. doi:10.1016/S0005-2736(01)00330-3

Provenzale, J.M., Mukundan, S., and Barboriak, D. (2006), Diffusion weighted and perfusion MR imaging for brain tumor characterization and assessment of treatment response, *Radiology*, 239 (3), 632–649. doi:10.1148/radiol.2393042031

Rabinovich-Guilatt, L., Couvreur, P., Lambert, G. et al. (2004), Extensive surface studies help to analyse zeta potential data: the case of cationic emulsions, *Chem. Phys. Lipids*, 131, 1–13. doi:10.1016/j.chemphyslip.2004.04.003

Redgrave, T.G., Rakic, V., Mortimer, B.-C. et al. (1992), Effects of sphingomyelin and phosphatidylcholine acyl chains on the clearance of triacylglycerol-rich lipoproteins from plasma. Studies with lipid emulsions in rats, *Biochim. Biophys. Acta*, 1126, 65–72. doi:10.1016/0005-2760(92)90218-k

Rensen, P.C.N., Van Dijk, M.C.M, Havenaar, E.C. et al. (1995), Selective liver targeting of antivirals by recombinant chylomicrons - a new therapeutic approach to hepatitis B, *Nat. Med.*, 1, 221–225. doi:10.1038/nm0395-221

Ribi, E.E. (1984), Refined detoxified endotoxin product, US Patent No. 4,436,727, March 13, 1984.

Ribi, E., Schwartzman, S.M., and Cantrell, J.L. (1984), Refined detoxified endotoxin product, US Patent No. 4,436,728A, March 13, 1984.

Rojanasakul, Y. and Robinson, J.R. (1989), Transport mechanisms of the cornea: characterization of barrier permselectivity, *Int. J. Pharm.*, 55 (2–3), 237–246. doi:10.1016/0378-5173(89)90047-1

Ross, R.J., Thompson, J.S., Kim, K. et al. (1982), Nuclear magnetic resonance imaging and evaluation of human breast tissue: preliminary clinical trials, *Radiology*, 143, 195–205. doi:10.1148/radiology.143.1.7063727

Rubin, M., Harell, D., Naor, N. et al. (1991), Lipid infusion with different triglyceride cores (long-chain vs medium-chain/long-chain triglycerides): effect on plasma lipids and bilirubin binding in premature infants, *J. Parenter. Enteral Nutr.*, 15, 642–646. doi:10.1177/0148607191015006642

Saba, T. M. (1970), Physiology and physiopathology of the reticuloendothelial system, *Arch. Intern. Med.*, 126 (6), 1031–1052. doi:10.1001/archinte.1970.00310120093014

Saftoiu, A. and Vilman, P. (2006), Endoscopic ultrasound elastography—a new imaging technique for the visualization of tissue elasticity distribution, *J. Gastrointestin. Liver Dis.*, 15 (2), 161–165.

Sakaeda, T. and Hirano, K. (1995), O/W lipid emulsions for parenteral drug delivery. II. Effect of composition on pharmacokinetics of incorporated drug, *J. Drug Target.*, 3 (3), 221–230. doi:10.3109/10611869509015949

Sakaeda, T. and Hirano, K. (1998), Effect of composition on biological fate of oil particles after intravenous injection of O/W lipid emulsions, *J. Drug Target.*, 6 (4), 273–284. doi:10.3109/10611869808996835

Sakaeda, T., Takahashi, K., Nishihara, Y. et al. (1994), O/W lipid emulsions for parenteral drug delivery. I. Pharmacokinetics of the oil particles and incorporated sudan II, *Biol. Pharm. Bull.*, 17 (11), 1490–1495. doi:10.1248/bpb.17.1490

Samama, J.P., Lee, K.M., and Biellmann, J.F. (1987), Enzymes and microemulsions. Activity and kinetic properties of liver alcohol dehydrogenase in ionic water-in-oil microemulsions, *Eur. J. Biochem.*, 163 (3), 609–617. doi:10.1111/j.1432-1033.1987.tb10910.x

Sarkeala, T., Heinavaara, S., and Anttila, A. (2008), Breast cancer mortality with varying invitational policies in organised mammography, *Br. J. Cancer*, 98 (3), 641–645. doi:10.1038%2Fsj.bjc.6604203

Sasaki, H., Yamamura, K., Nishida, K. et al. (1996), Delivery of drugs to the eye by topical application, *Prog. Retin. Eye Res.*, 15 (2), 583–620. doi:10.1016/1350-9462(96)00014-6

Scher, H.I., Halabi, S., Tannock I. et al. (2008), Design and end points of clinical trials for patients with progressive prostate cancer and castrate levels of testosterone: recommendations of the Prostate Cancer Clinical Trials Working Group, *J. Clin. Oncol.*, 26 (7), 1148–1159. doi:10.1200/JCO.2007.12.4487

Senior, J., Crawley, J.C.W., and Gregoriadis, G. (1985), Tissue distribution of liposomes exhibiting long half-lives in the circulation after intravenous injection, *Biochim. Biophys. Acta*, 839, 1–8. doi:10.1016/0304-4165(85)90174-6

Shils, M.E. (1998), Parentral nutrition, in: Shils, M.E., Olson, J.A., Shike, M.E., and Ross, O., Eds., *Modern Nutrition in Health and Disease*, 4th ed., Williams & Wilkins, Baltimore, MD.

Shim, S.S., Lee, K.S., Kim, B. et al. (2005), Accuracy of integrated PET/CT using fluorodeoxyglucose in preoperative staging on non-small cell lung cancer: a prospective comparison with standalone CT, *Radiology*, 236, 1011–1019.

Smith, J.J., Sorensen, A.G., and Thrall, J.H. (2003), Biomarkers in imaging: realizing radiology's future, *Radiology*, 227 (3), 633–638. doi:10.1148/radiol.2273020518

Song, Y.K., Liu, D., Maruyama, K.Z. et al. (1996), Antibody mediated lung targeting of long-circulating emulsions, *PDA J. Pharm. Sci. Technol.*, 50 (6), 372–377.

Sprecher, H. (2002), The roles of anabolic and catabolic reactions in the synthesis and recycling of polyunsaturated fatty acids, *Prostaglandins Leukot. Essent. Fatty Acids*, 67 (2–3), 79–83. doi:.10.1054/plef.2002.0402

Stroobants, S., Geominne, J., Seegers, M. et al. (2003), (18) FDG-Positron emission tomography for the early prediction of response in advanced soft tissue sarcoma treated with imatinib mesylate (Glivec), *Eur. J. Cancer*, 39 (14), 2012–2020. doi.10.1016/s0959-8049(03)00073-x

Szebeni, J. (2001), Complement activation-related pseudoallergy caused by liposomes, micellar carriers of intravenous drugs, and radiocontrast agents, *Crit. Rev. Ther. Drug Carrier Syst.*, 18 (6), 567–606.

Sznitowska, M., Janicki, S., Dabrowska, E.A. et al. (2002), Physicochemical screening of antimicrobial agents as potential preservatives for submicron emulsions, *Eur. J. Pharm. Sci.*, 15 (5), 489–495. doi:10.1016/S0928-0987(02)00034-9

Takakura, Y. and Hashida, M. (1995), Macromolecular drug carrier systems in cancer chemotherapy: macromolecular prodrugs, *Crit. Rev. Oncol. Hematol.*, 18 (3), 207–231. doi:10.1016/1040-8428(94)00131-C

Takayoshi, U., Yuen, S., Kasami, M. et al. (2007), Dynamic contrast enhanced screening of microcalcifications in the breast. Is there any value, *Breast Cancer Res. Treat.*, 103 (3), 269–281. doi:10.1007/s10549-006-9373-y

Takino, T., Konishi, K., Takakura, Y. et al. (1994), Long circulating emulsion carrier systems for highly lipophilic drugs, *Biol. Pharm. Bull.*, 17, 121–125. doi:10.1248/bpb.17.121

Takino, T., Koreeda, N., and Nomura, T. (1998), Control of plasma cholesterol-lowering action of probucol with various lipid carrier systems, *Biol. Pharm. Bull.*, 21, 492–497. doi:10.1248/bpb.21.492

Takino, T., Nakajima, C., Takakura, Y., et al. (1993), Controlled biodistribution of highly lipophilic drugs with various parenteral formulations, *J. Drug Targeting*, 1 (2), 117–124. doi:10.3109/10611869308996067

Tam, A. and Ahra, K. (2007), Palliative interventions for pain in cancer patients, *Semin. Intervent. Radiol.*, 24 (4), 419–429. doi:10.1055%2Fs-2007-992330

Tamilvanan, S. (2004), Oil-in-water lipid emulsions: implications for parenteral and ocular delivering systems, *Prog. Lipid Res.*, 43 (6), 489–533. doi:10.1016/j.plipres.2004.09.001

Tamilvanan, S. and Benita, S. (2004), The potential of lipid emulsion for ocular delivery of lipophilic drugs, *Eur. J. Pharm. Biopharm.*, 58 (2), 357–368. doi:10.1016/j.ejpb.2004.03.033

Teixeira, H., Dubernet, C., Puisieux, F. et al. (1999), Submicron cationic emulsions as a new delivery system for oligonucleotides, *Pharm. Res.*, 16, 30–36. doi:10.1023/A:1018806425667

Therasse, P., Arbuck, S.G., Eisenhauer, E.A. et al. (2000), New guidelines to evaluate the response to treatment in solid tumors (RECIST guidelines), *J. Natl. Cancer Inst.*, 92 (3), 205–216. doi:10.1093/jnci/92.3.205

Thike, A.A., Iqbal, J., Cheok, P.Y. et al. (2010), Triple negative breast cancer: outcome correlation with immunohistochemical detection of basal markers, *Am. J. Surg. Pathol.*, 34 (7), 956–964. doi:10.1097/PAS.0b013e3181e02f45

Tibell, A., Larsson, M., and Alvestrand, A. (1993), Dissolving intravenous cyclosporin A in a fat emulsion carrier prevents acute renal side effects in the rat, *Transplant Int. Off. J. Eur. Soc. Organ. Transplant*, 6 (2), 69–72. doi:10.1007/bf00336647

Tirosh, O., Barenholz, Y., Katzhendler, J. et al. (1998), Hydration of polyethylene glycol-grafted liposomes, *Biophys. J.*, 74 (3), 1371–1379. doi:10.1016%2FS0006-3495(98)77849-X

Tirucherai, G.S., Pezron, I., and Mitra, A.K. (2002), Novel approaches to nasal delivery of peptides and proteins. *STP Pharm. Sci.*, 12, 3–12.

Tromberg, B.J., Shah, N., Lanning, R. et al. (2000), Non-invasive in vivo characterization of breast tumors using photon migration spectroscopy, *Neoplasia*, 2 (1–2), 26–40. doi:10.1038%2Fsj.neo.7900082

Tsutsumi, M., Miyagawa, T., Matsumura, T. et al. (2007), The impact of real-time tissue elasticity imaging (elastography) on the detection of prostate cancer: clinicopathological analysis, *Int. J. Clin. Oncol.*, 12 (4), 250–255. doi:10.1007/s10147-007-0669-7

Ueda, K., Ishida, M., Inoue, T. et al. (2001), Effect of injection volume on the pharmacokinetics of oil particles and incorporated menatetrenone after intravenous injection as O/W lipid emulsions in rats, *J. Drug Target.*, 9 (5), 353–360. doi:10.3109/10611860108998771

Ueda, K., Yamazaki, Y., Noto, H. et al. (2003), Effect of oxyethylene moieties in hydrogenated castor oil on the pharmacokinetics of menatetrenone incorporated in O/W lipid emulsions prepared with hydrogenated castor oil and soybean oil in rats, *J. Drug Target.*, 11, 37–43. doi:10.1080/1061186031000086081

Van den Bent, M.J., Wefel, J.S., Schiff, D. et al. (2011), Response assessment in neuro-oncology (a report of the RANO group): assessment of outcome in trials of diffuse low-grade gliomas, *Lancet Oncol.*, 12 (6), 583–593. doi:10.1016/S1470-2045(11)70057-2

Viale, G., Rotmensz, N., Maisonneuve, P. et al. (2009), Invasive ductal carcinoma of the breast with the "triple-negative" phenotype: prognostic implications of EGFR immunoreactivity, *Breast Cancer Res. Treat.*, 116 (2), 317–328. doi:10.1007/s10549-008-0206-z

Werner, N. and Nickenig, G. (2006), Influence of cardiovascular risk factors on endothelial progenitor cells: limitations for therapy? *Arterioscler. Thromb. Vasc. Biol.*, 26, 257–266. doi:10.1161/01.ATV.0000198239.41189.5d

Wheeler, J.J., Wong, K.F., Ansell, S.M. et al. (1994), Polyethylene glycol modified phospholipids stabilize emulsions prepared from triacylglycerol, *J. Pharm. Sci.*, 83 (11), 1558–1564. doi:10.1002/jps.2600831108

Wisse, E. (1970), An electron microscopic study of the fenestrated endothelial lining of rat liver sinusoids, *J. Ultrastruct. Res.*, 31 (1–2), 125–150. doi:10.1016/S0022-5320(70)90150-4

Wolchok, J.D., Hoos, A., O'Day, S. et al. (2009), Guidelines for the evaluation of immune therapy activity in solid tumors: immune-related response criteria, *Clin. Cancer Res.*, 15 (23), 7412–7420. doi:10.1158/1078-0432.CCR-09-1624

Woodle, M.C. and Lasic, D.D. (1992), Sterically stabilized liposomes, *Biochim. Biophys. Acta*, 1113 (2), 171–199. doi:10.1016/0304-4157(92)90038-C

Wretlind, A. (1981), Parenteral nutrition, *Nutr. Rev.*, 39, 257–265.

Wu, N.Z., Da, D., Rudoll, T.L. et al. (1993), Increased microvascular permeability contributes to preferential accumulation of stealth liposomes in tumor tissue, *Cancer Res.*, 53 (16), 3765–3770.

Yamaguchi, T., Nishizaki, K., Itai, S. et al. (1995), Physicochemical characterization of parenteral lipid emulsion: influence of cosurfactants on flocculation and coalescence, *Pharm. Res.*, 12, 1273–1278. doi:10.1023/A:1016205203264

Yang, S.C. and Benita, S. (2000), Enhanced absorption and drug targeting by positively charged submicron emulsions, *Drug Dev. Res.*, 50, 476–486. doi:10.1002/1098-2299(200007/08)50:3/4%3C476::AID-DDR31%3E3.0.CO;2-6

Yeung, D.K.W., Ahuja, A.T., and Tse, G.M.K. (2002), Breast cancer: in vivo proton MR spectroscopy in the characterization of histopathologic subtypes and preliminary observations in axillary node metastases, *Radiology*, 225, 190–197. doi:10.1148/radiol.2243011519

Yi, S.W., Yune, Y., Kim, T.W. et al. (2000), A cationic lipid emulsion/DNA complex as a physically stable and serum-resistant gene delivery system, *Pharm. Res.*, 17, 314–320. doi:10.1023/A:1007553106681

Yuan, F., Dellian, M., Fukumura, D. et al. (1995), Vascular permeability in a human tumor xenograft: molecular size dependence and cutoff size, *Cancer Res.*, 55 (17), 3752–3756.

Yuan, F., Leunig, M., Huang, S.K. et al (1994), Microvascular permeability and interstitial penetration of sterically stabilized (stealth) liposomes in a human tumor xenograft, *Cancer Res.*, 54 (13), 3352–3356.

Yue, Y., Astvatsaturyan, K., Cui, X. et al. (2016), Stratification of prognosis of triple-negative breast cancer patients using combinatorial biomarkers, *PLoS One*, 11 (3), e0149661. doi:10.1371/journal.pone.0149661.t003

Zhang, M., Zhang, X., Zhao, S. et al. (2014), Prognostic value of surviving and EGFR protein expression in triple-negative breast cancer (TNBC) patients, *Target. Oncol.*, 9 (4), 349–357. doi:10.1007/s11523-013-0300-y

Figure 2.1. Freeze-dried emulsions using different cryo- or lyo-protectants and reconstitution of freeze-dried powder into nanosized emulsion.

Oil-in-Water Nanosized Emulsions for Drug Delivery and Targeting,
First Edition. Tamilvanan Shunmugaperumal.
© 2021 John Wiley & Sons, Inc. Published 2021 by John Wiley & Sons, Inc.

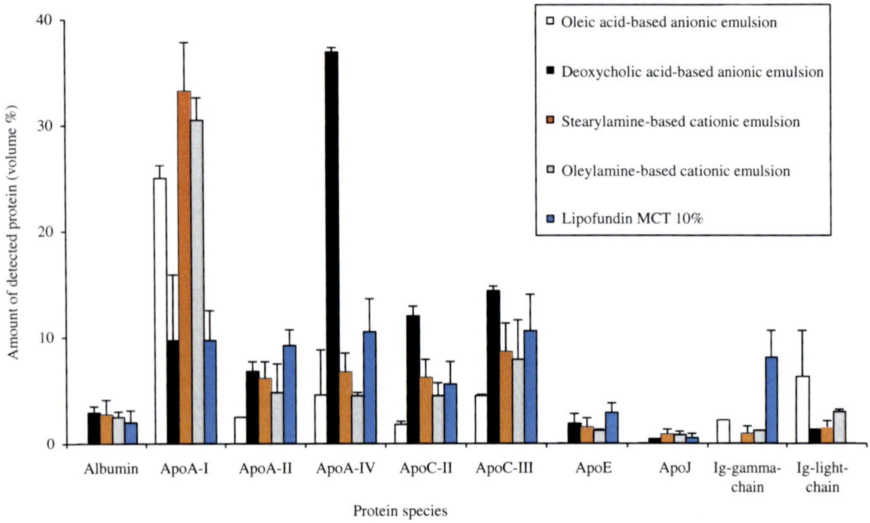

Figure 2.2. Amount of major proteins on the 2-D gels of plasma proteins adsorbed on emulsions with negative or positive surface charge in comparison with Lipofundin MCT 10%. [Taken with permission from Elsevier (Tamilvanan et al. 2005).]

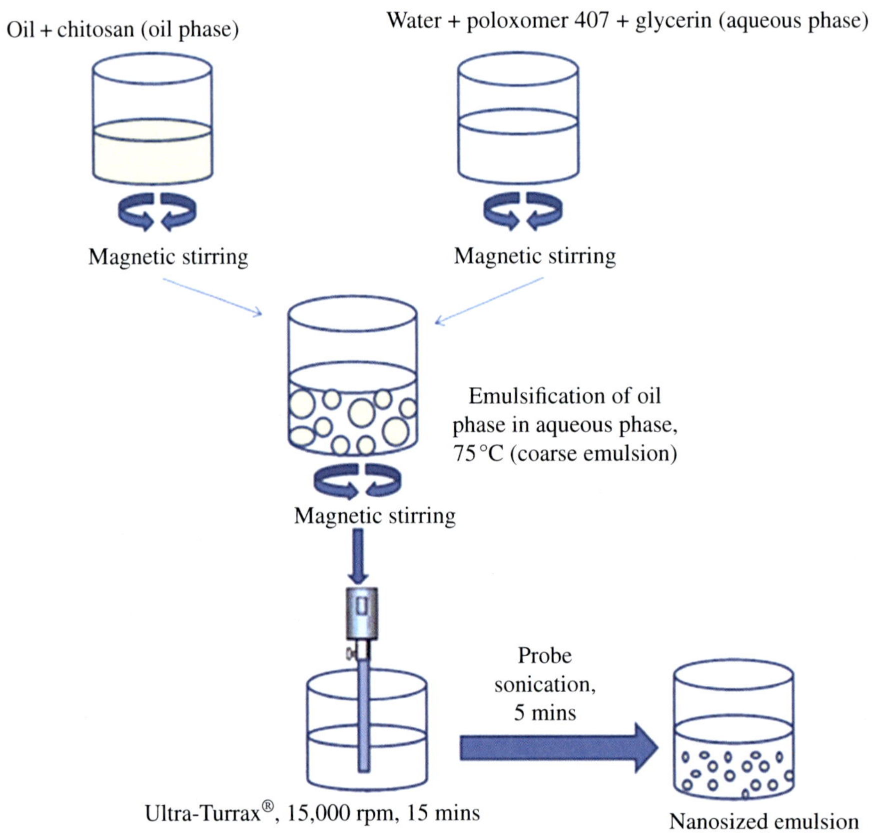

Figure 2.3. Schematic diagram for preparation of nanosized emulsion.

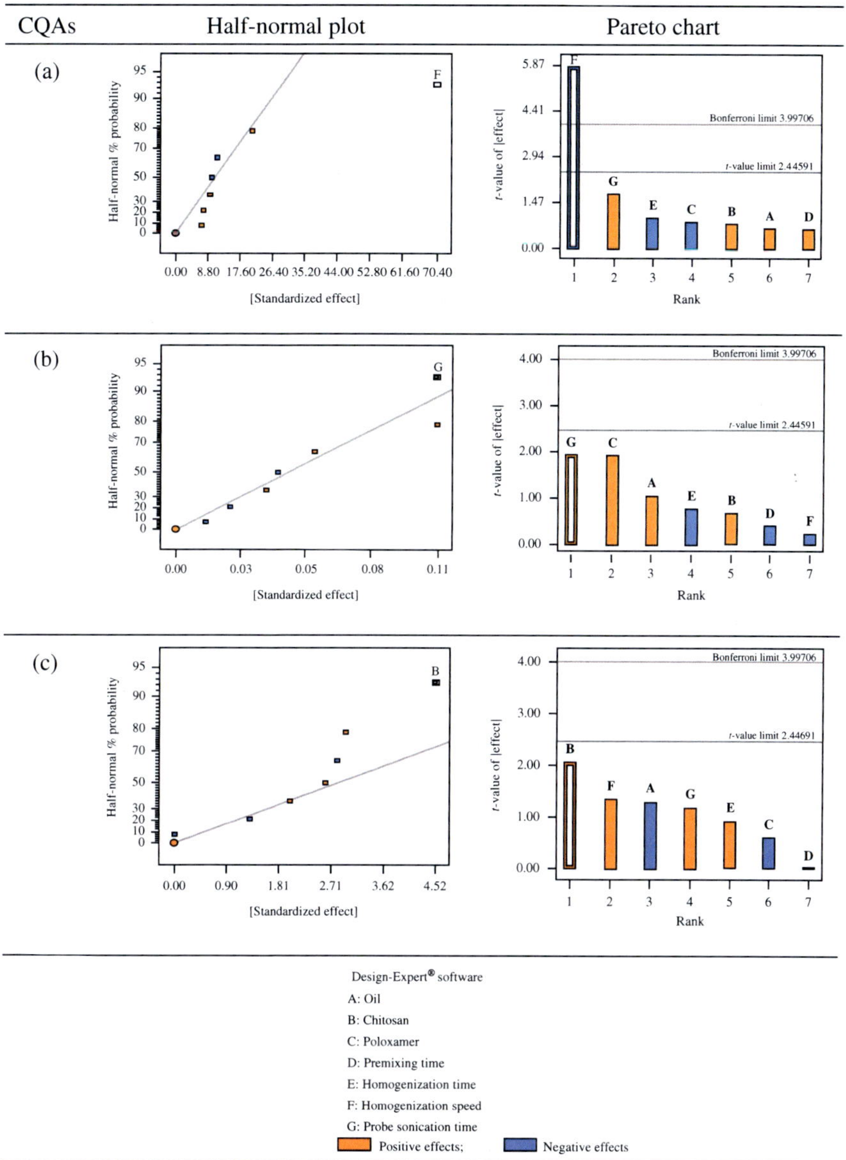

Figure 2.5. Half-normal and Pareto charts for screening of influential formulation and process variables as per Taguchi design using selected critical quality attributes (CQAs): (a) mean particle size, (b) polydispersity index, and (c) zeta potential.

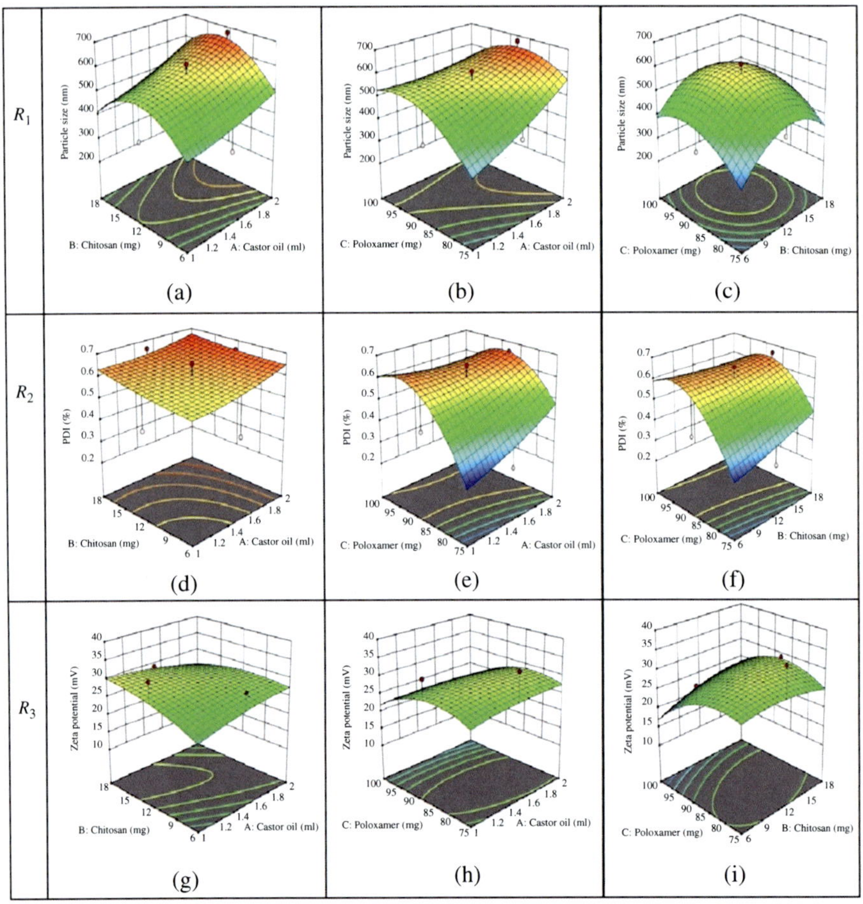

Figure 2.6. Response surface plots showing the interaction effects of castor oil and chitosan (a, d, g), castor oil and poloxamer (b, e, h), and chitosan and poloxamer (c, f, i) on response, mean particle size (R_1), polydispersity index (R_2), and zeta potential (R_3), respectively.

Critical quality attributes (CQAs)	Residual versus predicted plot	Actual versus predicted plot
R_1: MPS	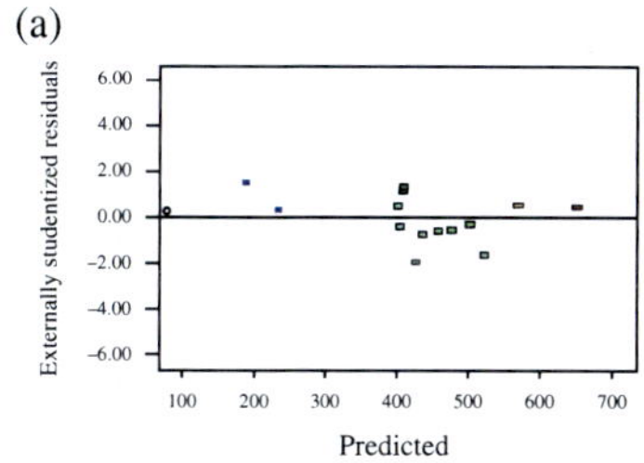	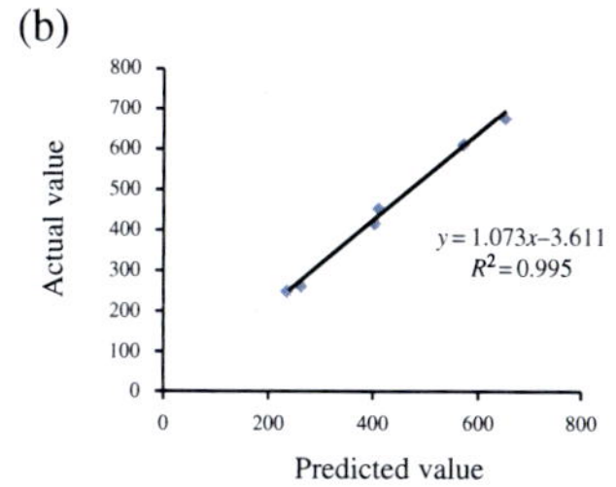
R_2: PDI	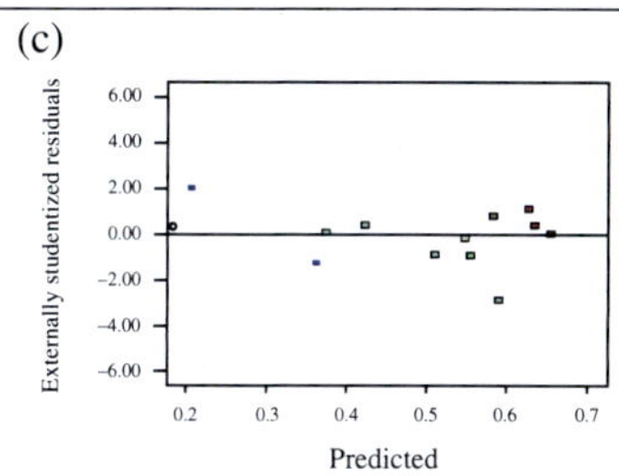	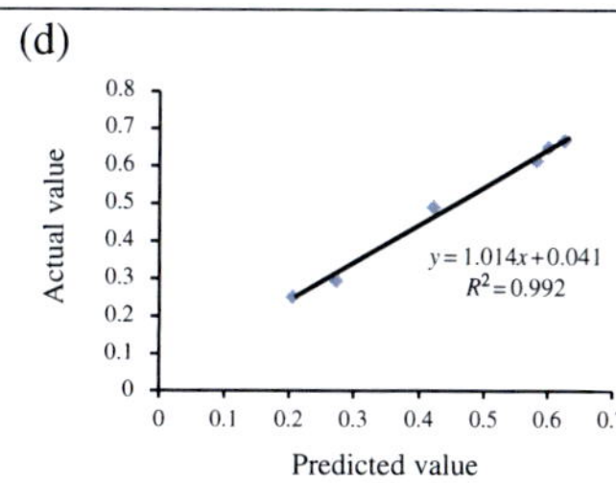
R_3: ZP	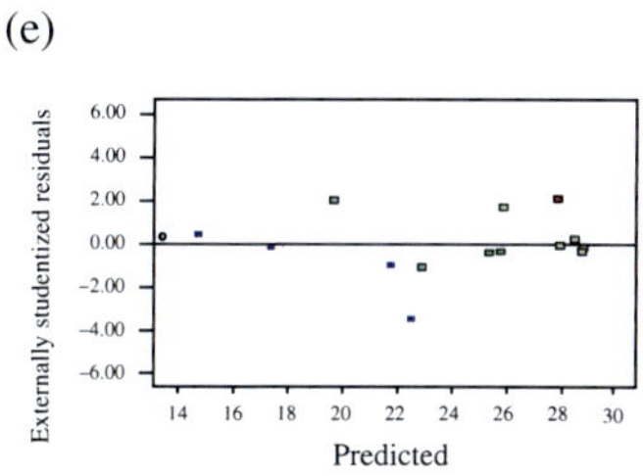	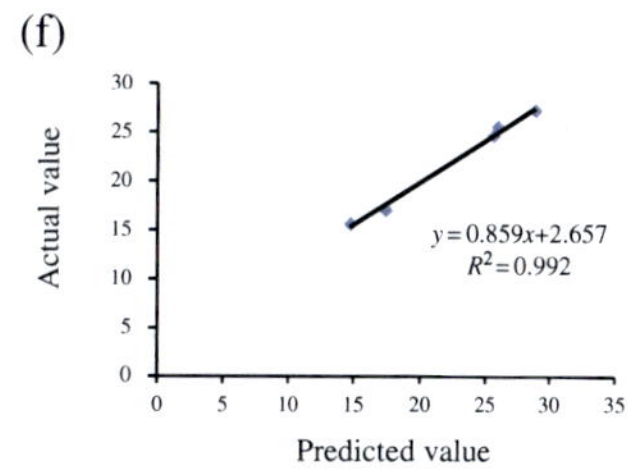
	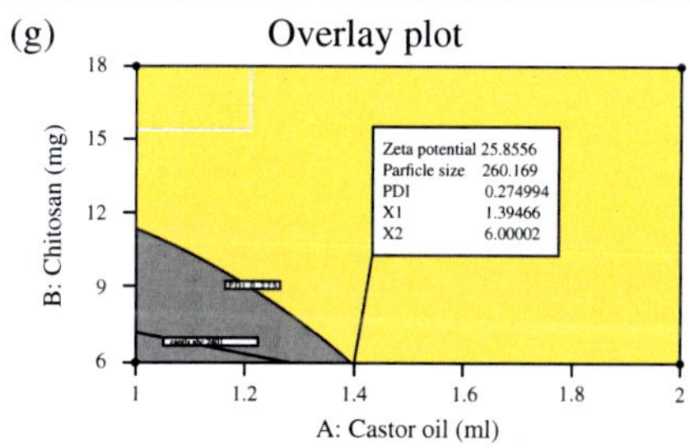	

Figure 2.7. Derived plots obtained from 3D-response surface model: residual versus predicted plot (a, c, e), actual versus predicted plot (b, d, f), and overlay plot (g).

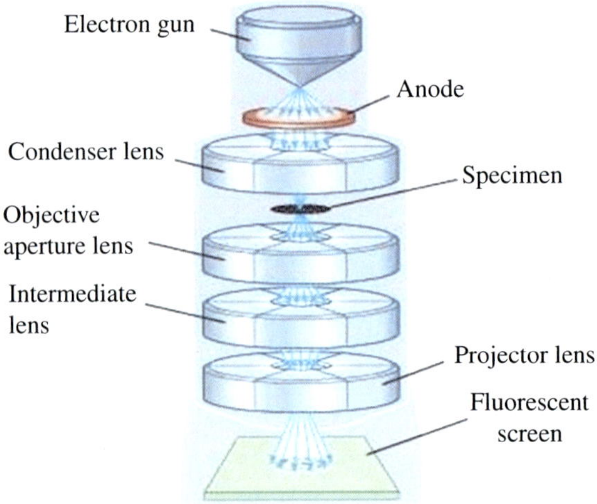

Figure 3.5. Essential parts of transmission electron microscope. [Adapted from https://www.britannica.com/technology/transmission-electron-microscope, accessed on November 28, 2019.]

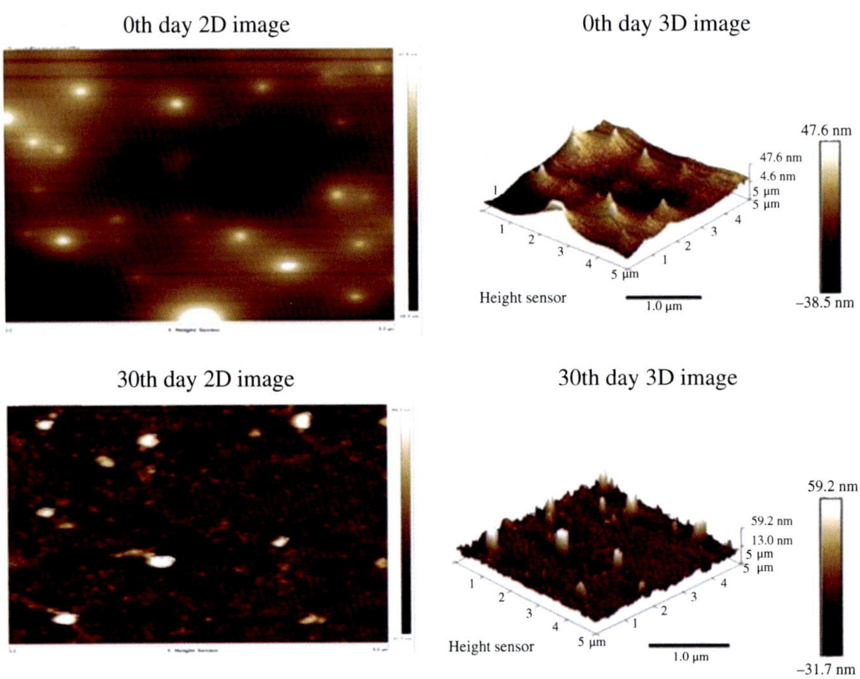

Figure 3.7. 2D and 3D AFM topography of seaweed oil nanosized emulsions. [Reproduced with permission from Sivagnanam et al. (2019).]

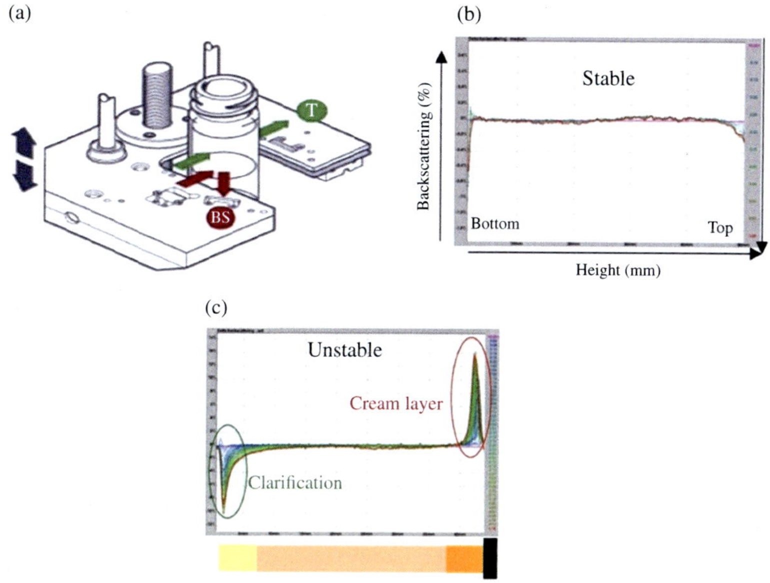

Figure 3.8. Turbiscan® assembly. Key: (a) Optical scanning analyser, Turbiscan®. (b) Superposition of scans with time for a stable sample. (c) Superposition of scans with time for an unstable sample (creaming). [Adapted from http://www.titanex.com.tw/doc/tecsupport/ANH-Turbiscan-application%20paper%20on%20%20pharmaceuticals.pdf, accessed on November 28, 2019.]

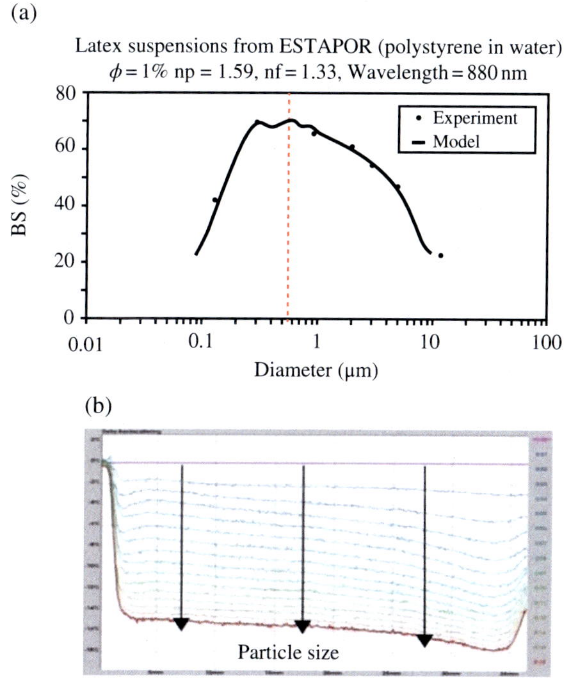

Figure 3.9. Assessment of colloidal dispersion using Turbiscan®. Key: (a) Backscattering level versus diameter for latex particles at 1%. (b) Typical profiles for flocculation phenomenon (initial size = 1 μm).

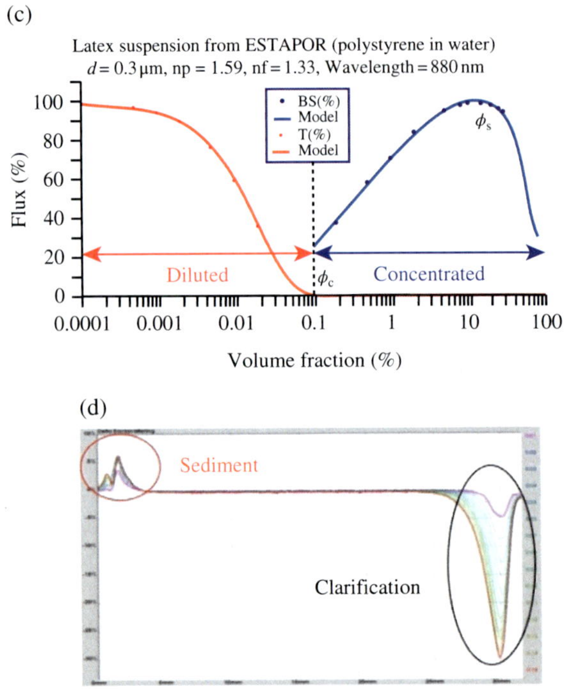

Figure 3.9. (c) Transmission (red) and backscattering (blue) levels versus volume fraction for latex particles of 0.3 μm. (d) Typical backscattering profiles for a sedimentation phenomenon. (Adapted from http://www.titanex.com.tw/doc/tecsupport/ANH-Turbiscan-application%20paper%20on%20%20pharmaceuticals.pdf, accessed on November 28, 2019.)

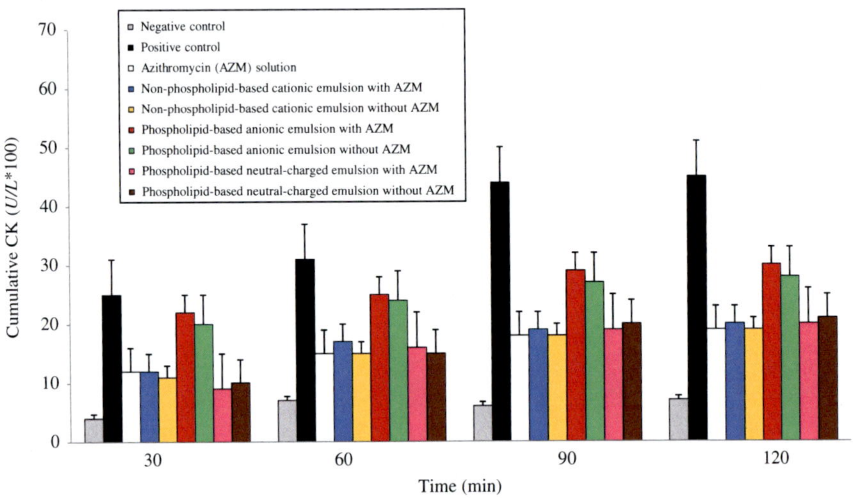

Figure 3.16. Cumulative creatine kinase (CK) release from 150 mg of the isolated rat extensor digitorum longus (EDL) muscle of male Sprague-Dawley rats after the treatment over a 120-min period with the various formulations containing equivalent amount of (100 μg) azithromycin (AZM) in comparison with the positive control (phenytoin) and the negative control (0.9% NaCl). [Adapted from Tamilvanan et al. (2013).]

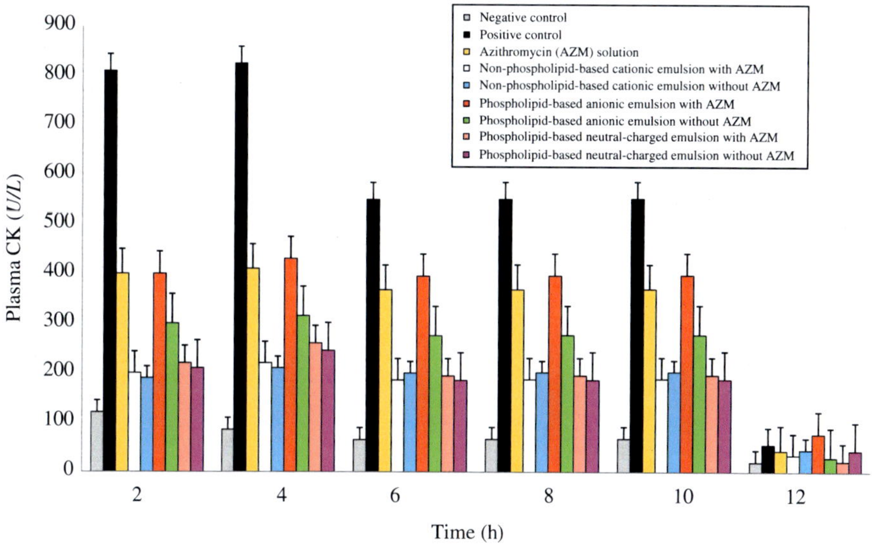

Figure 3.17. Mean plasma creatine kinase (CK) levels versus time following the intramuscular injection (0.3 ml) in the right thigh muscle of male Sprague-Dawley rats with various formulations containing equivalent amount of (100 μg) azithromycin (AZM), the positive control (phenytoin), and the negative control (0.9% NaCl).

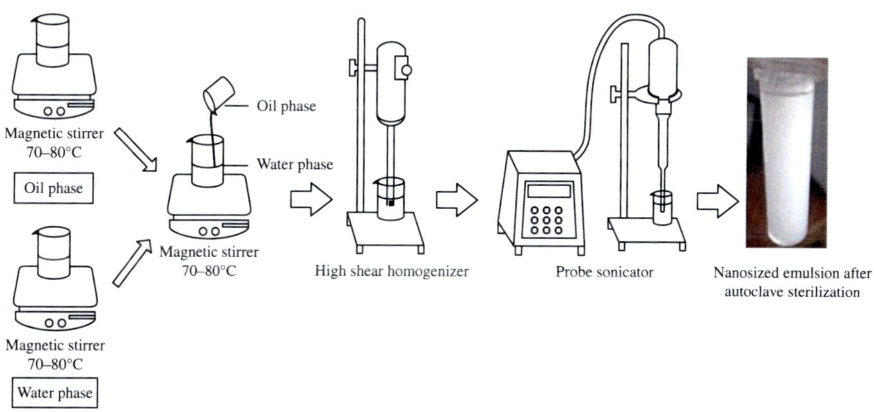

Figure 4.3. Oil-in-water nanosized emulsion preparation by *de novo* method.

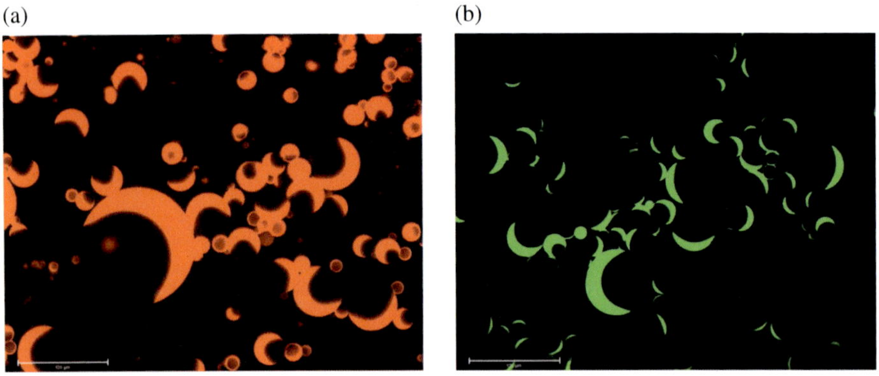

Figure 7.3.1. Formation of Janus structure on the dispersed oil droplets of oil-in-water emulsion. (a) Nile red-incorporated emulsion. (b) Coumarin 6-incorporated emulsion.

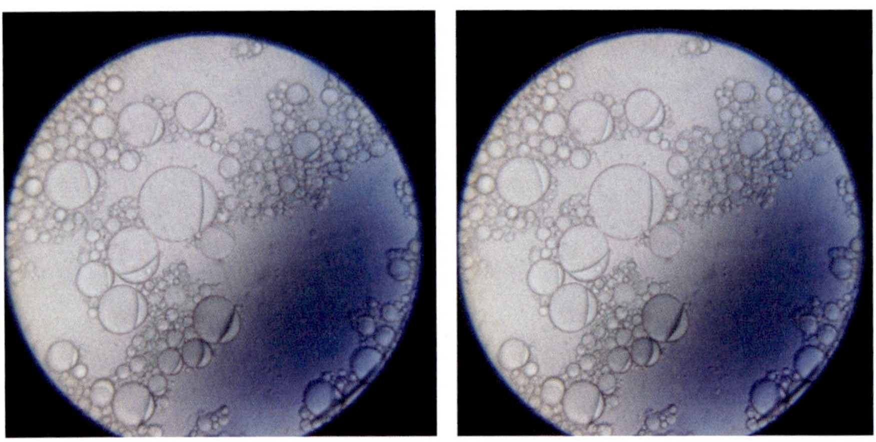

Figure 7.3.2. Optical microscopic picture showing the Janus structure on the dispersed oil droplets of oil-in-water macroemulsion.

Figure 7.3.3. Emulsion with Cerberus structure on the dispersed oil droplets of emulsion. [Adapted from Ge et al. (2017).]

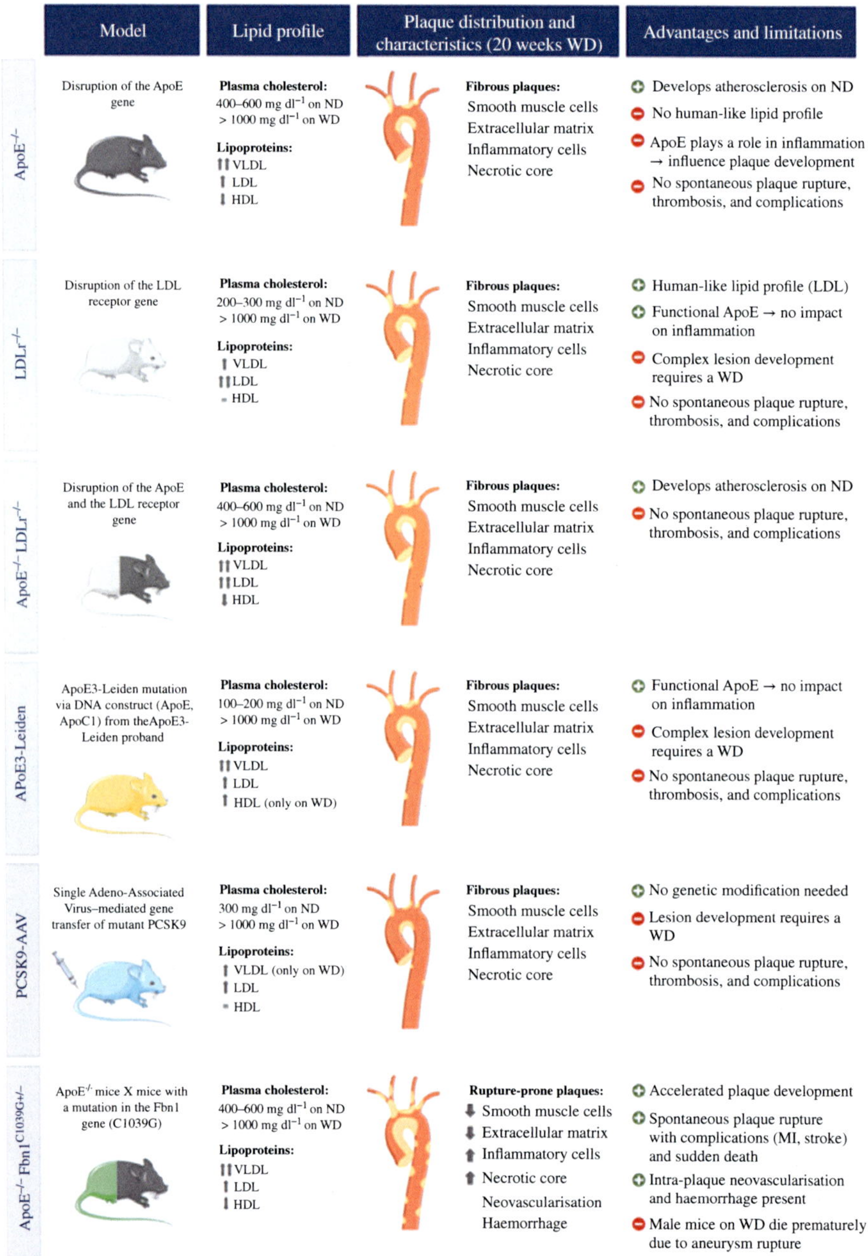

Figure 7.3.4. Overview of current mouse models of atherosclerosis. This figure describes the different models with their total plasma cholesterol levels on normal diet (ND) and Western-type diet (WD), lipoprotein profile, plaque characteristics, advantages, and limitations. The distribution of the plaques in the thoracic aorta and the complexity is shown for mice fed a WD for 20 weeks. The composition of the most complex lesions at that time point is displayed (usually found in the aortic root or brachiocephalic artery). [Reproduced with permission from Veseli et al. (2017).]

CHAPTER 5

BIOFATE OF NANOSIZED EMULSIONS

OLY KATARI*, SYED NAZRIN RUHINA RAHMAN*, AND TAMILVANAN SHUNMUGAPERUMAL

Department of Pharmaceutics, National Institute of Pharmaceutical Education and Research (NIPER), Guwahati, Assam, India

*The fundings provided by Department of Pharmaceuticals, Ministry of Chemicals and Fertilizers, Govt. of India to the author Oly Katari and Department of Biotechnology, Govt. of India to Syed Nazrin Ruhina Rahman are greatly acknowledged.

Oil-in-Water Nanosized Emulsions for Drug Delivery and Targeting,
First Edition. Tamilvanan Shunmugaperumal.
© 2021 John Wiley & Sons, Inc. Published 2021 by John Wiley & Sons, Inc.

EXPANSION OF ABBREVIATIONS

apos	apolipoproteins
AMD	age-related macular degeneration
API	active pharmaceutical ingredient
CDC	Centre for Disease Control
DHA	docosahexaenoic acid
2-D PAGE	two-dimensional polyacrylamide gel electrophoresis
EPA	eicosapentaenoic acid
GSD	globule size distribution
HDL	high-density lipoproteins
HSPG	heparan sunanosized emulsionsate proteoglycans
IEF	isoelectric focusing
LCT	long-chain triglyceride
LDL	low-density lipoprotein
LDLR	low-density lipoprotein receptor
LPL	lipoprotein lipase
LRP	LDL receptor-related protein
MCT	medium-chain triglyceride
NBT	nitroblue tetrazolium
NO	nitric oxide
O/W	oil-in-water
PC	phosphatidylcholine
PDA	piperazine diacrylamide
PFAT5	percentage of fat >5 μm
PN	parenteral nutrition
PNALD	parenteral nutrition-associated liver diseases
PPAR	peroxisome proliferator-activated receptor family
RPE	retinal pigment epithelium
SDS	sodium dodecyl sulfate
SM	sphingomyelin
SPOS	single-particle optical sensing
TG	triglyceride
TGRP	triglyceride-rich particles
TNA	total nutrient admixtures
TPN	total parenteral nutrition
VLDL	very low-density lipoprotein

5.1. INTRODUCTION

In recent years, the oil-in-water (o/w) nanosized emulsions have found therapeutic applications in all of the possible routes of administrations viz., nasal, ocular topical and intravitreal, oral, parenteral, topical skin, and vaginal. All of these active pharmaceutical ingredient (API)-administrative routes of human body do possess many constraints that may cause considerable and significant obstruction in the movement of API or API-loaded delivery systems inside the human body. As far as the o/w nanosized emulsion is concerned, this colloidal dispersion first of all considered as "foreign particles" by the human body following administration by using any of the above discussed API or API-loaded emulsion administrative routes. As a result, the parenterally administered API-loaded nanosized emulsion particles could be cleared from the human body as quickly as possible via different particle-clearance mechanism or simple metabolic reactions as shown in a tabular form in Chapter 4 (Table 4.3). Indirectly, we can say that the clearance/metabolism of administered nanosized emulsion particles by the human body is one of the hurdles to consider the o/w nanosized emulsion as controlled or sustained API delivery vehicle. However, by changing the formulation parameters and playing with the endogenous clearance pathway, it is somewhat possible to make the o/w nanosized emulsion as "a short-lived or transient" controlled or sustained API delivery vehicle. In this context, it is necessary to understand basically what are the issues faced by the o/w nanosized emulsion particles after their administration via different routes. In better words, understanding the issues of administered emulsion particles first will help to design the emulsion particles able to escape from—and even divert—the endogenous clearance pathway at least for short/transient period of time. Therefore, the main objective of this chapter is to elaborate possible issues faced by the administered emulsion particles after their entry into human body via various routes of administration. In biopharmaceutical term, the issues related to the administered emulsion particles may be termed as "biofate" of o/w nanosized emulsions. For our convenience, the biofate of o/w nanosized emulsions is broadly elaborated in this chapter based on the generations of emulsions (discussed in Chapter 4) and in particular, more focus is given on first-generation emulsions.

5.2. BIOFATE OF O/W NANOSIZED EMULSIONS

Literally, the interaction between emulsion particles and body tissues or plasma proteins can be covered under biofate of the o/w nanosized emulsions. Therefore, depending on the route of administration, the interaction cite of emulsion particles may vary. For instance, parenterally administered nanosized emulsions or emulsion particles entering into the blood stream from other routes of administration will face the interaction of plasma proteins as the blood stream-entered emulsions are considered as foreign particles. Similarly, the emulsions administered through topical application into the nasal cavity will face an intimate mixing problem with the nasal secretions and thus eventually most of the topically applied emulsion particles are squeezed out from the nasal cavity. Relatively, very less amount of the emulsion particles have chance to reach the alveolar region of the lungs. A more sensitive oozing/squeezing-out reaction can be faced by the topically instilled ophthalmic emulsions. Because, the ocular protective mechanism to prevent the entry of both liquid and solid particles is heavily occurring that results into the oozing/squeezing of 95% of the topically administered ophthalmic emulsion droplets from the ocular surface. Basically, the remaining 5% of the topically instilled ophthalmic emulsion particles are able to reach into the intraocular tissues of the eye via transcorneal or transconjunctival/scleral route. Here too, the biofate of ophthalmic topical emulsions coincide with the 5% of the emulsion particles entering into the intraocular tissues of the eye. Coming to the emulsions used in the topical skin application, the biofate depends on the contact time of topically applied emulsion particles with skin structure/surface, which is particularly controlled by residence time of emulsion on the skin along with other factors like dose applied on the skin, emulsifier concentration used to make the emulsion, spreading coefficient of the emulsion on the skin, etc. This chapter is particularly designed to showcase the biofate of o/w nanosized emulsions that were administered through intraocular injection or ocular topical application and parenteral route.

5.2.1. Issues Relevant to Ocular Nanosized Emulsions

For the eye, the method of API delivery is important. However, when nanosized emulsions are used as a vehicle for ocular API delivery purposes, both local/topical and intraocular routes of administration can be possible (though limited data concerning intraocular API delivery through nanosized emulsions is currently available). Nanosized emulsion provides a liquid-retentive carrier for ocular active agents particularly when topically instilled into the eye. In fact, in comparison with ocular inserts or implants and semisolid ocular preparations, the liquid-retentive nature gives impetus to investigate further the nanosized emulsion-based ophthalmic API delivery as it has a psychological attribution for comfortable use of ophthalmologists and of patients. In addition, through topical instillation of nanosized emulsions possessing ocular active substances, the delivering of API molecules even to the intraocular/posterior portion of the eye might be possible. In this context, the third-generation nanosized emulsions are being designed by adsorbing electrostatically the therapeutic oligonucleotides onto their surface for modulating functions of retinal pigment epithelium (RPE) cells effectively in order to treat blindness associated with age-related macular degeneration (AMD), proliferative vitreoretinopathy, retinal and choroidal neovascularization, and retinitis pigmentosa. To achieve this context, it becomes necessary to know first the ocular protective mechanisms and other concomitant factors to be faced by nanosized emulsions following ocular topical application. This point is further developed subsequently below.

5.2.1.1. Consequences of Nanosized Emulsions Following Ocular Topical Instillation From a medical point of view, nanosized emulsions for ophthalmic use aims at enhancing API bioavailability either by providing prolonged/

sustained delivery to the eye or by facilitating transcorneal/transconjunctival penetration. Following topical instillation into eye, the nanosized emulsion is likely to break due to tear fluid electrolytic action. Therefore, the water phase of the emulsion is drained off while the API-loaded oil phase of the emulsion remains in the cul-de-sac for a long time as a depot of the API (Sasaki et al. 1996). Since APIs includable in o/w-type nanosized emulsions are lipophilic in nature, the transconjunctival penetration appears to be more immense rather than classical transcorneal penetration following ocular topical instillation. Indeed, when compared with the cornea, API penetration through the conjunctiva has the advantage of larger surface area and higher permeability. Furthermore, the presence of API molecules in a considerable time in the *cul-de-sac* of the eye could result in a reservoir effect. However, following topical instillation of nanosized emulsions into eye, drainage and dilution are the two major events, which will occur in the precorneal area as a part of the protective mechanisms to ensure the proper functioning of the eye. Of these two events, the latter seems the least likely and as of yet has no experimental support. If the volume of instilled nanosized emulsions in the eye exceeds the normal lacrimal volume of 7–10 μl, then the portion of the instilled nanosized emulsions (one or two drops, corresponding to 50–100 μl) that is not eliminated by spillage from the palpebral fissure of conjunctiva is drained quickly via nasolacrimal system into nasopharynx and the gastrointestinal tract. In other words, the larger the instilled volume, the more rapidly the instilled nanosized emulsion is drained from the precorneal area. Hence the contact time of the nanosized emulsions with the absorbing surfaces (cornea, conjunctiva, and sclera) is reduced to a maximum of 2 min.

The so-called nonproductive absorption of nanosized emulsions occurs via the vessels of palpebral/scleral conjunctiva and concurs further in diminishing the nanosized emulsions/API-loaded nanosized emulsions absorption into the eye. In spite of a rapid removal of conjunctivally absorbed nanosized emulsions from the eye by local circulation and undergoing systemic absorption, direct transscleral access to some intraocular tissues cannot be excluded and this area needs however to be exploited further. Besides all above-described protective mechanisms of the eye, an unswept nanosized emulsion remaining in the precorneal area is subject to protein binding and to metabolic degradation in the tear film. In conjunction with blood plasma, although low, aqueous humor, tear film, and vitreous are also having varying amounts of relatively detectable proteins like albumin, globulin, and immunoglobulins (IgA, IgG, IgM, IgE, etc.) and the enzyme, lysozyme. Additional studies (at least *in vitro*) are necessary to understand clearly the nanosized emulsion interaction with these ocular fluid proteins. As expected, the destruction or depot action of nanosized emulsions might, indeed, result from their interaction with ocular fluid proteins.

In conclusion, the fluid dynamics (like blinking rate) in the precorneal area of the eye have a huge effect on ocular nanosized emulsion absorption and subsequently the nanosized emulsions incorporated API molecule disposition. The fluid dynamics may be altered by the physical and chemical properties of nanosized emulsions, which include interfacial charge, irritant ingredients, osmolality, pH, refractive index, tonicity, and viscosity. All these factors make the situation more complex particularly following topical instillation of nanosized emulsions into the eye. Thus, the formulation of ophthalmic API products must take into account not only the stability and compatibility of an API in the nanosized emulsions but also the influence of nanosized emulsions on precorneal fluid dynamics. All of the concepts exposed in this section may ultimately result in transcorneal/conjunctival absorption of 1–2% or less of the API applied topically through the nanosized emulsions. In summary, the rate of loss of API/nanosized emulsions from the eye can be 500–700 times greater than the rate of absorption into the eye. Thus, conventional topical delivery using nanosized emulsions cannot nevertheless achieve adequate intracellular concentrations of APIs or other substances such as oligonucleotides or genes for the treatment of endophthalmitis or other sight-threatening intraocular diseases, AMD.

5.2.1.2. Consequences of Nanosized Emulsions Following Intraocular Injection In order to achieve a high concentration of APIs within the eye using nanosized emulsion delivery vehicle, an approach enabling bypass of physiological and anatomical barriers (like blood-ocular) of eye is a more viable and attractive option. One such approach is to administer nanosized emulsions through direct intraocular injections such as intracameral, intracapsular, intravitreal, periocular (subconjunctival and sub-Tenon), or subretinal. Moreover, it is likely that intraocularly administered nanosized emulsions are able to both significantly increase API half-life and to minimize intraocular side effects appearing following intraocular injections of API alone. In general, APIs encapsulated within nanosized emulsions are less toxic than their free counterparts. Additionally, there is a possibility of obtaining slow API release from intraocularly injected nanosized emulsions. Taking into account the non-phagocytic character of neural retinal cells and ability of RPE cells to take up large molecules, including oligonucleotides, nanosized emulsions for intravitreal or subretinal injections are more likely to be a successful approach in future. Moreover, intravitreally administered API molecules are able to bypass the blood-ocular barrier to achieve constant therapeutic levels in the eye while minimizing systemic side effects. However, the hyalocytes, main cellular component of the vitreous, have been classified in at least one report (Maurice and Mishima 1984) as macrophages and thus could play a role in the uptake of intravitreally injected nanosized emulsions. It should be added that few studies are focused so far to inject nanosized emulsions intraocularly into eye and significant work should be devoted to generate this novel idea into a fruitful solution in ophthalmic API delivery applications.

5.2.2. Issues Relevant to Parenteral Nanosized Emulsions

Conceptually in a simple manner, what could happen inside a biosystem upon parenteral administration of nanosized emulsions will be discussed in more detail below. Interest in this concept has increased significantly with the greater understanding of the processes involved in parenteral nanosized emulsions both at the cellular and subcellular levels.

5.2.2.1. First-Generation Nanosized Emulsion Metabolism It is well established that when nanosized emulsions are mixed with blood, many plasma proteins associate with the emulsion droplet surfaces. Among the plasma proteins, the exchangeable apolipoproteins (apos) are particularly important because they can lead to disruption of the nanosized emulsions with concomitant release of any incorporated materials. In general, essentially all apos, except apo B, exchange between lipid surfaces and are attractive candidates for participation in nanosized emulsion catabolism.

5.2.2.1.1. Interaction between nanosized emulsions and plasma apos Traditionally, nanosized emulsions have been thought to be a part of clearance and targeting pathways similar to that of endogenous triglyceride-rich particles (TGRP), i.e., chylomicrons and very low-density lipoprotein (VLDL). Chylomicrons are triglyceride-rich lipoproteins synthesized in the enterocyte (intestinal mucosal cells) from the fats and other lipids ingested in the diet and absorbed by the intestine whereas VLDL are synthesized in the liver. These TGRP are secreted into extracellular fluid and mesenteric lymph and subsequently enter the blood circulation via the thoracic duct to undergo hydrolysis by lipoprotein lipase (LPL) on the capillary wall (or on the surface of endothelial cell) in order to generate small VLDL remnants or chylomicron remnants (monodiacylglycerols and fatty acids). However, to act on TGRP, LPL requires an activator, apo CII, a 79 amino acid residue peptide that is normally present in sufficient amount on the surface of the chylomicrons and VLDL particles. Apo CII possesses a domain that binds to the enzyme, LPL, and another domain that stimulates the enzymatic action. The remnants thus formed by the action of LPL are then sequestered into the space of Disse and taken up by the liver cells by various receptor mechanisms, mainly through either low-density lipoprotein (LDL) receptors or LDL receptor-related protein (LRP) (Cooper 1997). The space of Disse is rich in heparan sunanosized emulsionsate proteoglycans (HSPG) and contains an abundance of apo E secreted by hepatocytes and LPL presumably carried on the remnants. After entering the space of Disse, remnant particles become enriched in apo E, which enhance the binding of the remnants to HSPG on the surface of hepatocytes to undergo internalization of the particles via LDL receptors or LRP. Employing gene disruption techniques, it has been shown in mice that a deficiency in apo E results in massive accumulation of VLDL remnant, leading to severe hypercholesterolemia and premature atherosclerosis (Zhang et al. 1992). On the other hand, transgenic mice over expressing apo E manifest decreased plasma cholesterol levels on a chow diet and a marked resistance to diet-induced hypercholesterolemia. This study documents the direct relevance of apo E in modulating plasma lipoprotein metabolism, a property attributed to its ability to act as a ligand for the LDL receptors.

Although as many as 11 different exchangeable apos have been identified in human plasma (Segrest et al. 1992), apo E is nevertheless the main ligand of remnant particles to the hepatic receptors. However, it is interesting to see how the apo E would interact with remnant particles and in what the structural alterations took place on apo E during such interactions. A common feature of exchangeable apos is their ability to exist in a dual state: a lipid-free helix bundle state and a lipid-bound open state. The helix bundle organization of apos offers a unique way to sequester the hydrophobic protein interior in a lipid-free state until a metabolic need arises. Therefore, the appearance of hydrophobic lipidic surfaces like nanosized emulsions or remnant particles in blood circulation leads to recruitment of apos to the particle surface. Although devoid of apos, the nanosized emulsions on contact with plasma acquires exchangeable apos such as apo CII and apo E, within milliseconds. Consequently, the nanosized emulsion undergoes hydrolysis by LPL on the capillary wall, triggered by apo CII. Then, the resulting remnant particles are taken up by receptors in the liver parenchymal cells that recognize apo E (Floren and Nilsson 1987; Beisiegel et al. 1989), similar to that of TGRP, i.e., chylomicrons and VLDL. Alternatively, the remnant particles are also cleared via the LDL receptor (Redgrave and Maranhao 1985; Ishibashi et al. 1996), which is mediated by both apo B_{100} and apo E (Brown and Goldstein 1986). Considering nanosized emulsions only as fat-sorting lipoproteins or foreign particles for elimination from the blood circulation, formulation variables like particle size, oil phase, emulsifier, etc., on the biodistribution of nanosized emulsions was already reviewed very concisely by Nishikawa et al. (1998) in a book chapter. A synopsis of this book chapter and other conclusions about the contribution of formulation variables on the biodistribution of nanosized emulsions are briefly summarized in Table 4.3 (Chapter 4).

5.2.2.1.2. Apo E on nanosized emulsion metabolism Apo E, a 299-residue plasma apo, has seven amphipathic helical segments and is composed of two independently folded domains. The 10 kDa COOH-terminal domain (225–299) possesses high lipid affinity, while the 22 kDa NH_2-terminal domain (1–191) binds weakly to lipid (because of water-soluble character) and mediates receptor interactions (Weisgraber 1994). Lu et al. (2000) have shown that the 22 kDa NH_2-terminal structure of apo E reorganizes upon interaction with lipid. It has been proposed that the NH_2-terminal domain of apo E undergoes conformational changes, converting from a closed receptor-inactive conformation in the absence of lipid to an open receptor-active conformation in the presence of phospholipid.

A lipid-protein binding study employing synthetic nanosized emulsions and intact apo E4 (one of the isoform of apo E in humans relevant to Alzheimer's disease) supports this model (Saito et al. 2001). The binding enthalpies either between the nanosized emulsions and fragments of apo E4 (22 kDa NH_2-terminal fragments and 10 kDa COOH-terminal) or between nanosized emulsions and intact apo E4 were calculated depending on the surface concentrations of protein. At a low surface concentration of protein, the binding enthalpy of intact apo E4 was consistent with the sum of the enthalpies for the 22 and 10 kDa derivatives, indicating that both NH_2- and COOH-terminal domains bind to the emulsion surface. At saturation, however, the enthalpy for intact apo E4 was similar to that of the 10 kDa fragments, suggesting that only the COOH-terminal domain of intact apo E4 interacts with the nanosized emulsions or remnant particle surfaces. It has to be mentioned that the actual factor that determines hydrophobic/hydrophilic balance is the ΔG (free energy) of transfer from water to medium having lower dielectric coefficient such as ethanol, not the water solubility.

5.2.2.1.3. Apo E versus apo Cs, and LPL on nanosized emulsions metabolism Physiologically, the apo E-mediated hepatic uptake of plasma lipoproteins by cell receptors has, however, been argued to be governed by the ratio of apo E to apo Cs rather than the absolute apo E content (Jong et al. 1999). It is known that apo Cs inhibits the apo E-dependent recognition of TGRP by the LDL receptors and LRP (Sehayek and Eisenberg 1991). Oswald and Quarford (1987) showed that apo C-I, apo C-II, and apo C-III effectively inhibited the apo E-mediated uptake of nanosized emulsions by HepG2 cells. Perhaps, it has been shown that incorporation of sphingomyelin (SM) into the nanosized emulsion surface reduced the ratio of apo C-II and apo C-III to apo E (Arimoto et al. 1998). By combining these two studies, the effects of SM on apo E- and LPL-mediated cell uptake of SM-based nanosized emulsions have been investigated (Morita et al. 2003). SM in the nanosized emulsion surface enhanced the inhibitory effect of apo C-II and apo C-III on the apo E-mediated uptake of nanosized emulsions, presumably caused by intensified displacement of apo E. Furthermore, SM also inhibited the LPL-mediated uptake by decreasing the binding amount of LPL on the nanosized emulsion surface. LPL binds to the lipid surface without marked changes in the surface structure (MacPhee et al. 1997) and this suggests that a superficial binding of LPL to the phospholipid surface. Therefore, the lipid binding of LPL is presumed to depend on the structure of the head group region, which is significantly affected by the presence of SM in nanosized emulsions. It is further considered that SM could regulate lipolysis rates by several mechanisms: inhibiting LPL binding to nanosized emulsions, inhibiting catalytic activity of LPL, inhibiting apo C-II binding to nanosized emulsions, or a combination of the above.

In addition to apo E, there is strong evidence that the binding of VLDL to HSPG is also mediated by LPL (Olivecrona et al. 1977), by forming a bridge between the lipoproteins and HSPG. Furthermore, LPL enhances the uptake of remnants partially mediated by the ability of LPL to directly bind to LRP and LDL receptors (Medh et al. 1996). Taken together, these findings indicate that the balance between apo E and apo C II on the particle surface governs the hepatic uptake of remnant particles. Thus, apo E and LPL play key roles in the hepatic metabolism of VLDL and chylomicrons and in fact, display opposite effects on receptor binding.

5.2.2.1.4. Apo E and apo Cs versus apo As Various studies have reported the modulatory influence of apo As, particularly apo A-IV, on intravascular TGRP metabolism. Apo A-IV is an abundant 46 kDa plasma glycoprotein containing 6% carbohydrate and 376 amino acid residues (Weinberg and Scanu 1983). It is coded by a gene located on chromosome 11q and is clustered with apo A-I and apo C-III (Karathanasis et al. 1986). In humans, its synthesis is restricted to the enterocytes whereas in rodents, both the liver and small intestine produce it. Depending on the nutritional state, the plasma concentration of apo A-IV normally varies from 14 to 37 mg dl^{-1}. Structurally, apo A-IV contains 14-peptide repeats with the potential of forming amphipathic α-helices (Elshourbagy et al. 1986). The amphipathic nature of these peptide region repeats has led to the proposal that they are involved in lipid binding. Apo A-IV is initially incorporated into the surface of nascent chylomicrons and high-density lipoproteins (HDL) with a minor proportion in VLDL. An active exchange also occurs in the blood circulation between chylomicrons and HDL particles. The former provide apo A-IV and the latter donate apo C-II and apo E (Lenzo et al. 1988). By the way of facilitating the transfer of the activator apo C-II to TGRP, apo A-IV is able to mediate the activity of LPL (Goldberg et al. 1990). It was also shown to stimulate V_{max} of LPL in nanosized emulsions containing apo B and apo C-II (Sato et al. 2002), indicating further the influence of apo A-IV on this enzyme. The more recent advances in apo A-IV physiological functions and regulation were elaborately reviewed by Stan et al. (2003).

Another apo A of interest is apo A-I. Apo A-I has a single polypeptide chain containing 243 residues. These residues are divided into a 43-residue globular amino-terminal domain encoded by exon 3 of the apo A-I gene and a 200-residue lipid binding domain encoded by exon 4 of the apo A-I gene at the carboxy terminus (Luo et al. 1986). The secondary structure of the lipid binding domain is dominated by 10 helical repeats that have been classified as amphipathic α-helices. Eight of these repeats are 22 residues in length while two repeats are 11 residues in length with most of these amphipathic α-helical repeats punctuated by a proline residue at the beginning of the repeat (Segrest et al. 1992). Although apo A-I together with apo A-II are the two major proteins of HDL and are able to modulate the packing of phospholipid in the bilayers of liposome, thereby influencing the bilayer permeability characteristics (responsible for phospholipid-based liposomal structure disruption *in vivo*), both apo As role in nanosized emulsions metabolism is considered in general to be very limited. Moreover, apo A-I binds to phosphatidylcholine (PC) vesicles with higher affinity and lower capacity compared with triglyceride-PC emulsions (Saito et al. 1997). The binding of apo A-I with triolein-rich emulsion particles were studied *in vitro* as a function of increasing surface concentrations of oleic acid or cholesterol (Derksen and Small 1989; Derksen et al. 1989). No change in binding capacity of apo A-I occurred when the particle cholesterol content was increased from 1.1 to 3.7%. However, when the emulsion particles were saturated with cholesterol (above 3.7%), the protein binding capacity was decreased sharply without virtually unchanging the dissociation constant. This might be responsible for the impairment in the apo A-I redistribution and alteration in the metabolism of emulsion remnant particles (Derksen and Small 1989). Similarly, the effect of oleic acid contents in triolein-rich emulsion particles on apo A-I binding was studied and the binding capacity and dissociation constant values were linearly dependent on the surface oleic acid concentration (Derksen et al. 1989). These data show that oleic acid allows more apo A-I to bind with higher affinity to cholesterol-saturated emulsion particles.

5.2.2.1.5. Apo E isoforms' interaction with nanosized emulsions in vitro study In humans, apo E is a 34 kDa protein that exists as one of three predominant isoforms differing in amino acids at two positions (Weisgraber 1994). The three common isoforms of apo E are designated as apo E2 (C112/C158), apo E3 (C112/R158), and apo E4 (R112/R158). Apo E3, the most abundant isoform, contains a single cysteine at position 112 and an arginine at position 158. Apo E4 has arginines at positions 112 and 158 while apo E2 possesses cysteines at both the positions. Moreover, synthetic apo E3 was previously shown to have similar physical and biological properties to native human plasma apo E (Vogel et al. 1985). In addition, apo E in conjunction with apo C and apo A are soluble enough in water so that the interactions of the isolated proteins with nanosized emulsions can be studied by a range of physical–biochemical techniques. Studies of this type would provide an insight in the interaction between plasma apos and nanosized emulsions at molecular levels.

There is a very interesting recent study concerning the human apo E adsorption onto the dimyristoylglycerophosphocholine- and egg phosphatidylcholine-stabilized first-generation nanosized emulsions (Perugini et al. 2002). These authors prepared nanosized emulsions of well-defined composition but different particle size. The *in vitro* binding of human apo E isoforms to size-fractionated nanosized emulsions is characterized according to Langmuir adsorption isotherm. From this adsorption isotherm curve, the authors estimated the apparent dissociation constant and maximum binding capacity of different apo E depending on the size of nanosized emulsions. They have demonstrated that apo E3 binds preferentially to small nanosized emulsions whereas apo E4 exhibits a preference for large nanosized emulsion particles. Furthermore, although the apparent binding affinity for large nanosized emulsions is similar ($K_d \approx 0.5\,\mu M$), the maximum binding capacity for apo E4 is significantly higher than for apo E3 (3.0 and 1.8 amino acids per phospholipid, respectively).

5.2.2.1.6. Other mechanism on nanosized emulsion metabolism Recent studies have suggested that nanosized emulsions are cleared from blood with less lipolysis and release of free fatty acids than chylomicrons, and that substantial amounts of emulsions can be cleared as intact whole particles by different tissues (Hultin et al. 1995; Olivecrona and Olivecrona 1998). Of relevant interest, the metabolism of Intralipid™ has been studied by using electron microscopy and full-sized emulsion droplets were observed to be trapped in the space of Disse, and intact large emulsion droplets were present in hepatocytes next to mitochondria (Vilaro and Llobera 1988). This indicates that large particles, with little or no preceding lipolysis, can penetrate the fenestrated endothelium and enter liver cells intact for intracellular degradation. Also, Savonen et al. (1999) demonstrated that the core label of intact chylomicrons was distributed similarly to both liver and peripheral tissues in a mink model that lacks LPL as in normal mink. Thus, tissue uptake of intact whole particles with little hydrolysis can be a major pathway for nanosized emulsion clearance *in vivo*. Furthermore, a more recent study on mice by Qi et al. (2002) showed that the catabolism of fish-oil-based nanosized emulsions and the catabolism of soy oil or safflower oil-based nanosized emulsions occur via very different mechanisms. Removal of the soy or safflower oil-based nanosized emulsions is modulated by LPL, apo E, low-density lipoprotein receptor (LDL-R), and lactoferrin-sensitive pathways. In contrast, although clearance of fish-oil-based nanosized emulsions relies on LPL to a very minor extent and is independent of apo E, LDL-R, and lactoferrin-sensitive pathways, these emulsion particles were cleared faster from blood suggesting different catabolic pathways. Other possible mechanisms for this little or no preceding lipolysis of nanosized emulsions may include saturation of cellular uptake processes, depletion of adsorbed plasma apos from the emulsion droplet surfaces, or a combination of these events.

5.2.2.2. Newer Generation Nanosized Emulsion Metabolism For new-generation nanosized emulsions, with an exception of long circulation second- and third-generation nanosized emulsions, the *in vivo* studies focusing the metabolism aspect are not generally so extensive like the literatures that appeared for first-generation nanosized emulsions. However, an *in vitro* study is conducted on plasma protein adsorption onto the blank nanosized emulsion droplets (Harnisch and Müller 2000) to mimic the *in vivo* opsonization phenomenon responsible for the rapid clearance of the nanosized emulsion droplets from the blood. According to this author, the adsorption of many protein species like apo As, apo Cs, apo E, albumin, fibrinogen, and gamma globulin onto the emulsion droplet surfaces are detectable by two-dimensional polyacrylamide gel electrophoresis technique. However, it is still possible that the association of DNA/oligos/API with nanosized emulsions could change the structure of the nanosized emulsions in such a way as to adversely affect the adsorption of circulating apos and binding to the receptors. A more detailed biodistribution comparison study should be conducted between the nanosized emulsions alone and nanosized emulsions-DNA/oligos/API following parenteral administration in order to address the issue that association with DNA/oligos/API did not substantially change the uptake of the emulsion by the receptor mechanisms. Nonetheless, this type of classical comparison experiments in mice has recently been reported for carmustine (chemotherapeutic, lipophilic nitrosourea used in cancer treatment since the 1970s)-incorporated nanosized emulsions and nanosized emulsions alone (Maranhão et al. 2002). It was found that both nanosized emulsions carmustine and nanosized emulsions alone had an almost similar biodistribution in liver, spleen, kidney, lung, and brain.

The availability of transgenic and knockout animals will virtually allow investigators to address the roles of specific plasma apos in nanosized emulsion metabolism *in vivo*. To examine interactions of nanosized emulsions with lipoproteins or their receptors, there are now animal lines that overexpress or lack nearly the entire list of apo (A-I, A-IV, Cs, E, and B) as well as several receptors (LDL-R and LRP) and enzymes that act on lipoproteins (Breslow 1996). So, selecting the specific animals that lack or overexpress a particular apo would help to obtain the *in vivo* mechanism of the metabolism of nanosized emulsions alone or the metabolism of nanosized emulsions with adsorbed/incorporated biomolecules.

5.2.2.2.1. In vitro mimicking of opsonization process with surface (charged)-modified nanosized emulsions It is well known that the clearance of long-chain triglyceride (LCT) emulsions such as Intralipid® closely resembles the clearance of chylomicrons (Carpentier 1989) and is susceptible to the interaction of HDL acquiring plasma apos (Carpentier et al. 1987). Erkelens et al. (1979) reported that an anionic emulsion formulation (Intralipid®) captures apos within minutes after an infusion in man, facilitating its elimination. Once protein-free emulsions acquire apos along with other plasma proteins like albumin, fibrinogen, etc., the opsonins and other blood proteins can promote phagocytosis by forming a "bridge" between the emulsion particles and the phagocyte (Frank and Fries 1991). As a result, the emulsions will undergo the metabolic and transport pathways of lipoproteins and then be eliminated rapidly from the bloodstream. Studies (Wretlind 1981; Davis 1982) have shown that small changes in physical properties of fat emulsions can influence the elimination rate of these formulations from the blood. Indeed, an organ distribution study in male BALB/c mice of stearylamine-based cationic or deoxycholic acid-based anionic nanosized emulsion and Intralipid® was carried out (Klang et al. 1998; Yang and Benita 2000). In comparison with both of the anionic emulsions, the stearylamine-based cationic emulsion elicited a much longer retention time in the plasma, indicating clearly a long-circulating half-life for cationic emulsion in the blood. To enhance the API targeting efficacy of colloidal carriers like nanospheres and liposomes, a pegylation/cationization strategy is traditionally designed for the surface of these carriers (Barratt 2003). Whereas surface-pegylated colloidal carriers exhibit a prolonged plasma residence time through an escaping tendency from RES uptake following parenteral administration, surface-cationized colloidal carriers can facilitate the penetration of therapeutic agents into cell surface possibly via an endocytotic mechanism. The long-circulating effect of the cationic emulsion was attributed to the composition and conformation of the mixed emulsifiers' film at the o/w interface, namely Lipoid E 80, poloxamer 188, and oleylamine/stearylamine, which created an electrostatic and steric barrier at the oil–water interface. This combined effect is well established and has also been reported in an *in vitro* monolayer study (Korner et al. 1994).

Initially, surface (charge)-modified nanosized emulsions (anionic and cationic) were prepared following the well-established combined emulsification techniques and characterized for their droplet size distribution and surface charge. To further substantiate the previously observed *in vivo* behavior difference between the cationic and anionic emulsions, *in vitro* plasma protein adsorption patterns need to be tested onto these types of colloidal carriers to elucidate the anticipated influence of the various, mixed emulsifier combinations used while preparing these formulations. For this purpose, the two-dimensional polyacrylamide gel electrophoresis (2-D PAGE) technique, which was previously modified and successfully transferred to the marketed emulsion formulation (Harnisch and Muller 1998, 2000), was used. Following paragraphs discuss the results of *in vitro* plasma protein adsorption onto the surface (charge)-modified o/w nanosized emulsions (Tamilvanan et al. 2005).

The 2-D PAGE is a powerful tool for the simultaneous detection of up to 10,000 proteins at the same time. The 2-D PAGE separates the proteins according to two different parameters. In the first dimension, isoelectric focusing (IEF) is carried out for protein separation due to their isoelectric points. The second dimension, a sodium dodecyl sulfate (SDS)-PAGE, separates the proteins according to their molecular weights. This protocol enables the determination of proteins with an isoelectric point ranging from pH 4 to 10 and a molecular weight between 250 and 6 kDa. For the analysis of the *in vitro* plasma protein adsorption patterns on emulsion droplets with different surface charge, 2 ml of each emulsion were incubated with 6 ml citrate-stabilized human plasma. Incubation of the samples took place for 5 min at a temperature of 37°C. Emulsion droplets were separated from bulk plasma by centrifugation with 15,000 g (Cryofuge 20-3, Hereaus, Germany). Afterward, they were washed three times with phosphate buffer pH 7.4. Protein desorption was carried out with a SDS-solution according to a previously reported method (Bjellquist et al. 1993; Blunk et al. 1993). For the first dimension, immobilized nonlinear pH gradients ranging from 3.5 to 10 (Amershan Pharmacia, Sweden) were used. Sample entry was performed by in-sample-rehydration in a custom-made reswelling tray (Rabilloud et al. 1994; Sanchez et al. 1997).

The slab gels were cast with a Model 395 gradient former and multigel casting chamber containing 9–16% acrylamide gradient and 2.6% piperazine diacrylamide (PDA) as cross-linker.

All chemicals, devices, and running conditions are described elsewhere (Blunk et al. 1993; Harnisch and Muller 2000; Gessner et al. 2001). After 2-D PAGE, the gels were silver stained according to Blunk (1994) and scanned with an ImageScanner (Amershan Pharmacia, Sweden). Protein identification was carried out by matching the stained spots to a master map of human plasma, which is accessible on the Expasy server (Appel et al. 1993). Semiquantitative analysis of protein spots was possible with the Melanie III software (Bio-Rad, Germany).

The presence of poloxamer 188 in tested emulsions effectively eliminated the adsorption of the larger proteins like fibrinogen, immunoglobulins, etc. However, depending on the type of surface charges, the smaller proteins such as apos and albumin were almost completely adsorbed onto the nanosized emulsions. Indeed, when compared with marketed Lipofundin, medium-chain triglyceride (MCT) 10%- and deoxycholic acid-based anionic emulsions, the adsorption of apos, especially apoA-1, was approximately three times more on oleylamine- and stearylamine-based cationic emulsions and oleic acid-based anionic emulsions. In addition, the ratio between the apoA-1 to apoA-IV was found to be 1 for Lipofundin MCT 10% whereas it was about 0.26 for deoxycholic acid-based anionic emulsion and above 5 for oleic acid-based anionic emulsions and cationic emulsions. This indicates that emulsions having similar surface/interfacial charge imparted by different anion-forming stabilizers (oleic or deoxycholic acids) exhibited markedly different protein adsorption patterns.

Table 5.1 shows the plasma protein adsorption onto the surfaces of the developed anionic and cationic emulsions as well as the marketed Lipofundin MCT 10%. Figure 5.1 shows the 2-D PAGE gel of Lipofundin MCT 10%. The proteins designated in Fig. 5.1 represent more than 60% of the overall protein amount detected following adsorption on the emulsion droplets. Apos with a total amount of 45.6% were the most dominant protein species of the detected protein pattern, i.e., apoC-II (5.6%), apoC-III (10.6%), apoA-I (9.7%), apoA-II (9.2%), and apoA-IV (10.5%). Figure 5.2 compares the plasma protein adsorption patterns detected on the 2-D PAGE gels for both the anionic and cationic emulsions. From Figs. 5.1 and 5.2 and Table 5.1, it can be seen that with the exception of oleic acid-based anionic emulsion formulations, all other tested emulsions showed similar protein adsorption patterns when they were qualitatively compared. For oleic acid-based anionic emulsion formulation, no albumin is adsorbed and a reduction of apoA-IV is visible, whereas light chains of Ig are clearly detectable.

TABLE 5.1. Volume Percentage of Various Plasma Proteins Adsorbed onto the Surfaces of Developed Anionic and Cationic Emulsions and Commercial Lipofundin MCT 10% Obtained from 2-D PAGE Analyses

Various Emulsion Formulations Used		Volume Percentage of Various Adsorbed Plasma Proteins onto the Emulsion Surfaces Determined from 2-D PAGE Analyses									
		Albumin	ApoA-I	ApoA-II	ApoA-IV	ApoC-II	ApoC-III	ApoE	ApoJ	Ig-gamma-chain	Ig-light-chain
Anionic emulsions prepared based on	Deoxycholic acid	2.9 ± 0.6	9.7 ± 6.2	6.8 ± 0.9	36.9 ± 0.4	12.0 ± 0.9	14.4 ± 0.4	1.9 ± 0.9	0.4 ± 0.0	0.0 ± 0.0	1.3 ± 0.0
	Oleic acid	0.0 ± 0.0	25.0 ± 1.2	2.5 ± 0.0	4.6 ± 4.2	1.8 ± 0.3	4.5 ± 0.1	0.0 ± 0.0	0.0 ± 0.0	2.2 ± 0.0	6.3 ± 4.3
Cationic emulsions prepared based on	Oleylamine	2.5 ± 0.5	30.5 ± 2.1	4.8 ± 2.7	4.5 ± 0.3	4.5 ± 1.2	7.9 ± 3.7	1.2 ± 0.1	0.8 ± 0.3	1.2 ± 0.0	3.0 ± 0.2
	Stearylamine	2.7 ± 1.4	33.2 ± 4.6	6.1 ± 1.6	6.7 ± 1.8	6.2 ± 1.7	8.6 ± 2.7	1.5 ± 0.9	0.8 ± 0.5	0.9 ± 0.7	1.5 ± 0.6
Lipofundin MCT 10%		2.0 ± 1.1	9.7 ± 2.8	9.2 ± 1.5	10.5 ± 3.1	5.6 ± 2.1	10.6 ± 3.4	2.9 ± 0.9	0.5 ± 0.4	8.1 ± 2.5	0.0 ± 0.0

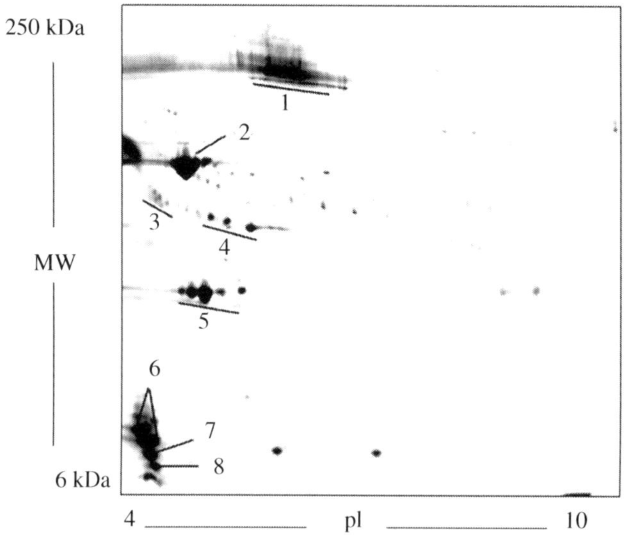

Figure 5.1. 2-D PAGE gel with protein adsorption pattern of Lipofundin MCT 10% after incubation in human plasma. Identification keys: (1) Albumin, (2) ApoA-IV, (3) ApoJ, (4) ApoE, (5) ApoA-I, (6) ApoC-II, (7) ApoC-III, (8) ApoA-II. [Reproduced with permission from Tamilvanan et al. (2005).]

Keeping the components of oil and water phases constant, the addition of acids like deoxycholic acid or oleic acid and primary amines like oleylamine or stearylamine allowed the formation of anionic and cationic emulsions, respectively, through an ionization process at the mixed o/w interfacial emulsifier film. Large proteins like gamma and light chains of Ig, fibrinogen, etc., were indeed prevented from being adsorbed on the emulsion formulations because of poloxamer hydrophilic conformation over the emulsion droplet surfaces. However, depending on the nature of electric charges of tested emulsions, smaller proteins such as various types of apoA and C, albumin, were adsorbed onto the emulsion droplet surfaces. Within the presented experimental data, the deoxycholic acid-based anionic emulsion with reported rapid clearance from plasma had by far the highest fraction of apos (>82%) adsorption patterns (Table 5.1). It should be emphasized that strong adsorption of apos does not automatically result in low recognition by the RES and cannot exclude regional effects from adsorbed lipoproteins. Moreover, the rate of elimination also depends on the ratio between apoC and apoE (Arimoto et al. 1998). The apos accounted for between 49 and 82% of the entire protein adsorption patterns onto all of the tested formulations except for the oleic acid-based anionic emulsion. Only a value of 38.4% apos adsorption was detectable on the oleic acid-based anionic emulsion formulation.

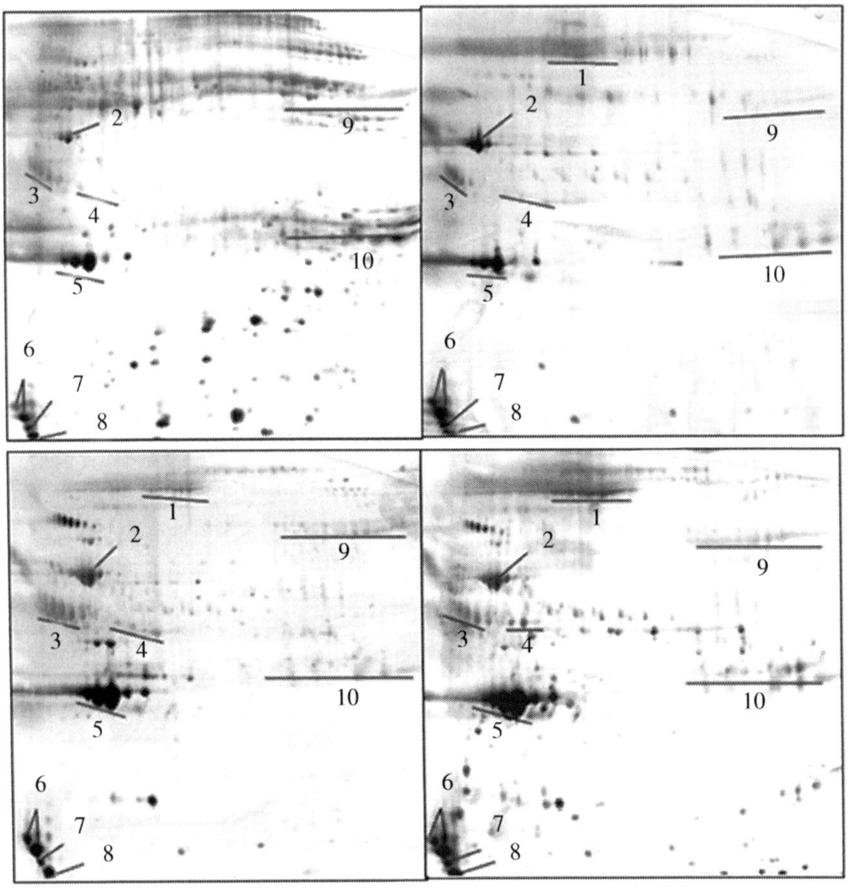

Figure 5.2. Silver-stained 2-D PAGE gels of emulsions with different surface charges after incubation in human plasma. Upper left: negative-charged formulation I (oleic acid-based anionic emulsion). Upper right: negative-charged formulation II (deoxycholic acid-based anionic emulsion). Lower left: positive-charged formulation I (stearylamine-based cationic emulsion). Lower right: positive-charged formulation II (oleylamine-based cationic emulsion). Identification keys: (1) Albumin, (2) ApoA-IV, (3) ApoJ, (4) ApoE, (5) ApoA-I, (6) ApoC-II, (7) ApoC-III, (8) ApoA-II, (9) IgG-gamma-chain, (10) Ig-light-chain. [Reproduced with permission from Tamilvanan et al. (2005).]

Based on Table 5.1, the observed threefold increase in the apoA-I protein adsorption on cationic emulsions in comparison with Lipofundin® MCT 10% and deoxycholic acid-based anionic emulsion formulation would be of clinical interest. In a previous report, apoA-I along with apoA-IV have been suggested to modulate the distribution of apoE between the different lipoprotein particles in the blood and thereby affect their clearance (Mahley and Hussain 1991). Furthermore, the attachment of apoE would greatly alter the *in vivo* distribution of fat emulsions since this protein is a ligand for the apoE-specific receptors on the liver parenchymal cells (Tenkanen and Ehnholm 1993). The higher the preferential adsorption of apoA-I onto the cationic emulsion droplets, the more intensified the displacement/redistribution of apoE would, therefore, be expected to occur on these types of cationic emulsion formulations in the blood. Indeed, the ratio of apoA-I to apoA-IV was very close to 1 for Lipofundin® MCT 10% whereas it was about 0.26 for deoxycholic acid-based anionic emulsion and above 5 for oleic acid-based anionic emulsion and both cationic emulsions. The binding of apoA-I with triolein-rich emulsion particles saturated with cholesterol was studied *in vitro* as a function of increasing surface concentrations of oleic acid (Derksen et al. 1989). The results indicated that oleic acid allows more apoA-I to bind with a higher affinity to cholesterol-saturated emulsion particles. As mentioned previously, the elimination of the anionic emulsion droplets was much faster than the elimination of the cationic emulsion droplets following intravenous administration to BALB/c mice (Klang et al. 1998; Yang and Benita 2000). It can further be deduced from the data depicted in Table 5.1 that the nature of the cationic lipid (oleylamine versus stearylamine) has no influence at all since almost identical protein adsorption patterns are observed. Based on this observation, the oleylamine-based cationic emulsion should possess a similar *in vivo* behavior of prolonged residence time in the blood as that of the stearylamine-based cationic emulsion following i.v. administration.

5.2.2.2.2. Influence of nanosized emulsions on monocyte-macrophages and neutrophils functions Since parenteral nanosized emulsions are eliminated from the bloodstream similarly to natural chylomicrons following the combined action of apos and LPL, the artificial nature of the nanosized emulsion-driven chylomicrons may lead to an undesirable interaction with the immune system. There is much evidence that phospholipid-stabilized first-generation nanosized emulsions also have an impact on mononuclear phagocytic function, particularly phagocytosis (Cleary and Pickering 1983; Waitzberg et al. 1996). Monocyte-macrophages are immune cells involved in antigen presentation, cytokine production, and phagocytosis. Flowchart 4.2 (Chapter 4) shows the mononuclear phagocyte system available in the body.

Experimental studies regarding the effects of nanosized emulsions on phagocytosis by circulating and resident monocyte-macrophages are controversial. Moreover, it seems that there are distinct macrophage responses to intravenous infusion of nanosized emulsions according to the anatomical origin of macrophages. Decreased phagocytosis and increased superoxide production by peritoneal macrophages, splenic macrophage proliferation, and bacterial translocation were seen following experimental total parenteral nutrition (TPN) infusion for a week (Shou et al. 1994). The addition of fish oil nanosized emulsions to LCT and MCT/LCT nanosized emulsions increased total liver and lung macrophage number and phagocytosis in rats after 96h of intravenous infusion (Cukier et al. 1999). The infusion of TPN with LCT nanosized emulsions into guinea pigs reduced phagocytosis of *Escherichia coli* by Kupffer cells and splenic macrophages (Hamawy et al. 1985). Wiernik et al. (1983) infused 20% soybean oil nanosized emulsions intravenously for 2 h into healthy volunteers. During the infusion, a significant increase in nitroblue tetrazolium (NBT) reduction by blood monocytes was noted. Preincubation of monocytes *in vitro* with LCT nanosized emulsions (20–100 mg ml^{-1}) for 30 min was found to increase the ability of the cells to migrate chemotactically and to phagocytose yeast particles. In a prospective, randomized, crossover clinical trial, malnourished gastric cancer patients received a 48 h TPN (40 kcal kg^{-1}) with 25% of energy provided as LCT or MCT/LCT nanosized emulsions (infused at the rate of 0.08 g kg^{-1} h^{-1}). No alteration in monocyte-macrophage function and phagocytosis was found with either nanosized emulsions compared with the appropriate controls (Waitzberg et al. 1997).

Since the initial function of neutrophils is to phagocytose and inactivate invasive agents, the effects of nanosized emulsions on the metabolic and microbial killing activities of neutrophils are investigated. Although an inhibitory effect of LCT and MCT/LCT nanosized emulsions on rat neutrophil phagocytosis *in vitro* was observed, no differences in *ex vivo* phagocytosis, chemotaxis, and bacterial killing were seen in healthy and septic rats when LCT nanosized emulsions, MCT/LCT nanosized emulsions, or saline was intravenously administered (Waitzberg et al. 1996). Similarly, the *in vitro* studies using human polymorphonuclear cells apparently showed a dose-dependent response to nanosized emulsions. While concentrations ranging from 15 to 100 mg ml^{-1} of soybean oil-based nanosized emulsions did show an inhibitory effect on neutrophil functions, concentrations higher than 100 mg ml^{-1} did not (Jarstrand et al. 1978; English et al. 1981; Wiernik et al. 1983; Bellinati-Pires et al. 1992). In contrast, a stimulatory effect was documented (Robin et al. 1989; Bellinati-Pires et al. 1992; Bellinati-Pires et al. 1993; Heine et al. 1999) when low doses of MCT/LCT nanosized emulsions were used. In a clinical trial with gastric cancer patients receiving TPN with LCT or MCT/LCT nanosized emulsions, bacterial killing by neutrophils was the only function reduced after LCT nanosized emulsions, although this function remained within the normal range values in 80% of the patients (Jarstrand et al. 1978; Waitzberg et al. 1997). Two studies showed no change in NBT reduction and superoxide anion production (Rasmussen et al. 1988; Waitzberg et al. 1997) while one study explored an inhibitory effect (Jarstrand et al. 1978).

The concerns about the association of nanosized emulsions with infectious morbidity were raised after a report of an increased risk of bacteriemia related to the use of nanosized emulsions in neonates (Freeman et al. 1990). However, in an extensive clinical trial in patients undergoing bone marrow transplantation while receiving TPN, there was no evidence that moderate doses of LCT nanosized emulsions were associated with the incidence of bacterial or fungal infections (Lenssen et al. 1998). In conclusion, LCT nanosized emulsions moderately decrease neutrophil bacterial killing. MCT containing nanosized emulsions, in a physical or structured way, may bring some advantages because of its positive effects on polymorphonuclear cells, macrophages, and cytokine production, particularly in critically ill or immunocompromised patients. However, there are no data available regarding the effect of *n*–3 fatty acid containing nanosized emulsions or olive oil emulsion on neutrophil function.

5.2.2.2.3. Immunomodulatory mechanism of nanosized emulsions Several mechanisms have been proposed to explain the different effects of nanosized emulsions on the immune system. Fatty acid incorporation and subsequent modification in the cell membrane composition is the main mechanism for immunomodulating effect of nanosized emulsions. Lipids can modulate the immune response and the impact of intravenously infused isolated fatty acids on the immune system has been studied since 1972 (DiLuzio 1972). In fact the physicochemical nature of the infused triacylglycerol, i.e., carbon chain length, saturation degree, size of the particle, and nature of fatty acids, may determine structural changes and alterations in the phagocytic cell activity. The incorporation of *n*–3 fatty acids into macrophage membranes occurs in 3–6 h *in vitro* and in 3 days *in vivo*. However, the effects of fatty acids on phagocytic activity *in vivo* and *in vitro* are controversial. The presence of dietary polyunsaturated fatty acids may inhibit microbicidal activity *in vivo* while the same fatty acids may stimulate this activity when directly added to cell culture *in vitro*. Possible explanation for this observed controversial result is provided by Mahoney et al. (1980). These authors calculated that, in normal conditions, the phagocytosis of an erythrocyte results in approximately 7% internalization of the macrophage surface area.

The incorporation of different fatty acids may influence cell permeability, dienoic eicosanoids secretion, membrane fluidity, phophatidylinositol production as well as membrane receptors, enzymes, and second messenger signaling processes. Furthermore, these effects are related to the membrane *n*–6:*n*–3 fatty acid ratio and to the total amount of *n*–3 and *n*–6 fatty acids infused. Although *n*–6 and *n*–3 polyunsaturated fatty acids are incorporated into cell membranes, it remains to be proved if MCT or saturated fatty acids also incorporated cell membranes when present in doses higher than in a normal diet. In addition, fatty acids may interfere with other immune-modulating mechanisms such as modulation of adhesion molecules, regulation of nitric oxide (NO) production, modulation of signal transduction pathways, and direct modulation of gene expression (Yaqoob 1998). Moreover, it has been shown that several unsaturated fatty acids bind directly to peroxisome proliferator-activated receptor family (PPAR—a group of key nuclear receptors involved in the regulation of lipid homeostasis) that regulates the expression of target genes by binding to DNA sequence elements. Through these actions, fatty acids could regulate expression of adhesion proteins, such as integrins and selectins, thus altering cell adhesion.

In spite of the reported different immunomodulating properties of isolated fatty acids, the first-generation nanosized emulsions, however, also contain other phospholipidic components, glycerol and α-tocopherol, apart from triglyceride. It is possible that the interaction of these distinct substances with each other may alter the above-described effects observed with isolated triglyceride (TG) on immunomodulatory functions (De la Fuente et al. 2000; Kruimel et al. 2000). The phagocytic (immunological) effects of TG-based nanosized emulsions particularly on monocyte-macrophages and neutrophils were reviewed by Waitzberg et al. (2002).

There is no reported immunological concern study using new-generation nanosized emulsions and this area has to be explored in near future. Although nanosized emulsions may not be an immunogenic, but following lipolysis of nanosized emulsions, the liberated fatty acids if they are in adequate (amount) enough could elucidate/produce immunomodulating properties, as they were described above.

5.3. CLINICAL ISSUES OF FIRST-GENERATION NANOSIZED EMULSIONS

The selection of various vegetable oils, such as those from cottonseed, soybean, or safflower plants, as the lipid sources for the early first-generation nanosized emulsions was principally based on providing a substrate that was high in the essential fatty acid, linoleic acid. According to the work of Holman et al. (1982), it was not until later that a pure safflower oil-based nanosized emulsion was demonstrated to supply insufficient amounts of linolenic acid, producing a clinical deficiency state involving significant neurologic abnormalities. Interestingly, both cottonseed oil and safflower oil contain very little linolenic acid, whereas soybean oil has approximately 10 times more than these per gram of oil (Bloch and Shils 2006). Consequently, the oil composition of certain commercial first-generation nanosized emulsions was modified by the addition of soybean oil, so as to increase the fraction of this essential fatty acid. Even today, with the newer emulsion products containing various mixtures of MCT or other LCT oils, soybean oil is routinely included because of the presence of relatively high amounts of essential fatty acids. Presently, only omega-6 LCT-based emulsions as either 100% soybean oil or as a 50:50 physical mixture (by weight) of safflower and soybean oils are available in the United States. Most of the serious adverse effects associated with these emulsions are a consequence of excessively high infusion rates. According to Klein and Miles (1994), LCT-based emulsions infused at rates exceeding 0.11 $g\,kg^{-1}\,h^{-1}$ underlie the major toxicities associated with their use in the clinical setting. Administration of lower amounts of LCTs, either by reducing the infusion rate or the dose of LCTs, or by employing mixtures of oils, such as those composed of LCTs and MCTs, to prepare first-generation nanosized emulsion, will decrease the risk of adverse events in susceptible patients. With the development of validated methods of fat globule analysis (Driscoll et al. 2001a, b), potential toxicity of first-generation nanosized emulsions from the infusion of unstable formulations has been an area of heightened research. The historical problems with ascertaining cause and effect of toxicity was largely the result of there being no reliable and quantifiable analytical technique to measure the presence of potentially embolic fat globules found in the large-diameter tail of the globule size distribution (GSD) (Driscoll 1997). Without such analytical capabilities, it is not possible to establish a dose–response relationship.

Conventional methods of assessing the GSD, such as laser diffraction, microscopy, electrical resistive pore techniques, and light scattering, have proved to be wholly inadequate in accurately quantifying the important large-diameter tail of injectable emulsions (Driscoll 2002). With the formal recognition of the single-particle (globule) optical sensing (SPOS) technique as the proposed stability-indicating assay capable of discerning the details of the large-diameter tail of emulsions by the USP (Chapter <729>, 2005), it is possible to explore this dose–response relationship in animal models. There have been three studies in animals that have used the SPOS technique in assessing the effects of infusing large fat globule-laden (i.e., unstable) first-generation nanosized emulsions (Driscoll et al. 2003, 2004, 2005). In the guinea pig model (Driscoll et al. 2005), 24-h total nutrient admixtures (TNA) infusions were compared with the stable group (n = 6) having a starting percentage of fat >5 μm (PFAT5) of 0.004% versus the unstable group with a starting PFAT5 of 2.4%. The lungs of the animals receiving the unstable emulsion showed evidence of oxidative stress. In the first 24-h TNA infusion study in rats, the starting PFAT5 levels for the stable group (n = 6) were 0.007%, whereas the unstable group (n = 7) had a PFAT5 = 0.682%. Infusion of the unstable emulsion was associated with oxidative stress in the liver (Driscoll et al. 2003). The second TNA study in rats was similar, but was conducted over 72 h with a stable group (PFAT5 = 0.004%) compared against an unstable group at a lower level of instability (PFAT5 = 0.117%) (Driscoll et al. 2004). In this study, the unstable group exhibited a similar degree of oxidative stress in the liver as the 24-h infusions, but this occurred at lower levels of instability, suggesting a cumulative adverse effect. The risk of infectious complications with the administration of first-generation nanosized emulsions from exogenous sources can occur in neonates and is likely a consequence of poor aseptic technique (i.e., during the preparation of syringes for lipid delivery) or excessively long infusion times. The Centre for Disease Control (CDC) (O'Grady et al. 2002) and USP Chapter 797 (2006) clearly recommend that as a separate infusion, lipid injectable emulsions not hang for a period exceeding 12 h. Despite this, some advocate 18- to 24-h infusion time in the neonatal ICU setting (Reiter 2002). A review of the risks suggests that adherence to the 12-h hang time is prudent and a safer method of lipid administration in infants (Sacks and Driscoll 2002). The clinical utility of first-generation nanosized emulsions goes beyond the provision of essential fatty acids and as a dense source of parenteral calories. The exogenous administration of the essential fatty acids, precursors to important second messengers influencing the metabolic response to injury (i.e., immune regulation, inflammation, pulmonary and renal function, and vagal tone) (Bistrian 2003), will have a profound effect on prostaglandin synthesis in cell membranes. Consequently, the inflammatory response can be greatly exaggerated, so that when essential fatty acids are given in sufficient quantities to critically ill patients, they may produce pathophysiologic effects (Driscoll et al. 2001a, b; Hasselmann and Reimund 2004; Lekka et al. 2004; McCowen and Bistrian 2005).

These adverse effects are largely mediated by the highly vasoactive prostaglandin-2 series. The omega-3 fatty acids, especially those of marine oil origin, including docosahexaenoic acid (DHA) and eicosapentaenoic acid (EPA), act principally through the less vasoactive prostaglandin-3 series and therefore downregulate or modulate the extent of the inflammatory response. Thus, beyond the provision of calories or supplements, certain lipids may exhibit favorable or even unfavorable pharmacologic-like effects under some clinical conditions. The etiology of parenteral nutrition-associated liver diseases (PNALD) may be because of the use of soybean-based emulsions, secondary to proinflammatory metabolites of omega-6 fatty acids (Grimminger et al. 1997), and decreased hepatic clearance of the parenteral lipid (Zaman et al. 1997). Soybean-derived lipids contain phytosterols (e.g., *b*-sitosterol, campesterol, and stigmasterol) that are linked with impairment of biliary secretion (Clayton et al. 1998). Past and very recent studies have suggested that phytosterols may be the "hepatotoxic" or "cholestatic" component of soybean-derived lipid emulsions, with recent molecular mechanisms of phytosterols being suggested (Carter and Shulman 2007; Carter et al. 2007). It has also been suggested that omega-6 fatty acids may contribute to impaired immunologic function (Wanten and Calder 2007). This multitude of factors results in a cholestatic, steatotic liver that is especially susceptible to inflammatory insults (e.g., bloodstream infections, surgery, and hepatotoxic medication) (Day and James 1998). In turn, repeated liver injury results in fibrosis, cirrhosis, and end-stage liver disease. Fish-oil-based emulsions address these problems on several fronts. Omega-fatty acid metabolites are less involved in the inflammatory response (Grimminger et al. 1997), and animal models have shown that parenteral fish oil does not impair biliary secretion and may prevent steatosis (Van Aerde et al. 1999; Araya et al. 2004; Alwayn et al. 2005). Hence, the liver is not predisposed to inflammatory insult, and liver injury can be prevented.

A popular work in a murine model of nonalcoholic fatty liver disease showed that administration of omega-3 fatty acids protected the liver against injury, whereas standard omega-6 fatty acids failed to do so (Alwayn et al. 2005). The data suggest that supplementation of omega-3 fatty acids might be beneficial in ameliorating parenteral nutrition (PN)-induced hepatic steatosis. Subsequently, the same group reports two infants with severe PN-associated liver disease where the clinical courses in both cases were reversed by the administration of parenteral omega-3 fatty acids (Gura et al. 2006). Although the data are preliminary and will require further study, the results are very promising as a potential treatment in this potentially life-threatening complication. Furthermore, the same authors (Gura et al. 2008) also compared safety and efficacy outcomes of a fish-oil-based fat emulsion in 18 infants with short-bowel syndrome who developed cholestasis (serum direct bilirubin level of >2 mg dl^{-1}) while receiving soybean emulsions with those from a historical cohort of 21 infants with short-bowel syndrome who also developed cholestasis while receiving soybean emulsions. The primary end point was time to reversal of cholestasis (three consecutive measurements of serum direct bilirubin level of ≤2 mg dl^{-1}). More importantly, the authors have not observed any deleterious adverse effects of treatment. These benefits may be because of the absence of soybean oil or because of the pharmacologic effects of fish oil. However, this hypothesis is difficult to test because of the need to provide essential fatty acids in parenterally fed patients. Ideally, a prospective randomized, controlled trial comparing fish-oil emulsions with soybean emulsions in the treatment of established parenteral nutrition-associated liver diseases should be conducted. But this type of study would be difficult to conduct, because some may consider it unethical to perform a study where children with preexisting parenteral nutrition liver injury could potentially be randomly assigned to a treatment group in which they would continue to receive a soybean oil-based parenteral lipid emulsion. A prospective, randomized trial is currently being underway at Children's Hospital Boston, Harvard Medical School, Boston, MA, USA, to assess the efficacy of fish-oil-based emulsions in the prevention of cholestasis in which infants who have never been exposed to either type of lipid emulsion are randomly assigned to either conventional soybean oil emulsion or fish-oil emulsion at the start of their parenteral nutrition course. In summary, from a clinical perspective, newer first-generation nanosized emulsions show great promise in certain patient settings, most notably in the intensive care unit in both adults and infants. The clinical use of alternative oils, such as fish oil, MCT, and olive oil show benefits over conventional soybean oil formulations. In adults, for example, the administration of omega-3 fatty acids via soybean oil-based lipids produces a heightened inflammatory response via production of 2-series prostaglandins, whereas substitution of a portion of the lipid with omega-3 fatty acids via fish oil can favorably dampen the inflammatory response. In infants, for example, the substitution of soybean oil with fish oil has recently been shown to reverse parenteral nutrition-associated liver disease. These advances should lead to safer infusion therapy in patients receiving lipid injectable emulsions. Furthermore, the current techniques for analyzing the pharmaceutical integrity of first-generation nanosized emulsions are now better defined, allowing standardization of these dosage forms and, ultimately, the formal establishment of Pharmacopoeial standards, and perhaps even compounding standards during the period of clinical use for these complex formulations.

5.4. CONCLUSION

The o/w nanosized emulsions are being categorized in terms of generations (Chapter 4) based on their performances in previous and present decades. First-generation nanosized emulsions are considered primarily as nutrient carriers to be administered via intravenous routes to bedridden patients. Upon intravenous injection, these nanosized emulsions mimic endogenous/natural TG-rich lipoprotein such as chylomicrons with respect to the rapid acquisition of apos, lipolysis by LPL, and uptake by the liver. Although the first-generation nanosized emulsions ostensibly contain commercially available lipids that are likely to closely mimic/resemble the metabolic fate of chylomicrons *in vivo*, these nanosized emulsions are however investigated extensively for immunological consequences after intravenous injections. Second-generation nanosized emulsions start initially as API carrier systems by solubilizing considerable amounts of lipophilic APIs at the oil phase or at the oil–water interface of the emulsion. This particular merit of nanosized emulsions is specifically exploited even commercially for both ocular- and parenteral-active APIs. Modifications made either in oil phase or at o/w interfacial film-forming emulsifier molecules allow the nanosized emulsions to be able to escape from lipolysis by LPL, apos adsorption, and liver uptake. Such surface-modified nanosized emulsions would prolong the circulation time in plasma and thereby an alteration in *in vivo* disposition of incorporated APIs following parenteral administration. Attachment of homing devices like antibody and apo E make the nanosized emulsions to deliver APIs selectively to target sites either a tumorized organ or hepatic system. Having together a positive charge and a steric stabilizing effect lead to the development of third-generation nanosized emulsions that contain a unique property: plasma half-life prolongation and electrostatic adhesion to ocular surface tissues after topical instillation into eye. Furthermore, the third-generation nanosized emulsions show a potential to carry a wide range of amphiphilic, lipophilic, and polyanionic compounds including DNA and oligonucleotides for ferrying across the cell membranes. Accumulating knowledge thus suggests that constant progress in the better understanding of principles and process governing the various issues related to nanosized emulsions has surely brought major improvements in the efficacy of parenteral or ophthalmic API delivery systems.

REFERENCES

Alwayn, I.P., Gura, K., Nosé, V. et al. (2005), Omega-3 fatty acid supplementation prevents hepatic steatosis in a murine model of nonalcoholic fatty liver disease, *Pediatr. Res.*, 57, 445–452. doi:10.1203/01.PDR.0000153672.43030.75

Appel, R.D., Sanchez, J.-C., Bairoch, A. et al. (1993), SWISS-2D PAGE: a database of two-dimensional gel electrophoresis images, *Electrophoresis*, 14, 1232–1238. doi:10.1002/elps.11501401185

Araya, J., Rodrigo, R., Videla, L.A. et al. (2004), Increase in long-chain polyunsaturated fatty acid n-6/n-3 ratio in relation to hepatic steatosis in patients with non-alcoholic fatty liver disease, *Clin. Sci.*, 106 (6), 635–643. doi:10.1042/CS20030326

Arimoto, I., Matsumoto, C., Tanaka, M. et al. (1998), Surface composition regulates clearance from plasma and triolein lipolysis of lipid emulsions, *Lipids*, 33, 773–779. doi:10.1007/s11745-998-0269-8

Barratt, G. (2003), Colloidal drug carriers: achievements and perspectives, *Cell. Mol. Life Sci.*, 60, 21–37. doi:10.1007/s000180300002

Beisiegel, U., Weber, W., Ihrke, G. et al. (1989), The LDL–receptor–related protein, LRP, is an apolipoprotein E-binding protein, *Nature*, 341, 162–164. doi:10.1038/341162a0.

Bellinati-Pires, R., Waitzberg, D.L., Salgado, M.M. et al. (1992), Effect of medium- and long-chain triglycerides on human neutrophil migration, *Braz. J. Med. Biol. Res.*, 25 (4), 369–373.

Bellinati-Pires, R., Waitzberg, D.L., Salgado, M.M. et al. (1993), Functional alterations of human neutrophils by medium-chain triglyceride emulsions: evaluation of phagocytosis, bacterial killing, and oxidative activity, *J. Leukoc. Biol.*, 53 (4), 404–410. doi:10.1002/jlb.53.4.404

Bistrian, B.R. (2003), Clinical aspects of essential fatty acid metabolism: Jonathan Rhoads lecture, *J. Parenter. Enter. Nutr.*, 27 (3), 168–175. doi:10.1177/0148607103027003168

Bjellquist, B., Pasquali, C., Ravier, F. et al. (1993), A nonlinear wide-range immobilized pH gradient for two-dimensional electrophoresis and its definition in a relevant pH scale, *Electrophoresis*, 14, 1357–1365. doi:10.1002/elps.11501401209

Bloch, A.S. and Shils, M.E. (2006), Part VIII: appendices A-21: average values for triglycerides, fatty acids and cholesterol, in: Shils, M.E., Shike, M., Ross, A.C., Caballero, B., and Cousins, R., Eds., *Modern Nutrition in Health and Disease*, Lippincott, Williams & Wilkins, Baltimore, MD.

Blunk, T. (1994), Plasmaproteinadsorption auf kolloidalen Arzneist-offträgern, Ph.D. Thesis, University of Kiel, Kiel, Germany.

Blunk, T., Hochstrasser, D.F., Sanchez, J.-C. et al. (1993), Colloidal carriers for intravenous drug targeting: plasma protein adsorption patterns on surface-modified latex particles evaluated by two-dimensional polyacrylamide gel electrophoresis, *Electrophoresis*, 14, 1382–1387. doi:10.1002/elps.11501401214

Breslow, J.L. (1996), Mouse models of atherosclerosis, *Science*, 272 (5262), 685–688. doi:10.1126/science.272.5262.685

Brown, M.S. and Goldstein, J.L. (1986), A receptor-mediated pathway for cholesterol homeostasis, *Science*, 232 (4746), 34–47. doi:10.1126/science.3513311

Carpentier Y.A. (1989), Intravascular metabolism of fat emulsions: the arvid wretlind lecture, espen 1988, *Clin. Nutr.*, 8 (3), 115–125. doi:10.1016/0261-5614(89)90060-5

Carpentier, Y.A., Richelle, M., Bihain, B.E. et al. (1987), Interactions between exogenous fat and plasma/lipoproteins, *Infusionstherapie*, 14, 29–32. doi:10.1159/000226178

Carter, B.A. and Shulman, R.J. (2007), Mechanisms of disease: update on the molecular etiology and fundamentals of parenteral nutrition associated cholestasis, *Nat. Clin. Pract. Gastroenterol. Hepatol.*, 4, 277–287. doi:10.1038/ncpgasthep0796

Carter, B.A., Taylor, O.A., Prendergast, D.R. et al. (2007), Stigmasterol, a soy lipid-derived phytosterol, is an antagonist of the bile acid nuclear receptor FXR, *Pediatr. Res.*, 62, 301–306. doi:10.1203/PDR.0b013e3181256492

Clayton, P.T., Whitfield, P., and Iyer, K. (1998), The role of phytosterols in the pathogenesis of liver complications of pediatric parenteral nutrition, *Nutrition*, 14, 158–164. doi:10.1016/S0899-9007(97)00233-5

Cleary, T.G. and Pickering, L.K. (1983), Mechanisms of intralipid effect on polymorphonuclear leukocytes, *J. Clin. Lab. Immunol.*, 11, 21–26.
Cooper, A.D. (1997), Hepatic uptake of chylomicron remnants, *J. Lipid Res.*, 38 (11), 2173–2192.
Cukier, C., Waitzberg, D.L., Logullo, A. et al. (1999), Lipid and lipid-free total parenteral nutrition: differential effects on macrophage phagocytosis in rats, *Nutrition*, 15 (11–12), 885–889. doi:10.1016/S0899-9007(99)00186-0
Davis, S.S. (1982), Emulsions systems for the delivery of drugs by the parenteral route, in: Bundgaard, H., Bagger Hansen, A., and Kofod, H., Eds., *Optimization of Drug Delivery*, Munksgaard, Copenhagen, pp. 333–346.
Day, C.P. and James, O.F. (1998), Steatohepatitis: a tale of two "hits"?, *Gastroenterology*, 114 (4), 842–845. doi:10.1016/S0016-5085(98)70599-2
De la Fuente, M., Carazo, M., Correa, R. et al. (2000), Changes in macrophage and lymphocyte functions in guinea-pigs after different amounts of vitamin E ingestion, *Br. J. Nutr.*, 84, 25–29. doi:10.1017/S0007114500001197
Derksen, A., Ekman, S., Small, D.M. et al. (1989), Oleic acid allows more apoprotein A-1 to bind with higher affinity to large emulsion particles saturated with cholesterol, *J. Biol. Chem.*, 264, 6935–6940.
Derksen, A. and Small, D.M. (1989), Interaction of apoA-1 and ApoE-3 with triglyceride-phospholipid emulsions containing increasing cholesterol concentrations. Model of triglyceride-rich nascent and remnant lipoproteins, *Biochemistry*, 28 (2), 900–906. doi:10.1021/bi00428a074
DiLuzio, N.R. (1972), Employment of lipids in the measurement and modification of cellular, humoral and immune responses, in: Paoletti, R., and Kritchevsky, D., Eds., *Advances in Lipid Research*, Academic Press, New York, pp. 43–48.
Driscoll, D.F. (1997), Physicochemical assessment of total nutrient admixture stability and safety: quantifying the risk, *Nutrition*, 13 (2), 166–167. doi:10.1016/s0899-9007(96)00394-2
Driscoll, D.F. (2002), The significance of particle-sizing measurements in the safe use of intravenous fat emulsions, *J. Dispers. Sci. Technol.*, 23, 679–687. doi:10.1081/DIS-120015371
Driscoll, D.F., Adolph, M., and Bistrian, B.R. (2001b), Lipid emulsions in parenteral nutrition, in: Rombeau, J.L. and Rolandelli, R.H., Eds., *Clinical Nutrition: Parenteral Nutrition*, W.B. Saunders and Co., Philadelphia, PA, pp. 35–59.
Driscoll, D.F. Etzler, F. Barber, T.A. et al. (2001a), Physicochemical assessments of parenteral lipid emulsions: light obscuration versus laser diffraction, *Int. J. Pharm.*, 219 (1–2), 21–37. doi:10.1016/s0378-5173(01)00626-3
Driscoll, D.F., Ling, P.R., and Bistrian, B.R. (2003), Hepatic indicators of inflammation following the infusion of pharmaceutically unstable all-in-one mixtures (AIO) in rats, *Clin. Nutr.*, 22, S16–S17. doi:10.1016/S0261-5614(03)80059-6
Driscoll, D.F., Ling, P.R., and Bistrian, B.R. (2004), Hepatic indicators of oxidative stress following a 72-h infusion of an unstable all-in-one (AIO) in rats, *Clin. Nutr.*, 23, 857.
Driscoll, D.F., Ling, P.R., Quist, W.C., and Bistrian, B.R. (2005), Pathological consequences from the infusion of unstable lipid emulsion admixtures in guinea pigs, *Clin. Nutr.*, 24, 105–113. doi:10.1016/j.clnu.2004.07.022
Elshourbagy, N.A., Walker, D.W., Boguski, M.S. et al. (1986), The nucleotide and derived amino acid sequence of human apolipoprotein A-IV mRNA and the close linkage of its gene to the genes of apolipoproteins A-I and C-III, *J. Biol. Chem.*, 261 (5), 1998–2002.

English, D., Roloff, J.S., Lukens, J.N. et al. (1981), Intravenous lipid emulsions and human neutrophil function, *J. Pediatr.*, 99 (6), 913–916. doi:10.1016/s0022-3476(81)80019-4

Erkelens, D.W., Brunzell, J.D., and Bierman, E.L. (1979), Availability of apolipoprotein CII in relation to the maximal removal capacity for an infused triglyceride emulsion in man, *Metabolism*, 28 (5), 495–501. doi:10.1016/0026-0495(79)90188-4

Floren, G.H. and Nilsson, A. (1987), Symptoms and endoscopic findings in the diagnosis of gastroesophageal reflux disease, *Scand. J. Gastroenterol.*, 22 (6), 714–718. doi:10.3109/00365528709011148

Frank, M.M. and Fries, L.F. (1991), The role of complement in inflammation and phagocytosis, *Immunol. Today*, 12 (9), 322–326. doi:10.1016/0167-5699(91)90009-I

Freeman, J., Goldmann, D.A., Smith, N.E. et al. (1990), Association of intravenous lipid emulsion and coagulase-negative staphylococcal bacteremia in neonatal intensive care units, *N. Engl. J. Med.*, 323, 301–308. doi:10.1056/NEJM199008023230504

Gessner, A., Olbrich, C., Schröder, W. et al. (2001), The role of plasma proteins in brain targeting: species dependent protein adsorption patterns on brain-specific lipid drug conjugate (LDC) nanoparticles, *Int. J. Pharm.*, 214 (1–2), 87–91. doi:10.1016/S0378-5173(00)00639-6

Goldberg, I.J., Scheraldi, C.A., Yacoub, L.K. et al. (1990), Lipoprotein ApoC-II activation of lipoprotein lipase. Modulation by apolipoprotein A-IV, *J. Biol. Chem.*, 265 (8), 4266–4272.

Grimminger, F., Wahn, H., Mayer, K. et al. (1997), Impact of arachidonic versus eicosapentaenoic acid on exotonin-induced lung vascular leakage: relation to 4-series versus 5-series leukotriene generation, *Am. J. Respir. Crit. Care Med.*, 155 (2), 513–519.

Gura, K.M., Duggan, C.P., Collier, S.B. et al. (2006), Reversal of parenteral nutrition associated liver disease in two infants with short bowel syndrome using parenteral fish oil: implications for future management, *Pediatrics*, 118, e197–e201. doi:10.1542/peds.2005-2662

Gura, K.M., Lee, S., Valim, C. et al. (2008), Safety and efficacy of a fish oil-based fat emulsion in the treatment of parenteral nutrition-associated liver disease, *Pediatrics*, 121 (3), e678–e686. doi:10.1542/peds.2007-2248

Hamawy, K.J., Moldawer, L.-L., Georgieff, M. et al. (1985), The effect of lipid emulsions on reticuloendothelial system function in the injured animal, *JPEN*, 9 (5), 559–565. doi:10.1177/0148607185009005559

Harnisch, S. and Muller, R.H. (1998), Plasma protein adsorption patterns on emulsions for parenteral administration: establishment of a protocol for two-dimensional polyacrylamide electrophoresis, *Electrophoresis*, 19 (2), 349–354. doi:10.1002/elps.1150190233

Harnisch, S. and Muller, R.H. (2000), Adsorption kinetics of plasma proteins on oil-in-water emulsions for parenteral nutrition, *Eur. J. Pharm. Biopharm.*, 49, 41–46. doi:10.1016/s0939-6411(99)00064-8

Hasselmann, M. and Reimund, J.M. (2004), Lipids in the nutritional support of critically ill patients, *Curr. Opin. Crit. Care*, 10 (6), 449–455. doi:10.1097/01.ccx.0000144767.81163.6d

Heine, J., Scheinichen, D., Jaeger, K. et al. (1999), In vitro influence of parenteral lipid emulsions on the respiratory burst of neutrophils, *Nutrition*, 15 (7–8), 540–545. doi:10.1016/s0899-9007(99)00112-4

Holman, R.T., Johnson, S.B., and Hatch, T.F. (1982), A case of human linolenic acid deficiency involving neurological abnormalities, *Am. J. Clin. Nutr.*, 35 (3), 617–623. doi:10.1093/ajcn/35.3.617

Hultin, M., Carneheim, C., Rosenquist, K. et al. (1995), Intravenous lipid emulsions: removal mechanisms as compared to chylomicrons, *J. Lipid Res.*, 36 (10), 2174–2184.

Ishibashi, S., Perrey, S., Chen, Z. et al (1996), Role of the low density lipoprotein (LDL) receptor pathway in the metabolism of chylomicron remnants. A quantitative study in knockout mice lacking the LDL receptor, apolipoprotein E, or both, *J. Biol. Chem.*, 271 (37), 22422–22427. doi:10.1074/jbc.271.37.22422

Jarstrand, C., Berghem, L., Lahnborg, G. (1978), Human granulocyte and reticuloendothelial system function during intralipid infusion, *J. Parenter. Enter. Nutr.*, 2 (5), 663–670. doi:10.1177/014860717800200508

Jong, M.C., Hofker, M.H., and Havekes, L.M. (1999), Role of ApoCs in lipoprotein metabolism: functional differences between ApoC1, ApoC2, and ApoC3, *Arterioscler. Thromb. Vasc. Biol.*, 19 (3), 472–484. doi:10.1161/01.atv.19.3.472

Karathanasis, S.K., Oettgen, P., Haddad, I.A. et al. (1986), Structure, evolution, and polymorphisms of the human apolipoprotein A4 gene (APOA4), *Proc. Natl. Acad. Sci. U. S. A.*, 83 (22), 8457–8461. doi:10.1073/pnas.83.22.8457

Klang, S.H., Parnas, M., and Benita, S. (1998), Emulsions as drug carriers-possibilities, limitations, and future perspectives, in: Muller, R.H., Benita, S., and Bohm, H.L., Eds., *Emulsions and Nanosuspensions for the Formulation of Poorly Soluble Drugs*, Medpharm, Stuttgart, pp. 31–65.

Klein, S. and Miles, J.M. (1994), Metabolic effects of long-chain and medium-chain triglyceride emulsions in humans, *J. Parenter. Enter. Nutr.*, 18 (5), 396–397. doi:10.1177/0148607194018005396

Korner, D., Benita, S., Albrecht, G. et al. (1994), Surface properties of mixed phospholipid-stearylamine monolayers and their interaction with a non-ionic surfactant (poloxamer), *Colloids Surf. B: Biointerfaces*, 3 (1–2), 101–109. doi:10.1016/0927-7765(93)01111-4

Kruimel, J.W., Naber, A.H., Curfs, J.H. et al (2000), With medium-chain triglycerides, higher and faster oxygen radical production by stimulated polymorphonuclear leukocytes occurs, *J. Parenter. Enter. Nutr.*, 24 (2), 107–112. doi:10.1177/0148607100024002107

Lekka, M.E., Liokatis, S., Nathanail, C. et al. (2004), The impact of intravenous fat emulsion administration in acute lung injury, *Am. J. Respir. Crit. Care Med.*, 169 (5), 638–644. doi:10.1164/rccm.200305-620OC

Lenssen, P., Bruemmer, B.A., Bowden, R.A. et al. (1998), Intravenous lipid dose and incidence of bacteremia and fungemia in patients undergoing bone marrow transplantation, *Am. J. Clin. Nutr.*, 67 (5), 927–933. doi:10.1093/ajcn/67.5.927

Lenzo, N.P., Martins, I., Mortimer, B.-C. et al. (1988), Effects of phospholipid composition on the metabolism of triacylglycerol, cholesteryl ester and phosphatidylcholine from lipid emulsions injected intravenously in rats, *Biochim. Biophys. Acta*, 960, 111–118. doi:10.1016/0005-2760(88)90016-1

Lu, B., Morrow, J.A., and Weisgraber, K.H. (2000), Conformational reorganization of the four-helix bundle of human apolipoprotein E in binding to phospholipid, *J. Biol. Chem.*, 275 (27), 20775–20781. doi:10.1074/jbc.M003508200

Luo, C.C., Li, W.H., Moore, M.N. et al. (1986), Structure and evolution of the apolipoprotein multigene family, *J. Mol. Biol.*, 187 (3), 325–340. doi:10.1016/0022-2836(86)90436-5

MacPhee, C.E., Chan, R.Y., Sawyer, W.H. et al. (1997), Interaction of lipoprotein lipase with homogeneous lipid emulsions, *J. Lipid Res.*, 38 (8), 1649–1659.

Mahley, R.W. and Hussain, M.M. (1991), Chylomicron remnant catabolism, *Curr. Opin. Lipidol.*, 2 (3), 170–176.

Mahoney, E.M., Scott, W.A., Landsberger, F.R. et al. (1980), Influence of fatty acyl substitution on the composition and function of macrophage membranes, *J. Biol. Chem.*, 255 (10), 4910–4917.

Maranhão, R.C., Graziani, S.R., Yamaguchi, N. et al. (2002), Association of carmustine with a lipid emulsion: in vitro, in vivo and preliminary studies in cancer patients, *Cancer Chemother. Pharmacol.*, 49 (6), 487–498. doi:10.1007/s00280-002-0437-3

Maurice, D.M. and Mishima, S. (1984), Ocular pharmacokinetics, in: Sears, M.L., Ed., *Pharmacology of the Eye*, Springer Verlag, New York, pp. 20–116.

McCowen, K.C. and Bistrian, B.R. (2005), Essential fatty acids and their derivatives, *Curr. Opin. Gastroenterol.*, 21 (2), 207–215. doi:10.1097/01.mog.0000153361.90653.cb

Medh, J.D., Bowen, S.L., Fry, G.L. et al. (1996), Lipoprotein lipase binds to low density lipoprotein receptors and induces receptor-mediated catabolism of very low density lipoproteins in vitro, *J. Biol. Chem.*, 271 (29), 17073–17080. doi:10.1074/jbc.271.29.17073

Morita, S., Okuhira, K., Tsuchimoto, N. et al. (2003), Effects of sphingomyelin on apolipoprotein E- and lipoprotein lipase-mediated cell uptake of lipid particles, *Biochim. Biophys. Acta*, 1631 (2), 169–176. doi:10.1016/s1388-1981(02)00365-7

Nishikawa, M., Takakura, Y., and Hashida, M. (1998), Biofate of fat emulsions, in: S. Benita, Ed., *Submicron Emulsions in Drug Targeting and Delivery*, Harwood Academic Publishers, Amsterdam, The Netherlands, pp. 99–118.

O'Grady, N.P., Alexander, M., Dellinger, E.P., et al. (2002), Guidelines for the prevention of intravascular catheter-related infections, *Infect. Control Hosp. Epidemiol.*, 23 (12), 759–769.

Olivecrona, G., Bengtsson, G., Marklund, S.E. et al. (1977), Heparin-lipoprotein lipase interactions, *Fed. Proc.*, 36, 60–65.

Olivecrona, G. and Olivecrona, T. (1998), Clearance of artificial triacylglycerol particles, *Curr. Opin. Clin. Nutr. Metab. Care*, 1 (2), 143–151. doi:10.1097/00075197-199803000-00002

Oswald, E. and Quarford, S. (1987), Effect of apoE on triglyceride emulsion interaction with hepatocyte and hepatoma G2 cells, *J. Lipid Res.*, 28 (7), 798–809.

Perugini, M.A., Schuck, P., Howlett, G.J. et al. (2002), Differences in the binding capacity of human apolipoprotein E3 and E4 to size-fractionated lipid emulsions, *Eur. J. Biochem.*, 269 (23), 5939–5949. doi:10.1046/j.1432-1033.2002.03319.x

Qi, K., Seo, T., Al-Haideri, M. et al. (2002), Omega-3 triglycerides modify blood clearance and tissue targeting pathways of lipid emulsions, *Biochemistry*, 41 (9), 3119–3127. doi:10.1021/bi015770h

Rabilloud, T., Valette, C., and Lawrence, J.J. (1994), Sample application by in-gel rehydration improves the resolution of two-dimensional electrophoresis with immobilized pH gradients in the first dimension, *Electrophoresis*, 15, 1552–1558. doi:10.1002/elps.11501501223

Rasmussen, A., Hessov, I., and Segel, E. (1988), The effect of intralipid® on polymorphonuclear leucocytes, *Clin. Nutr.*, 7, 37–41. doi:10.1016/0261-5614(88)90009-X

Redgrave, T.G. and Maranhao, R.C. (1985), Metabolism of protein-free lipid emulsion models of chylomicrons in rats, *Biochim. Biophys. Acta (BBA)—Lipids Lipid Metab.*, 835 (1), 104–112. doi:10.1016/0005-2760(85)90036-0

Reiter, P.D. (2002), Sterility of intravenous fat emulsion in plastic syringes, *Am. J. Health Syst. Pharm.*, 59 (19), 1857–1859. doi:10.1093/ajhp/59.19.1857

Robin, A.P., Arain, I., Phuangsab, A. et al. (1989), Intravenous fat emulsion acutely suppresses neutrophil chemiluminescence, *J. Parenter. Enter. Nutr.*, 13 (6), 608–613. doi:10.1177/0148607189013006608

Sacks, G.S. and Driscoll, D.F. (2002), Does lipid hang time make a difference? Time is of the essence, *Nutr. Clin. Pract.*, 17 (5), 284–290. doi:10.1177/0115426502017005284

Saito, H., Dhanasekaran, P., Baldwin, F. et al. (2001), Lipid binding-induced conformational change in human apolipoprotein E. Evidence for two lipid-bound states on spherical particles, *J. Biol. Chem.*, 276(44), 40949–40954. doi:10.1074/jbc.M106337200

Saito, H., Miyako, Y., Handa, T. et al. (1997), Effect of cholesterol on apolipoprotein A-I binding to lipid bilayers and emulsions, *J. Lipid Res.*, 38 (2), 287–294.

Sanchez, J.-C., Rouge, V., Pisteur, M. et al. (1997), Improved and simplified in-gel sample application using reswelling of dry immobilized pH gradients, *Electrophoresis*, 18 (3–4), 324–327. doi:10.1002/elps.1150180305

Sasaki, H., Yamamura, K., Nishida, K. et al. (1996), Delivery of drugs to the eye by topical application, *Prog. Retin. Eye Res.*, 15 (2), 583–620. doi:10.1016/1350-9462(96)00014-6

Sato, K., Takahashi, Y., Takahashi, T. et al. (2002), Identification of factors regulating lipoprotein lipase catalyzed hydrolysis in rats with the aid of monoacid-rich lipoprotein preparations(1), *J. Nutr. Biochem.*, 13 (9), 528–538. doi:10.1016/s0955-2863(02)00194-8

Savonen, R., Nordstoga, K., Christophersen, B. et al. (1999), Chylomicron metabolism in an animal model for hyperlipoproteinemia type I, *J. Lipid Res.*, 40 (7), 1336–1346.

Segrest, J.P., Jones, M.K., De Loof, H. et al. (1992), The amphipathic helix in the exchangeable apolipoproteins: a review of secondary structure and function, *J. Lipid Res.*, 33, 141–166.

Sehayek, E. and Eisenberg, S. (1991), Mechanisms of inhibition by apolipoprotein C of apolipoprotein E-dependent cellular metabolism of human triglyceride-rich lipoproteins through the low density lipoprotein receptor pathway, *J. Biol. Chem.*, 266 (27), 18259–18267

Shou, J., Lappin, J., Minnard, E.A. et al. (1994), Total parenteral nutrition, bacterial translocation, and host immune function, *Am. J. Surg.*, 167, 145–150. doi:10.1016/0002-9610(94)90065-5

Stan, S., Delvin, E., Lambert, M. et al. (2003), Apo A-IV: an update on regulation and physiologic functions, *Biochim. Biophys. Acta*, 1631 (2), 177–187. doi:10.1016/s1388-1981(03)00004-0

Tamilvanan, S., Schmidt, S., Müller R.H. et al. (2005), In vitro adsorption of plasma proteins onto the surface (charges) modified-submicron emulsions for intravenous administration, *Eur. J. Pharm. Biopharm.*, 59, 1–7. doi:10.1016/j.ejpb.2004.07.001

Tenkanen, H. and Ehnholm, C. (1993), Molecular biology of apolipoprotein A-IV, *Curr. Opin. Lipidol.*, 4 (2), 95–99.

USP Chapter <729> (2005), Globule size distribution in injectable lipid emulsions. *US Pharmacopoeial Forum*, 31 (5), 1448–1453.

Van Aerde, J.E., Duerksen, D.R., Gramlich, L. et al. (1999), Intravenous fish oil emulsion attenuates total parenteral nutrition-induced cholestasis in newborn piglets, *Pediatr. Res.*, 45 (2), 202–208. doi:10.1203/00006450-199902000-00008

Vilaro, S. and Llobera, M. (1988), Uptake and metabolism of intralipid by rat liver: an Electron-Microscopic Study. *J. Nutr.*, 118 (8), 932–940. doi:10.1093/jn/118.8.932

Vogel, T., Weisgraber, K.H., Zeevi, M.I. et al. (1985), Human apolipoprotein E expression in *Escherichia coli*: structural and functional identity of the bacterially produced protein with plasma apolipoprotein E, *Proc. Natl. Acad. Sci. U.S.A*, 82 (24), 8696–8700. doi:10.1073/pnas.82.24.8696

Waitzberg, D.L., Bellinati-Pires, R., Salgado, M.M. et al. (1997), Effect of total parenteral nutrition with different lipid emulsions on human monocyte and neutrophil functions, *Nutrition*, 13 (2), 128–132. doi:10.1016/S0899-9007(96)00386-3

Waitzberg, D.L., Bellinati-Pires, R., Yamaguchi, N. et al. (1996), Influence of medium-chain triglyceride-based lipid emulsion on rat polymorphonuclear cell functions, *Nutrition*, 12 (2), 93–99. doi:10.1016/0899-9007(96)90706-6

Waitzberg, D.L., Lotierzo, P.H., Logullo, A.F. et al. (2002), Parenteral lipid emulsions and phagocytic systems, *Br. J. Nutr.*, 87 (S1), S49–S57. doi:10.1079/bjn2001456

Wanten, G.J. and Calder, P.C. (2007), Immune modulation by parenteral lipid emulsions, *Am. J. Clin. Nutr.*, 85 (5), 1171–1184. doi:10.1093/ajcn/85.5.1171.

Weinberg, R.B. and Scanu, A.M. (1983), Isolation and characterization of human apolipoprotein A-IV from lipoprotein-depleted serum, *J. Lipid Res.*, 24, 52–59.

Weisgraber, K.H. (1994), Apolipoprotein E: structure-function relationships, *Adv. Protein Chem.*, 45, 249–302. doi:10.1016/s0065-3233(08)60642-7

Wiernik, A., Jarstrand, C., and Julander, I. (1983), The effect of intralipid on mononuclear and polymorphonuclear phagocytes, *Am. J. Clin. Nutr.*, 37 (2), 256–261. doi:10.1093/ajcn/37.2.256

Wretlind, A. (1981), Parenteral nutrition, *Nutr. Rev.*, 39 (7), 257–265. doi:10.1111/j.1753-4887.1981.tb06783.x

Yang, S.C. and Benita, S. (2000), Enhanced adsorption and drug targeting by positively charged submicron emulsions, *Drug Dev. Res.*, 50 (3–4), 476–486. doi:10.1002/1098-2299(200007/08)50:3/4%3C476::AID-DDR31%3E3.0.CO;2-6

Yaqoob, P. (1998), Lipids and the immune response, *Curr. Opin. Clin. Nutr. Metab. Care*, 1 (2), 153–161.

Zaman, N., Tam, Y.K., Jewell, L.D. et al. (1997), Effects of intravenous lipid as a source of energy in parenteral nutrition associated hepatic dysfunction and lidocaine elimination: a study using isolated rat liver perfusion, *Biopharm. Drug Dispos.*, 18 (9), 803–819. doi:10.1002/(sici)1099-081x(199712)18:9%3C803::aid-bdd65%3E3.0.co;2-s

Zhang, S.H., Reddick, R.L., Piedrahita, J.A. et al. (1992), Spontaneous hypercholesterolemia and arterial lesions in mice lacking apolipoprotein E, *Science*, 258 (5081), 468–471. doi:10.1126/science.1411543

CHAPTER 6

MEDICAL OR THERAPEUTICAL APPLICATIONS OF OIL-IN-WATER NANOSIZED EMULSIONS

OLY KATARI* AND TAMILVANAN SHUNMUGAPERUMAL

Department of Pharmaceutics, National Institute of Pharmaceutical Education and Research (NIPER), Guwahati, Assam, India

*The funding provided by Department of Pharmaceuticals, Ministry of Chemicals and Fertilizers, Govt. of India to the author Oly Katari is greatly acknowledged.

Oil-in-Water Nanosized Emulsions for Drug Delivery and Targeting,
First Edition. Tamilvanan Shunmugaperumal.
© 2021 John Wiley & Sons, Inc. Published 2021 by John Wiley & Sons, Inc.

EXPANSION OF ABBREVIATIONS

apo	apolipoprotein
AMD	age-related macular degeneration
API	active pharmaceutical ingredient
BBB	blood–brain barrier
CNS	central nervous system
CsA	cyclosporin A
EMLA	eutectic mixture of local anesthetics
Ig	immunoglobulin
IOP	intraocular pressure
IV	intravenous
MPS	mean particle size
NLC	nanostructured lipid carriers
NCE	new chemical entities
NSAIDs	nonsteroidal anti-inflammatory drugs
OTC	over-the-counter
O/W	oil-in-water
PEG	poly(ethylene glycol)
PUFAs	polyunsaturated fatty acids
RGCs	retinal ganglion cells
RPE	retinal pigment epithelium
SLN	solid lipid nanoparticles
TEER	transepithelial/transendothelial electrical resistance
TM	trabecular meshwork
TPN	total parenteral nutrition
USFDA	United States Food and Drug Administration
W/O	water-in-oil

6.1. INTRODUCTION

Aside from ancestral/conventional/traditional use of first-generation oil-in-water (o/w) nanosized emulsions as total parenteral nutrition (TPN), the developed second-, third-, and fourth-generation nanosized emulsions have already found commercial positioning for managing different diseases/infections either through parenteral or topical administration. In this perspective, Table 4.1 (Chapter 4) displays a non-exhaustive list of commercially exploited o/w nanosized emulsions administered through various routes into the human body. It is very sad to note that the o/w nanosized emulsion did not find commercial positioning, in particular, to the oral route. But, it has been shown in a number of studies that the incorporation of active pharmaceutical ingredient (API) in o/w nanosized emulsions significantly increased the absorption of the API when compared with the equivalent aqueous solution administered orally

(Ilan et al. 1996; Kimura et al. 1989; Myers and Stella 1992; Palin et al. 1986). However, the use of emulsions for oral application is limited since other attractive alternatives, such as self-emulsifying oil delivery systems, which are much less sensitive and easy to manufacture, are available (Charman et al. 1992; Gursoy and Benita 2004). Hence, the objective of this chapter is to cover the potential medical or therapeutical applications of o/w nanosized emulsions administered via parenteral and traditional non-parenteral topical routes such as nasal, ocular, and percutaneous. It should be added that the discussions provided about the targeting concepts of o/w nanosized emulsions in Chapter 4 also coincide and elaborate the objective of medical or therapeutical applications of nanosized emulsions. Only few research reports are available regarding the API-targeting to the central nervous system (CNS), especially the various regions of brain using o/w nanosized emulsions are discussed subsequently below. Similarly, the emulsions for intranasal delivery are also limited.

6.2. MEDICAL OR THERAPEUTICAL APPLICATIONS OF O/W NANOSIZED EMULSIONS

6.2.1. Medical or Therapeutical Applications of Parenteral O/W Nanosized Emulsions

Several publications have reported improved API exposures in the brain, following the systemic administration of API-encapsulating nanosized emulsions in rodent models (Dordevic et al. 2017; Ganta et al. 2010a, b; Prabhakar et al. 2013; Shah 2013; Tan et al. 2017; Vyas et al. 2008). Particularly, nanosized emulsions made with oils, such as fish oil and flaxseed oil, are rich in omega-3 and omega-6 polyunsaturated fatty acids (PUFAs) and have been shown to improve CNS API distribution and exposure in several preclinical experiments. PUFAs regulate various activities in the brain, such as neurotransmission, cell survival, and neuroinflammation; though saturated and monounsaturated fatty acids can be synthesized *de novo* within the brain, PUFAs enter the CNS from the blood (Bazinet and Laye 2014). Various pathways have been proposed to describe the transport of PUFAs across the blood–brain barrier (BBB). One theory proposes that fatty acids cross the BBB without specific transporters, crossing the luminal and transluminal leaflets of the endothelial cells by reversible flip-flop (Hamilton and Brunaldi 2007). Another theory indicates that fatty acids cross the BBB by facilitated diffusion; wherein fatty acid transporters attach to PUFA molecules on the luminal side of the BBB and diffuse across both the endothelial and neuronal cells (Edmond 2001; Liu et al. 2015). Additionally, PUFAs can also cause membrane fluidization at the BBB interface, thereby impacting tight junctions and enhancing CNS absorption (Navarro et al. 2011). Though the precise mechanism responsible for CNS transport of PUFAs remains unclear, it has been hypothesized that encapsulating APIs in oils containing PUFAs may result in enhanced CNS API delivery. A major assumption of this hypothesis is that the API remains encapsulated in the PUFA containing oil phase of the nanosized emulsion, and the intact API containing oil droplet diffuses across the BBB. Despite the absence of experimental validation of this assumption, mainly due to lack of bioanalytical approaches to distinguish between the free and oil-encapsulated API in biological matrices, several studies claim improved CNS API delivery following systemic dosing of nanosized emulsions formulated with oils rich in PUFAs. Selected examples that illustrate the potential of such nanosized emulsions include the work of Vyas et al. (2008) demonstrating that flaxseed oil nanosized emulsions improve CNS exposure of saquinavir (an anti-HIV protease inhibitor) subsequent to oral dosing in mice and the results of Shah et al. (2014) illustrating improved CNS exposure and analgesic effects of a neuroactive peptide encapsulated in a fish oil nanosized emulsion following systemic dosing in mice. Additionally, *in vivo* properties of these nanosized emulsions and sensitive formulation characteristics that enhance CNS API delivery are not completely understood. Improved CNS API delivery from nanosized emulsions may be the result of several factors that may vary significantly between free and nanosized emulsion-encapsulated APIs. These factors include the characteristics of systemic pharmacokinetics and biodistribution and the interactions at the cell membrane of the BBB interface (Ganta et al. 2010b). Increased circulation time of the nanosized emulsion-encapsulated API in the body often increases the time available for CNS penetration, and thus enhances the extent of CNS API delivery. Furthermore, the pharmacokinetic behavior of nanosized emulsions can be modified by varying the formulation approach, most commonly, this involves conjugation with poly(ethylene glycol) (PEG) (Ganta et al. 2010b; Hak et al. 2015). In addition, unlike the free API molecule, which must cross the BBB via passive diffusion across the cell membrane, nanosized emulsion-encapsulated APIs may penetrate the BBB via different transport processes. One such transport process relies on endocytosis of the nanosized emulsion-encapsulated API following fusion of the oil droplet with cell membrane of the BBB, which depends not only on the composition of the nanosized emulsion but also on the size and charge of the nanosized emulsion oil droplet. Despite potential advantages, the role of different transport processes and the interactions between various nanosized emulsion formulation attributes that are relevant for regulating API delivery to the CNS in a systemic environment remain unclear. For example, we do not know how the size of the oil droplet vs. the circulation half-life of the nanosized emulsion affects BBB permeability and resulting in CNS penetration of the nanosized emulsion-encapsulated API.

A new nanosized emulsion, incorporating the antiepileptic API carbamazepine for an API delivery to the brain, was developed by Madhusudhan et al. (2007). 1-*O*-alkylglycerol/lecithin blends were used as emulsifiers, e.g., 0.25%

w/w soy lecithin and 1.25% w/w 1-*O*-decylglycerol. Soybean oil (10% w/w) as the lipid phase dissolved the emulsifier and 0.1% w/w carbamazepine. After adding the water phase containing glycerol (2.5% w/w), the emulsion was formed by high-pressure homogenization. A tissue distribution study in mice showed a significant uptake in all tissues for all nanosized emulsions tested. 1-*O*-alkylglycerols are reported to improve the brain delivery, which was best shown using the nanosized emulsion containing 1-*O*-decylglycerol, which increased the brain availability 2.37 times. However, the tissue ranking was still lung > liver > spleen > brain > kidney > heart. A study that also targeted the brain was carried out by Wen et al. (2011), which compared solid lipid nanoparticles (SLN), nanostructured lipid carriers (NLC), and nanosized emulsions. These three categories were produced by using 100% cetyl palmitate, different ratios of cetyl palmitate and squalene, and 100% squalene as the lipid phase, respectively. To obtain a positive charge for all three categories of nanoparticles, Forestall, a cationic emulsifier/surfactant, was used. As a dye, sulforhodamine B was incorporated for imaging. An *in vivo* study in rats investigating the accumulation of the dye in the brain revealed that the retention time was most prolonged, from 20 to 50 min, using nanosized emulsions. It seems that due to the flexibility of nanosized emulsion's oil droplets, compared with more rigid NLC and SLN, the uptake into the brain overcoming the BBB is enhanced.

The o/w nanosized emulsion formulations of lipophilic APIs, such as propofol, etomidate, dexamethasone palmitate, and diazepam, were already developed and marketed (Table 4.1). Furthermore, various research groups across the world are currently undertaking projects to exploit the potential of o/w emulsions for parenteral delivery of a myriad of investigational APIs as well as other lipophilic APIs by receptor-mediated targeting to cancer cells. The important technical and clinical points to keep in mind before the use of the emulsion systems for this kind of work are given below.

It has to be clear that, once diluted and injected (or administered in ocular and other routes), the emulsion stability and fate are determined by three measurable parameters. The first is the partition coefficient of each emulsion component (including added APIs and agents) between the emulsion assembly and the medium. To some extent, this partition coefficient is related to oil–water and/or octanol–water partition coefficients. For example, it was well demonstrated that per component of which $\log P$ is lower than 8, the stability upon IV injection is questionable (Takino et al. 1994, 1995). The other two parameters are k_{off}, a kinetic parameter that describes the desorption rate of an emulsion component from the assembly, and k_c, the rate of clearance of the emulsion from the site of administration. This approach is useful to decide if and what application an API delivery system will have a chance to perform well (Barenholz and Cohen 1995). Stability in plasma is an important requirement for IV emulsions as flocculated droplets may result in lung embolism. It was found that tocol-based emulsions stabilized by sodium deoxycholate/lecithins flocculated strongly when mixed with mouse, rat, and sheep plasma and serum, whereas soya oil-based emulsions with the same emulsifiers did not (Han et al. 2004). It was hypothesized that this effect was caused by the adsorption of plasma proteins onto the tocol droplets (opsonization). Indeed, the steric stabilization of emulsions by incorporation of emulsifiers like poloxamer 188 or PEGylated phospholipids such as PEG5000 PE proved to be effective in the stabilization of tocol-based emulsions in plasma. Conversely, *in vitro* studies were conducted on plasma protein adsorption onto the blank second- and third-generation emulsion droplets (Harnisch and Müller 2000; Tamilvanan et al. 2005) to mimic the *in vivo* opsonization phenomenon responsible for the rapid clearance of the emulsion droplets from the blood. According to these authors, the adsorption of many protein species such as apolipoprotein (apo) As, apoCs, apoE albumin, fibrinogen, and gamma globulin onto the emulsion droplet surfaces is detectable by two-dimensional polyacrylamide gel electrophoresis.

6.2.2. Medical or Therapeutical Applications of Non-parenteral O/W Nanosized Emulsions

6.2.2.1. Intranasal Emulsions The intranasal route is a noninvasive, user-friendly, needle-free administration and does not require any trained personnel; even self-medication is possible, leading this route with possible patient compliance with a possibility of protein and peptide delivery. In recent years, a great focus has been given toward intranasal route as a promising and novel alternative to the invasive methods for administration of APIs for CNS because of its highly vascularized large surface area and ability to circumvent BBB providing direct and rapid delivery to the brain (Gadhave et al. 2019; Kim et al. 2018; Mahajan et al. 2014). Functionally, the nasal cavity plays an important protective role to filter, warm, and humidify the inhaled air before it reaches the lower airways. The unique anatomical connection between the nasal cavity and the CNS provides relatively quick and easy access to the brain for APIs delivered to the intranasal cavity (Alam et al. 2010; Pardeshi and Belgamwar 2013). It has been established that the nonpolar APIs, following intranasal administration, are diffused through the olfactory epithelium and reached to the brain by different pathways. Although all of the mechanisms involving nose to brain API transport are not clear, the involvement of neuronal pathways, cerebrospinal fluid, and nasal lymphatics was evidenced (Pardeshi and Belgamwar 2013).

The o/w nanosized emulsions by virtue of their lipophilic nature and low globule sizes are significantly absorbed by intranasal delivery due to increased uptake by nasal mucosa (Vyas et al. 2005). Moreover, the nanosized emulsion particles are modified into nanogelling systems, mucoadhesive systems, or coated particulate systems to overcome issues of fast nasal clearance and to improve mucosal absorption. An extensive literature survey performed by Chatterjee et al. (2019) indicated that out of 131 research articles published from 2016 to 2019 (February), about 27.48% of the studied intranasal API delivery system targeted to the brain are based on micro or nanosized emulsion platform.

In spite of numerous available proof of concept, described by several researchers, no intranasal nanosized emulsion formulations are successfully exploited commercially. Such unavailability, first triggered for the formulators to find the gap between "bench-to-bedside" of intranasal nanosized emulsion. The major limitation includes nasal-mucociliary clearance, which is a normal physiological phenomenon of the human body (Gänger et al. 2018). Mucociliary clearance is a defense mechanism that protects the lungs from inhaled harmful foreign matters, such as pathogens or microorganisms. Two different layers of mucous membranes are found on the nasal epithelium, where the average mucous flow rate is 5 mm min^{-1}, hence, mucous is regenerated in the nasal cavity every 15–20 min (Pardeshi and Belgamwar 2013). Mucociliary clearance depends on the viscoelastic nature of mucous, ciliary beat frequency, and coordination of cilia (Cao et al. 2009). These properties show human variation and hence variable duration of intranasal residence of delivery systems and therapeutic agents is considered a major challenge for nasal delivery. Enzymatic degradation at the nasal mucosa is another limitation especially for protein and peptide agents, where these agents found to be degraded by enzymes present in the lumen during crossing the epithelial barrier and incur poor bioavailability (Garg et al. 2015).

Other limitations for the nose to brain API delivery include nasal toxicity due to some therapeutic agents and absorption enhancers, problem to cross nasal mucosa for high molecular weight APIs, and chances of severe damage of nasal cavity by repeated API administration. Therefore, it becomes necessary to quickly review the influence/relationship of selected emulsion components on selected API and on the possible damage of nasal structure leading to the overall emulsion vehicle toxicity.

6.2.2.1.1. Lipid/oil component The o/w nanosized emulsions comprises of oil, emulsifier, and aqueous phase. It has been estimated that anywhere from 40 to as much as 70% of all new chemical entities (NCE) entering API development programs possess insufficient aqueous solubility to allow consistent gastrointestinal absorption of a magnitude sufficient to ensure therapeutic efficacy (Gorain et al. 2013; Gursoy and Benita 2004; Pandey et al. 2018). Therefore, the lipophilic APIs-incorporated o/w nanosized emulsions are capable of solubilizing the APIs in the oil phase and oil–water interface of the nanosized emulsion and the API solubility increase is partially depending on the increase in lipophilicity of the oils. As a thumb rule, the solubilizing capacity of oils would decrease in the order vegetable oils > medium-chain triglycerides > medium-chain mono- and di-glycerides (Sapra et al. 2013). However, the selection of oil does not depend only on the solubilization capacity; rather it also requires balance with the allowable oil concentration approved by the United States Food and Drug Administration (USFDA) for the o/w nanosized emulsion.

In general, the mean particle size (MPS) of o/w nanosized emulsions increases with an increase in oil concentration in the formulation (Choudhury et al. 2014). However, increased MPS reduces nasomucosal permeation of the API. Therefore, the optimal selection of oil quantity that is sufficient to dissolve the API is used. Sometimes, if the oil is used as a permeation enhancer, it can promote API permeation through the nasal mucosa. It is demonstrated by a study that the polar lipids present in butter oil used in their study played a key role in enhancing permeation of API (Quetiapine fumarate) across nasal mucosa, which could be attributed to both transcellular and paracellular pathways (Khunt et al. 2017).

The droplet size of nanosized emulsion is the most important factor for mucosal API permeation. Proposed mechanism of the nose to brain API transport is via olfactory or trigeminal pathways. The morphological analysis says that the average diameter of olfactory axons, in various species, is approximately 200 nm. In humans, this diameter ranged from 100 to 700 nm (Morrison and Costanzo 1992). By virtue of these size constraints, the nanocarriers should be below 200 nm, or even lower, to avail the transcellular pathways of neuronal absorption. In reported studies, the range increased. In a nice experimental study by Ahmad et al. (2017a, b), it was observed that nanosized emulsion with 100 nm average droplet size was transported to the olfactory bulb via olfactory and trigeminal nerve at higher rate and extent, compared with the nanosized emulsion with 900 nm average size. The authors explained that there was little distribution of intact nanocarrier at the brain, despite a higher concentration of dye that was delivered through the nanosized emulsion.

However, nanosized emulsions with droplet size above 200 nm were also shown to deliver the API effectively in the brain, when administered intranasally. For instance, the intranasally administered cyclosporin A (CsA)-loaded nanosized emulsion having the MPS value of 272 nm was more distributed to the brain than blood plasma (Yadav et al. 2015). Significantly higher CsA concentrations were observed for intranasally administered CsA-loaded nanosized emulsions in the olfactory bulb, midbrain, and hindbrain when compared to the API concentrations, which were observed in those tissues following IV administration of CsA-loaded nanosized emulsion. Moreover, the higher uptake in the midbrain region indicates that one of the roots of API uptake could be via olfactory epithelium and redistribution by the API through CSF.

Globule size also affects the retention time on the nasal mucosa, which is a significant factor in affecting API transport to the brain. The lesser the retention, the lower the API absorption. Ahmad et al. (2017a, b) has shown that higher droplet size tends to be cleared out more easily by nasomucosal clearance compared with higher droplet size. In their work, they observed almost all nanosized emulsion with average size >200 nm were cleared off the nasal mucosa after 4 h of intranasal administration, whereas nanosized emulsion with size 80 or 200 nm showed some degree of retention on nasal mucosa up to

16 and 12 h, respectively. Hence, from the above discussion, it can be concluded that droplet size is the most significant factor that influences API targeting to the brain after nasal administration.

6.2.2.1.2. Emulsifiers Emulsifiers are the integral component of o/w nanosized emulsion, which helps to reduce surface tension and prevent coalescence of dispersed oil droplets and thus the oil–water phase separation. An ideal emulsifier should have sufficient ability to dissolve a high quantity of the API to express enhanced API loading. Emulsifier helps to make nanosized emulsion stable as well as affect the MPS. Hence, it has a significant effect on nasomucosal permeation of API. It was evidenced that increased emulsifier ratio reduces MPS (Hosny and Banjar 2013). The lower the MPS, the higher the nasomucosal permeation. Higher nasomucosal permeation means a higher chance of brain targeting by the delivered API. Emulsifier is also treated as intranasal permeation enhancers. It can act either by altering the fluidity or membrane or weakening/damaging the tight junction of epithelial layers (Lin et al. 2007). Both actions can promote the API permeation into the major sections of nasal cavity such as nasal vestibule, which is the dilated opening of nostril; respiratory section, which refers to the passage for air travel; and olfactory section, which contains the olfactory receptors (Wang et al. 2018). It was observed that transepithelial/transendothelial electrical resistance (TEER) values were reduced for human nasal epithelial cell layer when incubated with different emulsifiers (sodium cholate, sodium taurocholate, Tween 80, and poloxamer F68) for 2 h (Lin et al. 2007). These authors found a linear relationship of TEER value reduction with an increase in transepithelial permeability of fexofenadine HCl. The quoted work also described that excluding poloxamer F68, other three emulsifiers also increased paracellular API uptake (Lin et al. 2007). The API permeation effect depends on the type and concentration of emulsifier. The alteration of membrane bilayer makes the use of emulsifier critically questionable in terms of possible toxicity. The damage of lipid bilayer could be irreversible above a certain concentration of emulsifiers. The study by Lin et al. (2007) also showed that the recovery of disrupted membrane barrier depends on the concentration and type of emulsifier (ionic or nonionic). Generally, the concentration of emulsifier is kept as low as possible to balance between API permeation and no or less toxic effect.

Zeta potential is an indicator of emulsion stability that says an absolute value of 30 is desirable for a stable emulsion (Win et al. 2017). Some studies indicate that zeta potential has also the role of API retention on the nasal mucosa. It is well known that positively charged droplet is better attached to a mucous layer containing negatively charged mucin (Samaridou and Alonso 2018). In his work, Ahmad et al. (2017a, b) have explained that positively charged chitosan in his formulation formed electrostatic bonding with negatively charged particle. However, most of the nanosized emulsions, reported for the nose to brain API delivery, have negative zeta potential. The mechanism of adhesion with respect to electrical charge has not been studied much using different zeta potential nanosized emulsions. Therefore, the effect of zeta potential on mucosal retention of the API should be studied in a reproducible manner. The low viscosity of nanosized emulsion formulations allowed quick clearance from the nasal cavity due to mucociliary movement, which ultimately decreases the absorption and transportation of APIs within the brain (Wen 2011). Therefore, current strategies of intranasal nanosized emulsion targeting CNS disorders mainly focus on the prolongation of contact time between nasal mucosa and the formulated dosage form to obtain complete absorption of the incorporated therapeutic agent. It had been proved by various researchers that if the retention time of the formulation within the nasal cavity be increased, then the API transportation to the brain will also be increased. Although partly overlapping, the strategies to improve the intranasal stay of the formulation can be subdivided into three approaches: nanosized emulsion with mucoadhesion, nanoemulgel, i.e. the gel form of nanosized emulsion, and nanosized emulsion forming *in situ* nasal gel (Chatterjee et al. 2019).

The nasal route is still receiving great attention due to a number of advantages over parenteral and oral administration (Hussain 1998), particularly when first-pass metabolism makes the API molecules ineffective. The approach of an o/w nanosized emulsion formulation of the API may increase absorption by solubilizing the API molecules in the inner phase of the emulsion and prolonging contact time between emulsion droplets and nasal mucosa.

One of the first examples for systemic delivery of peptides concerned a lipid-soluble rennin inhibitor (Kararli et al. 1992). The peptide was solubilized in the oil phase of an o/w emulsion containing membrane adjuvants such as oleic acid and mono- and di-glycerides. Emulsion formulations have been proposed to simultaneously increase the solubility of the peptide and to enhance membrane permeability through interaction between the membrane and the oil components. Enhanced and prolonged *in vivo* nasal absorption of the rennin inhibitor was observed in emulsion formulation compared to aqueous suspension. From morphological studies, the emulsions did not provoke any significant changes to the nasal mucosa (Kararli et al. 1992). Such a formulation approach was also used for the administration of a steroidal male sex hormone testosterone (Ko et al. 1998). The steroid was solubilized in the oil phase of the o/w emulsion and the ionic composition of the aqueous phase was modified in order to produce electrically positive, negative, and neutral droplets. Droplets with a surface charge led to better bioavailability than neutral droplets, but contrary to the above-described topical applicabilities of cationic emulsions over anionic emulsions, positively charged droplets did not provide the best results (Ko et al. 1998). However, the emulsion approach was advantageous since it helped to overcome the solubility problem of the hydrophobic API molecules.

In another study that does not involve peptide APIs, various emulsion formulations were prepared in order to modulate the partitioning of the API between the aqueous phase and the oil phase (Aikawa et al. 1998). The disappearance of an API from the nasal cavity was determined by an *in situ* perfusion technique. When the API was solubilized in the aqueous phase, the formulation did not have a significant effect on the API disappearance rate. However, partitioning of the API in the oil phase resulted in delaying absorption. It was suggested that oil droplets containing medium-chain triglycerides formed a pseudo-oily layer on the mucous membrane, which slowed down the API disappearance from the nasal cavity (Aikawa et al. 1998). Another interesting study reported nasal delivery of insulin formulated in both o/w and w/o emulsions (Mitra et al. 2000). As insulin partitions into the aqueous phase of the emulsion, the peptide is either incorporated within the continuous phase of the o/w emulsion or encapsulated in the aqueous droplets of the w/o emulsion. Following nasal perfusion experiments, plasma insulin concentration profiles showed enhanced insulin absorption when the peptide was formulated as an o/w emulsion compared to an aqueous solution. However, a w/o emulsion did not cause any significant increase in plasma insulin concentration. Delivery of insulin by administration of nasal drops also revealed a large dose-dependent increase in plasma insulin concentration. It also needs to be pointed out that the emulsifier mixture alone did not promote any absorption. It was suggested that insulin molecules probably were adsorbed at the surface of the oil droplets. Adhesion of the oil droplets on the mucosal membrane then induced a local increase in insulin concentration at the membrane surface. However, the number of droplets in contact with the surface had to be small enough. Otherwise, a stagnant oil layer is formed, which acts as an additional barrier to the transport, as was observed with the w/o emulsion (Aikawa et al. 1998).

Other recent applications of emulsion formulation involve mucosal gene and vaccine delivery (Benita et al. 2004; Jenkins 1999; Kim et al. 2000) and the preparation of polymeric microspheres by the w/o emulsification solvent extraction technique (El-Hameed and Kellaway 1997).

6.2.2.2. Ophthalmic Emulsions To manage/treat the ailments of ocular surfaces, the API delivery via oral route is not a best option due to the exposure of API molecules unnecessarily to the other tissues of the human body. The local/topical application of intended API molecule is the best option available for the ocular ailment management/treatment. However, delivering the API molecule in adequate concentration in front of the ocular surfaces is a cumbersome/herculean task.

6.2.2.2.1. Topical ophthalmic emulsions When o/w nanosized emulsions are used as a vehicle for ocular API delivery purposes, both topical/local and intraocular routes of administration can be possible (though no data concerning intraocular API delivery through emulsion are currently available). The o/w nanosized emulsions having both anionic and cationic charges provide a liquid-retentive carrier for ocular active agents, particularly when topically instilled into the eye. It is interesting to add here that thermodynamically stable and optically isotropic colloidal systems such as the water-in-oil (w/o) nanosized emulsion is also designed for ocular topical administration as it has the ability to incorporate considerable amounts of both hydrophilic and lipophilic APIs (Alany et al. 2006). In fact, in comparison to ocular inserts or implants and semisolid ocular preparations, the liquid-retentive nature gives impetus to investigating further the emulsion-based ophthalmic API delivery as it has the benefit of being comfortable to use for both ophthalmologists and patients. In addition, through topical instillation of emulsions possessing ocular active substances, the delivery of API molecules even to the posterior portion of the eye might be possible.

In this context, the third-generation emulsion is being designed by adsorbing electrostatically the therapeutic oligonucleotides onto its surface for modulating the functions of retinal pigment epithelium (RPE) cells effectively in order to treat blindness associated with age-related macular degeneration (AMD), proliferative vitreoretinopathy, retinal and choroidal neovascularization, and retinitis pigmentosa. To achieve this, it becomes necessary to know first the ocular protective mechanisms and other concomitant factors to be faced by emulsion droplets following ocular topical application. This point is further developed below.

Considerations of ocular API delivery are not detailed in this section. Pertinent information concerning factors affecting API permeation or retention as well as eye anatomy and physiology can be found in several reviews (Järvinen et al. 1995; Lee 1993; Ludwig 2005; Prausnitz and Noonan 1998; Stjernschantz and Astin 1993; Washington et al. 2001). From a medical point of view, o/w nanosized emulsions for ophthalmic use aim at enhancing API bioavailability either by providing prolonged delivery to the eye or by facilitating transcorneal/transconjunctival penetration. APIs incorporated in o/w nanosized emulsions are lipophilic in nature, and depending on the extent of lipophilicity, either the corneal or the conjunctival/scleral route of penetration may be favored (Chien et al. 1990). For the more lipophilic APIs, the corneal route was shown to be the predominant pathway for delivering APIs to the iris, whereas the less lipophilic APIs underwent conjunctival/scleral penetration for delivery into the ciliary body (Chien et al. 1990). Thus, transcorneal permeation has traditionally been the mechanism by which topically applied ophthalmic APIs are believed to gain access to the internal ocular structures. Relatively little attention has been given to alternate routes by which APIs may enter the eye. It was reported that APIs may be absorbed by the noncorneal route and appeared to enter certain intraocular tissues through the conjunctiva/sclera (Ahmed and Patton 1985; Dey et al. 2003; Kaur and Smitha 2002). Indeed, when compared to the cornea, API penetration through the conjunctiva has the advantage of a larger surface area and higher permeability, at least for APIs that are not highly lipophilic. Furthermore, the lasting presence of API molecules in the lower conjunctival *cul-de-sac* of the eye could result in a reservoir effect. Nevertheless, the o/w nanosized emulsions more or less physically resemble a simple aqueous-based eye drop dosage form since more than 90% of the external phase is aqueous irrespective of the formulation composition. Hence, following topical administration, nanosized emulsions would probably face almost similar ocular protective events as encountered with conventional eye drops into the eye. The o/w nanosized emulsions are likely to be destabilized by the tear fluid electrolytic and dynamic action. Because of constant eyelid movements, the basal tear flow rate ($1.2\,\mu l\,min^{-1}$) and the reflex secretion induced by instillation [up to $400\,\mu l\,min^{-1}$ depending on the irritating power of the topical ocular solutions (Sasaki et al. 1996)], topical eye drop dosage forms are known to be rapidly washed out from the eye. Therefore, the water phase of the emulsion is drained off while, probably, the oil phase of the emulsion remains in the *cul-de-sac* for a long period of time and functions as an API reservoir (Sasaki et al. 1996). If the volume of instilled emulsion in the eye exceeds the normal lachrymal volume of 7–10 μl, then the portion of the instilled emulsion (one or two drops, corresponding to 50–100 μl) that is not eliminated by spillage from the palpebral fissure of conjunctiva is drained quickly via the nasolacrimal system into the nasopharynx. In other words, the larger the instilled volume, the more rapidly the instilled emulsion is drained from the precorneal area. Hence, the contact time of the emulsion with the absorbing surfaces (cornea and conjunctiva) is estimated to be a maximum of a few minutes, well beyond the short residence time of conventional eye drops. In order to verify the extension of the residence time of the emulsion in the conjunctival sac, Beilin et al. (1995) added a fluorescent marker to the formulations. One minute after the topical instillation into eye, $39.9 \pm 10.2\%$ of the fluorescence was measured for the nanosized emulsions whereas only $6.8 \pm 1.8\%$ for regular eye drops. In addition, a study was carried out in male albino rabbits to compare the corneal penetration of indomethacin from Indocollyre (a marketed hydro-PEG ocular solution) to that of negatively and positively charged emulsions (Klang et al. 2000). By this comparison, it was intended to gain insightful mechanistic comprehension regarding the enhanced ocular penetration effect of the emulsion as a function of dosage form and surface charge. The contact angle of one droplet of the different dosage forms on the cornea was measured and found to be 70° for saline, 38° for the anionic emulsion, and 21.2° for the cationic emulsion. The values of the calculated spreading coefficient were −47, −8.6, and $-2.4\,mN\,m^{-1}$, respectively. It can clearly be deduced, owing to the marked low spreading coefficient values elicited by the emulsions, that both nanosized emulsions had better wettability properties on the cornea compared with saline. The emulsion may then prolong the residence time of the drop on the epithelial layer of the cornea, thereby enabling better API penetration through the cornea to the internal tissues of the eye, as confirmed by animal studies (Klang et al. 2000). It is therefore believed that API is not released from the oil droplet and equilibrates with the tears but rather partitions directly from the oil droplets to the cell membranes on the eye surface. Therefore, it is reasonable to consider that nanosized emulsions have a real advantage since they elicit an increased ocular residence time in comparison with conventional eye drops and will significantly improve the ocular API bioavailability (Aiache et al. 1997). In spite of a relatively rapid removal of conjunctivally absorbed emulsion from the eye by local circulation, direct transscleral access to some intraocular tissues cannot be excluded, especially if an electrostatic attraction does occur between the cationic oil droplets of emulsion and anionic membrane moieties of the sclera, as shown by some authors (Klang et al. 2000). There is no doubt that the API absorption from emulsion through the noncorneal route needs to be investigated further as it may elicit useful information on the potential of nanosized emulsions to promote API penetration to the posterior segment through a mechanism that bypasses the anterior chamber. In addition to the above-described protective and elimination mechanisms of the eye, nanosized emulsions remaining in the precorneal area may be subject to protein binding and to metabolic degradation in the tear film. In conjunction with blood plasma, although low, tear film, aqueous humor, and vitreous humor also have varying amounts of relatively detectable proteins such as albumin, globulin, and immunoglobulins (Ig) (e.g., IgA, IgG, IgM, and IgE), and the enzyme, lysozyme. Additional studies (at least *in vitro*) are necessary to understand clearly the nanosized emulsion interaction with the ocular fluid components. Although it is unlikely to happen because of the low emulsion volume remaining in the conjunctival sac, the fluid dynamics may be moderately altered by the physical and chemical properties of nanosized emulsions, which include tonicity, pH, refractive index, interfacial charge, viscosity, osmolality, and irritant ingredients. Thus, formulations of ophthalmic API products must take into account not only the stability and compatibility of an API in the emulsion but also the influence of the emulsion on precorneal fluid dynamics. All of the concepts exposed in this section may ultimately result in transcorneal/conjunctival absorption of 1–2% or less of the API applied topically through the emulsions. In summary, the rate of loss of API/emulsion from the eye can be 500–700 times greater than the rate of absorption into the eye. Thus, conventional topical delivery using emulsions cannot achieve adequate intracellular concentrations of APIs or other substances such as oligonucleotides or genes for the treatment of endophthalmitis or other sight-threatening intraocular diseases (e.g., AMD).

6.2.2.2.2. Non-topical ophthalmic emulsions In order to achieve a high concentration of API within the eye using an emulsion delivery vehicle, an approach that bypasses physiological and anatomical barriers (e.g., blood-ocular) of the eye is a more viable and attractive option. One such approach is to administer emulsion through direct intraocular injections such as periocular (subconjunctival and sub-Tenon), intracameral, intravitreal, intracapsular, or subretinal. Moreover, it is likely that intraocularly administered emulsion is able to both significantly increase API half-life and minimize intraocular side effects that appear following intraocular injection of API alone. In general, API encapsulated within emulsion is less toxic than its free counterparts. Additionally, there is a possibility of obtaining slow API release from an intraocularly injected emulsion. Taking into account the nonphagocytic character of neural retinal cells and the ability of RPE cells to take up large molecules, including oligonucleotides, the third-generation emulsion for intravitreal or subretinal injections is more likely to be a successful approach in future. Moreover, intravitreally administered API molecules are able to bypass the blood-ocular barrier to achieve constant therapeutic levels in the eye while minimizing systemic side effects. However, the hyalocytes, the main cellular components of the vitreous, have been classified in at least one report (Maurice and Mishima 1984) as macrophages and thus may play a role in the uptake of intravitreally injected emulsion. It should be added that no studies are focused so far on injecting emulsion intraocularly and significant work should be devoted to generate this novel idea into a fruitful solution in ophthalmic API delivery applications.

Over the last decade, o/w nanosized emulsions containing either anionic or cationic droplets have been recognized as interesting and promising ocular topical delivery vehicles for lipophilic APIs. Complete details are available elsewhere (Tamilvanan and Benita 2004). The *in vivo* data obtained from studies of early formulations confirm that o/w nanosized anionic emulsions can be effective topical ophthalmic API delivery systems (Muchtar et al. 1992) with a potential for sustained API release (Naveh et al. 2000).

Naveh et al. (1994) have also noted that the intraocular pressure (IOP)-reducing effect of a single, topically administered dose of pilocarpine-loaded anionic emulsion lasted for more than 29 h in albino rabbits whereas that of the generic pilocarpine lasted only 5 h. Zurowska-Pryczkowska et al. (1999) studied how nanosized emulsion as a vehicle influences the chemical stability of pilocarpine and the effect the API has on the physical stability of nanosized emulsions. In a subsequent paper (Sznitowska et al. 2001) from the same group on *in vivo* evaluation using normotensive rabbits, it was shown that the nanosized emulsion formulated with pilocarpine hydrochloride at pH 5.0 could be indicated as a preparation offering prolonged pharmacological action (miotic effect) together with satisfactory chemical stability. However, the ocular bioavailability arising from such a formulation is not significantly improved when compared to an aqueous solution. Calvo et al. (1996a, b) observed an improvement in indomethacin ocular bioavailability when the API was incorporated in an emulsion dosage form with respect to the commercial aqueous drops following topical application into rabbit eye.

In order to verify the extension of the residence time of the emulsion in the conjunctival sac, Beilin et al. (1995) added a fluorescent marker to the formulations, as mentioned previously. From that observation, it is reasonable to consider that an emulsion formulation has the real advantage of increasing ocular residence time in comparison to eye drops. Anselem et al. (1993) and Melamed et al. (1994) prepared a nanosized emulsion containing adaprolol maleate, a novel soft β-blocking agent, and observed a delayed IOP depressant effect in human volunteers. A similar pharmacological effect was also observed in human volunteers by Aviv et al. (1996) using pilocarpine base-loaded emulsion. Another randomized human trial conducted by Garty et al. (1994) compared the activity of the pilocarpine base-laden nanosized emulsion instilled twice daily with a generic dosage form instilled four times a day to 40 hypertensive patients for 7 days. No local side effects were observed. The IOP decreased 25% in both formulations during this time period. No significant difference was noticed between the two treatments. These results proved that the anionic emulsion extended the action of the API and two daily administrations have the same result as four instillations of regular eye drops.

A novel nanosized anionic emulsion incorporating the immunomodulatory agent CsA was developed and the clinical efficacy was investigated for the treatment of moderate-to-severe dry-eye disease in humans (Ding et al. 1995; Ding and Olejnik 1997; Acheampong et al. 1999). The novel CsA ophthalmic dosage form represents a real breakthrough in the formulation of a complex, highly lipophilic molecule such as CsA within an o/w nanosized emulsion. Following thorough validation of this formulation through several clinical studies in various countries (Acheampong et al. 1999; Brignole et al. 2001; Ding and Olejnik 1997; Ding et al. 1995; Galatoire et al. 2003; Kunert et al. 2000; Sall et al. 2000; Small et al. 2002; Stevenson et al. 2000; Tang-Liu and Acheampong 2005; Turner et al. 2000), an anionic o/w emulsion containing CsA 0.05% (Restasis, Allergan, Irvine, CA) was approved for the first time by the FDA, on December 23, 2002. In addition, this anionic emulsion having CsA is now available at pharmacies in the United States for the treatment of chronic dry-eye disease (available at http://ww.restasis.com and http://ww.dryeye.com). Furthermore, an over-the-counter (OTC) product that features an emulsion formula, Refresh Endura, is already launched in the U.S. market for eye-lubricating purposes in patients suffering from moderate-to-severe dry-eye syndrome.

The effect of Restasis on contact lens comfort and reducing dry-eye symptoms in patients with contact lens intolerance was evaluated in comparison with rewetting drops (carboxymethylcellulose 0.5%, Refresh Contacts) (Hom 2006). Both formulations were applied twice per day before and after lens wear. Symptoms were assessed by lens wear time, use of rewetting drops during lens wear, subjective evaluation of dryness, and completion of the ocular surface disease index questionnaire. The results of this pilot study indicate that CsA 0.05% is beneficial for contact lens wearers with dry eye and reduces contact lens intolerance (Hom 2006). Furthermore, Sall et al. (2006) have recently evaluated the efficacy of marketed artificial tears (Systane and Restasis) in relieving the signs and symptoms of dry eye when used as supportive therapy to a CsA-based ophthalmic emulsion (i.e., Restasis + Systane vs.

Restasis + Refresh). Significant differences were seen in favor of Restasis + Systane versus Restasis + Refresh for less ocular burning, stinging, grittiness, and dryness. Systane was better than Restasis + Refresh for less burning, dryness, and scratchiness. Results indicate that the choice of concomitant therapy used with Restasis has significant effects on outcome measures and both supportive therapies were compatible with Restasis (Sall et al. 2006).

When compared to either saline or anionic emulsions, the nanosized cationic emulsions were shown to enhance the ocular bioavailability of indomethacin (Klang et al. 2000), piroxicam (Klang et al. 1999), and CsA (Abdulrazik et al. 2001; Tamilvanan et al. 2001) following one single-drop dose instillation into the rabbit eye. A significant API reservoir effect was noted in the cornea and conjunctiva even for more than 8 h following the instillation (Abdulrazik et al. 2001). This long residence time may help reduce evaporation of the limited volume of natural tears present in patients with dry eye. This was probably due to the adhesion of the positively charged oil droplets to the negatively charged corneal surface moieties as a result of electrostatic attraction. This hypothesis was supported by data from an *ex vivo* study, which showed that cationic emulsion exhibited better wettability properties on albino rabbit eye cornea than either saline or anionic emulsion (Klang et al. 2000). Associated with Poloxamer and phospholipids, a cationic primary amine, stearylamine, has been used to obtain the above-described third-generation cationic emulsions. Additionally, a cationic emulsion based on an association of Poloxamer 188 and chitosan was prepared and also showed interesting physicochemical properties on stability and charge effects (Calvo et al. 1997; Jumaa and Müller 1999). Moreover, the stability and ocular tolerance following topical instillation into the eye of these cationic emulsion vehicles were investigated (Jumaa and Müller 1999; Klang et al. 1994). The overall studies hence stress the effectiveness of nanosized cationic emulsion, which promotes ocular API absorption via internalization possibly through an endocytic process (Yang and Benita 2000).

6.2.2.2.3. Future perspective of topical ophthalmic emulsions Apart from the above-discussed non-topical and topical ophthalmic emulsions, the development of targeted topical o/w nanosized emulsions is at the rudimentary stage wherein the API molecules can be targeted via topical ophthalmic application into human eye. This novel concept is highlighted below.

6.2.2.2.3.1. TARGETED TOPICAL OPHTHALMIC EMULSIONS To achieve an effective and precise medication delivery to the organs of human body, which really need the API action, the concept of API targeting is being introduced. The arrival of various nanoformulations such as liposomes, nanospheres, nanosized emulsions, etc., along with their success even at commercial level help a lot to achieve precisely the API targeting concept for managing the different oncological syndromes. However, it is well known that the IV-administered nanoformulations including the o/w nanosized emulsions will accumulate instantly in various mononuclear phagocytic systems (MPS, also known as the reticuloendothelial system or macrophage system)-rich organs such as bone, kidney, liver, lung, kidney, etc., due to their hydrophobic surfaces and this process is termed as passive targeting. If you decorate the surfaces of nanoformulation with any hydrophilic compounds, then the IV-administered particles stay in the blood stream for longer period of time without the immediate capturing by the MPS for elimination from the body. Due to the prolonged circulation of the IV-administered nanoformulations in the blood stream, the chances of the API-loaded particles to access or reach to their intended target cells or organs become imminent and this approach is termed as active targeting. Unfortunately, the API targeting concept is currently limited only to the parenterally administered API particularly through IV route. Coming to the topically administered ocular nanoformulations, the scenario of both active- and passive-targeting approaches is not practically happening. Instead of this, the topically applied nanoformulations should encounter and even need to travel against the situations of eye-protective mechanisms such as tear turnover rate, refractive index, spillage from the eye, etc., for reaching into the targeted intraocular tissues like choroid plexus, optic head, retina, sclera, and vitreous humor. In spite of the presence of eye-protective mechanisms, the topically instilled o/w nanosized emulsions were not only reaching into these intraocular tissues but also delivering the API in adequate therapeutic concentration in those tissues to treat few ocular ailments without any systemic API exposure and/or associated side effects (Tamilvanan and Benita 2004).

Since the topically applied o/w nanosized emulsions are able to reach the intraocular tissues for delivering the API, the development of targeted topical ophthalmic emulsions is currently being proposed for these pigmented dendritic cells near the optic disc, retinal ganglion cells (RGCs), and trabecular meshwork (TM) to manage the glaucoma.

6.2.2.3. Topical Dermal Emulsions Human skin consists of multiple layers with varied affinity and therefore limiting the entry/permeation of API molecules of interest applied topically onto the skin surface. Hence, many APIs exhibit low skin penetration, which results in poor therapeutic efficacy. In contrast to common chemical skin penetration enhancers, organic solvents that are generally associated to some degree with skin irritation, toxicity, and sensitization, a solvent-free topical vehicle based on API entrapment into the nanometer-sized dispersed oil droplets of emulsions is more efficacious in terms of percutaneous absorption and possibly devoid of adverse effects. In addition, the uniqueness of the large internal hydrophobic oil core of o/w nanosized emulsions allows high solubilization capacity for water-insoluble topically active APIs and also aids in carrying water, an excellent softener, to the skin.

Friedman et al. (1995) and Schwarz et al. (1995) proved initially the concept of using anionic nanosized emulsion vehicles for enhanced percutaneous absorption of nonsteroidal anti-inflammatory drugs (NSAIDs) and diazepam.

The NSAIDs and diazepam in a nanosized emulsion vehicle also demonstrated noticeable systemic activity. The o/w emulsion was tested for primary irritation in humans in a 48-h trial. The observed low irritancy and excellent human acceptance make the nanosized emulsions as attractive vehicle to be applied topically onto the human skin.

Even though emulsion vehicles increase dermal API delivery of lipophilic APIs in humans, one of the problems for topical API delivery is the difficulty of applying these vehicles to the skin because of their fluidity. Rheological properties are studied in transdermal formulations and different results are given. Realdaon and Ragazzi (1998) have investigated different mechanical emulsifying conditions on o/w nanosized emulsions containing methyl nicotinate. The influence of these procedures on rheological properties and *in vivo* availability of methyl nicotinate was evaluated. Even if various viscosities were obtained, differences between batches did not compromise API availability. On the contrary, Welin-Berger et al. (2000, 2001) concluded in their study on nanosized emulsions containing model compounds that both release and permeation rates decrease when the apparent yield stress (i.e., the macroviscosity) increases by addition of gelling polymers. Because a topical anesthetic agent will induce a pain-suppressing anesthesia, the eutectic mixture of local anesthetics (EMLA) has proven to be a useful medication for children. It is an emulsion containing a mixture of lidocaine and prilocaine. This cream gives an effective deep sedation and can be applied half an hour to 1 h prior to the procedure. Local side effects with this emulsion are very mild (Cordoni and Cordoni 2001; Dutta 1999). Systemic activity of nanosized emulsions containing diazepam was compared with regular creams or ointments by Schwarz et al. (1995). Their efficiency was tested on protection against pentamethylenetetrazole, which induces convulsive effects in mice. Diazepam applied topically in emulsion creams was strongly dependent on the MPS value of the emulsion. Furthermore, nanosized emulsions increased transdermal API delivery and prolonged protective activity for up to 6 h.

Many formulations for topical emulsions are available in the scientific literature, in patents, and on the market. Progresses made in the last years in this field are concentrated on the various aspects of API release and the influence of droplet size.

Third-generation cationic emulsions were suggested as API carriers for topical skin application. It was found that α-tocopherol-loaded cationic nanosized emulsion was able to prevent oxidative damage of cultured fibroblast cells (Ezra et al. 1996). In addition, the same cationic nanosized emulsion was able to protect rat skin against oxidative stress induced by ultraviolet irradiation significantly better than either the corresponding anionic nanosized emulsion or the cationic blank nanosized emulsion, as measured with a noninvasive evaluation of the lipid hydroperoxidation process of the rat skin. However, no difference was found between cationic or anionic nanosized emulsions of α-tocopherol as assessed by endogenous peroxyl radical scavenging ability. Taken together, these results suggest that the cationic nanosized emulsion allows the α-tocopherol to remain on the surface of the skin because of electrostatic interactions between the negatively charged sites of the superficial layers of the skin and the positively charged oil droplets. Although the extent of α-tocopherol incorporation into the skin is similar for both cationic and anionic nanosized emulsions, the prolonged skin surface residence time of the cationic nanosized emulsion allows an enhanced protective effect against oxidative stress. In contrast to these results, an *in vitro* percutaneous absorption study on hairless rat skin found that the antifungal APIs, econazole and miconazole nitrate, incorporated into a similar cationic nanosized emulsion formulation were more effective in terms of skin penetration than the corresponding anionic nanosized emulsion (Piemi et al. 1999). The enhanced rate of diffusion of these antifungal APIs through the skin by the cationic nanosized emulsion suggests a new approach for dermal API penetration enhancement (Piemi et al. 1999).

As far as the fourth-generation o/w nanosized emulsion is concerned, its therapeutic utility via topical skin route should not only be considered as API permeation promoting vehicle but also be utilized for imaging the infective microorganisms dwelling between the different layers of the skin to initiate skin infections/inflammations. It is possible because the fourth-generation nanosized emulsions to contain positive charge for interacting electrostatically with the negative charge of the skin tissues at physiological pH. The works toward this direction are currently underway in different laboratories across the world.

6.3. CONCLUSION

Initially considered as API-holding vehicle, the o/w nanosized emulsions depending on their administrative routes (nasal, ocular, parenteral, and skin) are able to find their therapeutic applications and thus commercial exploitation of few successful products. Further medical or therapeutical application potentials may rest on dual- or triple-API-loaded nanosized emulsions (as discussed in Chapter 7).

REFERENCES

Abdulrazik, M., Tamilvanan, S., Khoury, K. et al. (2001), Ocular delivery of cyclosporin A II. Effect of submicron emulsion's surface charge on ocular distribution of topical cyclosporine A, *STP Pharm. Sci.*, 11 (6), 427–432.

Acheampong, A.A., Shackleton, M., Tang-Liu, D.D.-S. et al. (1999), Distribution of cyclosporin A in ocular tissues after topical administration to albino rabbits and beagle dogs, *Curr. Eye Res.*, 18 (2), 91–103. doi:10.1076/ceyr.18.2.91.5381

Ahmad, E., Feng, Y., Qi, J. et al., (2017a), Evidence of nose-to-brain delivery of nanoemulsions: cargoes but not vehicles, *Nanoscale*, 9 (3), 1174–1183. doi:10.1039/c6nr07581a

Ahmad, N., Ahmad, R., Naqvi, A.A. et al. (2017b), Intranasal delivery of quercetin-loaded mucoadhesive nanoemulsion for treatment of cerebral ischaemia, *Artif. Cells Nanomed. Biotechnol.*, 46 (4), 717–729. doi:10.1080/21691401.2017.1337024

Ahmed, I. and Patton, T.F. (1985), Importance of the noncorneal absorption route in topical ophthalmic drug delivery, *Invest. Ophthalmol. Vis. Sci.*, 26 (4), 584–587.

Aiache, J.M., El Meski, S., Beyssac, E. et al. (1997), The formulation of drug for ocular administration, *J. Biomater. Appl.*, 11 (3), 329–248. doi:10.1177%2F088532829701100305

Aikawa, K., Matsumoto, K., Uda, H. et al. (1998), Prolonged release of drug from o/w emulsion and residence in rat nasal cavity, *Pharm. Dev. Technol.*, 3 (4), 461–469. doi:10.3109/10837459809028627

Alam, M.I., Beg, S., Samad, A. et al. (2010), Strategy for effective brain drug delivery, *Eur. J. Pharm. Sci.*, 40 (5), 385–403. doi:10.1016/j.ejps.2010.05.003

Alany, R.G., Rades, T., Nicoll, J. et al. (2006), W/O microemulsions for ocular delivery: evaluation of ocular irritation and precorneal retention, *J. Control. Release*, 111 (1–2), 145–152. doi:10.1016/j.jconrel.2005.11.020

Anselem, S., Beilin, M., and Garty, N. (1993), Submicron emulsion as ocular delivery system for adaprolol maleate, a soft β-blocker, *Pharm. Res.*, 10, S205.

Aviv, H., Friedman, D., Bar-Ilan, A. et al. (1996), Submicron emulsions as ocular drug delivery vehicles, U.S. Patent No. 5,496,811, March 5, 1996.

Barenholz, Y. and Cohen, R. (1995), Rational design of amphiphile-based drug carriers and sterically stabilized carriers, *J. Liposome Res.*, 5 (4), 905–932. doi:10.3109/08982109509012690

Bazinet, R.P. and Laye, S. (2014), Polyunsaturated fatty acids and their metabolites in brain function and disease, *Nat. Rev. Neurosci.*, 15 (12), 771–785. doi:10.1038/nrn3820

Beilin, M., Bar-Ilan, A., and Amselem, S. (1995), Ocular retention time of submicron emulsion (SME) and the miotic response to pilocarpine delivered in SME, *Invest. Ophthalmol. Vis. Sci.*, 36, S166.

Benita, M.B., Oudshoorn, M., Romeijn, S. et al. (2004), Cationic submicron emulsions for pulmonary DNA immunization, *J. Control. Release*, 100, 145–155. doi:10.1016/j.jconrel.2004.08.008

Brignole, F., Pisella, P.J., Saint-Jean, M.De. et al. (2001), Flow cytometric analysis of inflammatory markers in KCS: 6-month treatment with topical cyclosporin A, *Invest. Ophthalmol. Vis. Sci.*, 42, 90–95.

Calvo, P., Alonso, M.J., Vila-Jato, J.L. et al. (1996a), Improved ocular bioavailability of indomethacin by novel ocular drug carriers, *J. Pharm. Pharmacol.*, 48 (11), 1147–1152. doi:10.1111/j.2042-7158.1996.tb03911.x

Calvo, P., Remuñá-López, C., Vila-Jato, J.L. et al. (1997), Development of positively charged colloidal drug carriers: chitosan-coated polyester nanocapsules and submicron-emulsions, *Colloid Polym. Sci.*, 275, 46–53. doi:10.1007/s003960050050

Calvo, P., Vila-Jato, J.L., and Alonso, M.J. (1996b), Comparative in vitro evaluation of several colloidal systems, nanoparticles, nanocapsules, and nanoemulsions, as ocular drug carriers, *J. Pharm. Sci.*, 85 (5), 530–536. doi:10.1021/js950474+

Cao, S.L., Ren, X.W., Zhang, Q.Z. et al. (2009), In situ gel based on gellan gum as new carrier for nasal administration of mometasone furoate, *Int. J. Pharm.*, 365 (1–2), 109–115. doi:10.1016/j.ijpharm.2008.08.042

Charman, S.A., Charman, W.N., Rogge, M.C. et al. (1992), Self-emulsifying drug delivery systems: formulation and biopharmaceutical evaluation of an investigational lipophilic compound, *Pharm. Res.*, 9, 87–93. doi:10.1023/a:1018987928936

Chatterjee, B., Gorain, B., Mohananaidu, K. et al. (2019), Targeted drug delivery to the brain via intranasal nanoemulsion: available proof of concept and existing challenges, *Int. J. Pharm.*, 565, 258–268. doi:10.1016/j.ijpharm.2019.05.032

Chien, D.S., Homsy, J.J., Gluchowski, C. et al., (1990), Corneal and conjunctival/scleral penetration of *p*-aminoclonidine, AGN 190342, and clonidine in rabbit eyes, *Curr. Eye Res.*, 9 (11), 1051–1059. 10.3109/02713689008997579

Choudhury, H., Gorain, B., Karmakar, S. et al. (2014), Improvement of cellular uptake, in vitro antitumor activity and sustained release profile with increased bioavailability from a nanoemulsion platform, *Int. J. Pharm.*, 460 (1–2), 131–143. doi:10.1016/j.ijpharm.2013.10.055

Cordoni, A. and Cordoni, L.E. (2001), Eutectic mixture of local anesthetics reduces pain during intravenous catheter insertion in the pediatric patient, *Clin. J. Pain*, 17 (2), 115–118. doi:10.1097/00002508-200106000-00003

Dey, S., Anand, B.S., Patel, J. et al. (2003), Transporters/receptors in the anterior chamber: pathways to explore ocular drug delivery strategies, *Expert Opin. Biol. Ther.*, 3, 23–44. doi:10.1517/14712598.3.1.23.

Ding, S. and Olejnik, O. (1997), Cyclosporin ophthalmic oil/water emulsions. Formulation and characterization, *Pharm. Res.*, 14, S41.

Ding, S., Tien, W., and Olejnik, O. (1995) Nonirritating emulsions for sensitive tissue, U.S. Patent No. 5,474,979 (to Allergan), December 12, 1995.

Dordevic, S.M., Santrac, A., Cekic, N.D. et al. (2017), Parenteral nanoemulsions of risperidone for enhanced brain delivery in acute psychosis: physicochemical and in vivo performances, *Int. J. Pharm.*, 533 (2), 421–430. 10.1016/j.ijpharm.2017.05.051

Dutta, S. (1999), Use of eutectic mixture of local anesthetics in children, *Indian J. Pediatr.*, 66 (5), 707–715. doi:10.1007/BF02726260

Edmond, J. (2001), Essential polyunsaturated fatty acids and the barrier to the brain: the components of a model for transport, *J. Mol. Neurosci.*, 16 (2–3), 181–193. doi:10.1385/JMN:16:2-3:181

El-Hameed, M.D.A. and Kellaway, I.W. (1997), Preparation and in vitro characterisation of mucoadhesive polymeric microspheres as intra-nasal delivery systems, *Eur. J. Pharm. Biopharm.*, 44, 53–60. doi:10.1016/S0939-6411(97)00101-X

Ezra, R., Benita, S., Ginsberg, I. et al. (1996), Prevention of oxidative damage in fibroblast cell cultures and rat skin by positively-charged submicron emulsion of α-tocopherol, *Eur. J. Pharm. Biopharm.*, 42 (4), 291–298.

Friedman, D.I., Schwarz, J.S., and Weisspapir, M. (1995), Submicron emulsion vehicle for enhanced transdermal delivery of steroidal and nonsteroidal antiinflammatory drugs, *J. Pharm. Sci.*, 84 (3), 324–329. doi:10.1002/jps.2600840312

Gadhave, D., Choudhury, H., and Kokare, C. (2019), Neutropenia and leukopenia protective intranasal olanzapine-loaded lipid-based nanocarriers engineered for brain delivery, *Appl. Nanosci.*, 9, 151–168. doi:10.1007/s13204-018-0909-3

Galatoire, O., Baudouin, C., Pisella, P. J. et al. (2003), Flow cytometry in impression cytology during keratoconjunctivitis sicca: effects of topical cyclosporin A on HLA DR expression, *J. Fr. Ophtalmol.*, 26 (4), 337–343.

Gänger, S., Schindowski, K., Gänger, S. et al. (2018), Tailoring formulations for intranasal nose-to-brain delivery: a review on architecture, physico-chemical characteristics and mucociliary clearance of the nasal olfactory mucosa, *Pharmaceutics*, 10 (3), E116. doi:10.3390/pharmaceutics10030116.

Ganta, S., Deshpande, D., Korde, A. et al. (2010a), A review of multifunctional nanoemulsion systems to overcome oral and CNS drug delivery barriers, *Mol. Membr. Biol.*, 27 (7), 260–273. doi:10.3109/09687688.2010.497971

Ganta, S., Sharma, P., Paxton, J.W. et al. (2010b), Pharmacokinetics and pharmacodynamics of chlorambucil delivered in long-circulating nanoemulsion, *J. Drug Target.*, 18 (2), 125–133. doi:10.3109/10611860903244199

Garg, T., Rath, G., Murthy, R.R. et al. (2015), Current nanotechnological approaches for an effective delivery of bioactive drug molecules to overcome drug resistance tuberculosis, *Curr. Pharm. Des.*, 21 (22), 3076–3089. doi:10.2174/1381612821666150531163254

Garty, N., Lusky, M., Zalish, M. et al. (1994), Pilocarpine in submicron emulsion formulation for treatment of ocular hypertension: a phase II clinical trial, *Invest. Ophthalmol. Vis. Sci.*, 35, 2175.

Gorain, B., Choudhury, H., Biswas, E. et al. (2013), A novel approach for nanoemulsion components screening and nanoemulsion assay of olmesartan medoxomil through a developed and validated HPLC method, *RSC Adv.*, 27, 10887–10893. doi:10.1039/c3ra41452c

Gursoy, R.N. and Benita, S. (2004), Self-emulsifying drug delivery systems (SEDDS) for improved oral delivery of lipophilic drugs, *Biomed. Pharmacother.*, 58 (3), 173–182. doi:10.1016/j.biopha.2004.02.001

Hak, S., Garaiova, Z., Olsen, L.T. et al. (2015), The effects of oil-in-water nanoemulsion polyethylene glycol surface density on intracellular stability, pharmacokinetics, and biodistribution in tumor bearing mice, *Pharm. Res.*, 32 (4), 1475–1485. doi:10.1007/s11095-014-1553-6

Hamilton, J.A. and Brunaldi, K. (2007), A model for fatty acid transport into the brain, *J. Mol. Neurosci.*, 33, 12–17. doi:10.1007/s12031-007-0050-3

Han, J., Davis, S. S., Papandreou, C. et al. (2004), Design and evaluation of an emulsion vehicle for Paclitaxel. I. Physicochemical properties and plasma stability, *Pharm. Res.*, 21 (9), 1573–1580. doi:10.1023/b:pham.0000041451.70367.21

Harnisch, S. and Müller, R. H. (2000), Adsorption kinetics of plasma proteins on oil-in-water emulsions for parenteral nutrition, *Eur. J. Pharm. Biopharm.*, 49, 41–46. doi:10.1016/s0939-6411(99)00064-8

Hom, M. M. (2006), Use of cyclosporine 0.05% ophthalmic emulsion for contact lens-intolerant patients, *Eye Contact Lens*, 32 (2), 109–111. doi:10.1097/01.icl.0000175651.30487.58

Hosny, K.M. and Banjar, Z.M. (2013), The formulation of a nasal nanoemulsion zaleplon in situ gel for the treatment of insomnia, *Expert Opin. Drug Deliv.*, 10 (8), 1033–1041. doi:10.1517/17425247.2013.812069

Hussain, A.A. (1998), Intranasal drug delivery, *Adv. Drug Deliv. Rev.*, 29 (1–2), 39–49. doi:10.1016/S0169-409X(97)00060-4

Ilan, E., Amselem, S., Weisspapir, M. et al. (1996), Improved oral delivery of desmopressin via a novel vehicle: mucoadhesive submicron emulsion, *Pharm. Res.*, 13 (7), 1083–1087. doi:10.1023/a:1016023111248

Järvinen, K., Järvinen, T., and Urtti, A. (1995), Ocular absorption following topical delivery, *Adv. Drug Deliv. Rev.*, 16, 3–19. doi:10.1016/0169-409X(95)00010-5

Jenkins, P.G. (1999), Mucosal vaccine delivery, *Expert Opin. Ther. Pat.*, 9 (3), 255–262. doi:10.1517/13543776.9.3.255

Jumaa, M. and Müller, B.W. (1999), Physicochemical properties of chitosan–lipid emulsions and their stability during the autoclaving process, *Int. J. Pharm.*, 183 (2), 175–184. doi:10.1016/s0378-5173(99)00086-1

Kararli, T.T., Needham, T.E., Schoenhard, G. et al. (1992), Enhancement of nasal delivery of a renin inhibitor in the rat using emulsion formulations, *Pharm. Res.*, 9 (8), 1024–1028. doi:10.1023/a:1015850310589

Kaur, I.P. and Smitha, R. (2002), Penetration enhancers and ocular bioadhesives: two new avenues for ophthalmic drug delivery, *Drug Dev. Ind. Pharm.*, 28 (4), 353–369. doi:10.1081/ddc-120002997

Khunt, D., Shah, B., and Misra, M. (2017), Role of butter oil in brain targeted delivery of Quetiapine fumarate microemulsion via intranasal route, *J. Drug Deliv. Sci. Technol.*, 40, 11–20. doi:10.1016/j.jddst.2017.05.004

Kim, D., Kim, Y.H., and Kwon, S. (2018), Enhanced nasal drug delivery efficiency by increasing mechanical loading using hypergravity, *Sci. Rep.*, 8, 168. doi:10.1038/s41598-017-18561-x

Kim, T.W., Chung, H., Kwon, I.C. et al. (2000), In vivo gene transfer to the mouse nasal cavity mucosa using a stable cationic lipid emulsion, *Mol. Cell*, 10 (2), 142–147. doi:10.1007/s10059-000-0142-1

Kimura, T., Takeda, K., Kageyn, A. et al. (1989), Intestinal absorption of dolichol from emulsions and liposomes in rats, *Chem. Pharm. Bull.*, 37 (2), 463–466. doi:10.1248/cpb.37.463

Klang, S., Abdulrazik, M., and Benita, S. (2000), Influence of emulsion droplet surface charge on indomethacin ocular tissue distribution, *Pharm. Dev. Technol.*, 5 (4), 521–532. doi:10.1081/pdt-100102035

Klang, S.H., Frucht-Pery, J., Hoffman, A. et al. (1994), Physicochemical characterization and acute toxicity evaluation of a positively-charged submicron emulsion vehicle, *J. Pharm. Pharm.*, 46 (12), 986–993. doi:10.1111/j.2042-7158.1994.tb03254.xKlang, S.H., Siganos, C.S., and Benita, S. (1999), Evaluation of a positively-charged submicron emulsion of piroxicam in the rabbit corneum healing process following alkali burn, *J. Control. Release*, 57, 19–27. doi:10.1016/s0168-3659(98)00107-2

Ko, K.Y., Needham, T.E., and Zia, H. (1998), Emulsion formulations of testosterone for nasal administration, *J. Microencapsul.*, 15 (2), 197–205. doi:10.3109/02652049809006849

Kunert, K.S., Tisdale, A.S., Stern, M.E. et al. (2000), Analysis of topical cyclosporine treatment of patients with dry eye syndrome: effect on conjunctival lymphocytes, *Arch. Ophthalmol.*, 118 (11), 1489–1496. doi:10.1001/archopht.118.11.1489

Lee, V.H.L. (1993), Precorneal, corneal and postcorneal factors, in: Mitra, A.K., Ed., *Ophthalmic Drug Delivery System*, Marcel Dekker, New York, pp. 59–81.

Lin, H., Gebhardt, M., Bian, S. et al. (2007), Enhancing effect of surfactants on fexofenadine.HCl transport across the human nasal epithelial cell monolayer, *Int. J. Pharm.*, 330 (1–2), 23–31. doi:10.1016/j.ijpharm.2006.08.043

Liu, J.J., Green, P., John Mann, J. et al. (2015), Pathways of polyunsaturated fatty acid utilization: implications for brain function in neuropsychiatric health and disease, *Brain Res.*, 1597, 220–246. doi:10.1016/j.brainres.2014.11.059

Ludwig, A. (2005), The use of mucoadhesive polymers in ocular drug delivery, *Adv. Drug Deliv. Rev.*, 57 (11), 1595–1639. doi:10.1016/j.addr.2005.07.005

Madhusudhan, B., Rambhau, D., Apte, S.S. et al. (2007), 1-*O*-alkylglycerol stabilized carbamazepine intravenous o/w nanoemulsions for drug targeting in mice, *J. Drug Target.*, 15 (2), 154–161. doi:10.1080/10611860601141150

Mahajan, H.S., Mahajan, M.S., Nerkar, P.P. et al. (2014), Nanoemulsion-based intranasal drug delivery system of saquinavir mesylate for brain targeting, *Drug Deliv.*, 21 (2), 148–154. doi:10.3109/10717544.2013.838014

Maurice, D.M. and Mishima, S. (1984), Ocular pharmacokinetics, in: Sears, M.L., Ed., *Pharmacology of the Eye*, Springer Verlag, New York, pp. 20–116.

Melamed, S., Kurtz, S., Greenbaum, A. et al. (1994), Adaprolol maleate in submicron emulsion, a novel soft β-blocking agent, is safe and effective in human studies, *Invest. Ophthalmol. Vis. Sci.*, 35, 1387–1390.

Mitra, R., Pezron, I., Chu, W.A. et al. (2000), Lipid emulsions as vehicles for enhanced nasal delivery of insulin, *Int. J. Pharm.*, 205 (1–2), 127–134. doi:10.1016/s0378-5173(00)00506-8

Morrison, E.E. and Costanzo, R.M. (1992), Morphology of olfactory epithelium in humans and other vertebrates, *Microsc. Res. Tech.*, 23, 49–61. doi:10.1002/jemt.1070230105

Muchtar, S., Almog, S., Torracca, M.T. et al. (1992), A submicron emulsion as ocular vehicle for delta-8-tetrahydrocannabinol: effect on intraocular pressure in rabbits, *Ophthalmic Res.*, 24 (3), 142–149. doi:10.1159/000267160.

Myers, R.A. and Stella, V.J. (1992), Systemic bioavailability of penclomedine (NSC-338720) from oil-in-water emulsions administered intraduodenally to rats, *Int. J. Pharm.*, 78 (1–3), 217–226. doi:10.1016/0378-5173(92)90374-B

Navarro, C., Gonzalez-Alvarez, I., Gonzalez-Alvarez, M. et al. (2011), Influence of polyunsaturated fatty acids on cortisol transport through MDCK and MDCK-MDR1 cells as blood–brain barrier in vitro model, *Eur. J. Pharm. Sci.*, 42 (3), 290–299. doi:10.1016/j.ejps.2010.12.005

Naveh, N., Muchtar, S., and Benita, S. (1994), Pilocarpine incorporated into a submicron emulsion vehicle causes an unexpectedly prolonged ocular hypotensive effect in rabbits, *J. Ocul. Pharmacol.*, 10 (3), 509–520. doi:10.1089/jop.1994.10.509

Naveh, N., Weissman, C., Muchtar, S. et al. (2000), A submicron emulsion of HU-211, a synthetic cannabinoid, reduces intraocular pressure in rabbits, *Graefes Arch. Clin. Exp. Ophthalmol.*, 238 (4), 334–338. doi:10.1007/s004170050361

Palin, K.J., Phillips, A.J., and Ning, A. (1986), The oral absorption of cefoxitin from oil and emulsion vehicles in rats, *Int. J. Pharm.*, 33 (1–3), 99–104. doi:10.1016/0378-5173(86)90043-8

Pandey, M., Choudhury, H., Yeun, O.C. et al. (2018), Perspectives of nanoemulsion strategies in the improvement of oral, parenteral and transdermal chemotherapy, *Curr. Pharm. Biotechnol.*, 19 (4), 276–292. doi:10.2174/1389201019666180605125234

Pardeshi, C.V. and Belgamwar, V.S. (2013), Direct nose to brain drug delivery via integrated nerve pathways bypassing the blood–brain barrier: an excellent platform for brain targeting, *Expert Opin. Drug Deliv.*, 10 (7), 957–972. doi:10.1517/17425247.2013.790887

Piemi, M.P., Korner, D., Benita, S. et al. (1999), Positively and negatively charged submicron emulsions for enhanced topical delivery of antifungal drugs, *J. Control. Release*, 58 (2), 177–187. doi:10.1016/S0168-3659(98)00156-4

Prabhakar, K., Afzal, S.M., Surender, G. et al. (2013), Tween 80 containing lipid nanoemulsions for delivery of indinavir to brain, *Acta Pharm. Sin. B*, 3 (5), 345–353. doi:10.1016/j.apsb.2013.08.001

Prausnitz, M.R. and Noonan, J.S. (1998), Permeability of cornea, sclera and conjunctiva: a literature analysis for drug delivery to eye, *J. Pharm. Sci.*, 87 (12), 1479–1488. doi:10.1021/js9802594

Realdaon, N. and Ragazzi, E. (1998), Study of drug availability from o/w emulsions obtained in different manufacturing conditions, in: *Proceedings of 2nd World Meeting APGI/APV*, pp. 775–776.

Sall, K., Stevenson, O.D., Mundorf, T.K. et al. (2000), Two multicenter, randomized studies of the efficacy and safety of cyclosporine ophthalmic emulsion in moderate to severe dry eye disease. CsA Phase 3 Study Group, *Ophthalmology*, 107 (4), 631–639. doi:10.1016/s0161-6420(99)00176-1

Sall, K.N., Cohen, S.M., Christensen, M.T. et al. (2006), An evaluation of the efficacy of a cyclosporine-based dry eye therapy when used with marketed artificial tears as supportive therapy in dry eye, *Eye Contact Lens Sci. Clin. Pract.*, 32, 21–26. doi:10.1097/01.icl.0000170599.57728.29

Samaridou, E. and Alonso, M.J. (2018), Nose-to-brain peptide delivery—the potential of nanotechnology, *Bioorg. Med. Chem.*, 26 (10), 2888–2905. doi:10.1016/j.bmc.2017.11.001.

Sapra, B., Thatai, P., Bhandari, S. et al. (2013), A critical appraisal of microemulsions for drug delivery: part I, *Ther. Deliv.*, 4 (12), 1547–1564. doi:10.4155/tde.13.116.

Sasaki, H., Yamamura, K., Nishida, K. et al. (1996), Delivery of drugs to the eye by topical application, *Prog. Retin. Eye Res.*, 15 (2), 583–620. doi:10.1016/1350-9462(96)00014-6

Schwarz, J.S., Weisspapir, M.R., and Friedman, D.I. (1995), Enhanced transdermal delivery of diazepam by submicron emulsion (SME) creams, *Pharm. Res.*, 12, 687–692. doi:10.1023/A:1016255408348

Shah, L. (2013), Multifunctional Nanoemulsions for systemic delivery of analgesic peptides to the CNS, in: *Pharmaceutical Sciences School of Pharmacy: Northeastern University*.

Shah, L., Gattacceca, F., and Amiji, M.M. (2014), CNS delivery and pharmacokinetic evaluations of DALDA analgesic peptide analog administered in nano-sized oil-in-water emulsion formulation, *Pharm. Res.*, 31 (5), 1315–1324. doi:10.1007/s11095-013-1252-8

Small, D. S., Acheampong, A., Reis, B. et al. (2002), Blood concentrations of cyclosporin A during long-term treatment with cyclosporin A ophthalmic emulsions in patients with moderate to severe dry eye disease, *J. Ocul. Pharmacol. Ther.*, 18 (5), 411–418. doi:10.1089/10807680260362696

Stevenson, D., Tauber, J., and Reis, B.L. (2000), Efficacy and safety of cyclosporin A ophthalmic emulsion in the treatment of moderate-to-severe dry eye disease: a dose-ranging, randomized trial. The Cyclosporin A Phase 2 Study Group, *Ophthalmology*, 107 (5), 967–974. doi:10.1016/s0161-6420(00)00035-x

Stjernschantz, J. and Astin, M. (1993), Anatomy and physiology of the eye: physiological aspects of ocular drug therapy, in: Edman, P., Ed., *Biopharmaceutics of Ocular Drug Delivery*, CRC Press, Boca Raton, pp. 1–25.

Sznitowska, M., Janicki, S., Zurowska-Pryczkowska, K. et al. (2001), In vivo evaluation of submicron emulsions with pilocarpine: the effect of pH and chemical form of the drug, *J. Microencapsul.*, 18 (2), 173–181. doi:10.1080/02652040010000442

Takino, T., Konishi, K., Takakura, Y. et al. (1994), Long circulating emulsion carrier systems for highly lipophilic drugs, *Biol. Pharm. Bull.*, 17 (1), 121–125. doi:10.1248/Bpb.17.121

Takino, T., Nagahama, E., Sakaeda (neeKakutani), T. et al. (1995), Pharmacokinetic disposition analysis of lipophilic drugs injected with various lipid carriers in the single-pass rat liver perfusion system, *Int. J. Pharm.*, 114, 43–54. doi:10.1016/0378-5173(94)00211-M

Tamilvanan, S. and Benita, S. (2004), The potential of lipid emulsion for ocular delivery of lipophilic drugs, *Eur. J. Pharm. Biopharm.*, 58 (2), 357–368. doi:10.1016/j.ejpb.2004.03.033

Tamilvanan, S., Khoury, K., Gilhar, D. et al. (2001), Ocular delivery of cyclosporin A. I. Design and characterization of cyclosporin A-loaded positively charged submicron emulsion, *STP Pharm. Sci.*, 11, 421–426.

Tamilvanan, S., Schmidt, S., Muller, R.H. et al. (2005), In vitro adsorption of plasma proteins onto the surface (charges) modified-submicron emulsions for intravenous administration, *Eur. J. Pharm. Biopharm.*, 59, 1–7. doi:10.1016/j.ejpb.2004.07.001

Tan, S.F., Kirby, B.P., Stanslas, J. et al. (2017), Characterisation, in-vitro and in-vivo evaluation of valproic acid-loaded nanoemulsion for improved brain bioavailability, *J. Pharm. Pharmacol.*, 69 (11), 1447–1457. doi:10.1111/jphp.12800

Tang-Liu, D.D. and Acheampong, A. (2005), Ocular pharmacokinetics and safety of ciclosporin, a novel topical treatment for dry eye, *Clin. Pharmacokinet.*, 44, 247–261. doi:10.2165/00003088-200544030-00003

Turner, K., Pflugfelder, S.C., Ji, Z. et al. (2000), Interleukin-6 levels in the conjunctival epithelium of patients with dry eye disease treated with cyclosporine ophthalmic emulsion, *Cornea*, 19 (4), 492–496. doi:10.1097/00003226-200007000-00018

Vyas, T.K., Babbar, A.K., Sharma, R.K. et al. (2005), Intranasal mucoadhesive microemulsions of zolmitriptan: preliminary studies on brain-targeting, *J. Drug Target.*, 13 (5), 317–324. doi:10.1080/10611860500246217

Vyas, T.K., Shahiwala, A., and Amiji, M.M. (2008), Improved oral bioavailability and brain transport of Saquinavir upon administration in novel nanoemulsion formulations, *Int. J. Pharm.*, 347 (1–2), 93–101. doi:10.1016/j.ijpharm.2007.06.016

Wang, Q., Zhang, Y., Wong, C.-H. et al. (2018), Demonstration of direct nose-to-brain transport of unbound hiv-1 replication inhibitor DB213 Via intranasal administration by pharmacokinetic modeling, *AAPS J.*, 20 (1), 23. 10.1208/s12248-017-0179-0

Washington, N., Washington, C., and Wilson, C. (2001), Ocular drug delivery, in: Washington, N., Washington, C., and Wilson, C., Eds., *Physiological Pharmaceutics: Barriers to Drug Absorption*, Taylor & Francis, New York, pp. 250–270.

Welin-Berger, K. and Bergenstahl, B. (2000), Inhibition of Ostwald ripening in local anesthetic emulsions by using hydrophobic excipients in the disperse phase, *Int. J. Pharm.*, 200 (2), 249--260. doi:10.1016/s0378-5173(00)00395-1

Welin-Berger, K., Neelissen, J.A., and Bergenstahl, B. (2001), The effect of rheological behaviour of a topical anaesthetic formulation on the release and permeation rates of the active compound, *Eur. J. Pharm. Sci.*, 13 (3), 309–318. doi:10.1016/s0928-0987(01)00118-x

Wen, C.-J., Yen, T.-C., Al-Suwayeh, S.A. et al. (2011), In vivo real-time fluorescence visualization and brain-targeting mechanisms of lipid nanocarriers with different fatty ester:oil ratios, *Nanomedicine*, 6 (9), 1545–1559. doi:10.2217/nnm.11.46

Wen, M.M. (2011), Olfactory targeting through intranasal delivery of biopharmaceutical drugs to the brain, *Discov. Med.*, 11 (61), 497–503.

Win, T., Rajagopal, J., Manda, U.K. et al. (2017), Incorporation of carbopol to palm olein based analgesic cream: effect on formulation characteristics, *Latin Am. J. Pharm.*, 36 (11), 2144–2152.

Yadav, S., Gattacceca, F., Panicucci, R. et al. (2015), Comparative biodistribution and pharmacokinetic analysis of cyclosporine-A in the brain upon intranasal or intravenous administration in an oil-in-water nanoemulsion formulation, *Mol. Pharm.*, 12 (5), 1523–1533. doi:10.1021/mp5008376

Yang, S.C. and Benita, S. (2000), Enhanced absorption and drug targeting by positively charged submicron emulsions, *Drug Dev. Res.*, 50 (3–4), 476–486. doi:10.1002/1098-2299(200007/08)50:3/4%3C476::AID-DDR31%3E3.0.CO;2-6

Zurowska-Pryczkowska, K., Sznitowska, M., and Janicki, S. (1999), Studies on the effect of pilocarpine incorporation into a submicron emulsion on the stability of the drug and the vehicle, *Eur. J. Pharm. Biopharm.*, 47 (3), 255–260. doi:10.1016/s0939-6411(98)00098-8

PART I

OVERVIEW OF TOCOL-BASED EMULSIONS, OXYGEN-CARRYING EMULSIONS, EMULSIONS WITH DOUBLE OR TRIPLE CARGOS AND EMULSION-LIKE DISPERSIONS

CHAPTER 7

OVERVIEW OF TOCOL-BASED EMULSIONS, OXYGEN-CARRYING EMULSIONS, EMULSIONS WITH DOUBLE OR TRIPLE CARGOS AND EMULSION-LIKE DISPERSIONS

Oil-in-Water Nanosized Emulsions for Drug Delivery and Targeting,
First Edition. Tamilvanan Shunmugaperumal.

© 2021 John Wiley & Sons, Inc. Published 2021 by John Wiley & Sons, Inc.

CHAPTER 7.1

TOCOL-BASED NANOSIZED EMULSIONS

OLY KATARI* AND TAMILVANAN SHUNMUGAPERUMAL

Department of Pharmaceutics, National Institute of Pharmaceutical Education and Research (NIPER), Guwahati, Assam, India

*The funding provided by Department of Pharmaceuticals, Ministry of Chemicals and Fertilizers, Govt. of India to the author Oly Katari is greatly acknowledged.

Oil-in-Water Nanosized Emulsions for Drug Delivery and Targeting,
First Edition. Tamilvanan Shunmugaperumal.
© 2021 John Wiley & Sons, Inc. Published 2021 by John Wiley & Sons, Inc.

EXPANSION OF ABBREVIATIONS

APIs	active pharmaceutical ingredients
IV	intravenous
LCTs	long-chain triglycerides
MCT	medium-chain triglycerides
PEG	poly(ethyleneglycol)
PEG2000-DSPE	1,2-distearoyl-*sn*-glycero-3-phosphoethanolamine-*N*-[amino (polyethyleneglycol)2000]
VLBW	very low-birth-weight

7.1.1. INTRODUCTION

The term "tocols", broadly defined, indicates a family of tocopherols (vitamin E) and tocotrienols and their derivatives, including those common derivatives that are esterified at the 6-hydroxyl on the chroman ring (Packer and Fuchs 1993; Herrera 2001). In particular, the vitamin E family consists of eight naturally occurring isomers which are α-, β-, γ-, and δ-tocopherols as well as α-, β-, γ-, and δ-tocotrienols (Sylvester et al. 2010). Most of the studies involving vitamin E are reported on the α-tocopherol isomer since it is widely available in the market. However, the use of the term "tocols" is appropriate since all tocopherols and tocotrienols are fundamentally derivatives of the simplest tocopherol, 6-hydroxy-2-methyl-2-phytylchroman (sometimes referred to as "tocol"). Although both groups contain a chromanol ring along with a 15-carbon tail, tocopherols and tocotrienols differ in tail saturation and number and location of methyl groups on the chromanol ring. The presence of three trans-double bonds in the tail of tocotrienols highly distinguish them from tocopherols physiologically. Physically, vitamin E is a viscous liquid. It is insoluble in water but is highly soluble in cosolvents such as ethanol, oils, organic solvents such as acetone and chloroform as well as in surfactants. Among the various derivatives of vitamin E isomers, the most common derivatives are the tocopherol esters (acetate, succinate, tartrate, and nicotinate) and the polyethylene glycol succinate esters of vitamin E (Constantinides et al. 2006).

A number of studies reveal that α-tocopherol has been used in parenteral nanosized emulsions in small quantities (usually $<1\%$), to protect the encapsulated polyunsaturated fatty acids from oxidation (Collins-Gold et al. 1990). Tocopherol was also found to synergize with other antioxidants, such as green tea extract, rosmarinic acid, and vitamin C, in providing antioxidation effect (Carlotti et al. 1997; Panya et al. 2012; Yin et al. 2012). Bartels et al. (2004) reported that the free radicals induced in systemic and hepatic ischemia and reperfusion, which may be one of the related causes in liver peroxidation injury, could be improved by administering tocopherol emulsions. Therefore, tocopherol has been used in parenteral nutrition regimen in cases of oxidative stress (e.g., chronic inflammation, sepsis, lung distress syndrome, and organ failure) and was also found to be beneficial to the very low-birth-weight (VLBW) infants who need extra nutrition (Baeckert et al. 1988; Biesalski 2009). This section did not discriminate tocopherol and tocotrienol and therefore both of them are considered as tocol. Hence, in the following paragraphs, tocol is predominantly used as a prefix to indicate emulsions, i.e., tocol-based nanosized emulsions.

7.1.2. FORMULATION DEVELOPMENT USING TOCOL

Over the years, tocopherol (vitamin E) has been incorporated into many emulsions for parenteral administration to serve as an antioxidant. Recently, oils with potential biological activity, such as fish oil, olive oil, and vitamin E, have been used in parenteral emulsions to contribute therapeutic value in certain conditions (Hartman et al. 2009; Goulet et al. 2010). Vitamin E, for example, was found to protect against doxorubicin-induced cardiotoxicity, allowing for higher doxorubicin doses without an increase in toxicity (Geetha et al. 1990). A list of commercial products that contain vitamin E that are intended for parenteral administration is given in Table 7.1.1.

While great efforts were made to develop tocopherol emulsions, no commercial nanosized emulsions rich in tocopherol are currently available in the market. Due to the polarity and viscosity of tocopherol, it is extremely challenging to maintain the physical stability of emulsions when loaded with high

TABLE 7.1.1. Commercial Products for Parenteral Administration that Contain Vitamin E

Product	Manufacturer
Cernevit 2.02 mg (IV emulsion)	Baxter/Clintec Parenteral, France
Infuvite Adult 1 mg (IV emulsion)	Clintec/Baxter Healthcare, Illinois, USA
Infuvite PEDiatric 1.4 mg (IV emulsion)	Clintec/Baxter Healthcare, Illinois, USA
Intralipid 10% 6.5 μg (IV emulsion)	Fresenius Kabi, Bad Homburg, Germany
Intralipid 20% 13 μg (IV emulsion)	Fresenius Kabi, Bad Homburg, Germany
Intralipid 30% 19.5 μg (IV emulsion)	Fresenius Kabi, Bad Homburg, Germany
M.V.I. Adult 1 mg (IV emulsion)	Hospira, Illinois, USA
M.V.I.-12 1 mg (IV emulsion)	Hospira, Illinois, USA
M.V.I. Pediatric 1.4 mg (IV emulsion)	Hospira, Illinois, USA
Oliclinomel Trace (emulsion for infusion)	Baxter India Private Limited, Aurangabad, Maharashtra
Vitalipid N Adult 0.91 mg (oily injection)	Fresenius Kabi, Bad Homburg, Germany
Vitalipid N Infant 0.64 mg (oily injection)	Fresenius Kabi, Bad Homburg, Germany

This table shows the quantity of vitamin E in several commercial products that are intended to be used for parenteral administration.

concentrations of tocopherol (Alayoubi et al. 2012). Physical instability and coalescence of nanosized oil droplets may result in phase separation during storage or complications after administration such as lipid embolism (Hulman 1995). The presence of hydroxyl group on the aromatic chromanol ring of tocopherol makes it more polar than conventional oils and thus increases the solubility of phospholipids into it. This in turn reduces the amount of emulsifier available at the tocopherol/water interface (Constantinides et al. 2006). Phospholipids alone are therefore not sufficient to stabilize tocopherol in parenteral nanosized emulsions. It was found that the addition of coemulsifiers with high HLB values are needed to stabilize parenteral nanosized emulsions loaded with a high concentration of tocopherol (Jumaa and Müller 2002). It was found by Han et al. (2004) that a stable tocopherol emulsion could be obtained when a mixture containing lecithin and a combination of deoxycholic acid and sodium deoxycholate was used. A more thorough and systematic study was performed by Alayoubi et al. (2012, 2013c) on the effect of coemulsifiers on the stability of tocopherol emulsions. When sodium deoxycholate, Poloxamer® 188, and sodium oleate were evaluated for their stabilizing effect on emulsions loaded with up to 14% tocopherol, sodium oleate and sodium deoxycholate were found to have a significant hemolytic effect and were sensitive to the presence of electrolytes in the solution, which led to charge neutralization (Alayoubi et al. 2013c). In contrast, Poloxamer® at high concentrations (≥2.5%) was found to be sufficient to stabilize the emulsions by the steric effect of its bulky polyoxyethylene group (Alayoubi et al. 2013c).

In addition to its polarity, tocopherol is also viscous. Therefore, to lower its viscosity, tocopherol is often blended with medium-chain triglycerides (MCT) in the oil phase. The viscosity of binary blends at different vitamin E/MCT ratios was found to decrease from a high of 3,700 mPa·s for α-tocopherol or from 1,800 mPa·s for the tocotrienol-rich fraction of vitamin E to a low of 27 mPa·s for MCT alone (Alayoubi et al. 2012). Viscosity of the oil phase is thus a critical parameter for the emulsification and stability of nanosized emulsions. As a result, vitamin E to MCT ratio was identified as a primary factor that contributes to the long-term stability of these emulsions (Alayoubi et al. 2012). Blending vitamin E with MCT also helps lower the polarity of the oil phase and thereby increasing the ratio of emulsifiers at the oil/water interface.

7.1.3. NON- AND PRE-CLINICAL SAFETY STUDIES

7.1.3.1. Hemolytic Study

The hemolytic activity of formulations intended for parenteral administration may have severe consequences on patients' health (Banziger 1967). Evaluating the hemolytic activity of parenteral nanosized emulsions is therefore a critical *in vitro* toxicity screen and a reliable measure for estimating membrane damage caused by the formulation when administered *in vivo* (Jumaa et al. 1999). Phospholipids, for example, are known to cause erythrocyte hemolysis, and therefore direct contact between emulsion droplets and erythrocytes may lead to harmful effects (Ishii and Nagasaka 2004). A screening study was carried out by Alayoubi and coworkers (2013c) to compare the hemolytic effect of several tocopherol emulsions stabilized with different coemulsifiers. The emulsions under study were found to induce hemolysis, which significantly increased with time (Alayoubi et al. 2013c). Recent studies have shown a decrease in the extent of hemolysis when tocopherol emulsions are PEGylated with 1,2-distearoyl-*sn*-glycero-3-phosphoethanolamine-*N*-[amino(polyethyleneglycol)2000] (PEG2000-DSPE) and a significant increase in the bioavailability and circulation half-life of tocopherol in rats (Alayoubi et al. 2013a). In general, any hemolysis value below 10% is considered to be nonhemolytic, while values above 25% are considered to be hemolytic (Amin and Dannenfelser 2006). The erythrocyte lysis percentage, calculated from *in vitro* hemolysis study, of above 10% but below 25% simply indicates that the developed formulation was ascribed to nonhemolytic. The hemolysis may also be assessed indirectly through *in vivo* pharmacokinetic study in terms of circulation half-life value of the intravenously administered formulations. For example, plasma half-life of tocopherol after intravenous (IV) injection in tumor-free rats ranged from 1.7 h for the conventional formulation to 12.3 h for the PEGylated emulsion (Alayoubi et al. 2013a). Prolonged circulation time was attributed to the PEG density and its conformation on the surface of the droplets. This study also provided an early demonstration of a potentially viable "tocopherol"-rich preparation for parenteral administration.

7.1.4. THERAPEUTIC APPLICATIONS

7.1.4.1. Treatment to Critical or Intensive Care Unit Patients

An adequate intake of tocopherol is required by the VLBW infants to prevent tocopherol deficiency, which is characterized by hemolytic anemia (Brion et al. 2005). To address this issue, Greene et al. (1988) developed two formulations for the parenteral delivery of tocopherol; one in which synthetic emulsifiers like polysorbates (M.V.I. Pediatric™) were used and the other in which tocopherol and fractionated soybean oil were emulsified with egg phospholipids as in Vitalipid® N. M.V.I. Pediatric™ is a lyophilized powder that is used after reconstitution and dilution in IV infusions. Each 5 ml of reconstituted product provides 7 mg tocopherol (as DL-α-tocopheryl acetate) solubilized with 50 mg polysorbate 80 (Hospira 2013). Vitalipid® N is a multivitamin oily injection that must be diluted with Intralipid® 10 or 20% before use. It is available as Vitalipid® N Adult, which contains 9.1 mg (10 IU) DL-α-tocopherol per ampoule (10 ml), and Vitalipid® N Infant, which contains 0.64 mg (0.7 IU) DL-α-tocopherol per 1 ml (Medsafe 2013). Several preclinical and clinical trials were carried out to investigate the toxicity and health benefits of tocopherol in parenteral nutrition. A clinical study in 65 neonatal–perinatal training program centers that was carried out to determine the toxicity of IV tocopherol when given to VLBW infants (Brion et al. 2005) found that the smallest infants (<1.0 kg) were receiving excessive, potentially toxic doses of tocopherol.

7.1.4.2. As a Solvent, API Delivery Vehicle, and Therapeutic Agent

Apart from being a good antioxidant, tocopherol was also found to be a good solubilizer for many poorly water-soluble active pharmaceutical ingredients (APIs) like paclitaxel, cyclosporine, and itraconazole (Han et al. 2004). It was also reported that many APIs with poor solubility in long-chain triglycerides (LCTs) or MCTs have good solubility in tocopherol (Nielsen et al. 2001). Nielsen et al. (2001) also compared the solubility of various poorly soluble APIs in different oils and found that most APIs have a significantly higher solubility in tocopherol than in sesame oil. As a consequence, the Cremophor®-based commercial formulation was replaced by a tocopherol emulsion, TOCOSOL®, which was developed for the delivery of the chemotherapeutic API paclitaxel (Constantinides et al. 2000). TOCOSOL® was found to provide several advantages over the commercial Cremophor®-based formulation TAXOL®. These advantages included high paclitaxel loading (10 mg ml^{-1}) in a ready-to-use product, shorter infusion periods, and smaller dose volumes (Constantinides et al. 2000). These characteristics provided a better safety profile while tolerating the delivery of a greater "dose density", thereby resulting in greater antitumor efficacy. Unfortunately, while positive results were reported for TOCOSOL® in phase I and II clinical trials (Constantinides et al. 2004), it failed a pivotal phase III study that was conducted on breast cancer patients. In another example, a tocopherol injectable emulsion was manufactured for the delivery of amiodarone as an alternative to the current clinical formulation Cordarone® that contains 2% Tween® 80 (Kessler et al. 2002). The amiodarone-loaded tocopherol emulsion was tested in mice, rats, rabbits, and dogs. It showed a protective antiarrhythmic effect in an aconitine-induced arrhythmia model in rats and a reduction in venous irritation in rabbits (Kessler et al. 2002). It also improved activity in preventing the ischemia-induced decreases in left ventricular contractility in the ischemic-heart model in dogs (Kessler et al. 2002).

Another novel tocopherol emulsion was developed for the parenteral delivery of clarithromycin (Lu et al. 2008). It showed good physical stability and reduced irritation when examined by the rabbit ear vein irritation test. Other APIs such as resveratrol and vincristine were also formulated in parenteral tocopherol emulsions (Junping et al. 2003; Hung et al. 2007). Without tocopherol, emulsions loaded with vincristine were found to be unstable and resulted in the precipitation of white crystals.

In several studies, parenteral administration of tocopherol isomers was shown to have a potential therapeutic activity against various tumors, such as breast and colon cancers (Sen et al. 2006; Sylvester et al. 2010). The antitumor activity of tocopherol was mostly attributed to its tocotrienol isomers. Many studies reported on the antiproliferative activity of tocotrienols. Few studies, however, demonstrated the potential of nanosized emulsions in the delivery of tocopherol for cancer therapy. In a study, tocopherol nanosized emulsions loaded with either the tocopherol or tocotrienol isomers were tested for its antiproliferative activity against mammary gland/breast adenocarcinoma (MCF-7) and colorectal adenocarcinoma (SW-620) cells (Alayoubi et al. 2013c). No significant inhibition in viability was observed when the cells were treated with an emulsion loaded with the α-tocopherol isomer of tocopherol. Approximately 80 and 100% of MCF-7 and SW-620 cells remained viable, respectively, when treated with α-tocopherol at concentrations as high as 25 μM. On the other hand, a dose-dependent inhibition in cell growth was observed when cells were treated with an emulsion loaded with the tocotrienol isomers of tocopherol.

The tocotrienol emulsion was found to significantly reduce cell viability at tocotrienol concentrations as low as 15 μM against both cell lines (Alayoubi et al. 2015). The IC50 of the tocotrienol nanosized emulsion was 14 and 12 μM against MCF-7 and SW-620, respectively, whereas the IC50 of the α-tocopherol emulsion was approximately 69 and 78 μM against MCF-7 and SW-620, respectively (Alayoubi et al. 2013c). Based on these studies, it was concluded that fortifying parenteral nanosized emulsions with the tocotrienol fraction of tocopherol is a potentially viable formulation for use in cancer therapy. Tocotrienols were also found to potentiate the antitumor activity of several classes of APIs, such as statins (Wali and Sylvester 2007), COX-2 inhibitors (Shirode and Sylvester 2010), and tyrosine kinase inhibitors (Bachawal et al. 2010).

The IC50 of a 20% tocotrienol emulsion against the MCF-7 and MDA-MB-231 cells was 14 and 7 μM, respectively. It significantly decreased to 10 and 4.8 μM when simvastatin was added to the emulsions at a concentration of 9% w/w of the oil phase (Alayoubi et al. 2013b). The IC50 of simvastatin when used alone was 19 and 8 μM against MCF-7 and MDA-MB-231 cells, respectively. The synergistic anticancer activity between tocotrienols and many classes of APIs may allow for a reduction in the administered dose and consequently improvement in APIs' safety profile. For example, a lower acute toxicity and greater antitumor efficacy were reported for parenteral tocopherol emulsion containing aclacinomycin when compared to the free API (Wang et al. 2002). These findings suggested that developing a combination treatment of tocopherol with chemotherapeutic APIs may offer noteworthy health benefits in the treatment of cancer.

7.1.5. CONCLUSION

As one of the fat-soluble vitamin, the tocol (especially vitamin E/α-tocopherol) has been used in small quantities as an antioxidant in various pharmaceutical, nutritional, food, and cosmetic preparations for oral, topical, or parenteral administration to enhance product stability. However, a major shortcoming to tocol use is its low and erratic oral bioavailability due to its poor aqueous solubility. To overcome the limitations of its oral and topical application, IV delivery was considered a viable alternative route for tocol administration. Moreover, high tolerability of tocol was reported after daily IV administration especially of tocopherol to patients at doses up to 2,300 mg m^{-2} for 9 consecutive days. To prevent and supplement the tocol deficiency in VLBW infants, two commercial products were already in the market developed based on synthetic emulsifiers like polysorbates to produce nanosized emulsions (M.V.I. Pediatric™, Hospira, Illinois, USA) and the other one with egg phospholipids emulsifier-stabilized fractionated soybean oil nanosized emulsions (Vitalipid® N, Fresenius Kabi, Bad Homburg, Germany). Aside from the abovesaid applications, the tocol is itself used as solvent for synthetic APIs, API delivery vehicle, and therapeutic agent in recent years. More commercial success need to come from tocol as an API delivery vehicle for many water-insoluble APIs of potential therapeutic applications.

REFERENCES

Alayoubi, A., Abu-Fayyad, A., and Nazzal, S. (2015), Vitamin E in parenteral lipid emulsions, *Diet Nutr. Crit. Care*, 1861–1874. doi:10.1007/978-1-4614-7836-2_90

Alayoubi, A., Alqahtani, S., Kaddoumi, A. et al. (2013a), Effect of PEG surface conformation on anticancer activity and blood circulation of nanoemulsions loaded with tocotrienol rich-fraction of palm oil, *AAPS J.*, 15 (4), 1168–1179. doi:10.1208/s12248-013-9525-z

Alayoubi, A., Anderson, J.F., Satyanarayanajois, S.D. et al. (2013b), Concurrent delivery of tocotrienols and simvastatin by lipid nanoemulsions potentiates their antitumor activity against human mammary adenocarcinoma cells, *Eur. J. Pharm. Sci.*, 48 (3), 385–392. doi:10.1016/j.ejps.2012.12.011

Alayoubi, A., Kanthala, S., Satyanarayanajois, S.D. et al. (2013c), Stability and *in vitro* antiproliferative activity of bioactive "vitamin E" fortified parenteral lipid emulsions, *Colloids Surf. B Biointerfaces*, 103, 23–30. doi:10.1016/j.colsurfb.2012.10.003

Alayoubi, A., Nazzal, M., Sylvester, P.W. et al. (2012), "Vitamin E" fortified parenteral lipid emulsions: Plackett–Burman screening of primary process and composition parameters, *Drug Dev. Ind. Pharm.*, 39 (2), 363–373. doi:10.3109/03639045.2012.682223

Amin, K. and Dannenfelser, R. (2006), *In vitro* hemolysis: guidance for the pharmaceutical scientist, *Eur. J. Pharm. Sci.*, 95 (6), 1173–1176. doi:10.1002/jps.20627

Bachawal, S., Wali, V., and Sylvester, P. (2010), Enhanced antiproliferative and apoptotic response to combined treatment of gamma-tocotrienol with erlotinib or gefitinib in mammary tumor cells, *BMC Cancer*, 10, 84. doi:10.1186/1471-2407-10-84

Baeckert, P.A., Greene, H.L., Fritz, I. et al. (1988), Vitamin concentrations in very low birth weight infants given vitamins intravenously in a lipid emulsion: measurement of vitamins A, D, and E and riboflavin, *J. Pediatr.*, 113 (6), 1057–1065. doi:10.1016/S0022-3476(88)80583-3

Banziger, R. (1967), Hemolysis testing *in vivo* of parenteral formulations, *Bull. Parenter. Drug Assoc.*, 21 (4), 148–151.

Bartels, M., Biesalski, H.K., Engelhart, K. et al. (2004), Pilot study on the effect of parenteral vitamin E on ischemia and reperfusion induced liver injury: a double blind, randomized, placebo-controlled trial, *Clin. Nutr.*, 23 (6), 1360–1370. doi:10.1016/j.clnu.2004.05.003

Biesalski, H.K. (2009), Vitamin E requirements in parenteral nutrition, *Gastroenterology*, 137 (5), S92–S104. doi:10.1053/j.gastro.2009.07.073

Brion, L.P., Bell, E.F., and Raghuveer, T.S. (2005), Variability in the dose of intravenous vitamin E given to very low birth weight infants, *J. Perinatol.*, 25, 139–142. doi:10.1038/sj.jp.7211223

Carlotti, M.E., Gallarate, M., Gasco, M.R. et al. (1997), Synergistic action of vitamin C and amino acids on vitamin E in inhibition of the lipoperoxidation of linoleic acid in disperse systems, *Int. J. Pharm.*, 155 (2), 251–261. doi:10.1016/S0378-5173(97)00168-3

Collins-Gold, L.C., Lyons, R.T., and Bartholow, L.C. (1990), Parenteral emulsions for drug delivery, *Adv. Drug Deliv. Rev.*, 5 (3), 189–208. doi:10.1016/0169-409X(90)90016-L

Constantinides, P., Han, J., and Davis, S. (2006), Advances in the use of tocols as drug delivery vehicles, *Pharm. Res.*, 23, 243–255. doi:10.1007/s11095-005-9262-9

Constantinides, P.P., Lambert, K.J., Tustian, A.K. et al. (2000), Formulation development and antitumor activity of a filter-sterilizable emulsion of paclitaxel, *Pharm. Res.*, 17 (2), 175–182. doi:10.1023/A:1007565230130

Constantinides, P.P., Tustian, A., and Kessler, D.R. (2004), Tocol emulsions for drug solubilization and parenteral delivery, *Adv. Drug Deliv. Rev.*, 56 (9), 1243–55. doi:10.1016/j.addr.2003.12.005

Geetha, A., Sankar, R., Marar, T. et al. (1990), Alpha-tocopherol reduces doxorubicin-induced toxicity in rats—histological and biochemical evidences, *Indian J. Physiol. Pharmacol.*, 34 (2), 94–100.

Goulet, O., Antebi, H., Wolf, C. et al. (2010), A new intravenous fat emulsion containing soybean oil, medium-chain triglycerides, olive oil, and fish oil: a single-center, double-blind randomized study on efficacy and safety in pediatric patients receiving home parenteral nutrition, *J. Parenter. Enteral Nutr.*, 34 (5), 485–495. doi:10.1177/0148607110363614

Greene, H.L., Hambidge, M., Schanler, R. et al. (1988), Guidelines for the use of vitamins, trace elements, calcium, magnesium, and phosphorus in infants and children receiving total parenteral nutrition: report of the Subcommittee on Pediatric Parenteral Nutrient Requirements from the Committee on Clinical Practice Issues of the American Society for Clinical Nutrition, *Am. J. Clin. Nutr.*, 48 (5), 1324–1342. doi:10.1093/ajcn/48.5.1324

Han, J., Davis, S.S., Papandreou, C. et al. (2004), Design and evaluation of an emulsion vehicle for paclitaxel. I. Physicochemical properties and plasma stability, *Pharm. Res.*, 21, 1573–1580. doi:10.1023/B:PHAM.0000041451.70367.21

Hartman, C., Ben-Artzi, E., Berkowitz, D. et al. (2009), Olive oil-based intravenous lipid emulsion in pediatric patients undergoing bone marrow transplantation: a short-term prospective controlled trial, *Clin. Nutr.*, 28 (6), 631–635. doi:10.1016/j.clnu.2009.04.021

Herrera, B.C.E. (2001), Vitamin E: action, metabolism and perspectives, *J. Physiol. Biochem.*, 57, 43–56. doi:10.1007/BF03179812

Hospira(2013), M.V.I. Pediatric™ product insert, (Available at http://www.hospira.com/Images/EN-1593_32-5591_1.pdf) Accessed on January 10, 2020.

Hulman, G. (1995), The pathogenesis of fat embolism, *J. Pathol.*, 176, 3–9. doi:10.1002/path.1711760103

Hung, C.F., Fang, C.L., Liao, M.H. et al. (2007), The effect of oil components on the physicochemical properties and drug delivery of emulsions: tocol emulsion versus lipid emulsion, *Int. J. Pharm.*, 335 (1–2), 193–202. doi:10.1016/j.ijpharm.2006.11.016

Ishii, F. and Nagasaka, Y. (2004), Interaction between erythrocytes and free phospholipids as an emulsifying agent in fat emulsions or drug carrier emulsions for intravenous injections, *Colloids Surf. B Biointerfaces*, 37 (1–2), 43–47. doi:10.1016/j.colsurfb.2004.05.016

Jumaa, M., Kleinebudde, P., and Müller, B.W. (1999), Physicochemical properties and hemolytic effect of different lipid emulsion formulations using a mixture of emulsifiers, *Pharm. Acta Helv.*, 73 (6), 293–301. doi:10.1016/S0031-6865(99)00003-5

Jumaa, M. and Müller, B.W. (2002), Parenteral emulsions stabilized with a mixture of phospholipids and PEG-660-12-hydroxystearate: evaluation of accelerated and long-term stability, *Eur. J. Pharm. Biopharm.*, 54 (2), 207–212. doi:10.1016/S0939-6411(02)00057-7

Junping, W., Takayama, K., Nagai, T. et al. (2003), Pharmacokinetics and antitumor effects of vincristine carried by microemulsions composed of PEG-lipid, oleic acid, vitamin E and cholesterol, *Int. J. Pharm.*, 251 (1–2), 13–21. doi:10.1016/S0378-5173(02)00580-X

Kessler, D., Palepu, N., Tustian, A. et al. (2002), A novel amiodarone microemulsion injectable formulation, *AAPS PharmSci*, 4, W4122 (Abstract).

Lu, Y., Wang, Y., and Tang, X. (2008), Formulation and thermal sterile stability of a less painful intravenous clarithromycin emulsion containing vitamin E, *Int. J. Pharm.*, 346 (1–2), 47–56. doi:10.1016/j.ijpharm.2007.06.003

Medsafe (2013), Vitalipid® N datasheet, (Available at http://www.medsafe.govt.nz/profs/datasheet/v/vitalipidninj.pdf.) Accessed January 10, 2020.

Nielsen, P.B., Mullertz, A., Norling, T. et al. (2001), The effect of α-tocopherol on the *in vitro* solubilization of lipophilic drugs, *Int. J. Pharm.*, 222 (2), 217–224. doi:10.1016/S0378-5173(01)00701-3

Packer, L. and Fuchs, J., Eds. (1993), *Vitamin E in Health and Disease*, Marcel Dekker, New York.

Panya, A., Kittipongpittaya, K., Laguerre, M. et al. (2012), Interactions between α-tocopherol and rosmarinic acid and its alkyl esters in emulsions: synergistic, additive, or antagonistic effect? *J. Agric. Food Chem.*, 60 (41), 10320–10330. doi:10.1021/jf302673j

Sen, C.K., Khanna, S., and Roy, S. (2006), Tocotrienols: vitamin E beyond tocopherols, *Life Sci.*, 78 (18), 2088–2098. doi:10.1016/j.lfs.2005.12.001

Shirode, A.B. and Sylvester, P.W. (2010), Synergistic anticancer effects of combined γ-tocotrienol and celecoxib treatment are associated with suppression in Akt and NFκB signaling, *Biomed. Pharmacother.*, 64 (5), 327–332. doi:10.1016/j.biopha.2009.09.018

Sylvester, P.W., Kaddoumi, A., Nazzal, S. et al. (2010), The value of tocotrienols in the prevention and treatment of cancer, *J. Am. Coll. Nutr.*, 29 (Suppl. 3), 324S–333S. doi:10.1080/07315724.2010.10719847

Wali, V. and Sylvester, P. (2007), Synergistic antiproliferative effects of γ-tocotrienol and statin treatment on mammary tumor cells, *Lipids*, 42 (12), 1113–1123. doi:10.1007/s11745-007-3102-0

Wang, J., Maitani, Y., and Takayama, K. (2002), Antitumor effects and pharmacokinetics of aclacinomycin a carried by injectable emulsions composed of vitamin E, cholesterol, and PEG-lipid, *J. Pharm. Sci.*, 91 (4), 1128–1134. doi:10.1002/jps.10104

Yin, J., Becker, E.M., Andersen, M.L. et al. (2012), Green tea extract as food antioxidant. Synergism and antagonism with α-tocopherol in vegetable oils and their colloidal systems, *Food Chem.*, 135 (4), 2195–2202. doi:10.1016/j.foodchem.2012.07.025

CHAPTER 7.2

OXYGEN-CARRYING EMULSIONS

DATTA MAROTI PAWDE*, OLY KATARI*, AND TAMILVANAN SHUNMUGAPERUMAL

Department of Pharmaceutics, National Institute of Pharmaceutical Education and Research (NIPER), Guwahati, Assam, India

*The fundings provided by Department of Pharmaceuticals, Ministry of Chemicals and Fertilizers, Govt. of India to the author Oly Katari and National Mission on Himalayan Studies (NMHS), Ministry of Environment, Forest & Climate Change (MoEF & CC) Nodal and Serving hub with G.B. Pant National Institute of Himalayan Environment & Sustainable Development, Govt. of India to Datta Maroti Pawde are greatly acknowledged.

Oil-in-Water Nanosized Emulsions for Drug Delivery and Targeting,
First Edition. Tamilvanan Shunmugaperumal.
© 2021 John Wiley & Sons, Inc. Published 2021 by John Wiley & Sons, Inc.

EXPANSION OF ABBREVIATIONS

perflubron	perfluorooctyl bromide
API	active pharmaceutical ingredient
CO	carbon monoxide
F	perfluoro
MW	molecular weight
NO	nitric oxide
PFC	perfluorocarbons
QbD	quality-by-design
RBC	red blood cells
USFDA	United States Food and Drug Administration

7.2.1. INTRODUCTION

This section starts with the phrase "without oxygen, without life" signifying the importance of oxygen for human life. Supplying of oxygen to the tissues of internal organs in an uninterrupted manner ensures the live-condition of human. For healthy human beings, the supply of oxygen to the tissues of internal organs occurs via systemic blood circulation with red blood cells (RBCs), which carry the oxygen molecules obtained from the regular breathing exercise of the individual. If any disturbance occurred due to RBC breakage or blood loss, then the oxygen supply to the organ's tissues will be hampered that leads to loss of human life. The patients who arrive at hospital in heavily blood-loss condition require blood supplement or simply some oxygen carrier administration into their body. The whole blood transfusion is the mostly recommended method usually adapted in hospital setup to save the patient life in heavily blood-loss condition.

Ever since methods for the typing and storage of blood were developed early in the 20th century, blood transfusions have become an essential part of modern medicine (Giangrande 2000). The ability of donated RBCs to restore oxygen transport capacity is a life-saving measure for patients who have lost significant blood volume. The procedure is commonly employed in cases of acute trauma and during surgical procedures (Giangrande 2000; Napolitano et al. 2009). Currently, donated RBCs, particularly those from universal donor types, are a precious resource that have a limited shelf life under standard storage conditions, and cancellations of elective surgeries are common in times of short supply of blood (Goodnough and Shander 2012). Although efforts to lessen blood utilization by doctors and supply chain management improvements can help to reduce shortages (Tinmouth et al. 2005; Beliën and Forcé 2012), these methods alone are not expected to fully address the issue. As such, significant efforts have also been placed on finding alternative strategies to help reduce the demand for timely human donations (Henkel-Hanke and Oleck 2007; Chen et al. 2009; Buehler et al., 2010; Castro and Briceno 2010).

Alternatively, the promising products/platforms useful to correct blood shortage include hemoglobin and perfluorocarbons (PFCs). However, both have been met with significant challenges in terms of their clinical translation or human use. Currently, Fluosol-DA represents the only synthetic oxygen carrier to have been approved by the United States Food and Drug Administration (USFDA). However, it was taken off the market 5 years after approval in 1989 due to difficulties in its storage and use (Spence et al. 1994). In medical and pharmaceutical fields, the oxygen carriers like O_2-carrying emulsions have received a considerable interest to overcome the O_2 shortage to the tissues of internal organs of human body.

The objective of this chapter is to provide an insight on the oxygen-carrying emulsions developed so far for medical use.

7.2.2. RATIONALE FOR SELECTING PERFLUOROCARBON FOR *IN VIVO* OXYGEN TRANSPORT

The use of PFCs in biomedical research appeared in the 1960s and their unique properties have also been well characterized. The PFCs' use starts from triggered active pharmaceutical ingredient (API) delivery, to oxygen delivery and molecular imaging as extensively reviewed in many literatures (Ahrens and Zhong 2013; Chen et al., 2013, 2018; Yu 2013; Srivastava et al. 2015; Rapoport 2016; Chapelin et al. 2018; Jirak et al. 2019). The unique behavior of PFCs, including their high oxygen dissolving capacity, hydrophobic or lipophobic character, and extreme inertness, derived directly, in a predictable manner, from the electronic structure and spatial requirements of the fluorine atom (Lowe 1987, 2003; Riess 2005, 2006). Their low water solubility is key to the prolonged *in vivo* persistence of the now commercially available injectable microbubbles that serve as contrast agents for diagnostic ultrasound imaging. Riess (2005) developed Oxygent™, a stable, small-sized emulsion of a slightly lipophilic, rapidly excreted PFC, perfluorooctyl bromide (perflubron). Several such perfluorochemical-based contrast agents, namely Optison® (Amersham Health Corp.), SonoVue® (Bracco), Definity® (Bristol Myers Squib), and Imagent® (Alliance Pharmaceutical Corp.), have been licensed by the USFDA or EMEA in Europe in recent years and are now commercially available (Riess 2005).

By definition, PFCs represent a class of organic materials characterized by linear, branched, or cyclic carbon–carbon skeletons where hydrogens are replaced by fluorines (Lambert et al. 2019). As a result of their chemical structure, PFCs are highly hydrophobic and lowly reactive, giving them the capability to dissolve large amounts of gases such as oxygen and carbon dioxide. Compared with water, many PFCs have nearly 20 times the capacity for oxygen dissolution. As this is a physical process, a larger proportion of the carried oxygen is generally available for release to the tissues when compared with hemoglobin, which follows a sigmoidal dissociation curve (Lowe 2006).

The properties and behavior of PFCs [and of perfluoroalkylated (F-alkylated) compounds] are in essence of the same nature as those of regular organic (HC-derived) compounds. However, the exceptionally strong intramolecular binding and uniquely low intermolecular cohesiveness of liquid PFCs related to the low polarizability of fluorine result in properties that can become substantially different (exceeding the usual range) from those of H-analogs and, in practice, unique. Many fluorine-based compounds can reach a level of effectiveness in their performances that cannot be attained by HC compounds, leading to technological feats that just cannot be achieved with non-fluorinated materials. Compared to HCs, the PFCs are typically much more inert, have higher densities, compressibility, fluidity, spreading coefficients and gas-dissolving capacities, and lower refraction indexes, surface tensions, dielectric constants and water solubilities, and magnetic susceptibilities comparable to that of water. Moreover, the fluorine-based compounds offer unique combinations of properties that can make them irreplaceable and constitute the basis for further potential biomedical applications (Riess 2005). In addition, the PFCs are not metabolized in human body and they do not have effects on cell cultures either. This means that the PFCs possess the benefit of carrying the O_2 or CO_2.

7.2.2.1. PFC Suitability Assessment for *In Vivo* Oxygen Transfer

Using the PFC as a carrier for *in vivo* O_2 delivery assumes that a number of issues be resolved. Where chemistry is concerned, the principal difficulties are to decide which PFC (Krafft 1998; Riess and Krafft 1999; Riess 2001, 2002) to use and to develop a stable injectable emulsion of that PFC.

It has been determined that rapid excretion of the PFC requires a touch of lipid solubility, meaning that the PFC should not be too heavy in terms of molecular weight (MW). On the other hand, the emulsion stability requires low water solubility and hence the PFC not be too light. A good-candidate PFC should thus have relatively high lipid solubility and the lowest possible water solubility, two conditions that are difficult to satisfy simultaneously. Vapor pressure, which also depends on MW, is an important parameter. For example, too light a PFC can favor retention of air in the alveoli, resulting in increased pulmonary residual volume. In order to avoid this phenomenon, the vapor pressure of the PFC phase should not exceed about 10 torr.

Perfluoro(F)-octyl bromide offers the best combination of rapid excretion, ability to form stable emulsions with phospholipids, and easy of manufacture. Additionally, it has among the largest O_2 and CO_2 solubilities relative to its MW. F-octyl bromide has the advantage of being a totally synthetic material. Industrial access to perflubron is achieved via a simple bromination of a key intermediate, F-octyl iodide, on the route to commercial fluorosurfactants. Teflon also proceeds from the same manufacturing line. The telomerization process used for synthesizing F-octyl iodide from tetrafluoroethylene, CF_2CF_2, ensures that perflubron can be produced on a multiton scale in better than 99.9% purity (Riess 2005).

7.2.3. FORMULATION OF STABLE INJECTABLE PERFLUOROCARBON EMULSIONS

In formulation/product development point-of-view, the PFCs are immiscible with water and lipids, and for this reason, cannot be administered without further modification. Thus, the PFCs designed as nanometer-sized colloids, such as kinetically stable nanosized emulsions (droplet diameters $<\sim 500\,\text{nm}$), have been utilized extensively as injectable formulations. Furthermore, the PFCs can also be formulated into stable nanosized emulsions with unique capability of not only being a carrier for oxygen molecule but also helping the intracellular or extracellular delivery of API for therapeutic use and imaging agent for diagnostic purpose (Janjic et al. 2014; Herneisey et al. 2016). In spite of the multiple use, the PFC-based platforms generally have not experienced much clinical success, which can largely be attributed to issues such as difficulty of storage and adverse immune reactions (Castro and Briceno 2010).

In conjunction with the first-generation oil-in-water nanosized emulsions, which are simply called as parenteral nutrition or fat emulsions, the PFC emulsions do contain the same emulsifier, egg yolk phospholipids, and the same production process of/through high-shear homogenization and terminal heat sterilization. However, many formulation-related hurdles might arise during the development of stable PFC emulsions. Emulsifying the PFCs is difficult a priori, because of the high interfacial tension, γ_i, of $50–60\,\text{mN m}^{-1}$ that opposes the dispersion of PFCs in water. The target PFC droplets are in the 0.1–0.2 mm range and are typically covered by a monomolecular film of the phospholipids. Practical PFC emulsions need to remain stable for several years without significant changes in particle sizes and particle size distribution. Frozen storage, thawing, and reconstitution are clearly impractical and not acceptable. Over the period of time, the submicronic PFC droplets grow, not through droplet coalescence, but as a result of molecular diffusion (also known as Ostwald ripening). In this process, individual PFC molecules leave the smaller droplets, where the chemical potential is higher, to join larger droplets, where curvature and, consequently, chemical potential are smaller (Fig. 7.2.1). Molecular diffusion is characterized by a linear increase of the average droplet volume over time and by a time-invariant droplet size distribution function. Droplet growth by molecular diffusion follows the Lifshitz–Slezov equation:

$$\frac{\mathrm{d}r^3}{\mathrm{d}t} = \omega = \frac{8V_\mathrm{m}CD\gamma_\mathrm{i}}{9RT} f(\Phi) \tag{7.2.1}$$

where the average droplet volume r^3 in a given emulsion increases over time t that is directly proportional to the water/PFC interfacial tension γ_i, molar volume of the particle material V_m, and to the solubility and diffusibility, C and D, respectively, of the PFC in the aqueous phase. By chance, phospholipids (the emulsifier used in Intralipid and other pharmaceuticals, including

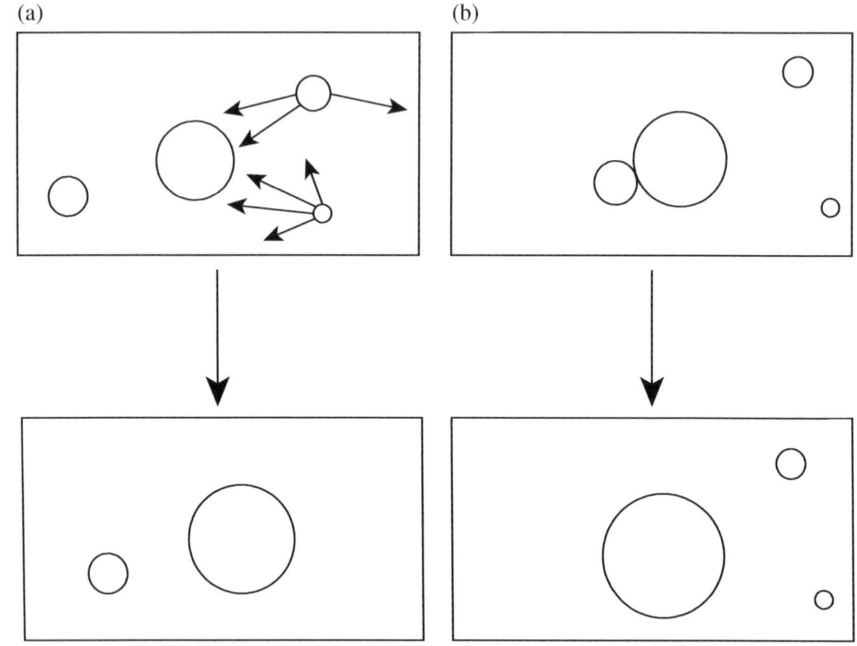

Figure 7.2.1. Schematic representation of droplet size growth over time in fluorocarbon emulsions. (a) Molecular diffusion (joining of PFC molecules from smaller droplets to larger droplets, Ostwald ripening). (b) Droplet coalescence (formation of bigger droplets at the expense of smaller droplets).

liposome preparations) are particularly apt at reducing γ_i. Additionally, the solubility of the PFC phase in water diminishes rapidly when a "heavier" (higher MW) PFC is added. The longer organ retention of higher MW PFCs can be mitigated by using a somewhat lipophilic PFC. This is the case when F-decyl bromide is selected as the heavier PFC. The droplet growth in an F-octyl bromide emulsion can be very effectively reduced by adding just a small percentage of its higher homologue, F-decyl bromide. Table 7.2.1 summarizes the properties of PFC emulsions that are related to their being a dispersion of droplets.

In practice, the manufacturing of small-sized, narrowly dispersed emulsions requires a good deal of know-how. Extensive formulation and process optimization led to Oxygent™ AF0144, a 60% w/v concentrated PFC emulsion, that is heat sterilized, has an average droplet size of 160 nm after terminal heat sterilization, and a viscosity around 4 cP, that is, slightly above that of water. Its pH and osmolarity were adjusted to 7.1 and 304 mOsm, respectively. It is stable for 2 years at 5–10°C and is ready for use (Riess 2005). Table 7.2.2 shows the selected commercial examples of perfluorocarbon (PFC)-based emulsions and few selected examples illustrating the current diversity of PFC nanosized

TABLE 7.2.1. Characteristics of Perfluorocarbon (PFC) Emulsions Related to the Dispersed PFC(s)

Desired Properties	Observation in PFC Emulsion
Adjustable viscosity	Preferably close to blood
Dissolved O_2	Readily and immediately available; high extraction ratio
Foreign particles—RES clearance	Can activate macrophages (flu-like symptoms); short intravascular persistence
Linear O_2 vs. partial oxygen pressure (pO_2) uptake	No saturation
Mechanical resistance	Resistance against the mechanical forces of pumps and filters
No metabolism	Respiratory excretion
Numerous particles	Facilitates O_2 diffusion
Passive delivery	No binding with carbon monoxide (CO), nitric oxide (NO)
O_2 dissolution	Increases when temperature decreases
CO_2 dissolution and removal	Yes
Small sizes/RBC (0:15–0:2 mm vs. 7 mm)	Yet no extravasation
Straightforward large-scale manufacturing and heat sterilization	Yes

emulsion composition in oxygen delivery applications with respect to the choice of PFC and surfactants.

There is no doubt that PFCs dissolve, transport, and deliver O_2 *in vivo* (Riess 2001; Spahn et al. 2002; Gladysz et al. 2006). A randomized, multicenter, European, Phase III clinical evaluation of Oxygent in general surgery patients has established the ability of the emulsion to significantly reduce and avoid RBC transfusion. The trial was conducted using an augmented acute normovolemic hemodilution with PFC emulsion protocol. In the protocol-defined target population (330 subjects with blood loss $\geq 20\,ml\,kg^{-1}$ body weight), significantly greater avoidance of any RBC transfusion, as compared with controls, was maintained through day 21 or day of hospital discharge ($P<0.05$). There was also a significant reduction in the number of units of blood transfused ($P<0.001$). From the clinical data collected, the hemoglobin equivalency, in terms of added O_2-delivering capacity, of a 1 g PFC/kg body weight dose was estimated to be around 1.5 g hemoglobin. Perflubron still stands out as the best candidate PFC for intravascular O_2 delivery. Oxygent is a well-designed, stable emulsion, although by now, even more stable and smaller-sized emulsions have become feasible. A further objective is to prolong intravascular persistence. There is little doubt that improved products are feasible. On the other hand, the voluntary suspension of a cardiopulmonary bypass surgery trial with Oxygent because of side effects most likely due to an inadequate clinical

TABLE 7.2.2. Selected Commercial Examples of Perfluorocarbon (PFC)-Based Emulsions and Few Select Examples Illustrating the Current Diversity of PFC Nanoemulsion Composition in Oxygen Delivery Applications with Respect to the Choice of PFC and Surfactants

Commercial Examples of Perfluorocarbon (PFC)-Based Emulsions			
Product Name	Amount of PFC	Emulsifiers	Particle Size Range (nm)
Oxycyte	60% w/v Perfluoro (*t*-butylcyclohexane)	Purified egg phospholipids, NF, or egg lecithin	200–400
Oxygent	60% w/v Perfluorooctyl bromide	Egg yolk phospholipids (2–3.6% w/v) Perfluorodecyl bromide (additive, optional)	160–180

Composition Diversity in Perfluorocarbons Formulations for Oxygen Delivery Applications				
Ftoremulsion III, Perftoran Plus (Perfluorodecalin)				
PFC	PFC Content	Emulsifiers and Additives	Particle Size Range (nm)	References
Perfluorodecaline	20% w/v	Proxanol 268, perfluoromethylcyclohexylpiperidine	NA	Vorob'ev et al. (2009)
	88% w/v	Proprietary surfactants (undisclosed)	NA	Torres et al. (2014)
	5–40% v/v	Pluronic F68 (0.125–2% w/v) + Pluronic F127 (0.125–2% w/v) blend	206.6–259.8 nm	Fraker et al. (2012)
	50% w/v	Pluronic F-68 (5% w/v) Span 20 (5% w/v) L-α-Phosphatidylcholine (5% w/v)	NA	Freire et al. (2005)
	20% w/v	Pluronic F68 (2.7% w/v) Yolk phospholipids (0.4% w/v)	NA	Engelman et al. (1981)
	40% w/v	Egg yolk lecithin (4% w/v)	200 nm	Deuchar et al. (2018)
	13.3% v/v	Globin (0.33% w/v) and red blood cells membrane	220 nm	Yu et al. (2019)

Perfluorotributylamine	18.75% v/v	NA	150 nm	Zhou et al. (2018)
	10.7% v/v	Human serum albumin, red blood cell membranes; indocyanin green; DiD	180 nm	Ren et al. (2017)
	33.3% v/v	Lecithin E80 (9% w/v)	NA	Benjamin et al. (2013)
	40% v/v	Egg yolk lecithin (9% w/v)	1–2 μm	Goh et al. (2010)
	40% v/v	Egg yolk lecithin (9.5% w/v)	NA	Ma et al. (2013)
Perfluoro-15-crown-5-ether	7% v/v	Human serum	160 nm	Song et al. (2016)
	NA	Lutrol F68	164 nm	Wang et al. (2013)
Perfluorooctyl bromide	10% v/v	F8TAC11	177 nm	Lorton et al. (2018)
		Tris(hydroxymethyl)aminomethane head, hydrophobic perfluorinated tail	259 nm	
	NA	Span 80, PLGA-PEG polymer	302 nm	Yao et al. (2015)
	Variable	Red blood cell membranes	170 nm	Noveck et al. (2000)

NA, not applicable.

protocol that resulted in overly aggressive autologous blood harvesting in the treatment group prior to bypass is a serious setback in the development of PFC-based O_2-carriers. Further efforts will undoubtedly improve our understanding of their *in vivo* behavior, teach us more about how to perform clinical trials, and, more generally, about how to use the O_2 delivering capacity of PFC emulsions optimally (Riess 2005).

7.2.4. MULTIFUNCTIONAL PERFLUOROCARBON-BASED EMULSIONS

The PFC-based emulsions can also be formulated with unique capability of not only oxygen carriage but also intracellular or extracellular delivery of an API (therapeutic) or imaging agent (diagnostic) (Janjic et al. 2014; Herneisey et al. 2016). In juxtaposition with the *in vivo* behavior of parenterally administered HC-based nanosized emulsions, which showed high affinity for serum protein binding and thus the lesser bioavailability or half-life of incorporated APIs, the parenterally administered PFC-based emulsions also showed the similar serum protein binding properties, which may be dependent on the mean particle size values of dispersed PFC droplets (Keipert et al. 1994).

The targeting concept using the nanosized emulsions can be accomplished either actively, by tagging the nanosized emulsion surface with homing ligands (Anton and Vandamme 2011), or passively, via the enhanced permeability and retention effect (Fang et al. 2011). Consequently, API dose and dosing frequency can be reduced. Additionally, tagging a nanosized emulsion with imaging reagent (e.g., near-infrared dyes, fluorescence biosensors, and metal chelators) facilitates tracking biodistribution of the system and monitoring of pathogenesis of disease (Patel et al. 2013; Patrick et al. 2013; Janjic et al. 2014; Jahromi et al. 2018).

^{19}FMRI has emerged as a unique noninvasive quantitative molecular imaging technique where ^{19}F nuclei provide highly selective and specific signals for the biodistribution of immune cells labeled with PFC nanosized emulsions (Srinivas et al. 2009; Hitchens et al. 2011). ^{19}F is extremely selective for labeled cells, and in addition to visualization of the cell location *in vivo*, cell numbers can be quantified directly from the spin-density-weighted ^{19}F MR images. The concept of "*in vivo* cytometry" has been developed by labeling cells with PFC *in vivo* or *ex vivo* for cell detection using ^{19}F MRI (Srinivas et al. 2007, 2009, 2010; Janjic et al. 2008). This approach was extensively utilized for imaging monocyte-derived cells as a means of quantitatively and qualitatively assessing inflammation in a variety of preclinical models (Weise et al. 2011; Stoll et al. 2012; Balducci et al. 2013; Patel et al. 2015) and as a measure of response to treatment (Balducci et al. 2012; Zhong et al. 2015). Noninvasive imaging of inflammation with PFCs can provide invaluable insights into organ rejection and therapeutic transplant outcomes by providing unbiased long-term monitoring. Furthermore, PFCs are particularly useful for regenerative medicine applications as noninvasive stable cell tracking agents for sensitive cell populations such as progenitor or stem cells (Helfer et al. 2013; Srivastava et al. 2015; Richard et al. 2019).

The clinical trials with ^{19}F MRI have successfully demonstrated that PFCs administered to patients can be unambiguously imaged with ^{19}F MRI because low levels of naturally occurring fluorine result in extremely high contrast-to-noise ratio and specificity (Ahrens et al. 2014; Chapelin et al. 2018). The chemical inertness of PFCs allows time-course studies using ^{19}F MRI. In the case of oxygen-carrying PFCs, molecular oxygen concentrations can be monitored with ^{19}F MR spectroscopy (Zhao et al. 2004). Extensive efforts are underway to overcome known problems with ^{19}F MRI such as long imaging times and low sensitivity by redesigning PFC nanosized emulsions and imaging sequences (Chapelin et al. 2018; Jahromi et al. 2018). The technology to synthesize PFCs in bulk exists and is well understood. The same can be said about manufacturing nanosized emulsions in bulk (Liu et al. 2015). Additionally, many of the materials used to manufacture a nanosized emulsion product are generally regarded as safe and approved by the USFDA. The success in clinical use of PFCs as imaging agents supports the notion of their safety for use elsewhere. This is important for obtaining regulatory approval of the nanosized emulsion platforms in organ preservation or transplant application. Thus, designing the clinically translatable PFCs will be a breakthrough advance to solve one of the major problems in regenerative medicine for organ and tissue preservations.

7.2.5. GENERATIONS AND HURDLES OF PERFLUOROCARBON-BASED NANOSIZED EMULSIONS AS OXYGEN CARRIERS

In comparison to the natural oxygen carrier such as RBCs, the nanosized emulsions can distribute into microcirculation and offer an abundant surface area for gas exchange with an extent/order of one-tenth the scale of erythrocytes. Despite a large body of literature evidence, and demonstrated preclinical safety, there exists limited number of clinically approved PFC products for respiratory gas transport. A handful of PFC-based oxygen therapeutics in the form of PFC-in-water nanosized emulsions have been commercialized and entered clinical trials (Vercellotti et al. 1982; Millard 1994; Paxian et al. 2003; Kozhura et al. 2005; Liu et al. 2015). One way or another, the trials were terminated prematurely, or the products were not sustainable and withdrawn from the market after the trials. Reasons for failure include off-target effects and insufficient colloidal stability (Castro and Briceno 2010). Although the development of PFC-based nanosized emulsions as oxygen therapeutics commenced in the 1970s, many products still remain in a state of development today. Even the developmental stages have been roughly broken down into “first-generation-” and “second-generation-” products. This type of generation-wise classification of PFC-based nanosized emulsions is in principle similar to the generation-wise classification made on HC-based nanosized emulsions as shown in Chapter 4. A trademark of the "first-generation" PFC nanosized emulsions was the surfactant component. These were comprised of a PFC phase-stabilized with synthetic nonionic surfactant, Pluronic F68. One of the most prevalent first-generation products is Fluosol (Green Cross Corp., Osaka, Japan). The USFDA initially approved this product for the treatment option in percutaneous transluminal coronary artery balloon angioplasty (Lowe 1994). Not long after approval, the FDA retracted Fluosol due to insufficient long-term stability and handling limitations (Cabrales and Intaglietta 2013). The development of Perftoran (first-generation, Ftorosan, OJCS SPF Perftoran Russian, Moscow, Russia) was likewise discontinued due to limited stability and long organ half-life (Castro and Briceno 2010). Similarly, Oxypherol (Green Cross Corp.), initially named Fluosol-43, was discontinued due to excessive organ retention and long clearance times (Riess 2001).

The challenges that first-generation products met were insurmountable at the time, but some experimental findings were of value. In 1982, a report suggested F68 is responsible for activation of complement particles inside the biosystems (Vercellotti et al. 1982).

Consequently, "second-generation" products utilized egg-based phospholipids in place of F68 as the surfactant and increased PFC concentrations (Kuznetsova 2003; Castro and Briceno 2010). The products such as Oxyfluor (HemaGen/PFC, MO, USA) and Oxygent (Alliance Pharmaceutical Corp., CA, USA) were two of the more prevalent second-generation PFC-based nanosized emulsions. Similar to the first-generation products, second-generation PFC-based nanosized emulsions were also hindered by clinical side effects. Most importantly, both the Oxygent and Oxyfluor products caused flu-like symptoms, which were attributed to macrophage uptake of the injected emulsion and subsequent downstream effects related to arachidonic acid conversion to prostaglandins (Flaim 1994). A combination of Alliance Pharmaceutical's financial situation and a rise in the stroke prevalence in the Oxygent group during Phase II trials led the trial to a halt (Niiler 2002). Double-Crane Pharmaceutical Co. Ltd (Beijing) resumed clinical trials of Oxygent, but this trial was again discontinued after Phase II (Castro and Briceno 2010). Although adding a small amount of safflower oil to Oxyfluor's composition led to innovation to enhance the colloidal stability, the experimental data were not strong enough to indicate a clear advantage. Oxycyte, another second-generation product, gathered attention and entered Phase II clinical trials for severe non-penetrating traumatic brain injury. However, the trial was terminated by 2014 (Paxian et al. 2003).

In 2018, clinical trials for Oxycyte, which is currently under the code name ABL-101 (Aurum Biosciences Ltd, Glasgow, UK), were relaunched in the United Kingdom for acute ischemic stroke after showing promise in preclinical setting (Lowe 1994; Deuchar et al. 2018; Zhuang et al. 2018). This clinical trial is novel because it uses the PFC-based nanosized emulsion as a theranostic platform. It can monitor metabolic status of tissues under oxidative stress to identify salvageable tissue zones (diagnostic) while simultaneously mitigating the ischemic insult to tissues through oxygen delivery (therapeutic).

The problems observed with PFC-based nanosized emulsions extend beyond the scope of clinical product performance. For instance, a recent review identified fatal problems centered on subpar manufacturing standards. Perftech (KEM Laboratories, Tijuana, Mexico), a product previously available in Mexico, was suspended due to noncompliance with Good Manufacturing Practice standards (Bialas et al. 2019). Two of the most extensively studied and commercially available PFC-based nanosized emulsion formulations, Oxygent and Oxycyte, are described in more detail in Table 7.2.2, which includes use as oxygen therapeutics across diverse preclinical and clinical applications (Nolte et al. 2000).

7.2.6. PROPOSED SOLUTIONS TO OVERCOME THE HURDLES OF PERFLUOROCARBON-BASED NANOSIZED EMULSIONS USED AS OXYGEN CARRIERS

Overall, the safety, feasibility, and efficacy of PFC nanosized emulsion-based oxygen therapeutics in mitigating hypoxia and improving tissue preservation has been confirmed across a wide range of experimental studies or preclinical animal models. They have also been shown to be effective across the blood–brain barrier. However, preservation of ischemic tissues and protection from effects of reperfusion injury demand capabilities that go beyond respiratory gas transport. A technology capable of co-delivering oxygen and additional therapeutic entity (e.g., NO scavenger and antioxidant) during preservation and after replantation would be of considerable value. Commonly reported PFC-based nanosized emulsion structures are not ideal for carrying APIs because of the lipophobic and hydrophobic nature of PFCs. To solve this problem, Lambert et al. (2019) have focused on triphasic systems for API delivery with internal phase composed of two oils, PFC and hydrocarbon, in which they demonstrated that triphasic systems can be produced on large scale and with high control of quality attributes (Janjic et al., 2014, 2018; Liu et al. 2015; Herneisey et al. 2016). They also demonstrate that quality-by-design (QbD) approaches can significantly improve triphasic systems development and allow for in-depth understanding of the impact of the composition of the nanosized emulsion on manufacturing and final product quality (Herneisey et al. 2019). Based on these findings, it is proposed that the future of oxygen delivery lies in the more complex PFC nanosystems such as triphasic nanosized emulsions, which offer high stability, pharmaceutical development-friendly processes for scale-up, and overall high level of quality control.

7.2.7. CONCLUSION

In principle, any ideal oxygen carrier must be completely safe and perform at par or better than whole blood, without any formulation hurdles, especially the colloidal stability. Although the generation of such a "blood substitute" remains an ambitious goal for the next decade of research and development in oxygen carriers, the new advances and innovation in the field of PFC colloid-based oxygen therapeutics such as the encapsulation of PFCs in RBC membranes are promising but require significant validation in translational models prior to clinical use. On the other hand, the quality control metrics should be central to characterizing colloidal stability of PFC-based nanosized emulsions. Either by changing the PFC components or by altering the additives/surfactants, the development of PFC-based nanosized emulsions with improved stability characteristics is an attainable near-term goal. In addition, the PFC-based nanosized emulsions are not merely oxygen carriers but are considered as multifunctional agents, that is, imaging capable across multimodality platforms. Collectively, the features of multifunctional PFC-based nanosized emulsions can help inform *in vitro* formulation, *in vivo* validation, therapeutic monitoring, and decision-making, and improve safety, efficacy, and feasibility of use across a wide range of applications necessitating mitigation of the deleterious effects of ischemia and reperfusion injury.

REFERENCES

Ahrens, E.T., Helfer, B.M., O'Hanlon, C.F. et al. (2014), Clinical cell therapy imaging using a perfluorocarbon tracer and fluorine-19 MRI, *Magn. Reson. Med.*, 72 (6), 1696–1701. doi:10.1002/mrm.25454

Ahrens, E.T. and Zhong, J. (2013), In vivo MRI cell tracking using perfluorocarbon probes and fluorine-19 detection, *NMR Biomed.*, 26 (7), 860–871. doi:10.1002/nbm.2948

Anton, N. and Vandamme, T.F. (2011), Nano-emulsions and micro-emulsions: clarifications of the critical differences, *Pharm. Res.*, 28 (5), 978–985. doi:10.1007/s11095-010-0309-1

Balducci, A., Helfer, B.M., Ahrens, E.T. et al. (2012), Visualizing arthritic inflammation and therapeutic response by fluorine-19 magnetic resonance imaging (^{19}F MRI), *J. Inflamm.*, 9 (1), 24. doi:10.1186/1476-9255-9-24

Balducci, A., Wen, Y., Zhang, Y. et al. (2013), A novel probe for the non-invasive detection of tumor-associated inflammation, *Oncoimmunology*, 2 (2), e23034. doi:10.4161/onci.23034

Beliën, J. and Forcé, H. (2012), Supply chain management of blood products: a literature review, *Eur. J. Oper. Res.*, 217 (1), 1–16. doi:10.1016/j.ejor.2011.05.026

Benjamin, S., Sheyn, D., Ben-David, S. et al. (2013), Oxygenated environment enhances both stem cell survival and osteogenic differentiation, *Tissue Eng. Part A*, 19 (5–6), 748–758. doi:10.1089/ten.TEA.2012.0298

Bialas, C., Moser, C., and Sims, C.A. (2019), Artificial oxygen carriers and red blood cell substitutes: a historic overview and recent developments toward military and clinical relevance, *J. Trauma Acute Care Surg.*, 87 (1S), S48–S58. doi:10.1097/TA.0000000000002250

Buehler, P., D'Agnillo, F., and Schaer, D.J. (2010), Hemoglobin-based oxygen carriers: from mechanisms of toxicity and clearance to rational drug design, *Trends Mol. Med.*, 16 (10), 447–457. doi:10.1016/j.molmed.2010.07.006

Cabrales, P. and Intaglietta, M. (2013), Blood substitutes: evolution from non-carrying to oxygen and gas carrying fluids, *ASAIO J.*, 59 (4), 337–354. doi: 10.1097/MAT.0b013e318291fbaa

Castro, C.I. and Briceno, J.C. (2010), Perfluorocarbon-based oxygen carriers: review of products and trials, *Artif. Organs*, 34 (8), 622–634. doi:10.1111/j.1525-1594.2009.00944.x

Chapelin, F., Capitini, C.M., and Ahrens, E.T. (2018), Fluorine-19 MRI for detection and quantification of immune cell therapy for cancer, *J. Immunother. Cancer*, 6 (1), 105. doi:10.1186/s40425-018-0416-9

Chen, J., Pan, H., Lanza, G.M. et al. (2013), Perfluorocarbon nanoparticles for physiological and molecular imaging and therapy, *Adv. Chronic Kidney Dis.*, 20 (6), 466–478. doi:10.1053/j.ackd.2013.08.004

Chen, J.-Y., Scerbo, M., and Kramer, G. (2009), A review of blood substitutes: examining the history, clinical trial results, and ethics of hemoglobin-based oxygen carriers, *Clinics*, 64 (8), 803–813. doi:10.1590/S1807-59322009000800016

Chen, Q., Yu, J., and Kim, K. (2018), optically-triggered phase-transition droplets for photoacoustic imaging, *Biomed. Eng. Lett.*, 8 (2), 223–229. doi:10.1007/s13534-018-0069-0

Deuchar, G.A., Brennan, D., Holmes, W.M. et al. (2018), Perfluorocarbon enhanced Glasgow Oxygen Level Dependent (GOLD) magnetic resonance metabolic imaging identifies the penumbra following acute ischemic stroke, *Theranostics*, 8 (6), 1706–1722. doi: 10.7150/thno.21685

Engelman, R.M., Rousou, J.H., and Dobbs, W.A. (1981), Fluosol-DA: an artificial blood for total cardiopulmonary bypass, *Ann. Thorac. Surg.*, 32 (6), 528–535. doi:0.1016/S0003-4975(10)61793-6

Fang, J., Nakamura, H., and Maeda, H. (2011), The EPR effect: unique features of tumor blood vessels for drug delivery, factors involved, and limitations and augmentation of the effect, *Adv. Drug Deliv. Rev.*, 63 (3), 136–151. doi:10.1016/j.addr.2010.04.009

Flaim, S.F. (1994), Pharmacokinetics and side effects of perfluorocarbon-based blood substitutes, *Artif. Cells Blood Sub. Biotechnol.*, 22 (4), 1043–1054. doi:10.3109/10731199409138801

Fraker, C.A., Mendez, A.J., Inverardi, L. et al. (2012), Optimization of perfluoro nano-scale emulsions: the importance of particle size for enhanced oxygen transfer in biomedical applications, *Colloids Surf. B Biointerfaces*, 98, 26–35. doi:10.1016/j.colsurfb.2012.04.011.

Freire, M., Dias, A., Coutinho, J. et al. (2005), Enzymatic method for determining oxygen solubility in perfluorocarbon emulsions, *Fluid Phase Equilibr.*, 231 (1), 109–113. doi:10.1016/j.fluid.2005.01.008

Giangrande, P.L. (2000), The history of blood transfusion, *Br. J. Haematol.*, 110 (4), 758–767. doi:10.1046/j.1365-2141.2000.02139.x

Gladysz, J.A., Curran, D.P., and Horváth, I.T. (2006), *Handbook of Fluorous Chemistry*, Wiley, Hoboken, NJ.

Goh, F., Gross, J.D., Simpson, N.E. et al. (2010), Limited beneficial effects of perfluorocarbon emulsions on encapsulated cells in culture: experimental and modeling studies, *J. Biotechnol.*, 150 (2), 232–239. doi:10.1016/j.jbiotec.2010.08.013

Goodnough, L.T. and Shander, A. (2012), Patient blood management, *Anesthesiology*, 116 (6), 1367–1376. doi:10.1097/ALN.0b013e318254d1a3

Helfer, B.M., Balducci, A., Sadeghi, Z. et al. (2013), ^{19}F MRI tracer preserves in vitro and in vivo properties of hematopoietic stem cells, *Cell Transplant.*, 22 (1), 87–97. doi:10.3727/096368912X653174

Henkel-Hanke, T. and Oleck, M. (2007), Artificial oxygen carriers: a current review, *AANA J.*, 75 (3), 205–211.

Herneisey, M., Liu, L., Lambert, E. et al. (2019), Development of theranostic perfluorocarbon nanoemulsions as a model non-opioid pain nanomedicine using a quality by design (QbD) approach, *AAPS PharmSciTech*, 20 (2), 65. doi:10.1208/s12249-018-1287-6

Herneisey, M., Williams, J., Mirtic, J. et al. (2016), Development and characterization of resveratrol nanoemulsions carrying dual-imaging agents, *Ther. Deliv.*, 7 (12), 795–808. doi:10.4155/tde-2016-0050

Hitchens, T.K., Ye, Q., Eytan, D.F. et al. (2011), ^{19}F MRI detection of acute allograft rejection with in vivo perfluorocarbon labeling of immune cells, *Magn. Reson. Med.*, 65 (4), 1144–1153. doi:10.1002/mrm.22702

Jahromi, A.H., Wang, C., Adams, S.R. et al. (2018), Fluorous-soluble metal chelate for sensitive fluorine-19 magnetic resonance imaging nanoemulsion probes, *ACS Nano*, 13 (1), 143–151. doi:10.1021/acsnano.8b04881

Janjic, J.M., Shao, P., Zhang, S. et al. (2014), Perfluorocarbon nanoemulsions with fluorescent, colloidal and magnetic properties, *Biomaterials*, 35 (18), 4958–4968. doi:10.1016/j.biomaterials.2014.03.006

Janjic, J.M., Srinivas, M., Kadayakkara, D.K. et al. (2008), Self-delivering nanoemulsions for dual fluorine-19 MRI and fluorescence detection, *J. Am. Chem. Soc.*, 130 (9), 2832–2841. doi: 10.1021/ja077388j

Janjic, J.M., Vasudeva, K., Saleem, M. et al. (2018), Low-dose NSAIDs reduce pain via macrophage targeted nanoemulsion delivery to neuroinflammation of the sciatic nerve in rat, *J. Neuroimmunol.*, 318, 72–79. doi:10.1016/j.jneuroim.2018.02.010

Jirak, D., Galisova, A., Kolouchova, K. et al. (2019), Fluorine polymer probes for magnetic resonance imaging: quo vadis?, *Mag. Reson. Mater. Phys. Biol. Med.*, 32 (1), 173–185. doi:10.1007/s10334-018-0724-6

Keipert, P., Otto, S., Flaim, S. et al. (1994), Influence of perflubron emulsion particle size on blood half-life and febrile response in rats, *Artif. Cells Blood Sub. Biotechnol.*, 22 (4), 1169–1174. doi:10.3109/10731199409138812

Kozhura, V.L., Basarab, D.A., Timkina, M.I. et al. (2005), Reperfusion injury after critical intestinal ischemia and its correction with perfluorochemical emulsion "perftoran," *World J. Gastroenterol.*, 11 (45), 7084–7090. doi:10.3748/wjg.v11.i45.7084

Krafft, M.P., Riess, J.G., and Weers, J.G. (1998), The design and engineering of oxygen-delivering fluorocarbon emulsions, in: Benita, S., Ed., Submicron Emulsions in Drug Targeting and Delivery, Taylor & Francis, London, pp. 235–333. doi:10.1201%2F9780367810528-10

Kuznetsova, I. (2003), Perfluorocarbon emulsions: stability in vitro and in vivo (a review), *Pharm. Chem. J.*, 37 (8), 415–420. doi:10.1023/A:1027355913348

Lambert, E., Gorantla, V.S., and Janjic, J.M. (2019), Pharmaceutical design and development of perfluorocarbon nanocolloids for oxygen delivery in regenerative medicine, *Nanomedicine*, 14 (20), 2697–2712. doi:10.2217/nnm-2019-0260

Liu, L., Bagia, C., and Janjic, J.M. (2015), The first scale-up production of theranostic nanoemulsions, *BioRes. Open Access*, 4 (1), 218–228. doi:10.1089/biores.2014.0030

Lorton, O., Hyacinthe, J.-N., Desgranges, S. et al. (2018), Molecular oxygen loading in candidate theranostic droplets stabilized with biocompatible fluorinated surfactants: particle size effect and application to in situ ^{19}F MRI mapping of oxygen partial pressure, *J. Magn. Reson.*, 295, 27–37. doi:10.1016/j.jmr.2018.07.019

Lowe, K.C. (1987), Perfluorocarbons as oxygen-transport fluids, *Comp. Biochem. Physiol. A Physiol.*, 87 (4), 825–838. doi:10.1016/0300-9629(87)90001-6

Lowe, K.C. (1994), Perfluorochemicals in vascular medicine, *Vasc. Med. Rev.*, 5, 15–32. doi:10.1177%2F1358863X9400500103

Lowe, K.C. (2003), Engineering blood: synthetic substitutes from fluorinated compounds, *Tissue Eng.*, 9 (3), 389–399. doi:10.1089/107632703322066570

Lowe, K.C. (2006), Blood substitutes: from chemistry to clinic, *J. Mater. Chem.*, 16 (43), 4189–4196. doi:10.1039/B604923K

Ma, T., Wang, Y., Qi, F. et al. (2013), The effect of synthetic oxygen carrier-enriched fibrin hydrogel on Schwann cells under hypoxia condition in vitro, *Biomaterials*, 34 (38), 10016–10027. doi:10.1016/j.biomaterials.2013.09.047

Millard, R.W. (1994), Oxygen solubility, rheology and hemodynamics of perfluorocarbon emulsion blood substitutes, *Artif. Cells Blood Substit. Biotechnol.*, 22 (2), 235–244. doi:10.3109/10731199409117417

Napolitano, L.M., Kurek, S., Luchette, F.A. et al. (2009), Clinical practice guideline: red blood cell transfusion in adult trauma and critical care, *Crit. Care Med.*, 37 (12), 3124–3157. doi:10.1097/CCM.0b013e3181b39f1b

Niiler, E. (2002), Setbacks for blood substitute companies, *Nat. Biotechnol.*, 20 (10), 962–963. doi:10.1038/nbt1002-962

Nolte, D., Pickelmann, S., Lang, M. et al. (2000), Compatibility of different colloid plasma expanders with perflubron emulsion: an intravital microscopic study in the hamster, *Anesthesiology*, 93 (5), 1261–1270. doi:10.1097/00000542-200011000-00020

Noveck, R.J., Shannon, E., Leese, P.T. et al. (2000), Randomized safety studies of intravenous perflubron emulsion. II. Effects on immune function in healthy volunteers, *Anesth. Anal.*, 91 (4), 812–822. doi:10.1097/00000539-200010000-00009

Patel, S.K., Beaino, W., Anderson, C.J. et al. (2015), Theranostic nanoemulsions for macrophage COX-2 inhibition in a murine inflammation model, *Clin. Immunol.*, 160 (1), 59–70. doi:10.1097/00000539-200010000-00009

Patel, S.K.K., Patrick, M.J., Pollock, J.A. et al. (2013), Two-color fluorescent (near-infrared and visible) triphasic perfluorocarbon nanoemulsions, *J. Biomed. Opt.*, 18 (10), 101312. doi:10.1117/1.JBO.18.10.101312

Patrick, M.J., Janjic, J.M., Teng, H. et al. (2013), Intracellular pH measurements using perfluorocarbon nanoemulsions, *J. Am. Chem. Soc.*, 135 (49), 18445–18457. doi:10.1021/ja407573m

Paxian, M., Keller, S.A., Huynh, T.T. et al. (2003), Perflubron emulsion improves hepatic microvascular integrity and mitochondrial redox state after hemorrhagic shock, *Shock*, 20 (5), 449–457. doi:10.1097/01.shk.0000090601.26659.87

Rapoport, N. (2016), Drug-loaded perfluorocarbon nanodroplets for ultrasound-mediated drug delivery, *Adv. Exp. Med. Biol.*, 880, 221–241. doi:10.1007/978-3-319-22536-4_13

Ren, H., Liu, J., Li, Y. et al. (2017), Oxygen self-enriched nanoparticles functionalized with erythrocyte membranes for long circulation and enhanced phototherapy, *Acta Biomater.*, 59, 269–282. doi:10.1016/j.actbio.2017.06.035

Richard, J.P., Hussain, U., Gross, S. et al. (2019), Perfluorocarbon labeling of human glial-restricted progenitors for ^{19}F magnetic resonance imaging, *Stem Cells Transl. Med.*, 8 (4), 355–365. doi:10.1002/sctm.18-0094

Riess, J.G. (2001), Oxygen carriers ("blood substitutes") Raison d'Etre, chemistry, and some physiology Blut ist ein ganz besondrer Saft, *Chem. Rev.*, 101 (9), 2797–2920. doi:10.1002/sctm.18-0094

Riess, J.G. (2002), Fluorous micro-and nanophases with a biomedical perspective, *Tetrahedron*, 58 (20), 4113–4131. doi:10.1016/S0040-4020(02)00262-4
Riess, J.G. (2005), Understanding the fundamentals of perfluorocarbons and perfluorocarbon emulsions relevant to in vivo oxygen delivery, *Artif. Cells Blood Substit. Biotechnol.*, 33 (1), 47–63. doi:10.1081/BIO-200046659
Riess, J.G. (2006), Perfluorocarbon-based oxygen delivery, *Artif. Cells Blood Substit. Biotechnol.*, 34 (6), 567–580. doi:10.1080/10731190600973824
Riess, J.G. and Krafft, M.P. (1999), Fluorocarbons and fluorosurfactants for in vivo oxygen transport (blood substitutes), imaging, and drug delivery, *MRS Bull.*, 24 (5), 42–48. doi:10.1557/S0883769400052313
Song, X., Feng, L., Liang, C. et al. (2016), Ultrasound triggered tumor oxygenation with oxygen-shuttle nanoperfluorocarbon to overcome hypoxia-associated resistance in cancer therapies, *Nano Lett.*, 16 (10), 6145–6153. doi:10.1021/acs.nanolett.6b02365
Spahn, D.R., Waschke, K.F., Standl, T. et al. (2002), Use of perflubron emulsion to decrease allogeneic blood transfusion in high-blood-loss non-cardiac surgery, *Anesthesiology*, 97 (6), 1338–1349. doi:10.1097/00000542-200212000-00004
Spence, R., Norcross, E., Costabile, J. et al. (1994), Perfluorocarbons as blood substitutes: the early years: experience with fluosol DA-20% in the 1980s, *Artif. Cells Blood Substit. Biotechnol.*, 22 (4), 955–963. doi:10.3109/10731199409138794
Srinivas, M., Heerschap, A., Ahrens, E.T. et al. (2010), ^{19}F MRI for quantitative in vivo cell tracking, *Trends Biotechnol.*, 28 (7), 363–370. doi:10.1016/j.tibtech.2010.04.002
Srinivas, M., Morel, P.A., Ernst, L.A. et al. (2007), Fluorine-19 MRI for visualization and quantification of cell migration in a diabetes model, *Magn. Reson. Med.*, 58 (4), 725–734. doi:10.1002/mrm.21352
Srinivas, M., Turner, M.S., Janjic, J.M. et al. (2009), In vivo cytometry of antigen-specific t cells using ^{19}F MRI, *Magn. Reson. Med.*, 62 (3), 747–753. doi:10.1002/mrm.22063
Srivastava, A.K., Kadayakkara, D.K., Bar-Shir, A. et al. (2015), Advances in using MRI probes and sensors for in vivo cell tracking as applied to regenerative medicine, *Dis. Model. Mech.*, 8 (4), 323–336. doi:10.1242/dmm.018499
Stoll, G., Basse-Lüsebrink, T., Weise, G. et al. (2012), Visualization of inflammation using ^{19}F-magnetic resonance imaging and perfluorocarbons, *Wiley Interdiscip. Rev. Nanomed. Nanobiotechnol.*, 4 (4), 438–447. doi:10.1002/wnan.1168
Tinmouth, A., MacDougall, L., Fergusson, D. et al. (2005), Reducing the amount of blood transfused: a systematic review of behavioral interventions to change physicians' transfusion practices, *Arch. Intern. Med.*, 165 (8), 845–852. doi:10.1001/archinte.165.8.845
Torres, L.N., Spiess, B.D., and Torres Filho, I.P. (2014), Effects of perfluorocarbon emulsions on microvascular blood flow and oxygen transport in a model of severe arterial gas embolism, *J. Surg. Res.*, 187 (1), 324–333. doi:10.1016/j.jss.2013.08.011
Vercellotti, G.M., Hammerschmidt, D.E., Craddock, P.R. et al. (1982), Activation of plasma complement by perfluorocarbon artificial blood: probable mechanism of adverse pulmonary reactions in treated patients and rationale for corticosteroids prophylaxis, *Blood*, 59 (6), 1299–1304.
Vorob'ev, S.I., Moiseenko, O.M., Belyaev, B.L. et al. (2009), Colloid-chemical and medico-biological characteristics of the perfluorocarbon Ftoremulsion III, *Pharm. Chem. J.*, 43 (5), 267. doi:10.1007/s11094-009-0278-z
Wang, Y.-G., Kim, H., Mun, S. et al. (2013), Indocyanine green-loaded perfluorocarbon nanoemulsions for bimodal ^{19}F-magnetic resonance/nearinfrared fluorescence imag-

ing and subsequent phototherapy, *Quant. Imaging Med. Surg.*, 3 (3), 132–140. doi:10.3978%2Fj.issn.2223-4292.2013.06.03

Weise, G., Basse-Luesebrink, T., Wessig, C. et al. (2011), In vivo imaging of inflammation in the peripheral nervous system by ^{19}F MRI, *Exp. Neurol.*, 229 (2), 494–501. doi: 10.1016/j.expneurol.2011.03.020

Yao, Y., Zhang, M., Liu, T. et al. (2015), Perfluorocarbon-encapsulated PLGA-PEG emulsions as enhancement agents for highly efficient reoxygenation to cell and organism, *ACS Appl. Mater. Interfaces*, 7 (33), 18369–18378. doi:10.1021/acsami.5b04226

Yu, P., Han, X., Yin, L. et al. (2019), Artificial red blood cells constructed by replacing heme with perfluorodecalin for hypoxia-induced radioresistance, *Adv. Therap.*, 2 (6), 1900031. doi:10.1002/adtp.201900031

Yu, Y.B. (2013), Fluorinated dendrimers as imaging agents for ^{19}F MRI, *Wiley Interdiscip. Rev. Nanomed. Nanobiotechnol.*, 5 (6), 646–661. doi:10.1002/wnan.1239

Zhao, D., Jiang, L., and Mason, R.P. (2004), Measuring changes in tumor oxygenation, *Methods Enzymol.*, 386, 378–418. doi:10.1016/S0076-6879(04)86018-X

Zhong, J., Narsinh, K., Morel, P.A. et al. (2015), In vivo quantification of inflammation in experimental autoimmune encephalomyelitis rats using fluorine-19 magnetic resonance imaging reveals immune cell recruitment outside the nervous system, *PLoS One*, 10 (10), e0140238. doi:10.1371/journal.pone.0140238

Zhou, Z., Zhang, B., Wang, H. et al. (2018), Two-stage oxygen delivery for enhanced radiotherapy by perfluorocarbon nanoparticles, *Theranostics*, 8 (18), 4898–4911. doi:10.7150/thno.27598

Zhuang, J., Ying, M., Spiekermann, K. et al. (2018), Biomimetic nanoemulsions for oxygen delivery in vivo, *Adv. Mater.*, 30 (49), e1804693. doi:10.1002/adma.201804693

CHAPTER 7.3

NANOSIZED EMULSIONS FOR MULTIPLE MEDICAMENT LOADINGS, IMAGING, AND THERANOSTIC PURPOSES

DATTA MAROTI PAWDE*, OLY KATARI*, AND TAMILVANAN SHUNMUGAPERUMAL

Department of Pharmaceutics, National Institute of Pharmaceutical Education and Research (NIPER), Guwahati, Assam, India

*The fundings provided by Department of Pharmaceuticals, Ministry of Chemicals and Fertilizers, Govt. of India to the author Oly Katari and National Mission on Himalayan Studies (NMHS), Ministry of Environment, Forest & Climate Change (MoEF & CC) Nodal and Serving hub with G.B. Pant National Institute of Himalayan Environment & Sustainable Development, Govt. of India to Datta Maroti Pawde are greatly acknowledged.

Oil-in-Water Nanosized Emulsions for Drug Delivery and Targeting,
First Edition. Tamilvanan Shunmugaperumal.
© 2021 John Wiley & Sons, Inc. Published 2021 by John Wiley & Sons, Inc.

EXPANSION OF ABBREVIATIONS

AP	atherosclerotic plaques
API	active pharmaceutical ingredient
BODIPY	boron dipyrromethene
CA	contrast agent
CT	computed tomography
EGFR	epidermal growth factor receptor
ETPTA	ethoxylated trimethylolpropane triacrylate
^{19}F	fluorine-19
FL	fluorescence
ICG	indocyanine green
MI	myocardial infarction
MR	magnetic resonance
MRI	magnetic resonance imaging
ND	normal diet
NIR	near-infrared
O/W	oil-in-water
PA	photoacoustic
PDT	photodynamic therapy
PEG	poly(ethylene glycol)
PEG-F_{54}-BODIPY	poly(ethylene glycol)-boron dipyrromethene amphiphile with 54 fluorine-19
PFC	perfluorocarbon
PFH	perfluorohexane
PS	photosensitizer
QDs	quantum dots
SO	silicone oil
SPIONs	superparamagnetic iron oxide nanoparticles
SPIT	sterol-based post-insertion
VO	vegetable oil
WD	western-type diet

7.3.1. INTRODUCTION

Currently, it is highly imperative that diagnostic approaches are helping pharmaceutical and biotechnology sponsors to streamline their clinical trial process. Hence, both molecular assays and diagnostic imaging are routinely being used to stratify patients for treatment, monitor disease, and provide reliable early clinical-phase assessments. Moreover, the importance of diagnostic approaches in active pharmaceutical ingredient (API) development is highlighted by the rapidly expanding global cancer diagnostics market and the emergent attention of regulatory agencies worldwide, who are beginning to offer more structured platforms and guidance for this area. Furthermore, the medical strategies also utilize rhythm-based medical treatments (i.e., chronopharmaceutics) as well as newly introduced medical strategy, which is called theranosis, theragnosis, or pharmacodiagnostics.

Medical imaging comprehensively provides information about internal structures of organs in a noninvasive manner. Basically the imaging may be of two types: anatomical/physiological imaging and biomedical imaging. Whereas the anatomical/physiological imaging helps to visualize the internal structures of various organs of human body, the biomedical imaging plays a vital role to explore the essential information regarding management of several diseases. Specifically for management of cancer, the imaging techniques help to derive estimation of morphological characters, structure, and metabolic and functional information about the site to be examined. The purpose of the detection techniques are to make the therapy effective and to reduce the mortality rate by grabbing the exact cause along with origin of the disease like cancer (Fass 2008).

The currently available noninvasive biomedical technologies for macroscopically visualizing the tumors comprise of both electromagnetic and optical-based imaging techniques such as X-ray computed tomography (CT) scans, magnetic resonance imaging (MRI) scans, positron emission tomography scans, single photon emission CT, ultrasound scans, and optical imaging. It is unfortunate to specify that all the abovesaid imaging techniques have limitations like sensing of the unusual growth of tissue in a nomothetic manner, i.e., microscopic.

In order to achieve information at the molecular level, many biomarkers have also been associated with the imaging systems. These biological agents can selectively target the tumors beside amplification of imaging signals; it also helps to identify the stage and aggressiveness of the tumor (Mishra and Verma 2010). Understanding of the imaging and response requires spatial mapping at the level of gene expression, which makes the techniques not only complicated but it also leads to extraction of extraordinary information regarding stages of the disease (Smith et al. 2003). Many of the advanced imaging techniques are based upon the interaction of the electromagnetic waves, ultrasonic waves, microwaves, X-rays, and other radiation either with the body tissues- or fluid-surrounding cells. Sometimes, the hybrid of the techniques is also employed to collect all the desired information about cell proliferation, gene expression, angiogenesis, and vascular infunctionalities. The selection of techniques for imaging depends upon the sensitivity, specificity, targeting, and availability of the imaging agent (Fass 2008).

As all advanced techniques, the MRI and X-ray CT both suffer from inherent drawbacks. For instance, X-ray CT is cost-effective and has short processing time but has very low sensitivity and exposes tissue to ionizing radiation. MRI is much more sensitive and provides detailed images without ionizing radiation but it has a lower resolution and remains expensive to use.

Consequently, MRI can complete X-ray CT information to deliver accurate diagnostics. In order to limit the discomfort to patient, a single multimodal CA should be the best alternative in case of multimodal imaging needs. To perform the multiple incorporation of several imaging components into a single formulation, nanoparticles and nanocarriers seem the perfect match as shown by many researches (if no interaction between imaging components occurs) (Jarzyna et al. 2010; Hahn et al. 2011; Niu et al. 2011; Oh et al. 2011; Kircher and Willmann 2012; Lee et al. 2012; Dong et al. 2013; Key and Leary 2014; Xue et al. 2014).

The objective of this section is to inculcate the utility of oil-in-water (o/w) nanosized emulsions as an imaging agent (and also as theranostics) through briefly reviewing the research reports published in recent years. This in turn laid a foundation stone on the concept of theragnostics using the multifunctional o/w nanosized emulsions. Through an example of atherosclerosis, the theranostic concept is eloquently discussed with the recently observed various topological structures in oil droplets of the o/w nanosized emulsions.

7.3.2. BRIEF REVIEW OF RESEARCH REPORTS ON NANOSIZED EMULSIONS FOR IMAGING AND THERANOSTIC PURPOSE

Table 7.3.1 displays a non-comprehensive list of o/w nanosized emulsions used in imaging and theranostic purposes.

Soft nanocarriers, namely lipid nanoparticles and nanosized emulsions, have emerged as multivalency nanovehicle platforms. Beside their biocompatibility coming from the use of nontoxic and biodegradable materials, the main advantage of soft nanocarriers is their high loading capability, making an important reservoir of contrasting element. Increasing the contrast agent concentration per particle, along with a specificity of the particle accumulation in the desired site (i.e., passive or active targeting) has the direct consequence on reducing the administrated dosage, therefore decreasing potential side effects and toxicity. A condition enhancing the nanocarriers' targeting is the time in contact with the targeted tissue, i.e., increasing the chance of potential interactions. This contact time is generally increased with the circulation time in blood pool, and enhanced by specific nanocarrier surface properties. Thanks to a neutral, stealth, and hydrophilic PEGylated surface made onto the soft nanocarriers. As a result, premature recognition by the immune system is decreased through a hindrance poly(ethylene glycol) (PEG) shell ("chameleon effect"), thus preventing opsonins adsorption and increasing their colloidal stability and residence time in bloodstream (Storm et al. 1995; Aggarwal et al. 2009; Hallouard et al. 2010). It is noteworthy that PEG is not the only family of macromolecules that can be used for improving the stealth properties of nanocarriers, but other types of hydrophilic polymers, like polyglycerols and polysaccharides (Hadjesfandiari and Parambath 2018), can be used for that purpose.

TABLE 7.3.1. Non-comprehensive List of Oil-in-Water Nanosized Emulsions Used in Imaging and Theranostic Purposes

Research Paper Title	Brief Description	Reference
A versatile theranostic nanoemulsion for architecture-dependent multimodal imaging and dually augmented photodynamic therapy	To design a clinically translatable nanomedicine for photodynamic theranostics, the ingredients should be carefully considered. In this work, a poly(ethylene glycol)-boron dipyrromethene amphiphile (PEG-F_{54}-BODIPY) with 54 fluorine-19 (^{19}F) is synthesized and employed to emulsify perfluorohexane (PFH) into a theranostic nanosized emulsion (PFH@PEG-F_{54}-BODIPY). The as-prepared PFH@PEG-F_{54}-BODIPY can perform architecture-dependent fluorescence/photoacoustic/^{19}F magnetic resonance multimodal imaging, providing more information about the *in vivo* structure evolution of nanomedicine. Importantly, this nanosized emulsion significantly enhances the therapeutic effect of BODIPY through both the high oxygen dissolving capability and less self-quenching of BODIPY molecules. More interestingly, PFH@PEG-F_{54}-BODIPY shows high level of tumor accumulation and long tumor retention time, allowing a repeated light irradiation after a single-dose intravenous injection	Zhang et al. (2019)
Near-infrared absorbing nanoemulsions as nonlinear ultrasound contrast agents for cancer theranostics	Recently, the development of perfluorocarbon (PFC) nanosized emulsions has led to the ability of these nanoparticles to encapsulate therapeutic agents and vaporize into microbubbles and nanobubbles leading to cancer cell death. The PFC bubbles formed from nanosized emulsions can serve as ultrasound contrast agents that can be used to verify the delivery of the nanosized emulsion payload. In this work, perfluorohexane nanosized emulsions were synthesized for use as theranostic agents, by both encapsulating these nanoparticles with different chemotherapeutic agents (i.e., paclitaxel, doxorubicin) as well as to use the laser-induced PFH bubbles to enhance the nonlinear ultrasound signals from cells for treatment monitoring. Since nonlinear ultrasound contrast arises specifically from the nonlinear scattering from oscillating microbubbles and nanobubbles and does not account for linear scattering from tissue, these PFH nanosized emulsions have the potential to effectively locate the tumor region for treatment monitoring	Fernandes and Kolios (2019)

TABLE 7.3.1. ***Continued***

Research Paper Title	Brief Description	Reference
Magnetite and iodine-containing nanoemulsion as dual-modal contrast agent for X-ray/MR imaging	Noninvasive diagnostic by imaging combined with contrast agent (CA) is by now the most used technique to get an insight inside human bodies. X-ray and magnetic resonance imaging (MRI) are widely used technologies providing complementary results. Nowadays, it seems clear that bimodal CAs could be an emerging approach to increase the patient compliance, accessing the different imaging modalities with a single CA injection. The efficient superparamagnetic iron oxide nanoparticles (SPIONs) loaded into iodinated nanosized emulsions was investigated as dual-modal injectable CAs for X-ray imaging and MRI. The strength of this new CA lies not only in its dual-modal contrasting properties and biocompatibility but also in the simplicity of the nanoparticulate assembling: iodinated oily core was synthesized by triiodo-benzene group grafting on vitamin E (41.7% of iodine) via esterification and SPIONs were produced by thermal decomposition during 2, 4, and 6 h to generate SPIONs with different morphologies and magnetic properties. SPIONs with most anisotropic shape and characterized by the highest ratio of transverse (r2) and longitudinal (r1) relaxivity parameters (i.e., r2/r1) once encapsulated into iodinated nanosized emulsions were used for animal experimentation. The *in vivo* investigation showed an excellent contrast modification due to the presence of the selected nanosized emulsions, for both imaging technics explored, i.e., MRI and X-ray imaging	Wallyn et al. (2018)
Modified nanoemulsions with iron oxide for magnetic resonance imaging	Biocompatible iron oxide nanoparticles have been used as a unique imaging approach to study the dynamics of cells or molecular migration. To study the uptake of nanosized emulsions and track them *in vivo*, iron oxide nanoparticles were synthesized and dispersed in soybean oil to make iron oxide-modified nanosized emulsions. *In vitro* experiments demonstrated that iron oxide-modified nanosized emulsions can affect uptake by TC-1 cells (a murine epithelial cell line) and reduce the intensity of magnetic resonance (MR) images by shortening the T2 time. Most importantly, *in vivo* studies demonstrated that iron oxide-modified nanosized emulsions could be detected in mouse nasal septum by both transmission electron microscopy and MR imaging	Fan et al. (2016)

Noninvasive imaging of early venous thrombosis by ^{19}F MRI using targeted perfluorocarbon nanoemulsions	A novel technique for the sensitive and specific identification of developing thrombi making use of background-free ^{19}F MRI together with α_2-antiplasmin peptide (α2AP) targeted PFCs nanosized emulsions as contrast agent, which is cross-linked to fibrin by active factor XIII. Ligand functionality was ensured by mild coupling conditions using sterol-based post-insertion (SPIT). Developing thrombi with a diameter smaller than 0.8 mm could be unequivocally visualized in the murine vena cava inferior as hot spots *in vivo* by simultaneous acquisition of anatomical matching ^{1}H and ^{19}F MR images at 9.4 T with both excellent signal- and contrast-to-noise ratio (71 ± 22 and 17 ± 5, respectively). Furthermore, α2AP-PFC could be successfully applied for diagnosis of experimentally induced pulmonary thromboembolism. In line with the reported half-life of FXIIIa, application of α2AP-PFCs later than 60 min after thrombus induction did not any longer result in detectable ^{19}F MRI signals. Corresponding results were obtained in *ex vivo* generated human clots. Thus, α2AP-PFCs can visualize freshly developed thrombi, which might still be susceptible to pharmacologic intervention	Temme et al. (2015)
Development of EGFR-targeted nanoemulsion for imaging and novel platinum therapy of ovarian cancer	Platinum-based chemotherapy is the treatment of choice for malignant epithelial ovarian cancers, but generalized toxicity and platinum resistance limits its use. Theranostic nanosized emulsion with a novel platinum prodrug, myrisplatin, and the pro-apoptotic agent, C_6-ceramide, were designed to overcome these limitations. Ovarian cancer cells *in vitro* efficiently took up the non-targeted and *epidermal growth factor* receptor (EGFR)-targeted nanosized emulsions and improved cytotoxicity was observed for the these nanosized emulsions with the latter showing a 50-fold drop in the IC_{50} in SKOV3 cells as compared with cisplatin alone. The addition of gadolinium did not affect cell viability *in vitro*, but showed relaxation times comparable to Magnevist® (a clinically approved gadopentetate dimeglumine contrast agent at a 10 mM Gd concentration). The myrisplatin/C_6-ceramide nanosized emulsion synergistically enhanced *in vitro* cytotoxicity. An EGFR binding peptide (synthesized at lab scale) addition further increased *in vitro* cytotoxicity in EGFR positive cancer cells. The diagnostic version showed MR imaging similar to the clinically relevant Magnevist® and may be suitable as a theranostic for ovarian cancer	Ganta et al. (2014)

TABLE 7.3.1. *Continued*

Research Paper Title	Brief Description	Reference
Highly enhanced optical properties of indocyanine green/perfluorocarbon nanoemulsions for efficient lymph node mapping using near-infrared and magnetic resonance imaging	The near-infrared (NIR) fluorescence probe has better tissue penetration and lower autofluorescence. Indocyanine green (ICG) is an NIR organic dye for extensive biological application, and it has been clinically approved for human medical imaging and diagnosis. However, application of this dye is limited by its numerous disadvantageous properties in aqueous solution, including its concentration-dependent aggregation, poor aqueous stability *in vitro*, and low quantum yield. Its use in molecular imaging probes is limited because it loses fluorescence after binding to nonspecific plasma proteins, leading to rapid elimination from the body with a half-life of 2–4 min. In this study, the multifunctional PFC/ICG nanosized emulsions were investigated with the aim of overcoming all of these limitations. The PFC/ICG nanosized emulsions as a new type of delivery vehicle for contrast agents have both NIR optical imaging and ^{19}F-MR imaging moieties. These nanosized emulsions exhibited less aggregation, increased fluorescence intensity, long-term stability, and physicochemical stability against external light and temperature compared with free aqueous ICG. Also, the PFC/ICG bimodal nanosized emulsions allow excellent detection of lymph nodes *in vivo* through NIR optical imaging and ^{19}F-MR imaging. This result showed the suitability of the proposed nanosized emulsions for noninvasive lymph node mapping as they enable long-time detection of lymph nodes	Bae et al. (2014)
CT/fluorescence dual-modal nanoemulsion platform for investigating atherosclerotic plaques	Macrophages have become widely recognized as a key target for atherosclerosis imaging, since they contribute significantly to the progression of atherosclerosis. Dual-modal imaging contrast agents with unique X-ray computed tomography (CT) and optical imaging capabilities have great potential in disease diagnosis because of complementary combination of the high spatial resolution of CT with the high sensitivity of optical imaging. Here, a kind of quantum dots (QDs)-iodinated oil nanosized emulsion of 80 nm was developed as a CT/fluorescence dual-modal contrast agent. Hydrophobic QDs were embedded in iodinated oil, which subsequently dispersed in water to form the oil-in-water (o/w) nanosized emulsions. The morphology and hydrodynamic size of the nanosized emulsion were characterized, CT values and fluorescence properties were detected. Its cytotoxicity and affinity to three different cells were determined *in vitro* by MTT assay. *In vitro* Micro-CT and confocal microscopy cell imaging abilities of the nanosized emulsion were confirmed by co-incubating with murine macrophage cells and human liver cells. Then, *in vivo* accumulation of this nanosized emulsion in macrophages in atherosclerotic rabbits was investigated with clinic CT and fluorescence imagings. The results not only indicated the nanosized emulsion could be served as a dual-modal contrast agent, but revealed it could specifically target to macrophages and visualize atherosclerotic plaques	Ding et al. (2013)

Nevertheless, PEG remains the most reported component to minimize interactions with plasma proteins or antibodies. *In vivo* fate of nanocarriers including their pharmacokinetic profile, biodistribution, colloidal stability in physiological environment, and their retention time (Torchilin and Trubetskoy 1995) can be controlled by their composition, surface decoration, and also by their size (Caruso 2001; Torchilin 2002; Mornet et al. 2004; Torchilin 2006; Cormode et al. 2010; Hallouard et al. 2010; Jarzyna et al. 2010; Bae et al. 2011; Hahn et al. 2011; Parveen et al. 2012; Lusic and Grinstaff 2013; Anton and Vandamme 2014). In this context, Table 5.1 (Chapter 5) presents the various ways to overcome the plasma protein adsorption onto the parenterally administered o/w nanosized emulsions and thus to prolong the systemic circulation time of the carrier.

Due to all the advantages aforementioned, multimodality imaging nanocarriers are nowadays a target in research as an alternative to clinical contrast agents (CAs), balanced by the human health concerns due to their nanoparticulate constitution (Singh et al. 2010). Preclinical studies in rodents prior to translation to the clinic presents two main advantages: (1) mice have small blood volume providing a good contrast even for small administered volumes due to the high payload in contrasting materials (Badea et al. 2008; Hallouard et al. 2010) and (2) the multimodality will provide enough details of mice anatomic and physiologic functions and consequently details of nanocarriers' effects on animal health and model pathologies.

The X-ray CT and MRI are the common imaging techniques at the clinical stage for diagnosing the diseases as well as for inside observation of human body in conjecture with pharmaceuticals for contrast enhancement. They both aim at exploring and checking-up the morphology, the anatomy and physiologic functions via reconstructed 2D or 3D images without invasive gesture to patient (Torchilin 2007; Cormode et al. 2010; Hahn et al. 2011; Key and Leary 2014). The X-ray CT scanner is one of the oldest modality and uses the ability of electron-dense body tissues to absorb X-rays. The X-ray CT images can provide details about abdominal region, cardiorespiratory system, head, tumors, and skeleton due to their different opacification (Yu and Watson 1999; Hallouard et al. 2010; Anton and Vandamme 2014; Cormode et al. 2014; De La Vega and Häfeli 2014). Early diagnosis of soft tissues pathology remains so far quite challenging due to a lack of clear delineation between healthy and altered tissue, and very close X-ray attenuation properties. Contrast enhancer pharmaceutics, or CAs, based on small hydrosoluble iodinated compounds are used in clinic. However, owing to their small size, they undergo a fast renal clearance once in the bloodstream requiring the use of large doses of CAs causing serious adverse effects and poor image quality (Widmark 2007; Lee et al. 2013; Lusic and Grinstaff 2013; Idée and Guiu 2013; De La Vega and Häfeli 2014).

Initially inspired by nuclear magnetic resonance (Damadian 1971), MRI provides images from water molecule hydrogens using magnetic field and radiofrequency pulses with a contrast influenced by tissue relaxation properties (longitudinal T1 and transversal T2 relaxation time) (Grover et al. 2015). Typical CAs for MRI are based on gadolinium chelates for T1-weighted MRI and superparamagnetic iron oxide nanoparticles (SPIONs) for T2-weigthed MRI (Blanco-Andujar et al. 2016). It is noteworthy that MRI can also use T2*-weighted MRI more sensitive to magnetic field inhomogeneites than T2-weighted MRI and therefore more sensitive to SPIONs-induced susceptibility effects (Wang et al. 2001; Pankhurst et al. 2003; Na et al. 2009; Li et al. 2015).

In general, the combination of diagnosis and therapy (theranostics) offers a promising approach for the battle against cancer (Chen et al. 2017; Fan et al. 2017b; Wang et al. 2017). Among various theranostic systems, photodynamic theranostics integrates photodynamic therapy (PDT) with real-time diagnosis into a nanoplatform, thus providing an attractive option for personalized medicine (Lovell et al. 2010; Tong et al. 2016). The PDT has advantages over conventional tumor treatment modalities (including surgery, chemotherapy, and radiotherapy), such as noninvasiveness, low toxicity, repeatability, and avoidance of multidrug resistance. Many photodynamic theranostic nanoplatforms that encapsulate both diagnostic agents and photosensitizers (PSs) into a nanocarrier have achieved simultaneous cancer diagnosis and treatment (Kievit and Zhang 2011; Huang et al. 2012; Li et al. 2014; Cheng et al. 2015). However, the nanocarrier is just an excipient for API/gene delivery and the addition of diagnostic agents typically reduces the API-to-carrier ratio, resulting in a poor loading capability. Moreover, the incorporation of multiple theranostic agents into nanocarrier remarkably complicates the preparation process and increases the metabolic load. Therefore, an ideal theranostic nanoplatform is in desperate need whereby the carrier per se exerts most of the therapeutic and diagnostic functions.

The PDT kills cancer cells by converting tumor oxygen into reactive singlet oxygen using a PS (Fan et al. 2017a; Yun and Kwok 2017). However, the hypoxic tumor microenvironment and continuous oxygen consumption during the PDT process can result in an inadequate oxygen supply, which in turn hampers photodynamic efficacy (Qian et al. 2016). Two main strategies have been reported to solve this problem: (1) increasing the oxygen concentration around the PSs via higher oxygen supply (Cheng et al. 2015; Song et al. 2016) and (2) enlarging the distance of PSs in a nanosystem to prevent PSs from self-quenching by π–π stacking (Ren et al. 2017). To simultaneously address the above issues, we integrate multimodal imaging signals and PDT function into a poly(ethylene glycol)-boron dipyrromethene amphiphile (PEG-F_{54}-BODIPY) with 54 fluorine-19 (^{19}F), as an "all-in-one" nanocarrier. To enhance the PDT efficacy of boron dipyrromethene (BODIPY), perfluorohexane (PFH) is emulsified by the amphiphile to form an oxygen self-supplied nanosized emulsion for fluorescence/photoacoustic/magnetic resonance (FL/PA/MR) multimodal imaging-guided photodynamic therapy. Compared with traditional polymeric API carriers, the PEG-F_{54}-BODIPY can provide versatile theranostic functions. First, BODIPY is one of the most popular fluorescent chromophores due to its high fluorescence quantum yield and narrow absorption bands (Ulrich et al. 2008; Boens et al. 2012; Fan et al. 2015; Kowada et al. 2015). Second, through BODIPY core modifications, the absorption peak of BODIPY redshifts from ≈550 nm to the near-infrared (NIR) range (≈685 nm), allowing for efficient photoacoustic imaging. Third, the symmetrical ^{19}F with a single ^{19}F-MR signal in the amphiphile can be used for sensitive ^{19}F MRI, which creates much less background noise than ^{1}H MRI (Bo et al. 2015). Finally, BODIPY has emerged as a new class of PSs for PDT over the past decade (Kamkaew et al. 2013; Guo et al. 2016; Huang et al. 2017; Yang et al. 2017).

7.3.3. VARIOUS TOPOLOGIES OBSERVED IN DISPERSED OIL DROPLETS OF THE EMULSIONS

Flowchart 7.3.1 portrays the classification of o/w nanosized emulsions based on the internal structure of dispersed oil droplets.

7.3.3.1. Janus Architecture

Being one of the nanoparticulate API delivery systems, the o/w nanosized emulsions consist of multifunctional activities and even the emulsions are commercially exploited as API solubility enhancer, API delivery carrier, API targeting system, and recently as theranostic vector (Tamilvanan 2009; Vecchione et al. 2017). But the commercially available emulsions are simply stabilized against the non-stabilization forces like coalescence, dispersed droplet aggregation, etc., by using phospholipid-based emulsifying agent as one of the stabilizer molecule. The presence of phospholipid along with other emulsifying agents over the period of time always directs to the generation of free fatty acid either inside the emulsion or particularly at the oil–water interface of the emulsion leading to the microclimate pH change. This change in microclimate pH of the emulsion is very harmful for emulsion-incorporated API molecules, which are specifically sensitive to the change of pH unit over the storage time period. To eliminate the phospholipid molecule from the emulsion systems, we have developed a non-phospholipid-based emulsion containing

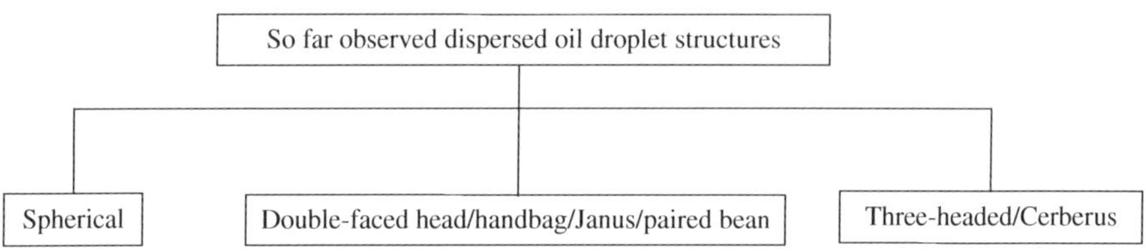

Flowchart 7.3.1. Classification of o/w nanosized emulsions based on internal structure of dispersed oil droplets.

dispersed oil droplets stabilized by either chitosan-poloxamer or chitosan-Tween 80 emulsifier film (Tamilvanan and Kumar 2011).

Among the gamut of so far designed colloidal dispersion system, such as liposome, nanospheres, nanocapsules, niosomes, transferosomes, ethosomes, cubosomes, etc., another heterogeneous dispersion system called 'oil-in-water emulsion' receives a considerable interest to make the successful commercial pharmaceutical formulations for hydrophobic API moiety (Barkat et al. 2011). However, it could be of interest to find a way for incorporating both the hydrophilic and hydrophobic API molecules together in a single emulsion dispersion system. One of the ways that was found during the year 2000 is to develop positively charged submicron emulsion based on medium-chain triglyceride stearylamine, Lipoid-E80, glycerin, poloxamer-188, vitamin E, and double distilled water (Teixeira et al. 2000). This emulsion dispersion system, when stearylamine molecule was incorporated, shows an unexpected bicompartmental structure, termed as 'handbag'. The handbag structure formation is believed due to the association of lipoid-E80, stearylamine, and poloxamer-188 at the oil–water interface to prevent the random collision of dispersed oil droplets, coalescence, and Ostwald ripening. While performing this activity by covering the dispersed oil droplets, with mono- or multi-molecular emulsifiers' film, the presence of cationic lipidic stearylamine promotes the elevation of polyoxyethylene structure of the hydrophilic poloxamer-188 toward the water dispersion medium of o/w emulsion system. These two competing driving forces exerted by stearylamine and poloxamer-188 ultimately lead to the formation of unexpected handbag structure or two compartmental structures within the emulsion dispersion system.

Rarely, some peculiar structure in the dispersed oil droplets of the emulsion is evident along with the regular round-shaped dispersed oil droplet structure. The bicompartmental or paired beans or handbag or simply double-faced head structure in the dispersed oil droplets of the emulsion is the direct instance of peculiar structure (Fig. 7.3.1). The presently developed emulsions, however, differ significantly from the work reported by Teixeira et al. (2000) in many ways. Firstly, two different nonvolatile oils (silicone and olive oils), one-single emulsifier molecule (Tween 80), and double distilled water will be used to make o/w emulsions. Secondly, the emulsions will be produced without phospholipid emulsifier molecules. Rather, simple nonionic surfactant (Tween 80) with or without cation-producing excipient, preferably chitosan, will be enough to produce the stable emulsions. The emulsions thus produced having the mean particle size values at nanometer levels could be termed as nanosized emulsions. The unique feature of this presently developed olive- and silicone-oils-based nanosized emulsions is the presence of the double-faced head structure like the one, which was shown after incorporation of cationic lipidic stearylamine emulsifier moiety into the phospholipid-based nanosized emulsion developed by Teixeira et al. (2000). In the literature, it is envisaged that the presence of double-faced head

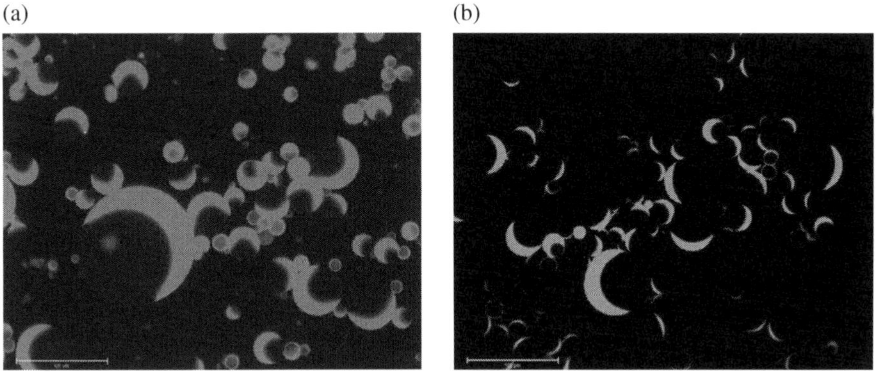

Figure 7.3.1. Formation of Janus structure on the dispersed oil droplets of oil-in-water emulsion. (a) Nile red-incorporated emulsion. (b) Coumarin 6-incorporated emulsion. (See color insert.)

structure within the emulsion dispersion system is capable of entrapping both hydrophilic and hydrophobic API molecules together in a confined manner. In the recent time, the term, anisotropic, is ostensibly being used to describe the Janus structure that has different surface features on the two sides. However, in the present study, the term "paired beans (Janus) structure" will also analogously be used to denote the "double-faced oily head structure" and/or "unexpected bicompartmental handbag structure." It needs to be mentioned that the selected hydrophilic API moieties should have also contained a measurable solubility in the surrounding continuous water medium of the o/w nanosized emulsion.

To substantiate whether or not the Janus or bicompartmental structure is really formed in the dispersed oil droplets of the emulsion, the two different oil-soluble dyes, Nile red and 6-cumarin, were added separately in the oil phase of the nanosized emulsion during its preparation. If oil droplets did not have Janus architecture, the entrapped two different-colored dyes should occupy whole space of the oil droplets. Interestingly, both of these tested two dyes occupied nearly half of the Janus architecture of the dispersed oil droplets indicating that the prepared emulsion indeed consists of the Janus structure in its dispersed oil droplets. Figure 7.3.1 shows the emulsion prepared by adding the Nile red and 6-cumarin dyes.

Ge et al. (2014) observed droplet topology by image observation of a Janus emulsion in a vegetable oil (VO)/silicone oil (SO)/Tween 80 aqueous solution prepared in a one-step high-energy mixing.

In recent year, the emulsion having dispersed oil droplets in Janus or bicompartmental structure was reported (Puri et al. 2019). Since the developed Janus oil droplets possessed the mean droplet size range of micron, the emulsion is termed as "Janus macroemulsion." In the literature, the presence of double-faced oily head structure is simply called as "Janus particles" honoring the roman God

"Janus." The Janus or bicompartmental structure of the dispersed oil droplets of emulsion is capable of entrapping two different API molecules that consist of similar pharmacological activity but dissimilar physicochemical properties. Figure 7.3.2 shows the Janus macroemulsion having the dispersed oil droplets with Janus architecture to confer dual API delivery.

It is being further speculated that

1. The mean droplet size of Janus dispersed oil droplet of emulsion might be further size-reduced up to nanometer range, so that the finally developed liquid-retentive and non-phospholipid-based multifunctional o/w nanosized Janus emulsion can be considered for parenteral administration.
2. The Janus or bicompartmental structure of the dispersed oil droplets of emulsion could accommodate one water-soluble API and one lipid-soluble tracer or diagnostic molecule for atherosclerosical application so that the complete nanoparticulate system can be termed as theranostic o/w nanosized Janus emulsion.
3. The liquid-retentive and non-phospholipid-based multifunctional o/w nanosized Janus emulsion can further be converted into solid dry powder for reconstitution through lyophilization or spray drying technique for improving the storage stability of emulsion.

Until now, the targeting moiety or diagnostic agent is just physically adsorbed or chemically bonded or non-covalently conjugated onto the surfaces of dispersed oil droplets of the multifunctional nanosized emulsion. Because the architectural structure of the dispersed oil droplets of the multifunctional o/w nanosized emulsion is in such a way that it can carry a single cargo into it at a

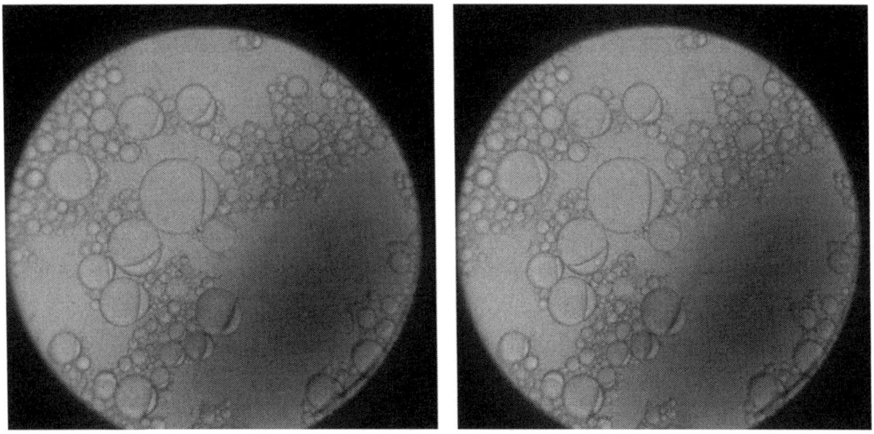

Figure 7.3.2. Optical microscopic picture showing the Janus structure on the dispersed oil droplets of oil-in-water macroemulsion. (See color insert.)

time. Longtime back, it was felt that due to the liquid-retentive nature of o/w nanosized emulsion, the possibility of entrapping two APIs in a single nanoparticulate system and thus the use of emulsion as dual API delivery carrier seems to be difficult or even impossible. Furthermore, the dual API delivery from a single dosage form is the good case for only to solid single- and multiple-unit dosage forms such as capsules, granules, microparticles, and tablets. However, with the advent of Janus or bicompartmental architectural structure onto the dispersed oil droplets of the liquid-retentive multifunctional o/w nanosized emulsion, it is now reasonably possible or plausible that either two different API molecules or one API molecule and one targeting moiety or diagnostic molecule might simultaneously be entrapped into the nanosized Janus oil droplets of emulsion at the same time. It should be added that both the API and diagnostic molecule, in selection, may be for atherosclerosical applications.

7.3.3.2. Cerberus Architecture

Multiple emulsions with droplets consisting of three singlets of mutually immiscible oils were first reported by Fryd and Mason (2013). The dispersed oil droplets of o/w nanosized emulsions possessing the three-singlet configuration is now termed as 'Cerberus'. The Cerberus signifies the three-head dog that guards the gates of Hades in Roman mythology. A report by Fryd and Mason (2014), however, demonstrated that Janus emulsions are formed by adding a solution of a cationic surfactant to an emulsion stabilized using an anionic surfactant under moderate agitation. Finally, Zarzar et al. (2015) established both Janus and Cerberus structures could be made onto the surfaces of dispersed oil droplets of emulsions by varying the emulsifier molecules and the emulsion preparation methods. The Cerberus emulsion with a specific "three room" morphology is attractive in both fundamental scientific research and future real-world applications because it processes the superiorities of "different compositions," "separate sub-room," and "sharply contoured space" compared with traditional single-oil emulsions. Basically, the mutual solubility of the selected three different oils such as silicone, fluorocarbon HFE-7200, and ethoxylated trimethylolpropane triacrylate (ETPTA) create the Cerberus structure on the surface of oil droplets during emulsification with aqueous phase containing Pluronic F127 (Ge et al. 2017). The observed dispersed oil droplets of the emulsion with Cerberus structure is shown in Fig. 7.3.3.

7.3.4. POSSIBLE THERAPEUTIC UTILITY OF JANUS EMULSION

7.3.4.1. Status of Recent Research and Development in Janus Structure

Janus particles are the elementary case of non-centro symmetrical particles, one of the important fundamental physics, which have profoundly linked with numerous material applications. These particles are known after the Roman

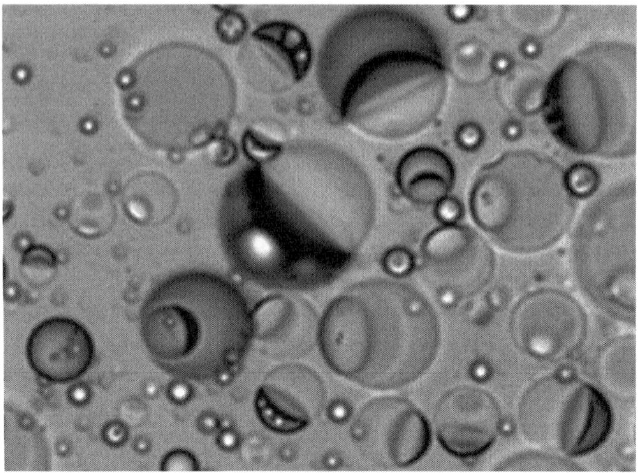

Figure 7.3.3. Emulsion with Cerberus structure on the dispersed oil droplets of emulsion. [Adapted from Ge et al. (2017).] (See color insert.)

god Janus, who is represented by double-faced head. In the same manner, Janus particles are also represented as two faces with different physical and chemical properties. Various latest reviews have done a tremendous job that summarizes the synthesis method and properties of Janus particles. Janus particles are microscopic particles in dimension ranging between 1 and 100 nm with two different sides. It is an analogue to the surfactant molecule, which has a hydrophilic head group and hydrophobic tail. Due to its fundamental as well as commercial application, Janus particles have collectively received considerable attention in the API delivery system due to their advanced surface-active properties with a uniform wettability. Binks and Fletcher (2001) investigated the wettability of Janus beads at the interface between oil and water and conclude that Janus particles are both surface-active and amphiphilic and are more stable compared with homogeneous particles that are only surface-active.

Emulsion is referred to as dispersion of droplets of one fluid within second immiscible fluid. These involve shearing of two immiscible fluids, leading to the fragmentation of one phase into another phase. These types of emulsions are intrinsically unstable because of large interfacial area within the system. To obtain stable emulsion, it involves the incorporation of another third component that is the emulsifying agent such as surfactant and colloidal particles that act onto the interface between the two immiscible fluids and further stabilize the emulsion (Gangwal et al. 2008). Such emulsions have a wide range of activity from food stuffs to pharmaceutical to personal care products to agricultures. Surfactants that are used in emulsion are commonly referred to as surface-active agents that are used to segregate at the interface and stabilize the emulsion. The colloidal particles have property to attach strongly to the interface in order to stabilize the emulsion. The surface activity of the surfactants originated from their amphiphilic structure. It has been demonstrated that Janus particles too have amphiphilic character that are used in order to stabilize the emulsion thermodynamically (Walther and Müller 2013).

Tu and Lee (2014) investigated the "One-Step Encapsulation and Triggered Release Based on Janus Particle-Stabilized Multiple Emulsions" in which they carried out the one-step formation of stable multiple emulsions using stimuli-responsive amphiphilic Janus particle as an emulsifier. This work is proven as the strong demonstration of Janus particle versatility as solid surfactant.

Fujii et al. (2013) has investigated the "Micrometer-Sized Gold–Silica Janus Particles as Particulate Emulsifiers" in which they worked on the preparation of gold-silica Janus particles (micrometer-sized), which act as an effective stabilizer for o/w interface and hence producing the pickering emulsion. The prepared emulsion collaborated to have both hydrophobic and hydrophilic head due to the presence of gold and silica surface. In opposite to this, only silica particles do not have properties to stabilize the emulsion and complete de-emulsification takes place within 2 h. Also, the particle's adsorption energy is three times as compared with silica particles and the toxicity profile proven to be lower compared with the conventional molecular level surfactant. By using vacuum evaporation, they fabricated the gold-silica Janus particles. This investigation opens a new way to prepare the wide range of hybrid microspheres (organic–inorganic).

Li et al. (2019) showed a simple and large-scale fabrication approach to prepare Janus emulsions that can be controlled in size, geometry, and stability. The role played by interfacial tensions in creating the Janus topology, and then the use of colloidal particles to investigate the origin of the stability for Janus emulsions was also systematically shown with experimental results.

A facile technique of "single-step solvent emulsion method" was used to seamlessly fabricate polymeric-based hybrid Janus particles for theranostic applications with little complexity by Lim et al. (2019). They showed a selective encapsulation of therapeutic and diagnostic agents in a very streamlined manner into different "faces" of the Janus structure.

Through a review article of "Janus Particles Recent and Novel Approach in Drug Delivery: An Overview," Indalkar et al. (2013) illustrated that the Janus particles are microscopic particles composed of two different sides having a hydrophilic head group and hydrophobic tail. The Janus particles possessed the surface-active properties and showed considerably the high wettability. Synthesis of Janus particles by fractionalization leads to the formation of fragments by breaking of the particle and gives automatic Janus character to the fragments. Considerable attention is given firstly on how to manage the Janus particles' geometry (their double-sided relative areas), and secondly on how to produce the production of Janus particles in large quantities. This review article highlighted the various applications of Janus particles in the field of biological sciences especially for diagnosing, imaging, and thus the theranostic potentials.

Using VO/SO/Tween 80 aqueous solution, Ge et al. (2014) observed dispersed oil droplet's topology and found the Janus structure.

Puri et al. (2019) very recently demonstrated the development of Janus macroemulsion to entrap fenugreek seed powder or ginger rhizome powder or both for managing the pain produced at primary dysmenorrheal condition in a mice model.

The development of Janus nanoparticles prepared either preformed- or *in situ* forming-polymers for achieving the dual API delivery is already going on different laboratories. The main focus is to generate the polymer-based Janus nanoparticles for API targeting purpose rather than to explore their theranostic potentials. Moreover, the generation of polymeric Janus nanoparticles utilizes the structural modification of the polymer network skeleton. Thus, this strategy to make polymer-based Janus nanoparticles is tedious and even the produced so-called "me-too" new polymeric skeleton should be considered as a new material, thus, requiring a fresh approval from regulatory authorities before releasing the Janus nanoparticles for human use. Using the atherosclerosis as the target disease, the injectable atherosclerosical theranostic dry powder for reconstitution made from the liquid-retentive and non-phospholipid-based multifunctional o/w nanosized emulsions having the dispersed oil droplets with Janus architecture can be projected to confer dual API delivery. More importantly, the produced Janus emulsion thus do not need any fresh regulatory approval to use in human body as it is not involved in any structural modification of polymeric network skeleton.

The possible theranostic utility of o/w nanosized emulsions is exemplified with initial experimental proof in the next section using atherosclerosis as the target disease.

7.3.4.2. Janus Emulsion for Atherosclerosis Theragnosis

Atherosclerosis is a progressive inflammatory disease occurring due to accumulation of lipids in the arterial vessel wall, which starts probably early in life. Disease progression leads to buildup of atherosclerotic plaques (AP) that cause narrowing of the arterial lumen to create hypertension in blood flow. AP often remain stable for years, but can rapidly become unstable, rupture, and trigger thrombus formation. Accordingly, in addition to restriction of the vessel lumen, the presence of AP is linked to an increased risk of acute cardiovascular events such as myocardial infarction (MI) and stroke. Incidentally, eating of junk or fast food made from the oil containing much bad cholesterol that allows the forced deposition of the chylomicron into the inner wall of arteries and thus the narrowing down of the blood flow lane. This condition is collectively named as AP. An early detection of AP condition in the individuals will certainly help to caution them not to use bad cholesterol-containing food or adjust the ailment with regular medicine intake. In this context, the detection of AP condition and simultaneous management of at least one condition of AP like hypertension is crucial. In recent years, the water insoluble APIs, obtained from both natural and synthetic sources, are conveniently being grouped under either brick dust or grease ball molecules (Lipinski 2005; Merisko-Liversidge and Liversidge 2011; Müller and Keck 2012). The API with brick dust molecular arrangements is suitable to convert into nanocrystals or API nanosuspensions to overcome their inherent dissolution limitation. However, the API with grease ball molecular characteristics is rightly fitted for entrapping into preformed lipid, oil or polymer-based nanoparticulate systems, so that the intrinsic solubility restriction of this kind of API can be defeated.

The approach of theragnosis with API-loaded Janus nanosized emulsion looks interesting/lucrative since the use of dispersed Janus oil droplets will be able to encapsulate both API and imaging agent or targeting moiety together in the nanosized emulsion. Furthermore, this technology will also be used to detect the level of hardening inside the blood vessel. Therefore, the development of injectable atherosclerosical theranostic solid dry powder for reconstitution from Janus nanosized emulsion might accord alarm for individual people about the blood bad lipid profile and its consequences like possible heart attack. Thus, the untimely death of people due to heart attack could drastically be reduced. Theragnosis or image-guided API delivery systems using the nanoformulations such as multifunctional o/w nanosized emulsion is one of the best options available currently to manage AP.

The commercially exploited emulsions are prepared based on phospholipid emulsifier molecule, which eventually generates free fatty acid over the time period resulting in microclimate pH change. Recently, the o/w nanosized emulsions were prepared by omitting the phospholipid emulsifier but stabilized with chitosan-Tween 80 or chitosan-poloxamer combination. Antihypertensive API along with targeting moiety and imaging agent are attached by using either adsorption or conjugation onto the oil droplets' surface of the o/w emulsion. Such anchoring or attachment was found to produce a non-satisfactory API targeting efficiency at *in vivo* conditions. This propels the formulation scientists to look for alternative ways to enhance the API targeting efficiency using the o/w nanosized emulsion. A sudden observation of the presence of bicompartmental or handbag or double-faced head or Janus structure in the dispersed oil droplets makes the possibility to use this peculiar Janus structure for entrapping the API and the remaining two cargos (targeting moiety and imaging agent) in a single API delivery carrier for AP theragnosis. The developed emulsion could be termed as liquid-retentive and non-phospholipid-based Janus nanosized emulsion. Worryingly, the liquid-retentive Janus nanosized emulsions are stable for 90 days at 25°C and after that the thick oily layer is coming up to the emulsion's surface, which does not redisperse into emulsion upon shaking. To further improve the storage stability of liquid-retentive Janus nanosized emulsion, the conversion into injectable solid dry powder for reconstitution is possible. Figure 7.3.4 displays the overview of currently available mouse models of atherosclerosis to perform the *in vivo*

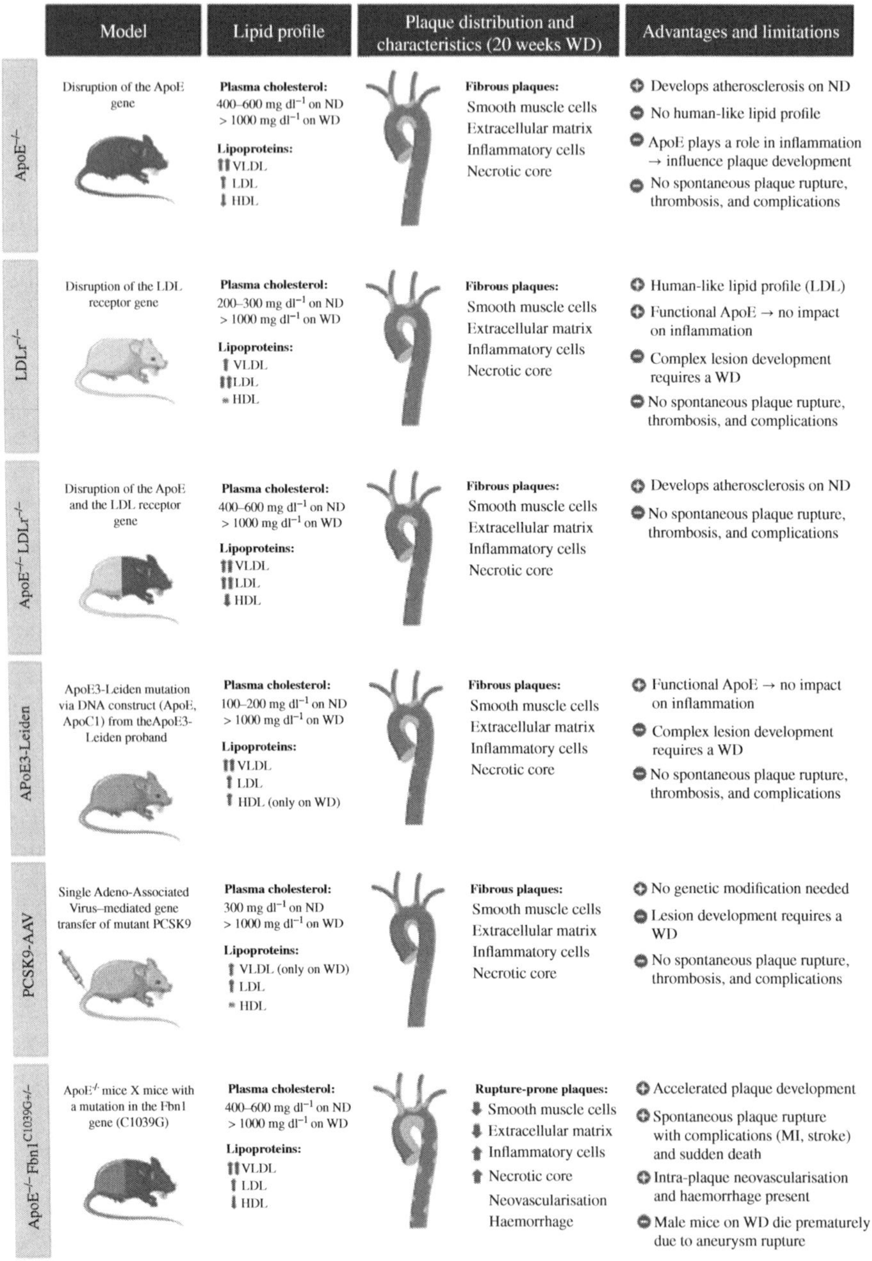

Figure 7.3.4. Overview of current mouse models of atherosclerosis. This figure describes the different models with their total plasma cholesterol levels on normal diet (ND) and Western-type diet (WD), lipoprotein profile, plaque characteristics, advantages, and limitations. The distribution of the plaques in the thoracic aorta and the complexity is shown for mice fed a WD for 20 weeks. The composition of the most complex lesions at that time point is displayed (usually found in the aortic root or brachiocephalic artery). [Reproduced with permission from Veseli et al. (2017).] (See color insert.)

theranostic proof using liquid-retentive and solid-dry powder Janus nano-sized emulsions containing API, imaging agent, and targeting moiety (Veseli et al. 2017).

7.3.5. CONCLUSION

Keeping multiple medicament loadings in mind with the help of bicompartmental structure of the dispersed oil droplets present in Janus emulsion, the hypothesis of the atherosclerosis theranostic concept is that whether the optimized, liquid-retentive, and non-phospholipid-based Janus emulsion having API or imaging agent or both API and imaging agent attempted using the Janus emulsion in reconstitutable solid dry powder form possesses more atherosclerosical theranostic potential. This initial excerpt should convert into most fruitful commercial exploitation for diagnosing, targeting, and treating other life-compromising/threatening diseases like malignancy and rheumatoid arthritis. The newborn baby of Cerberus architecture in emulsion system must also meet the demands for acceptable storage stability in order to further exploit its potential in multiple medicament loadings, imaging, and theranostic purposes.

REFERENCES

Aggarwal, P., Hall, J.B., McLeland, C.B. et al. (2009), Nanoparticle interaction with plasma proteins as it relates to particle biodistribution, biocompatibility and therapeutic efficacy, *Adv. Drug Deliv. Rev.*, 61 (6), 428–437. doi:10.1016/j.addr.2009.03.009

Anton, N. and Vandamme, T.F. (2014), Nanotechnology for computed tomography: a real potential recently disclosed, *Pharm. Res.*, 31, 20–34. doi:10.1007/s11095-013-1131-3

Badea, C.T., Drangova, M., Holdsworth, D.W. et al. (2008), In vivo small-animal imaging using micro-CT and digital subtraction angiography, *Phys. Med. Biol.*, 53 (19), R319.

Bae, K.H., Chung, H.J., and Park, T.G. (2011), Nanomaterials for cancer therapy and imaging, *Mol. Cell*, 31 (4), 295–302. doi:10.1007/s10059-011-0051-5

Bae, P.K., Jung, J., and Chung, B.H. (2014), Highly enhanced optical properties of indocyanine green/perfluorocarbon nanoemulsions for efficient lymph node mapping using near-infrared and magnetic resonance imaging, *Nano Converg.*, 1, 6. doi:10.1186/s40580-014-0006-6

Barkat, A.K., Naveed, A., Haji, M.S.K. et al. (2011), Basics of pharmaceutical emulsions: a review, *Afr. J. Pharm. Pharmacol*, 5 (25), 2715–2725. doi:10.5897/AJPP11.698

Binks, P. and Fletcher, P.D.I. (2001), Particles adsorbed at the oil-water interface: a theoretical comparison between spheres of uniform wettability and "Janus" particles, *ACS Publ.*, 17 (16), 4708–4710, doi:10.1021/la0103315

Blanco-Andujar, C., Walter, A., Cotin, G. et al. (2016), Design of iron oxide-based nanoparticles for MRI and magnetic hyperthermia, *Nanomedicine*, 11 (14), 1889–1910. doi:10.2217/nnm-2016-5001

Bo, S., Song, C., Li, Y. et al. (2015), Design and synthesis of fluorinated amphiphile as ^{19}F MRI/fluorescence dual-imaging agent by tuning the self-assembly, *J. Org. Chem.*, 80 (12), 6360–6366. doi:10.1021/acs.joc.5b00810

Boens, N., Leen, V., and Dehaen, W. (2012), Fluorescent indicators based on BODIPY, *Chem. Soc. Rev.*, 41 (3), 1130–1172. doi:10.1039/C1CS15132K

Caruso, F. (2001), Nanoengineering of particle surfaces, *Adv. Mater.*, 13, 11–22. doi:10.1002/1521-4095(200101)13:1%3C11::AID-ADMA11%3E3.0.CO;2-N

Chen, H., Zhang, W., Zhu, G. et al. (2017), Rethinking cancer nanotheranostics, *Nat. Rev. Mat.*, 2, 17024. doi:10.1038/natrevmats.2017.24

Cheng, Y., Cheng, H., Jiang, C. et al. (2015), Perfluorocarbon nanoparticles enhance reactive oxygen levels and tumour growth inhibition in photodynamic therapy, *Nat. Commun.*, 6, 8785. doi:10.1038/ncomms9785

Cormode, D.P., Naha, P.C., and Fayad, Z.A. (2014), Nanoparticle contrast agents for computed tomography: a focus on micelles, *Contrast Media Mol. Imaging*, 9 (1), 37–52. doi:10.1002/cmmi.1551

Cormode, D.P., Skajaa, T., Fayad, Z.A. et al. (2010), Nanotechnology in medical imaging: probe design and applications, *Arterioscler. Thromb. Vasc. Biol.*, 29 (7), 992–1000. doi:10.1161/ATVBAHA.108.165506

Damadian, R. (1971), Tumor detection by nuclear magnetic resonance, *Science*, 171 (3976), 1151–1153. doi:10.1126/science.171.3976.1151

De La Vega, J.C. and Häfeli, U.O. (2014), Utilization of nanoparticles as X-ray contrast agents for diagnostic imaging applications, *Contrast Media Mol. Imaging*, 10 (2), 81–95. doi:10.1002/cmmi.1613

Ding, J., Wang, Y., Ma, M. et al. (2013), CT/fluorescence dual-modal nanoemulsion platform for investigating atherosclerotic plaques, *Biomaterials*, 34, 209–216. doi:10.1016/j.biomaterials.2012.09.025

Dong, W., Li, Y., Niu, D. et al. (2013), A simple route to prepare monodisperse Au NP-decorated, dye-doped, superparamagnetic nanocomposites for optical, MR, and CT trimodal imaging, *Small*, 9 (15), 2500–2508. doi:10.1002/smll.201202649

Fan, G., Lin, Y.X., Yang, L. et al. (2015), Co-self-assembled nanoaggregates of BODIPY amphiphiles for dual colour imaging of live cells, *Chem. Commun.*, 51 (62), 12447–12450. doi:10.1039/C5CC04757A

Fan, W., Lu, N., Xu, C. et al. (2017a), Enhanced afterglow performance of persistent luminescence implants for efficient repeatable photodynamic therapy, *ACS Nano*, 11 (6), 5864–5872. doi:10.1021/acsnano.7b01505

Fan, W., Yung, B., Huang, P. et al. (2017b), Nanotechnology for multimodal synergistic cancer therapy, *Chem. Rev.*, 117 (22), 13566–13638. doi:10.1021/acs.chemrev.7b00258

Fan, Y., Guo, R., Shi, X. et al. (2016), Modified nanoemulsions with iron oxide for magnetic resonance imaging, *Nanomaterials*, 6 (12), 223. doi:10.3390/nano6120223

Fass, L. (2008), Imaging and cancer: a review, *Mol. Oncol.*, 2 (2), 115–152. doi:10.1016/j.molonc.2008.04.001

Fernandes, D.A. and Kolios, M.C. (2019), Near-infrared absorbing nanoemulsions as nonlinear ultrasound contrast agents for cancer theranostics, *J. Mol. Liq.*, 287, 110848. doi:10.1016/j.molliq.2019.04.125

Fryd, M.M. and Mason, T.G. (2013), Cerberus nanoemulsions produced by multidroplet flow-induced fusion, *Langmuir*, 29 (51), 15787–15793. doi:10.1021/la403817a

Fryd, M.M. and Mason, T.G. (2014), Self-limiting droplet fusion in ionic emulsions, *Soft Matter*, 10 (26), 4662–4673. doi:10.1039/C4SM00453A

Fujii, S., Yokoyama, Y., Miyanari, Y. et al. (2013), Micrometer-sized gold–silica Janus particles as particulate emulsifiers, *Langmuir*, 29 (18), 5457–5465. doi:10.1021/la400697a

Gangwal, S., Cayre, O.J., and Velev, O.D. (2008), Diectrophoretic assembly of metallodielectric Janus particles in AC electric fields, *Langmuir*, 24 (23), 13312–13320. doi:10.1021/la8015222

Ganta, S., Singh, A., Patel, N.R. et al. (2014), Development of EGFR-targeted nanoemulsion for imaging and novel platinum therapy of ovarian cancer, *Pharm. Res.*, 31 (9), 2490–2502. doi:10.1007/s11095-014-1345-z

Ge, L., Li, J., Zhong, S. et al. (2017), Single, Janus, and Cerberus emulsions from the vibrational emulsification of oils with significant mutual solubility, *Soft Matter*, 13 (5), 1012–1019. doi:10.1039/C6SM02690G

Ge, L., Lu, S., and Guo, R. (2014), Janus emulsions formed with a polymerizable monomer, silicone oil, and Tween 80 aqueous solution, *J. Colloid Interface Sci.*, 423, 108–112. doi:10.1016/j.jcis.2014.02.023

Grover, V.P., Tognarelli, J.M., Crossey, M.M. et al (2015), Magnetic resonance imaging: principles and techniques: lessons for clinicians, *J. Clin. Exp. Hepatol.*, 5 (3), 246–255. doi:10.1016/j.jceh.2015.08.001

Guo, Z., Zou, Y., He, H. et al. (2016), Bifunctional platinated nanoparticles for photoinduced tumor ablation, *Adv. Mater.*, 28 (46), 10155. doi:10.1002/adma.201602738

Hadjesfandiari, N. and Parambath, A. (2018), Stealth coatings for nanoparticles: polyethylene glycol alternatives, in: Parambath, A., Ed., *Engineering of Biomaterials for Drug Delivery Systems*, Woodhead Publishing, Elsevier, pp. 345–361. doi:10.1016/B978-0-08-101750-0.00013-1

Hahn, M.A., Singh, A.K., Sharma, P. et al. (2011), Nanoparticles as contrast agents for in-vivo bioimaging: current status and future perspectives, *Anal. Bioanal. Chem.*, 399, 3–27. doi:10.1007/s00216-010-4207-5

Hallouard, F. Anton, N. Choquet, P. et al. (2010), Iodinated blood pool contrast media for preclinical X-ray imaging applications—a review, *Biomaterials*, 31 (24), 6249–6268. doi:10.1016/j.biomaterials.2010.04.066

Huang, L., Li, Z., Zhao, Y. et al. (2017), Enhancing photodynamic therapy through resonance energy transfer constructed near-infrared photosensitized nanoparticles, *Adv. Mater.*, 29 (28), 1604789. doi:10.1002/adma.201604789

Huang, P., Lin, J., Wang, X. et al. (2012), Light-triggered theranostics based on photosensitizer-conjugated carbon dots for simultaneous enhanced-fluorescence imaging and photodynamic therapy, *Adv. Mater.*, 24 (37), 5104–5110. doi:10.1002/adma.201200650

Idée, J.M. and Guiu, B. (2013), Use of lipiodol as a drug-delivery system for transcatheter arterial chemoembolization of hepatocellular carcinoma: a review, *Crit. Rev. Oncol. Hematol.*, 88 (3), 530–549. doi:10.1016/j.critrevonc.2013.07.003

Indalkar, Y.R., Gaikwad, S.S., and Ubale, A.T. (2013), Janus particles recent and novel approach in drug delivery: an overview, *Curr. Pharma Res.*, 3 (4), 1031–1037.

Jarzyna, P.A., Gianella, A., Skajaa, T. et al. (2010), Multifunctional imaging nanoprobes, *Wiley Interdiscip. Rev.: Nanomed. Nanobiotechnol.*, 2 (2), 138–150. doi:10.1002/wnan.72

Kamkaew, A., Lim, S.H., Lee, H.B. et al. (2013), BODIPY dyes in photodynamic therapy, *Chem. Soc. Rev.*, 42 (1), 77–88. doi:10.1039/C2CS35216H

Key, J. and Leary, J.F. (2014), Nanoparticles for multimodal in vivo imaging in nanomedicine, *Int. J. Nanomed.*, 9, 711–726. doi:10.2147%2FIJN.S53717

Kievit, F.M. and Zhang, M. (2011), Cancer nanotheranostics: improving imaging and therapy by targeted delivery across biological barriers, *Adv. Mater.*, 23 (36), H217–H247. doi:10.1002/adma.201102313

Kircher, M.F. and Willmann, J K. (2012), Molecular body imaging: MR imaging, CT, and US. Part II. Applications, *Radiology*, 264 (2), 349–368. doi:10.1148/radiol.12111703

Kowada, T., Maeda, H., and Kikuchi, K. (2015), BODIPY-based probes for the fluorescence imaging of biomolecules in living cells, *Chem. Soc. Rev.*, 44 (14), 4953–4972. doi:10.1039/C5CS00030K

Lee, N., Cho, H.R., Oh, M.H. et al. (2012), Multifunctional Fe_3O_4/TaO x core/shell nanoparticles for simultaneous magnetic resonance imaging and X-ray computed tomography, *J. Am. Chem. Soc.*, 134 (25), 10309–10312. doi:10.1021/ja3016582

Lee, N., Choi, S.H., and Hyeon, T. (2013), Nano-sized CT contrast agents, *Adv. Mater.*, 25 (19), 2641–2660, doi:10.1002/adma.201300081

Li, J., Cheng, F., Huang, H. et al. (2015), Nanomaterial-based activatable imaging probes: from design to biological applications, *Chem. Soc. Rev.*, 44 (21), 7855–7880. doi:10.1039/C4CS00476K

Li, T., Schofield, A.B., Chen, K. et al. (2019), Particle-stabilized Janus emulsions that exhibit pH-tunable stability, *Chem. Commun.*, 55 (41), 5773–5776. doi: 10.1039/c8cc09842e

Li, Y., Lin, T.Y., Luo, Y. et al. (2014), A smart and versatile theranostic nanomedicine platform based on nanoporphyrin, *Nat. Commun.*, 5, 4712. doi:10.1038/ncomms5712

Lim, Y.G.J., Poh, K.C.W., and Loo, S.C.J. (2019), Hybrid Janus microparticles achieving selective encapsulation for theranostic applications via a facile solvent emulsion method, *Macromol. Rapid Commun.*, 40 (7), e1800801. doi: 10.1002/marc.201800801

Lipinski, C.A. (2005), Solubility in water and DMSO: issues and potential solutions, in: Borchardt, R., Kerns, E., Lipinski, C., Thakker, D., and Wang, B., Eds., *Pharmaceutical Profiling in Drug Discovery for Lead Selection*, AAPS Press, Arlington, TX, pp. 93–125.

Lovell, J.F., Liu, T.W.B., Chen, J. et al. (2010), Activatable photosensitizers for imaging and therapy, *Chem. Rev.*, 110 (5), 2839–2857. doi:10.1021/cr900236h

Lusic, H. and Grinstaff, M.W. (2013), X-ray computed tomography contrast agents, *Chem. Rev.*, 113 (3), 1641–1666. doi:10.1021/cr200358s

Merisko-Liversidge, E. and Liversidge, G.G. (2011), Nanosizing for oral and parenteral drug delivery: a perspective on formulating poorly-water soluble compounds using wet media milling technology, *Adv. Drug Deliv. Rev.*, 63 (6), 427–440. doi:10.1016/j.addr.2010.12.007

Mishra, A. and Verma, M. (2010), Cancer biomarkers: are we ready for the prime time, *Cancers*, 2, 190–208. doi:10.3390/cancers2010190

Mornet, S., Vasseur, S., Grasset, F. et al. (2004), Magnetic nanoparticle design for medical diagnosis and therapy, *J. Mater. Chem.*, 14, 2161–2175. doi:10.1039/B402025A

Müller, R.H. and Keck, C.M. (2012), Twenty years of drug nanocrystals: where are we, and where do we go?, *Eur. J. Pharm. Biopharm.*, 80, 1–3. doi:10.1016/j.ejpb.2011.09.012

Na, H.B., Song, I.C., and Hyeon, T. (2009), Inorganic nanoparticles for MRI contrast agents, *Adv. Mater.*, 21 (21), 2133–2148. doi:10.1002/adma.200802366

Niu, D., Liu, X., Li, Y. et al. (2011), Fabrication of uniform, biocompatible and multifunctional PCL-b-PAA copolymer-based hybrid micelles for magnetic resonance imaging, *J. Mater. Chem.*, 21 (36), 13825–13831. doi:10.1039/C1JM10929D

Oh, M.H., Lee, N., Kim, H. et al. (2011), Large-scale synthesis of bioinert tantalum oxide nanoparticles for X-ray computed tomography imaging and bimodal image-guided sentinel lymph node mapping, *J. Am. Chem. Soc.*, 133 (14), 5508–5515. doi:10.1021/ja200120k

Pankhurst, Q.A., Connolly, J., Jones, S.K. et al. (2003), Applications of magnetic nanoparticles in biomedicine, *J. Phys. D. Appl. Phys.*, 36 (13), R167–R181.

Parveen, S., Misra, R., and Sahoo, S.K. (2012), Nanoparticles: a boon to drug delivery, therapeutics, diagnostics and imaging, *Nanomedicine*, 8 (2), 147–166. doi:10.1016/j.nano.2011.05.016

Puri, D., Khatik, G.L., and Shunmugaperumal, T. (2019), Studies on olive-and silicone-oils-based Janus macroemulsions containing ginger to manage primary dysmenorrheal pain, *Mater. Sci. Eng. C*, 100, 276–285. doi:10.1016/j.msec.2019.01.137

Qian, C., Yu, J., Chen, Y. et al. (2016), Light-activated hypoxia-responsive nanocarriers for enhanced anticancer therapy, *Adv. Mater.*, 28 (17), 3313–3320. doi:10.1002/adma.201505869

Ren, H., Liu, J., Su, F. et al. (2017), Relighting photosensitizers by synergistic integration of albumin and perfluorocarbon for enhanced photodynamic therapy, *ACS Appl. Mater. Interfaces*, 9 (4), 3463–3473. doi:10.1021/acsami.6b14885

Singh, N., Jenkins, G.J., Asadi, R. et al. (2010), Potential toxicity of superparamagnetic iron oxide nanoparticles (SPION), *Nano Rev.*, 1 (1), 5358. doi:10.3402/nano.v1i0.5358

Smith, J.J., Sorensen, A.G., and Thrall, J.H. (2003), Biomarkers in imaging: realizing radiology's future, *Radiology*, 227 (3), 633–638. doi:10.1148/radiol.2273020518

Song, X., Feng, L., Liang, C. et al. (2016), Ultrasound triggered tumor oxygenation with oxygen-shuttle nanoperfluorocarbon to overcome hypoxia-associated resistance in cancer therapies, *Nano Lett.*, 16 (10), 6145–6153. doi:10.1021/acs.nanolett.6b02365

Storm, G., Belliot, S.O., Daemen, T. et al. (1995), Surface modification of nanoparticles to oppose uptake by the mononuclear phagocyte system, *Adv. Drug Deliv. Rev.*, 17 (1), 31–48. doi:10.1016/0169-409X(95)00039-A

Tamilvanan, S. (2009), Formulation of multifunctional oil-in-water nanosized emulsions for active and passive targeting of drugs to otherwise inaccessible internal organs of the human body, *Int. J. Pharm.*, 381, 62–76. doi:10.1016/j.ijpharm.2009.08.001

Tamilvanan, S. and Kumar, B.A. (2011), Influence of acetazolamide loading on the (*in vitro*) performances of non-phospholipid-based cationic nanosized emulsion in comparison to phospholipid-based anionic and neutral-charged nanosized emulsions, *Drug Dev. Ind. Pharm.*, 37 (9), 1003–1015. doi:10.3109/03639045.2011.555407

Teixeira, H., Dubernet, C., Rosilio, V. et al. (2000), New bicompartmental structures are observed when stearylamine is mixed with triglyceride emulsions, *Pharm. Res.*, 17 (10), 1329–1332. doi:10.1023/A:1026416208482

Temme, S., Grapentin, C., Quast, C. et al. (2015), Noninvasive imaging of early venous thrombosis by ^{19}F magnetic resonance imaging with targeted perfluorocarbon nanoemulsions, *Circulation*, 131 (16), 1405–1414. doi:10.1161/circulationaha.114.010962

Tong, H., Chen, Y., Li, Z. et al. (2016), Photodynamic theranostics: glutathione activatable photosensitizer-conjugated pseudopolyrotaxane nanocarriers for photodynamic theranostics, *Small*, 12 (45), 6223–6232. doi:10.1002/smll.201670234

Torchilin, V.P. (2002), PEG-based micelles as carriers of contrast agents for different imaging modalities, *Adv. Drug Deliv. Rev.*, 54 (2), 235–252. doi:10.1016/s0169-409x(02)00019-4

Torchilin, V.P. (2006), Multifunctional nanocarriers, *Adv. Drug Deliv. Rev.*, 58 (14), 1532–1555. doi:10.1016/j.addr.2006.09.009

Torchilin, V.P. (2007), Targeted pharmaceutical nanocarriers for cancer therapy and imaging. *AAPS J.*, 9 (2), E128–E147. doi:10.1208/aapsj0902015

Torchilin, V.P. and Trubetskoy, V.S. (1995), Which polymers can make nanoparticulate drug carriers long-circulating?, *Adv. Drug Deliv. Rev.*, 16 (2–3), 141–155. doi:10.1016/0169-409X(95)00022-Y

Tu, F. and Lee, D. (2014), One-step encapsulation and triggered release based on Janus particle-stabilized multiple emulsions, *Chem. Commun.*, 50 (98), 15549–15552. doi:10.1039/C4CC07854C

Ulrich, G., Ziessel, R., and Harriman, A. (2008), The chemistry of fluorescent bodipy dyes: versatility unsurpassed, *Angew. Chem. Int. Ed.*, 47 (7), 1184–1201. doi:10.1002/anie.200702070

Vecchione, R., Quagliariello, V., Giustetto, P. et al. (2017), Oil/water nano-emulsion loaded with cobalt ferrite oxide nanocubes for photo-acoustic and magnetic resonance dual imaging in cancer: in vitro and preclinical studies, *Nanomed. Nanotechnol. Biol. Med.*, 13 (1), 275–286. doi: 10.1016/j.nano.2016.08.022.

Veseli, B.E., Perrotta, P., De Meyer, G.R.A., et al. (2017), Animal models of atherosclerosis, *Eur. J. Pharmacol.*, 816, 3–13. doi:10.1016/j.ejphar.2017.05.010

Wallyn, J., Anton, N., Mertz, D. et al. (2018), Magnetite and iodine-containing nanoemulsion as dual modal contrast agent for X-ray/MR imaging, *ACS Appl. Mater. Interfaces*, 11, 403–416. doi:10.1021/acsami.8b19517

Walther, A. and Müller, A.H. (2013), Janus particles: synthesis, self-assembly, physical properties, and application, *Chem. Rev.*, 113 (7), 5194–5261. doi:10.1021/cr300089t

Wang, S., Lin, J., Wang, Z. et al. (2017), Core–satellite polydopamine-gadolinium-metallofullerene nanotheranostics for multimodal imaging guided combination cancer therapy, *Adv. Mater.*, 29 (35), 1701013. doi:10.1002/adma.201701013

Wang, Y.X.J., Hussain, S.M., and Krestin, G.P. (2001), Superparamagnetic iron oxide contrast agents: physicochemical characteristics and applications in MR imaging, *Eur. Radiol.*, 11 (11), 2319–2331. doi:10.1007/s003300100908

Widmark, J.M. (2007), Imaging-related medications: a class overview, *Proc. (Baylor Univ. Med. Cent.)*, 20 (4), 408–417. doi:10.1080/08998280.2007.11928336

Xue, S., Wang, Y., Wang, M. et al. (2014), Iodinated oil-loaded, fluorescent mesoporous silica-coated iron oxide nanoparticles for magnetic resonance imaging/computed tomography/fluorescence trimodal imaging, *Int. J. Nanomed.*, 9, 2527–2538. doi:10.2147%2FIJN.S59754

Yang, T., Liu, L., Deng, Y. et al. (2017), Ultrastable near-infrared conjugated-polymer nanoparticles for dually photoactive tumor inhibition, *Adv. Mater.*, 29 (31), 1700487. doi:10.1002/adma.201700487

Yu, S.B. and Watson, A.D. (1999), Metal-based X-ray contrast media, *Chem. Rev.*, 99 (9), 2353–2378. doi:10.1021/cr980441p

Yun, S.H. and Kwok, S.J.J. (2017), Light in diagnosis, therapy and surgery, *Nat. Biomed. Eng.*, 1, 0008. doi:10.1038/s41551-016-0008

Zarzar, L. D., Sresht, V., Sletten, E. M. et al. (2015), Dynamically reconfigurable complex emulsions via tunable interfacial tensions, *Nature*, 518, 520–524. doi:10.1038/nature14168

Zhang, Y., Bo, S., Feng, T. et al. (2019), A versatile theranostic nanoemulsion for architecture dependent multimodal imaging and dually augmented photodynamic therapy, *Adv. Mater.*, 31 (21), e1806444. doi:10.1002/adma.201806444

CHAPTER 7.4

EMULSION-LIKE DISPERSIONS

ABHINAB GOSWAMI*, OLY KATARI*, AND TAMILVANAN SHUNMUGAPERUMAL

Department of Pharmaceutics, National Institute of Pharmaceutical Education and Research (NIPER), Guwahati, Assam, India

*The fundings provided by Department of Pharmaceuticals, Ministry of Chemicals and Fertilizers, Govt. of India to the authors Abhinab Goswami and Oly Katari are greatly acknowledged.

Oil-in-Water Nanosized Emulsions for Drug Delivery and Targeting, First Edition. Tamilvanan Shunmugaperumal.
© 2021 John Wiley & Sons, Inc. Published 2021 by John Wiley & Sons, Inc.

EXPANSION OF ABBREVIATIONS

API	active pharmaceutical ingredient
ATPS	aqueous two-phase systems
CD	cyclodextrin
HEC	hydroxyethylcellulose
O/W	oil-in-water
O/W/W	oil-in-water-in-water
STL	slope of tie-line
TLL	tie-line length
W/O	water-in-oil
W/W	water-in-water emulsions

7.4.1. INTRODUCTION

Emulsions are heterogeneous dispersions of oil-in-water (o/w) or water-in-oil (w/o) (Lu and Gao 2010). The o/w nanosized emulsions comprise of a single oil or oil combinations stabilized by a single emulsifier molecule or combination of emulsifier molecules and dispersed in surrounding aqueous medium. The o/w nanosized emulsions thus produced appear as bluish-white in color at room temperature conditions. In pharmaceutical research, however, emulsion-like dispersions may also be produced without the inclusions of a single oil or oil combinations and therefore, the produced bluish white-colored dispersion is termed as oil-less emulsions. Similarly, the emulsion-like dispersions may also be made without the inclusion of single emulsifier molecule or combination of emulsifier molecules, and these dispersions are called as emulsifier/surfactant-free emulsions. There is an interesting emulsion-like dispersion that may be formulated without both the inclusion of oil(s) and emulsifier(s) molecules and the produced dispersion is called as emulsifier/surfactant- and oil-free emulsions. Flowchart 7.4.1 displays the classification of emulsion-like dispersions.

The objective of this section is to show the manufacturing and possible application potentials of various emulsion-like dispersions.

7.4.2. FORMULATION OF EMULSION-LIKE DISPERSIONS

The name or nomenclature allotted to each one of the emulsion-like dispersions is explained in the following sections on the basis of projecting the other FDA-approved excipients like β-cyclodextrin (CD), eutectic-liquid formers, and hydrophilic macromolecules.

7.4.2.1. Projection of β-Cyclodextrin as Emulsifier

Due to the amphiphilic characteristics of most of the emulsifiers (Eccleston 2007), they are capable of adsorbing/concentrating at the oil–water interface

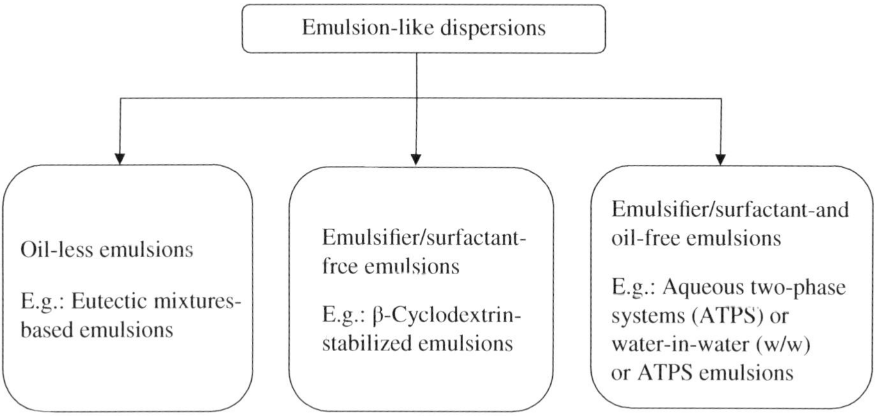

Flowchart 7.4.1. Classification of emulsion-like dispersions.

and to form mono- or multi-molecular films onto the dispersed oil droplets of the emulsions. These emulsifiers can be of natural or synthetic origin. However, the synthetic emulsifier is found to be associated with various intestinal, cardiovascular, and metabolic diseases leading to a paradigm shift from employing synthetic emulsifiers to natural ones (Halmos et al. 2019). Natural emulsifiers include phospholipids, biosurfactants, proteins, polysaccharides, and natural colloidal particles (McClements et al. 2017). Among the various carbohydrate emulsifiers, β-CD finds its application in the manufacturing of emulsifier/surfactant-free emulsions (Inoue et al. 2010).

The β-CD is an oligosaccharide of either dextrose or dextrose derivatives joined by α-1,4-linkages. Natural CD composed of 6, 7, and 8 glucose units and are referred to as α, β, and γ-CD, respectively. CDs have a central hydrophobic core, which is capable of forming stable inclusion complex especially with hydrophobic guest molecules (Frömming and Szejtli 1993). In particular, the β-CD is used not only as solubility enhancer for poorly water-soluble active pharmaceutical ingredient (API) via complex formation but also enhancing the physicochemical properties and chemical stability of API, thus, increasing the driving force of diffusion required to cross the biological membrane for API (Swiech et al. 2012; Sierpe et al. 2015). The emulsifying property of β-CD can be attributed to its capability of forming a dense film at the oil–water interface and a three-dimensional structural network in the continuous phase by precipitated complexes (Inoue et al. 2008). This type of emulsion, which is stabilized by solid particles, is known as Pickering emulsions (Pickering 1907; Abend et al. 1998). The advantages of Pickering emulsion include (1) reduction in the use of environmental pollution caused by certain surfactants, (2) less influence of pH, salt, or temperature on emulsion stability, and (3) control of the viscoelastic property of the emulsion (Inoue et al. 2009). Thus, the emulsion-like dispersions, which are prepared devoid of emulsifier/surfactant or emulsifier/surfactant combination, get the terminology as emulsifier/surfactant-free emulsions.

7.4.2.2. Projection of Eutectic Mixture as Oil Phase

Traditionally, the o/w nanosized emulsions are developed by using emulsifiers, oil, and water combination (Moreira-Oliveira et al. 2018). The presence of oil or oil combinations in emulsions possesses the tendency to undergo an oxidation reaction and thus initiating the degradation of the incorporated API. This can be prevented by incorporating suitable antioxidant molecules at low concentrations into the emulsions. Alternative approaches include the use of hydrotropic mixture systems such as eutectic mixture, which can be made from eutectic-forming excipients and may be used to dissolve lipophilic API.

The eutectic mixture is a mixture of components that do not interact to form a new compound but when mixed together at certain ratios inhibits the crystallization process of one another resulting in a system with a lower melting point than either of the component (Stott et al. 1998). There are reports that indicate the use of eutectic mixtures in pharmaceutical to increase API solubility, permeation, and absorption (Stott et al. 1998; Stott et al. 2001; Yong et al. 2004; Lazerges et al. 2010) and also as oil phase in emulsion (Nazzal et al. 2002).

The peculiar property of eutectic liquid is the hydrophobic nature and thus able to dissolve the sufficient amount of plant-derived active constituents (coriander) like the traditional oil or oil combination does. These eutectic mixtures are a very attractive but less expensive replacement of the oil phase to prepare finally the emulsion-like dispersions. The o/w nanosized emulsions prepared devoid of oil or oil combination is termed as oil-less emulsions (Tamilvanan et al. 2019).

7.4.2.2.1. Case Study Using Oil-Less Emulsions Table 7.4.1 shows the composition of the oil-less emulsion containing coriander seed powder.

The oil-less emulsion was manufactured by mixing the coriander constituents containing oil-less phase with the water phase using the mechanical stirrer followed by electric stirrer at 800 rpm for over the period of 30 min and then keeping the crude emulsion formed in an ultrasonicator for another 10 min. The final emulsion formed looks like milky white with minimum or no amount of suspended coriander seed powder as it was completely dispersed or dissolved in the eutectic liquid (oil-less phase) before emulsification with water phase.

TABLE 7.4.1. Composition of Oil-Less Emulsion

Oil-Less Phase	Water Phase
Camphor	Aqueous extract of coriander leaves
Coriander seed powder	Tween 80
Hydroxyethylcellulose	
Menthol	

The developed coriander-loaded oil-less emulsions had a mean particle diameter of 71.227 μm. Other characterization works performed are the API content, *in vitro* quantitative and qualitative antifungal activities, and *in vitro* API diffusion/permeation. The coriander seed powder is initially mixed with the eutectic liquid. To impart viscosity to the formulation, the oil-less emulsion was deliberately mixed with a fixed amount of hydroxyethylcellulose (HEC) during the emulsification step. The viscosity of the formulation is enhanced intentionally by keeping the point in mind that the final product should consist of required resistance against the eroding effect produced by the salivary fluid (for managing oral candidiasis condition) after the formulation application (Tamilvanan et al. 2019).

7.4.2.3. Projection of β-Cyclodextrin and Eutectic Mixture as Emulsifier and Oil Phase

By combining the concept of emulsifier/surfactant-free emulsions and oil-less emulsions together, new emulsion-like dispersions emerged out, which conveniently be termed as "emulsifier/surfactant- and oil-free emulsions." Keeping the point of topical application potential of eutectic mixture-based oil-less emulsions particularly onto the broken skin structure, the use of eutectic mixture (having the properties of counterirritant and rubefacient) may compromise the patient compliance because the further exacerbation of the already affected skin structures are likely to occur (Rangel et al. 2018; Rohilla et al. 2018). At the same time, the chances of so-called irritation effect produced by the eutectic liquid are also likely to be less because the eutectic liquid is simply dispersed in the continuous aqueous medium with the coverage of mono- or multi-molecular emulsifier film. In addition, the question of droplet–droplet coalescence and subsequent breakage of dispersed eutectic liquid droplets of the oil-less emulsion are not likely to occur due to the emulsifier film coverage around the dispersed eutectic liquid droplets of the oil-less emulsion.

To eliminate the possible patient noncompliance-producing properties (counterirritant and rubefacient) of eutectic mixture-based oil-less emulsions, it is better to eliminate the eutectic liquid as one of the components from the production of emulsion-like dispersions. By doing so, the emulsion-like dispersions can also be made without eutectic mixture but with β-CD. This type of emulsion-like dispersions produced should get another nomenclature in the pharmaceutical formulation literature. The term thus introduced should also indicate the word emulsion because of the bluish white-color appearance of the final product. Taking this point into consideration, the β-CD and other hydrophilic macromolecules are dispersed in water to produce two different aqueous dispersions. Following mixing of these two different aqueous dispersions, the obtained multicomponent multiphase system is called as "aqueous two-phase systems (ATPS) or water-in-water (w/w) emulsions" (Hatti-kaul 2001; Atefi et al. 2016; Esquena 2016) and sometimes, denoted as "ATPS emulsions" (Capron et al. 2001; Sagis 2007).

7.4.2.4. Projection of β-Cyclodextrin and Other Hydrophilic Macromolecules as Components to Make Multicomponent Multiphase System

7.4.2.4.1. Water-in-Water Emulsion A water-in-water (w/w) emulsion is defined as a multicomponent multiphase system that consists of aqueous droplets dispersed in a continuous aqueous phase. Such dispersion systems are composed of at least two hydrophilic macromolecules that are thermodynamically incompatible with each other in solution thus generating two phases (Piculell and Lindman 1992; Kasapis et al. 1993; Grinberg and Tolstoguzov 1997). These aqueous systems are also denoted as "Aqueous two-phase system" (Hatti-kaul 2001; Atefi et al. 2016) abbreviated as ATPS (Bridges et al. 2007; Onghena et al. 2015) or "Aqueous bi-phase system." However, the phenomenon of phase separation in solutions containing polymer mixtures is very common. Most hydrophilic polymers are incompatible with each other in aqueous solution thus yielding two coexisting phases in equilibrium with each other (Beijerinck 1896). Water–water phase separation can occur in a large variety of systems, which includes aqueous mixtures of polymer, polymers and surfactants, polymers and electrolytes, and surfactant solutions. The phase separation in aqueous two-phase systems is very fast, which can vary from seconds and a few hours. However, the colloidal stability of droplets in such systems is poor because of the lack of significant inner droplet repulsion forces, and therefore emulsions in many aqueous biphasic systems are unstable. The colloidal stability can be greatly controlled by the addition of particles and thus improving the stability of the w/w emulsions (Esquena 2016).

The w/w emulsions have a wide range of applications that are greatly expanding over the years primarily due to their all-water nature and absence of both oil and surfactants. The w/w emulsions can be used for the design of novel encapsulation and delivery systems for labile molecules (Sagis 2008, 2009). The w/w emulsions can be prepared using food-grade components, thus ensuring its biocompatibility. Biopolymers, such as polysaccharides and proteins, can be used in a wide range of delivery systems for encapsulating, protecting, and controlled delivery of APIs (Matalanis et al. 2011). These delivery systems are easily being formulated with the help of simple processing operations such as mixing, homogenizing, and thermal processing (Scholten et al. 2006; Sagis 2009). The w/w emulsions can be used as microreactors for the synthesis of products of high added value (Dewey et al. 2014). This can be attributed to their ability to provide reaction-relevant internal environments, while allowing entry/exit of substrates and product in mild conditions (Oh et al. 2008; Fernández-Barbero et al. 2009). Among all the possible applications, the most studied application of w/w emulsions is their application as template for the preparation of microgel particles with controlled particle size. These micro/nanogels are crosslinked polymer particles, which usually have the ability to respond to the external stimuli (pH, temperature, and ionic strength) (Esquena 2016). The w/w emulsions can allow the design of smart particles for the encapsulation of APIs. In a broader sense, the w/w emulsions should include the formation of oil-water-water (o/w/w) emulsions. These o/w/w particles might be useful for encapsulation and delivery of lipophilic components in the food, cosmetics, and pharmaceutical industries (Matalanis et al. 2010).

The two different types of phase separation usually occur in w/w emulsions and the following section discusses about them.

7.4.2.4.2. Types of Phase Separation in Water-in-Water Emulsions The mixture of hydrophilic polymer in aqueous solution is rather very complex (Piculell and Lindman 1992; Matalanis et al. 2011). The two types of phase separation that can occur in w/w emulsion are segregative and associative phase separation. These two separations are depending upon the repulsion and attraction forces between the two different hydrophilic polymers and also the hydration interaction between the polymer and water (Esquena 2016).

1. **Segregative phase separation:** Segregative phase separation may occur in many mixtures of two hydrophilic components. At low concentration, the mixture remains as a single phase with two components in the solution. However, with the increase in the concentration above a certain value, the components separate into two different phases that are in thermal equilibrium. Each phase is enriched in one hydrophilic component and is saturated with others. Segregation may arise due to negative entropy of mixing, which is highly dependent on molecular conformation. The occurrence of repulsion between the polymer chains can be attributed to the free conformational movement of the polymers. The conformational movement largely depends on the molecular weight, and short molecules tend to restrict the free conformation of long linear chain molecules. Therefore, the repulsive interactions expect to present in a mixture of nonionic polymer with clearly distinct molecular weight (Esquena 2016).

   The segregative phase behavior is shown schematically in the ternary phase diagram (Fig. 7.4.1), in which the immiscibility region is set by a binodal line. Compositions that are mutually in equilibrium are indicated by tie-lines, which converge into a critical point, where the two immiscible phases merge to become one single phase. This behavior is commonly observed in mixtures of nonionic polymers, or the case of one ionic polymer mixed with a nonionic one. This segregative phase separation can also occur in the mixture of a surfactant and a polymer (Piculell and Lindman 1992).
2. **Associative phase separation:** Associative phase separation occurs where mixtures of two components and water separate in a solid-like precipitate, which contains a high concentration of the two hydrophilic polymers, and a supernatant liquid solution that only has a residual low

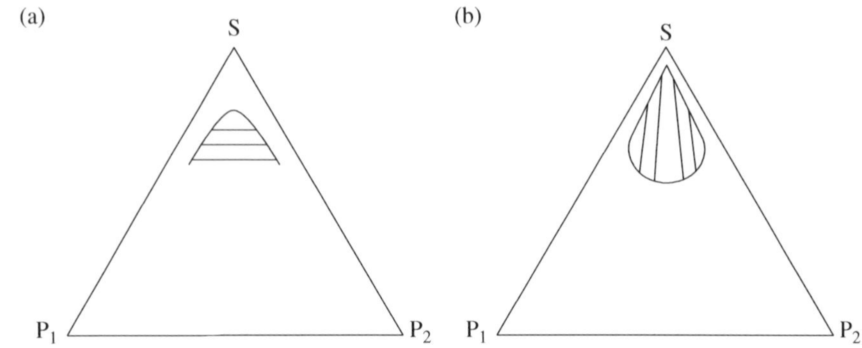

Figure 7.4.1. Schematic ternary phase diagrams, which illustrate (a) segregative and (b) associative phase separation in mixed hydrophilic polymer systems. Key: P_1, Polymer 1; P_2, Polymer 2; S, Solvent (water). [Adapted from Esquena (2016).]

number of polymers (Schmitt and Turgeon 2011). The tie-lines represented in the ternary phase diagram (Fig. 7.4.1) is almost vertical (Piculell and Lindman 1992). This type of phase separation is commonly produced by electrostatic complexation, due to the presence of oppositely charged polymers. The precipitate consists of amorphous noncrystalline particles, which are either solid-like particles or highly viscous droplets, made of two oppositely charged macro-ions held together by strong electrostatic attractions (Esquena 2016).

7.4.2.5. Formulation Realities of ATPS

Segregative phase separation can be prepared by applying mechanical agitation in the two-aqueous phase systems. The phase with smaller volume fraction becomes the internal phase, and the phase with smaller volume fraction becomes the internal phase and phase inversion occurs approximately at composition in which the volume fraction of the two phases is approximately equal. At composition near to the inversion point at 50 : 50 volume ratio, the bicontinuous emulsion can be formed (Norton and Frith 2001). This is illustrated in the phase diagram (shown in Fig. 7.4.2), which is like a fingerprint to a system under specific conditions such as temperature and pH (Raja et al. 2011).

The *x*-axis of the phase diagram indicates % w/w of bottom phase constituents while the *y*-axis shows % w/w of top phase constituents. The phase diagram provides a set of information like the concentration of components for the two-phase formation and their concentration in the top and bottom phases. The diagram shows a binodal curve representing the points, TCB (denotes the condition at which bottom and top distinct phases may coexist), which divides the region of the component's concentration. This curve divides the concentration that forms two immiscible aqueous phases (biphasic region) from those that make

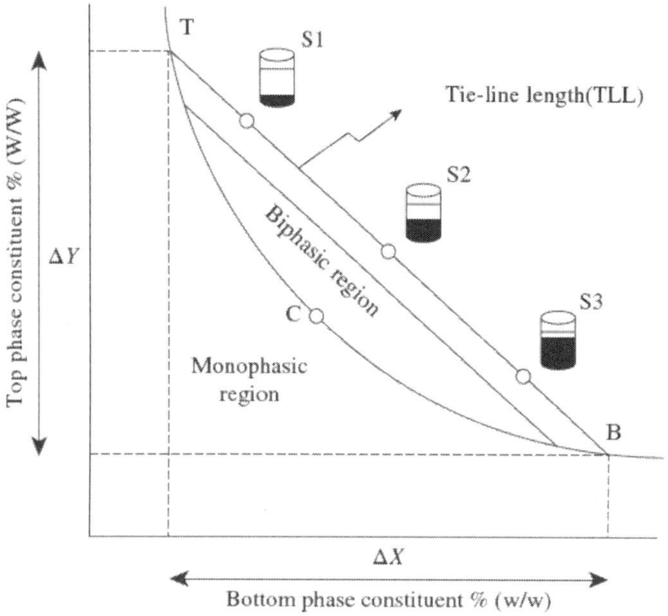

Figure 7.4.2. Schematic representation of the phase diagram. Concentrations above the binodal curve (TCB) form the aqueous two-phase system. [Adapted from Iqbal et al. (2016).]

one phase (monophasic region), which is below the binodal curve. In simple words, the phase that is above the TCB curve is the biphasic region and the phase that is below the TCB curve is the monophasic region. The line TB is the tie-line and it connects two nodes (T and B), which lie on the binodal curve. The tie-line can be estimated by using the weight ratio as shown in the following equation:

$$\frac{V_t \rho_t}{V_b \rho_b} = \frac{SB}{ST} \tag{7.4.1}$$

where V and ρ stand for volume and density of top (t) and bottom (b) phase while SB and ST are segments' length as shown in the phase diagram or by analysis of top and bottom phases, which is the more precise method.

All the potential systems S1, S2, and S3 have the same top and bottom phase equilibrium composition because of being on the same tie-line. Point C on the binodal curve is the critical point, just above this point the volume of both phases is theoretically equal. At point C, the value of tie-line length (TLL) is equal to zero [Eq. (7.4.2)].

$$\text{TLL} = \sqrt{\Delta X^2 + \Delta Y^2} \tag{7.4.2}$$

The tie-lines are commonly straight and the slope of tie-line (STL) can be calculated with the help of Eq. (7.4.3). This is also helpful in the construction of further tie-lines.

$$\mathrm{STL} = \frac{\Delta Y}{\Delta X} \tag{7.4.3}$$

The binodal curve can be determined by three methods: turbidometric, titration method; cloud point method; and node determination method.

The various types of ATPS and their great potential for the extraction, separation, purification, and enrichment of proteins, membranes, viruses, enzymes, nucleic acids, and other biomolecules both in industry and academia have been recently reviewed by Iqbal et al. (2016) along with optimization of the formula for ATPS. Current research shows that the ATPS has also been successfully used in the detection of veterinary drug residues in food, separation of precious metals, sewage treatment, and a variety of other purposes. Table 7.4.2 displays the various applications of ATPS.

7.4.2.5.1. ATPS Types Before proceeding with issues of ATPS stability, it becomes pertinent to discuss the building blocks of ATPS. Table 7.4.3 depicts the ATPS types along with their building blocks/compositions. The most common biphasic systems are formed by two polymers [usually polyethylene glycol (PEG) and dextran] or a polymer and a salt (e.g., phosphate, sulfate, or citrate). Other types include ionic liquids and short-chain alcohols (Van Berlo et al. 1998; Hatti-Kaul 2000; Hatti-Kaul 2001; Asenjo and Andrews 2012; Ruiz-Ruiz et al. 2012; Molino et al. 2013; Grilo et al. 2016). In addition to this, ionic and/or nonionic surfactants are used for the formation of micellar and reverse micellar ATPSs (Liu et al. 1998; Xiao et al. 2000; Ruiz-Ruiz et al. 2012). Polymer–polymer/salt systems are also studied.

Coming to the polymer–polymer ATPS system, PEGs are most widely used as one of the polymer components. In fact, PEGs are available in both liquid and solid forms depending on their grades that are being segregated based on the molecular weight (Rowe et al. 2009). Whereas PEG grades 200–600 are liquids at ambient temperatures, grades 1,000 and above appear as solids possessing the consistency, which range from pastes to waxy flakes. However, grades of PEG 6000 and above are available as free-flowing milled powders. The second polymer of interest is dextran. Dextran is a synthetic colloid consisting of a mixture of glucose polymers derived from the action of *Leuconostoc mesenteroides* on sucrose and have the high water-binding capacity. It is currently available in 10% dextran 40 (40 kDa) and 6% dextran 70 (70 kDa) (Winkler 2019). Due to the possession of low ionic strength, the polymer–polymer systems are preferably used for the separation, recovery, and purification of solutes sensitive to the ionic environment.

TABLE 7.4.2. Application of Aqueous Two-Phase Systems (ATPS)

Application of ATPS	Molecule of Interest	Composition of ATPS	Outcome	Reference
Purification of proteins and peptides	Bovine serum albumin, insulin, and protein A	Poly(ethylene glycol)–phosphate buffer and poly (ethylene glycol)–dextran	The separation process utilizes a microfluidic aqueous two-phase system platform along fluorescence microscopy for determination of partition coefficient value. The microfluid system coupled with fluorescence microscopy was found to fast and low material consumption as compared to its macroscale approach	Silva et al. (2017)
	Trypsin from bovine pancreas	Poly(ethylene glycol)–sodium citrate	The yield was found to be 99.7% in the top phase with a purification factor of 2.555	Pérez et al. (2015)
	Sso7d-Taq polymerase	Poly (ethylene glycol)–sodium sulphate	SSo7d-Taq polymerase was purified using a single-step aqueous, two-phase system. Extraction was found to be >95% pure and active	Sundarrajan et al. (2018)
Templates for microgel preparation	Human liver carcinoma HepG2 cells	Poly(ethylene glycol)–dextran	Encapsulated cells show high viability and growth potential	Liu et al. (2017)
	Lysozyme and bovine serum albumin	Poly(ethylene glycol)–dextran	The prepared microgel has loading efficiencies of 51.5–73.6%. The release of encapsulated agents was sustained over a period of 6–14 days. The poly(ethylene glycol)microgel particles showed a marked reduction in phagocytosis rates compared with the polystyrene particles, thus increasing the local and systemic bioavailability of encapsulated biological molecules	Graf et al. (2019)

TABLE 7.4.2. ***Continued***

Application of ATPS	Molecule of Interest	Composition of ATPS	Outcome	Reference
Separation and enrichment of drug from water, food of animal origin, and herbs	Ciprofloxacin from milk, egg, and shrimp	Poly(ethylene glycol-ran-propylene glycol) EOPOL31–K_2HPO_4	The average extraction efficiency in the first extraction was found to be 99.7 and 85.6%, in the second extraction with a detection limit of 6.8 ng g^{-1}.	Chen et al. (2014)
	Sulphonamide from milk	1-Butyl-3-methylimidazolium. tetrafluoraborate–$C_6H_5Na_3O_7{\cdot}2H_2O$	The detection limit of sulphonamide was found to be in the range of 2.04–2.84 ng l^{-1}. The recovery rate of sulphonamide was found to be in the range of 72.32–108.96%	Shao et al. (2014)
	Tetracycline, oxytetracycline, and chloramphenicol from honey	1-Octyl-3-methylimidazolium bromide–sodium dodecyl sulfate	The detection limit of tetracycline and oxytetracycline and chloramphenicol was found to be 5.8, 8.2, and 4.2 μg kg^{-1}, respectively. The recovery % was found to be in the range of 85.5–110.9	Yang et al. (2014)
	Chloramphenicol from shrimp	Polyoxyethylene lauryl ether (POELE10)–NaH_2PO_4	The average extraction efficiency was found to be 99.42% with a detection limit of 0.8 μg kg^{-1}. The recovery % of chloramphenicol was in the range of 98–100.4	Lu et al. (2016)
Purification of whole cell and their organelles	Isolation of extracellular vesicle	Poly(ethylene glycol)–dextran	The method provides a high extracellular vesicle recovery efficiency of ~70% in a short duration of time (15 min)	Shin et al. (2015)
Purification of low weight compounds	Phenylethanoid glycoside (echinacoside and acetoside)	Ethanol and ammonium sulphate	An increase of 2.46 and 2.58 folds were obtained by using ultrasound-assisted aqueous two-phase extraction	Dong et al. (2015)

Extraction of DNA, RNA, and other nucleic acid-based biomolecules	Plasmid DNA (pDNA) and proteins	Poly(ethylene glycol) (MW 600–4000 Da) and sodium sulphate	Partitioning of plasmid DNA and protein in cell lysis solution was investigated at different operational conditions in poly(ethylene glycol)–sodium sulphate systems. It was found that with an increase in molecular weight of the poly(ethylene glycol), the affinity of pDNA and proteins increases toward the salt phase. A similar result was observed when the pH of the system was increased	Nazer et al. (2017)
Purification of monoclonal antibodies	Monoclonal antibodies (immunoglobulin type G, IgG)	Poly(ethylene glycol) 400, poly(ethylene glycol) 1,450, sodium phosphate monobasic anhydrous, and potassium phosphate citrate	Aqueous two-phase-mediated extraction from the cultivation broth in combination with the new membrane-based phase separation led to a monoclonal antibody yield of 78% with a simultaneous reduction of DNA and host cell protein load	Kruse et al. (2019)
Aqueous two-phase system-assisted formation of micro/nanoparticles	Formation of self-assembled poly(lactic-co-glycolic acid) microparticles	Pluronic F127–dextran	The particle size of the prepared poly(lactic-co-glycolic acid) microparticles were found to be in the range of 2–10 μm with average size of around 7 μm. Both hydrophilic and hydrophobic molecules can be incorporated into the microparticles owing to the amphiphilic macromolecule composition. The ATPS-based microparticle formation can serve as a novel platform for poly(lactic-co-glycolic acid)/polymer-based tunable micro/nanoparticles and polymersome development	Yeredla et al. (2016)

TABLE 7.4.3. ATPS Types Along with Composition

Types of ATPS	Composition of ATPS	Reference
Polymer–polymer	PEG–dextran	Chen and Lee (1995)
	PEG–dextran	Helfrich et al. (2005)
Polymer–salt	PEG–K_2HPO_4	Benavides and Rito-Palomares (2006)
	PEG 4000-sulfate + 8.8% NaCl	Schmidt et al. (1994)
Alcohol–salt	2-Propanol–K_2HPO_4	Ooi et al. (2009)
	Ethanol–K_2HPO_4	Jiang et al. (2009)
Micellar/reverse micellar ATPS	*n*-Decyl tetra (ethylene oxide)	Liu et al. (1998)
	Isooctane/ethylhexanol/methyltrioctyl ammoniumchloride	Streitner et al. (2007)
Ionic liquids (ILs)-based ATPS	1-Butyl-3-methylimidazolium chloride–salt	Li et al. (2005)
	Imidazolium–K_2HPO_4	Shu et al. (2016)

Taking the ionic strength as the sensitive point for solute separation, the presence of high ionic strength becomes the only disadvantage of polymer–salts system (Albertsson 1961). However, the study by Schmidt et al. (1994) has shown an interesting result, which is being briefly described below. The enzyme α-amylase is widely used, particularly in the food industry. Bacterial α-amylase is added to barley in breweries to improve the liquefaction of the starch. The α-amylase of *B. subtilis* has several advantages over other α-amylases, as it is the one that most resembles the malt α-amylase during mashing conditions (Wieg 1987). The partition of pure α-amylase in ATPS composed of PEG/phosphate, PEG/sulfate, and PEG/dextran was studied along with the effect of PEG molecular weight (MW), PEG concentration, pH, enzyme concentration, and the addition of NaC1 on α-amylase partition. The addition of NaCl had the greatest effect on the partition coefficient (K) of pure α-amylase in the studied ATPS at the order shown below:

PEG/sulfate (1,000-fold increase), PEG/phosphate (150-fold increase), and PEG/dextran (8-fold increase) systems. However, the highest value of K was obtained in the PEG/phosphate system ($K > 300$). The value of K for α-amylase strongly depends on the MW of PEG in PEG/phosphate systems. The lower the PEG MW, the higher is the K value achieved for α-amylase.

Alcohol–salt ATPS are inexpensive as compared to polymers and copolymers. These systems are also characterized by low viscosity, easy constituent recovery, and reduced settling times, but a major drawback of using this type of ATPS is that many proteins are not compatible with alcohol-rich phase (Louwrier 1998; Jiang et al. 2009). ATPSs composed of an organic solvent and a salt solution has been used to isolate proteins, amino acids, and other natural products (Louwrier 1999; Tianwei et al. 2002; Zhi and Deng 2006). The advantages of this system include its low viscosity, high polarity, and the ease with which the alcohol can be recovered by evaporation. Furthermore, the system is inexpensive and of low toxicity to the environment (Tianwei et al. 2002; Zhi and Deng 2006). Ethanol can form a stable and adjustable aqueous two-phase system with inorganic salt solutions, which might be due to salting-out and the low solubility of inorganic salt in ethanol. When an aqueous two-phase system is formed, the top phase is rich in ethanol and the bottom phase is rich in inorganic salt. Ethanol/phosphate aqueous two-phase system is effective for the purification of biomolecules, with advantages of low cost and convenient operation (Jiang et al. 2009).

The aqueous micellar two-phase system was first introduced by Bordier for the separation of integral membrane proteins (Bordier 1981). These systems are also useful for ionic environment-sensitive solutes as nonionic surfactants can be used for the formation of these systems. Mixed micellar systems are becoming popular for showing selectivity features (Lye et al. 1995). Berthod et al. (2008) developed ionic liquids (ILs)-based ATPSs. Poly-phase systems (three or four polymer phases) also have been constructed for the separation of biomolecules (Hatti-Kaul 2001). One-polymer ATPSs have also been reported, which utilize only one polymer for the formation of ATPS in water (Johansson et al. 1999).

7.4.2.5.2. Stability Issues with ATPS The interfacial tension at the water–water interface is very low, which is often below the range of 10^{-2} mN m^{-1} (Ding et al. 2002; Antonov et al. 2004; Scholten et al. 2004; Buzza et al. 2013; Vis et al. 2015b). This value becomes even low near the critical point (Buzza et al. 2013). The value of interfacial tension at the water–water interface is as low as 10^{-5} mN m^{-1} as measured by the rheo-optical technique (Antonov et al. 2004). Moreover, the electric charge at the w/w interface, in case of a mixture of polyelectrolytes and neutral polymers, phase separation reduces even more than already extremely low interfacial tension, which can be explained by the Poisson–Boltzmann theory (Vis et al. 2015b).

The major drawback of the w/w emulsion is their usual lack of stability, especially in the composition near the critical point. Fast coalescence or flocculation tends to occur, thus leading to rapid and irreversible phase separation. The w/w interface has a length scale larger than the correlation length of the polymer solution (Nguyen et al. 2015). Small hydrophilic molecules, therefore, do not encounter an interface when they move from one polymer phase to the other. This causes the small molecule to not adsorb on w/w interfaces. The stability of w/w emulsion can be improved by using either latex particles or globular proteins (Nguyen et al. 2013). These particles adsorb onto the water–water interface despite low interfacial tension. The adsorption capacity of globular proteins (β-lactoglobulins) is more as compared to that of the latex particles. Apart from the abovementioned stabilizer, other reported stabilizers are bi-block and tri-block copolymers, nanoplates (gibbsite nanoplates) (Vis et al. 2015a), hydrophilic nanorods (cellulose nanocrystal), and liposomes; moreover, o/w emulsions can also improve the stability by adsorbing on the w/w interface, soft microgel particle of 150 nm in size (Esquena 2016). The w/w emulsion stabilized by soft microgel particles is sensitive to pH and temperature, thus the emulsion can be designed to be stimuli-responsive (Nguyen et al. 2015).

7.4.3. CONCLUSION

The emulsion-like dispersions are at the rudimentary stage of development, especially when a therapeutic application is concerned. The oil-less emulsions try to find place in the topical skin application. The emulsifier/surfactant-free emulsion-like dispersions are already in use for cosmetic purposes. On the other hand, the emulsifier/surfactant- and oil-free emulsion-like dispersions look to have a brighter future in medical and pharmaceutical applications. Because of the elimination of both emulsifier/surfactants and oil from the traditional o/w emulsions, the so-called ATPS emulsions are very good in terms of reduction in the patient-related complications. Although the ATPS has already proven in the separation of high-value products from crude raw mixtures, its potential to use as a delivery vehicle for API is not yet explored. Overall, more and more focus should be made on emulsion-like dispersions to extract their therapeutical utility in many possible syndrome management.

REFERENCES

Abend, S., Bonnke, N., Gutschner, U. et al. (1998), Stabilization of emulsions by heterocoagulation of clay minerals and layered double hydroxides, *Colloid Polym. Sci.*, 276 (8), 730–737. doi:10.1007/s003960050303

Albertsson, P.A. (1961), Fractionation of particles and macromolecules in aqueous two-phase systems, *Biochem. Pharmacol.*, 5, 351–358. doi:10.1016/0006-2952(61)90028-4

Antonov, Y.A., Van Puyvelde, P., and Moldenaers, P. (2004), Interfacial tension of aqueous biopolymer mixtures close to the critical point, *Int. J. Biol. Macromol.*, 34 (1–2), 29–35. doi:10.1016/j.ijbiomac.2004.01.001

Asenjo, J.A. and Andrews, B.A. (2012), Aqueous two-phase systems for protein separation: phase separation and applications, *J. Chromatogr. A*, 1238, 1–10. doi:10.1016/j.chroma.2012.03.049

Atefi, E., Fyffe, D., Kaylan, K.B. et al. (2016), Characterization of aqueous two-phase systems from volume and density measurements, *J. Chem. Eng. Data*, 61 (4), 1531–1539. doi:10.1021/acs.jced.5b00901

Beijerinck, M.W. (1896), About emulsion formation when mixing aqueous solutions of certain gelatinizing colloids, *J. Chem. Ind. Colloids*, 16–20, 1910. doi:10.1007/BF01503954

Benavides, J. and Rito-Palomares, M. (2006), Simplified two-stage method to B-phycoerythrin recovery from porphyridium cruentum, *J. Chromatogr. B Anal. Technol. Biomed. Life Sci.*, 844 (1), 39–44. doi:10.1016/j.jchromb.2006.06.029

Berthod, A., Ruiz-Angel, M.J., and Carda-Broch, S. (2008), Ionic liquids in separation techniques, *J. Chromatogr. A*, 1184 (1–2), 6–18. doi:10.1016/j.chroma.2007.11.109

Bordier, C. (1981), Phase separation of integral membrane proteins in Triton X-114 solution, *J. Biol. Chem.*, 256 (4), 1604–1607.

Bridges, N.J., Gutowski, K.E., and Rogers, R.D. (2007), Investigation of aqueous biphasic systems formed from solutions of chaotropic salts with kosmotropic salts (salt-salt ABS), *Green Chem.*, 9 (2), 177–183. doi:10.1039/b611628k

Buzza, D.M.A., Fletcher, P.D., Georgiou, T.K. et al. (2013), Water-in-water emulsions based on incompatible polymers and stabilized by triblock copolymers-templated polymersomes, *Langmuir*, 29 (48), 14804–14814. doi:10.1021/la403356j

Capron, I., Costeux, S., and Djabourov, M. (2001), Water in water emulsions: phase separation and rheology of biopolymer solutions, *Rheol. Acta*, 40 (5), 441–456. doi:10.1007/s003970100161

Chen, B., Han, J., Wang, Y. et al. (2014), Separation, enrichment and determination of ciprofloxacin using thermoseparating polymer aqueous two-phase system combined with high performance liquid chromatography in milk, egg, and shrimp samples, *Food Chem.*, 148, 105–111. doi:10.1016/j.foodchem.2013.10.011

Chen, J.-P. and Lee, M.-S. (1995), Enhanced production of *Serratia marcescens* chitinase in PEG/dextran aqueous two-phase systems, *Enzym. Microb. Technol.*, 17 (11), 1021–1027. doi:10.1016/0141-0229(95)00030-5

Dewey, D.C., Strulson, C.A., Cacace, D.N. et al. (2014), Bioreactor droplets from liposome-stabilized all-aqueous emulsions, *Nat. Commun.*, 5 (4670), 1–9. doi:10.1038/ncomms5670

Ding, P., Wolf, B., Frith, W. J. et al. (2002), Interfacial tension in phase-separated gelatin/dextran aqueous mixtures, *J. Colloid Interface Sci.*, 253 (2), 367–376. doi:10.1006/jcis.2002.8572

Dong, B., Yuan, X., Zhao, Q. et al. (2015), Ultrasound-assisted aqueous two-phase extraction of phenylethanoid glycosides from *Cistanche deserticola* Y.C. Ma stems, *J. Sep. Sci.*, 38 (7), 1194–1203. doi:10.1002/jssc.201401410

Eccleston, G.M. (2007), Emulsions and microemulsions, *Encycl. Pharm. Technol.*, 2, 1066–1085.

Esquena, J. (2016), Water-in-water (w/w) emulsions, *Curr. Opin. Colloid Interface Sci.*, 25, 109–119. doi:10.1016/j.cocis.2016.09.010

Fernández-Barbero, A., Suárez, I.J., Sierra-Martín, B. et al. (2009), Gels and microgels for nanotechnological applications, *Adv. Colloid Interface Sci.*, 147–148, 88–108. doi:10.1016/j.cis.2008.12.004

Frömming, K.H. and Szejtli, J. (1993), Cyclodextrin inclusion complexes, in: *Cyclodextrins in Pharmacy*, Vol. 5, Springer, Dordrecht, pp. 45–82. doi:10.1007/978-94-015-8277-3_4

Graf, M., Ziegler, C. E., Gregoritza, M. et al. (2019), Hydrogel microspheres evading alveolar macrophages for sustained pulmonary protein delivery, *Int. J. Pharm.*, 566, 652–661. doi:10.1016/j.ijpharm.2019.06.019

Grilo, A.L., Aires-Barros, M.R., and Azevedo, A.M. (2016), Partitioning in aqueous two-phase systems: fundamentals, applications and trends, *Sep. Purif. Rev.*, 45 (1), 68–80. doi:10.1080/15422119.2014.983128

Grinberg, V.Y. and Tolstoguzov, V.B. (1997), Thermodynamic incompatibility of proteins and polysaccharides in solutions, *Food Hydrocoll.*, 11 (2), 145–158. doi:10.1016/S0268-005X(97)80022-7

Halmos, E.P., Mack, A., and Gibson, P.R. (2019), Emulsifiers in the food supply and implications for gastrointestinal disease, *Aliment. Pharmacol. Ther.*, 49 (1), 41–50. doi:10.1111/apt.15045

Hatti-Kaul, R. (2000), *Aqueous Two-Phase Systems: Methods and Protocols*, Springer Science & Business Media, Berlin.

Hatti-Kaul, R. (2001), Aqueous two-phase systems, *Mol. Biotechnol.*, 19 (3), 269–277. doi:10.1385/MB:19:3:269

Helfrich, M.R., El-Kouedi, M., Etherton, M.R. et al. (2005), Partitioning and assembly of metal particles and their bioconjugates in aqueous two-phase systems, *Langmuir*, 21 (18), 8478–8486. doi:10.1021/la051220z

Inoue, M., Hashizaki, K., Taguchi, H. et al. (2008), Emulsion preparation using β-cyclodextrin and its derivatives acting as an emulsifier, *Chem. Pharm. Bull.*, 56 (9), 1335–1337. doi:10.1248/cpb.56.1335

Inoue, M., Hashizaki, K., Taguchi, H. et al. (2009), Preparation and characterization of n-alkane/water emulsion stabilized by cyclodextrin, *J. Oleo Sci.*, 58 (2), 85–90. doi:10.5650/jos.58.85

Inoue, M., Hashizaki, K., Taguchi, H. et al. (2010), Emulsifying ability of β-cyclodextrins for common oils, *J. Dispers. Sci. Technol.*, 31 (12), 1648–1651. doi:10.1080/01932690903297058

Iqbal, M., Tao, Y., Xie, S. et al. (2016), Aqueous two-phase system (ATPS): an overview and advances in its applications, *Biol. Proced. Online*, 18, 1–18. doi:10.1186/s12575-016-0048-8

Jiang, B., Li, Z.-G., Dai, J.-Y. et al. (2009), Aqueous two-phase extraction of 2, 3-butanediol from fermentation broths using an ethanol/phosphate system, *Process Biochem.*, 44 (1), 112–117. doi:10.1016/j.procbio.2008.09.019

Johansson, H.-O., Persson, J., and Tjerneld, F. (1999), Thermoseparating water/polymer system: a novel one-polymer aqueous two-phase system for protein purification, *Biotechnol. Bioeng.*, 66 (4), 247–257. doi:10.1002/(sici)1097-0290(1999)66:4%3C247::aid-bit6%3E3.0.co;2-5

Kasapis, S., Morris, E.R., Norton, I.T. et al. (1993), Phase equilibria and gelation in gelatin/maltodextrin systems-part II: polymer incompatibility in solution, *Carbohydr. Polym.*, 21 (4), 249–259. doi:10.1016/0144-8617(93)90056-A

Kruse, T., Schmidt, A., Kampmann, M. et al. (2019), Integrated clarification and purification of monoclonal antibodies by membrane based separation of aqueous two-phase systems, *Antibodies*, 8 (3), 1–17. doi:10.3390/antib8030040

Lazerges, M., Rietveld, I.B., Corvis, Y. et al. (2010), Thermodynamic studies of mixtures for topical anesthesia: lidocaine–salol binary phase diagram, *Thermochim. Acta*, 497 (1–2), 124–128. doi:10.1016/j.tca.2009.08.016

Li, S., He, C., Liu, H. et al. (2005), Ionic liquid-based aqueous two-phase system, a sample pretreatment procedure prior to high-performance liquid chromatography of opium alkaloids, *J. Chromatogr. B Anal. Technol. Biomed. Life Sci.*, 826 (1–2), 58–62. doi:10.1016/j.jchromb.2005.08.005

Liu, C., Kamei, D.T., King, J.A. et al. (1998), Separation of proteins and viruses using two-phase aqueous micellar systems, *J. Chromatogr. B Anal. Technol. Biomed. Life Sci.*, 711 (1–2), 127–138. doi:10.1016/s0378-4347(98)00013-9

Liu, Y., Nambu, N.O., and Taya, M. (2017), Cell-laden microgel prepared using a biocompatible aqueous two-phase strategy, *Biomed. Microdevices*, 19 (3), 1–7. doi:10.1007/s10544-017-0198-8

Louwrier, A. (1998), Model phase separations of proteins using aqueous/ethanol components, *Biotechnol. Tech.*, 12 (5), 363–365. doi:10.1023/A:1008818229903

Louwrier, A. (1999), Model isolations of nucleic acids from prokaryotic and eukaryotic sources using an organic/aqueous biphasic system, *Biotechnol. Tech.*, 13 (5), 329–330. doi:10.1023/A:1008911104349

Lu, G.W. and Gao, P. (2010), Emulsions and microemulsions for topical and transdermal drug delivery, in: Kulkarni, V.S., Ed., *Handbook of Non-invasive Drug Delivery Systems*, William Andrew Publishing, Elsevier, UK, pp. 59–94. doi:10.1016/B978-0-8155-2025-2.10003-4

Lu, Y., Yao, H., Li, C. et al. (2016), Separation, concentration and determination of trace chloramphenicol in shrimp from different waters by using polyoxyethylene lauryl ether-salt aqueous two-phase system coupled with high-performance liquid chromatography, *Food Chem.*, 192, 163–170. doi:10.1016/j.foodchem.2015.06.086

Lye, G.J., Asenjo, J.A., and Pyle, D.L. (1995), Extraction of lysozyme and ribonuclease-a using reverse micelles: limits to protein solubilization, *Biotechnol. Bioeng.*, 47 (5), 509–519. doi:10.1002/bit.260470502

Matalanis, A., Jones, O.G., and McClements, D.J. (2011), Structured biopolymer-based delivery systems for encapsulation, protection, and release of lipophilic compounds, *Food Hydrocoll.*, 25 (8), 1865–1880. doi:10.1016/j.foodhyd.2011.04.014

Matalanis, A., Lesmes, U., Decker, E.A. et al. (2010), Fabrication and characterization of filled hydrogel particles based on sequential segregative and aggregative biopolymer phase separation, *Food Hydrocoll.*, 24 (8), 689–701. doi:10.1016/j.foodhyd.2010.04.009

McClements, D.J., Bai, L., and Chung, C. (2017), Recent advances in the utilization of natural emulsifiers to form and stabilize emulsions, *Annu. Rev. Food Sci. Technol.*, 8, 205–236. doi:10.1146/annurev-food-030216-030154

Molino, J.V.D., Marques, V., de Araújo, D. et al. (2013), Different types of aqueous two-phase systems for biomolecule and bioparticle extraction and purification, *Biotechnol. Prog.*, 29 (6), 1343–1353. doi:10.1002/btpr.1792

Moreira-Oliveira, S., Amaral-Machado, L., de Oliveira, W. et al. (2018), Buccal bullfrog (Rana Catesbeiana Shaw) oil emulsion: a mucoadhesive system intended for treatment of oral candidiasis, *Pharmaceutics*, 10 (4), E257. doi:10.3390/pharmaceutics10040257

Nazer, B., Dehghani, M. R., and Goliaei, B. (2017), Plasmid DNA affinity partitioning using polyethylene glycol–sodium sulfate aqueous two-phase systems, *J. Chromatogr. B Anal. Technol. Biomed. Life Sci.*, 1044–1045, 112–119. doi:10.1016/j.jchromb.2017.01.002

Nazzal, S., Smalyukh, I.I., Lavrentovich, O.D. et al. (2002), Preparation and in vitro characterization of a eutectic based semisolid self-nanoemulsified drug delivery system (SNEDDS) of ubiquinone: mechanism and progress of emulsion formation, *Int. J. Pharm.*, 235 (1–2), 247–265. doi:10.1016/s0378-5173(02)00003-0

Nguyen, B.T., Nicolai, T., and Benyahia, L. (2013), Stabilization of water-in-water emulsions by addition of protein particles, *Langmuir*, 29 (34), 10658–10664. doi:10.1021/la402131e

Nguyen, B.T., Wang, W., Saunders, B.R. et al. (2015), pH-responsive water-in-water pickering emulsions, *Langmuir*, 31 (12), 3605–3611. doi:10.1021/la5049024

Norton, I.T., and Frith, W.J. (2001), Microstructure design in mixed biopolymer composites, *Food Hydrocoll.*, 15 (4–6), 543–553. doi:10.1016/S0268-005X(01)00062-5

Oh, J.K., Drumright, R., Siegwart, D.J. et al. (2008), The development of microgels/nanogels for drug delivery applications, *Prog. Polym. Sci.*, 33 (4), 448–477. doi:10.1016/j.progpolymsci.2008.01.002

Onghena, B., Opsomer, T., and Binnemans, K. (2015), Separation of cobalt and nickel using a thermomorphic ionic-liquid-based aqueous biphasic system, *Chem. Commun.*, 51 (88), 15932–15935. doi:10.1039/C5CC06595J

Ooi, C.W., Tey, B.T., Hii, S.L. et al. (2009), Purification of lipase derived from *Burkholderia pseudomallei* with alcohol/salt-based aqueous two-phase systems, *Process Biochem.*, 44 (10), 1083–1087. doi:10.1016/j.procbio.2009.05.008

Pérez, R.L., Loureiro, D.B., Nerli, B.B. et al. (2015), Optimization of pancreatic trypsin extraction in PEG/citrate aqueous two-phase systems, *Protein Expr. Purif.*, 106, 66–71. doi:10.1016/j.pep.2014.10.014

Pickering, S.U. (1907), CXCVI.-emulsions, *J. Chem. Soc. Trans.*, 91 (1907), 2001–2021. doi:10.1039/CT9079102001

Piculell, L. and Lindman, B. (1992), Association and segregation in aqueous polymer/polymer, polymer/surfactant, and surfactant/surfactant mixtures: similarities and differences, *Adv. Colloid Interf. Sci.*, 41, 149–178. doi:10.1016/0001-8686(92)80011-L

Raja, S., Murty, V. R., Thivaharan, V. et al. (2011), Aqueous two phase systems for the recovery of biomolecules—a review, *Sci. Technol.*, 1 (1), 7–16. doi:10.5923/j.scit.20110101.02

Rangel, M.D.L., Aquino, S.G.D., Lima, J.M.D. et al. (2018), In vitro effect of *Cinnamomum zeylanicum* Blume essential oil on *Candida* spp. involved in oral infections, *Evid. Based Complement. Alternat. Med.*, 2018, 1–13. doi:10.1155/2018/4045013

Rohilla, S., Bhatt, D.C., and Gupta, A. (2018), Therapeutic potential of phytomedicines and novel polymeric strategies for significant management of candidiasis, *Curr. Pharm. Des.*, 24 (16), 1748–1765. doi:10.2174/1381612824666180524102933

Rowe, R.C., Sheskey, P.J., and Quinn, M.E. (2009), *Handbook of Pharmaceutical Excipients*, 6th ed., Pharmaceutical Press, London, pp. 517–522.

Ruiz-Ruiz, F., Benavides, J., Aguilar, O. et al. (2012), Aqueous two-phase affinity partitioning systems: current applications and trends, *J. Chromatogr. A*, 1244, 1–13. doi:10.1016/j.chroma.2012.04.077

Sagis, L.M. (2007), Fluctuations of permeable interfaces in water-in-water emulsions, *Phys. Rev. Lett.*, 98 (6), 066105. doi:10.1103/PhysRevLett.98.066105

Sagis, L.M. (2008), Dynamics of controlled release systems based on water-in-water emulsions: a general theory, *J. Control. Release*, 131 (1), 5–13. doi:10.1016/j.jconrel.2008.07.010

Sagis, L. M. (2009), Dynamics of encapsulation and controlled release systems based on water-in-water emulsions: liposomes and polymersomes, *Phys A Stat. Mech. Appl.*, 388 (13), 2579–2587. doi:10.1016/j.physa.2009.03.024

Schmidt, A., Ventom, A., and Asenjo, J. (1994), Partitioning and purification of α-amylase in aqueous two-phase systems, *Enzyme Microb. Technol.*, 16 (2), 131–142. doi:10.1016/0141-0229(94)90076-0

Schmitt, C. and Turgeon, S.L. (2011), Protein/polysaccharide complexes and coacervates in food systems, *Adv. Colloid Interf. Sci.*, 167 (1–2), 63–70. doi:10.1016/j.cis.2010.10.001

Scholten, E., Sagis, L.M., and van der Linden, E. (2006), Effect of bending rigidity and interfacial permeability on the dynamical behavior of water-in-water emulsions, *J. Phys. Chem. B*, 110 (7), 3250–3256. doi:10.1021/jp056528d

Scholten, E., Visser, J.E., Sagis, L.M. et al. (2004), Ultralow interfacial tensions in an aqueous phase-separated gelatin/dextran and gelatin/gum arabic system: a comparison, *Langmuir*, 20 (6), 2292–2297. doi:10.1021/la0351919

Shao, M., Zhang, X., Li, N. et al. (2014), Ionic liquid-based aqueous two-phase system extraction of sulfonamides in milk, *J. Chromatogr. B*, 961, 5–12. doi:10.1016/j.jchromb.2014.04.054

Shin, H., Han, C., Labuz, J.M. et al. (2015), High-yield isolation of extracellular vesicles using aqueous two-phase system, *Sci. Rep.*, 5 (13103), 1–11. doi:10.1038/srep13103.

Shu, Y., Gao, M., Wang, X. et al. (2016), Separation of curcuminoids using ionic liquid based aqueous two-phase system coupled with in situ dispersive liquid-liquid microextraction, *Talanta*, 149, 6–12. doi:10.1016/j.talanta.2015.11.009

Sierpe, R., Lang, E., Jara, P. et al. (2015), Gold nanoparticles interacting with β-cyclodextrin–phenylethylamine inclusion complex: a ternary system for photothermal drug release, *ACS Appl. Mater. Interfaces*, 7 (28), 15177–15188. doi:10.1021/acsami.5b00186

Silva, D.F.C., Azevedo, A.M., Fernandes, P. et al. (2017), Determination of partition coefficients of biomolecules in a microfluidic aqueous two-phase system platform using fluorescence microscopy, *J. Chromatogr. A*, 1487, 242–247. doi:10.1016/j.chroma.2016.12.036

Stott, P.W., Williams, A.C., and Barry, B.W. (1998), Transdermal delivery from eutectic systems: enhanced permeation of a model drug, ibuprofen, *J. Control. Release*, 50 (1–3), 297–308. doi:10.1016/S0168-3659(97)00153-3

Stott, P.W., Williams, A.C., and Barry, B.W. (2001), Mechanistic study into the enhanced transdermal permeation of a model β-blocker, propranolol, by fatty acids: a melting point depression effect, *Int. J. Pharm.*, 219 (1–2), 161–176. doi:10.1016/s0378-5173(01)00645-7

Streitner, N., Voß, C., and Flaschel, E. (2007), Reverse micellar extraction systems for the purification of pharmaceutical grade plasmid DNA, *J. Biotechnol.*, 131 (2), 188–196. doi:10.1016/j.jbiotec.2007.06.011

Sundarrajan, S., Parambath, S., Suresh, S. et al. (2018), Novel properties of recombinant Sso7d-Taq DNA polymerase purified using aqueous two-phase extraction: utilities of the enzyme in viral diagnosis, *Biotechnol. Rep.*, 19, 1–36. doi:10.1016/j.btre.2018.e00270

Swiech, O., Mieczkowska, A., Chmurski, K. et al. (2012), Intermolecular interactions between doxorubicin and β-cyclodextrin 4-methoxyphenol conjugates, *J. Phys. Chem. B*, 116 (6), 1765–1771. doi:10.1021/jp2091363

Tamilvanan, S., Gill, S., Kaur, I., Rahman, S.N.R. et al. (2019), Candidiasis management: current status of allopathic dugs and utility of coriander-based oil-less emulsions, *Lett. Appl. NanoBioSci.*, 8 (3), 586–590. doi:10.33263/LIANBS83.586590

Tianwei, T., Qing, H., and Qiang, L. (2002), Purification of glycyrrhizin from *Glycyrrhiza uralensis* Fisch with ethanol/phosphate aqueous two phase system, *Biotechnol. Lett.*, 24, 1417–1420. doi:10.1023/A:1019850531640

Van Berlo, M., Luyben, K.C.A., and van der Wielen, L.A. (1998), Poly(ethylene glycol)-salt aqueous two-phase systems with easily recyclable volatile salts, *J. Chromatogr. B Anal. Technol. Biomed. Life Sci.*, 711 (1–2), 61–68. doi:10.1016/S0378-4347(97)00627-0

Vis, M., Opdam, J., and Van't Oor, I.S. et al. (2015a), Water-in-water emulsions stabilized by nanoplates, *ACS Macro Lett.*, 4 (9), 965–968. doi:10.1021/acsmacrolett.5b00480

Vis, M., Peters, V. F., Blokhuis, E. M. et al. (2015b), Decreased interfacial tension of demixed aqueous polymer solutions due to charge, *Phys. Rev. Lett.*, 115 (7), 0783031–0783035. doi:10.1103/PhysRevLett.115.078303

Wieg, A.J. (1987), Barley brewing, in: Pollock, J.R.A., Ed., *Brewing Science*, Vol 3, Academic Press Ltd, London, pp. 538–550.

Winkler, A.M. (2019), Albumin and related products, in: Shaz, B.H, Hillyer, C.D. and Gil, M.R., Eds., *Transfusion Medicine and Hemostasis*, Elsevier, UK, pp. 229–233. doi:10.1016/b978-0-12-813726-0.00038-6

Xiao, J.X., Sivars, U., and Tjerneld, F. (2000). Phase behavior and protein partitioning in aqueous two-phase systems of cationic-anionic surfactant mixtures, *J. Chromatogr. B Anal. Technol. Biomed. Life Sci.*, 743 (1–2), 327–338.

Yang, X., Zhang, S., Yu, W. et al. (2014), Ionic liquid-anionic surfactant based aqueous two-phase extraction for determination of antibiotics in honey by high-performance liquid chromatography, *Talanta*, 124, 1–6. doi:10.1016/j.talanta.2014.02.039.

Yeredla, N., Kojima, T., Yang, Y. et al. (2016), Aqueous two phase system assisted self-assembled PLGA microparticles, *Sci. Rep.*, 6 (27736), 1–8. doi:10.1038/srep27736

Yong, C.S., Oh, Y.K., Jung, S.H. et al. (2004), Preparation of ibuprofen-loaded liquid suppository using eutectic mixture system with menthol, *Eur. J. Pharm. Sci.*, 23 (4–5), 347–353. doi:10.1016/j.ejps.2004.08.008

Zhi, W. and Deng, Q. (2006), Purification of salvianolic acid B from the crude extract of *Salvia miltiorrhiza* with hydrophilic organic/salt-containing aqueous two-phase system by counter-current chromatography, *J. Chromatogr. A*, 1116 (1–2), 149–152. doi:10.1016/j.chroma.2006.03.036

PART II

SELECTED CASE STUDIES

CHAPTER 8

SELECTED CASE STUDIES

Oil-in-Water Nanosized Emulsions for Drug Delivery and Targeting,
First Edition. Tamilvanan Shunmugaperumal.

© 2021 John Wiley & Sons, Inc. Published 2021 by John Wiley & Sons, Inc.

CHAPTER 8.1

CASE STUDY 1 CATIONIC NANOSIZED EMULSIONS: NARRATION OF COMMERCIAL SUCCESS

OLY KATARI* AND TAMILVANAN SHUNMUGAPERUMAL

Department of Pharmaceutics, National Institute of Pharmaceutical Education and Research (NIPER), Guwahati, Assam, India

*The funding provided by Department of Pharmaceuticals, Ministry of Chemicals and Fertilizers, Govt. of India to the author Oly Katari is greatly acknowledged.

Oil-in-Water Nanosized Emulsions for Drug Delivery and Targeting,
First Edition. Tamilvanan Shunmugaperumal.
© 2021 John Wiley & Sons, Inc. Published 2021 by John Wiley & Sons, Inc.

EXPANSION OF ABBREVIATIONS

API	active pharmaceutical ingredients
BAC	benzalkonium chloride
CFS	corneal fluorescein staining
CKC	cetalkonium chloride
CsA	cyclosporin A
DES	dry eye syndrome
DOTAP	(*N*-(1-(2,3-dioleoyloxy)propyl)-*N*,*N*,*N*-trimethylammonium) chloride
DOPE	dioleoyl phosphatidylethanolamine
PEI	poly(ethylenimine)
PLL	poly-L-lysine
VAS	visual analogue scale

8.1.1. INTRODUCTION

Ophthalmic diseases are most commonly treated by topical eye-drop instillation of aqueous products. These formulations, however, raise technical problems like preservation, solubility, and stability as well as clinical issues such as efficacy, local toxicity, and compliance. The conventional aqueous solutions are limited to water-soluble active pharmaceutical ingredients (APIs) and over 80% of the product is expected to be eliminated within 2 min after instillation via the nasolacrimal drainage system limiting ocular penetration of the API to less than 1% of the administered dose (Shell 1984).

As a consequence, the pharmaceutical companies have faced the obstacles of developing a formulation for ocular topical administration, which would expand the range of potential APIs, remain/reside longer time on the ocular surface, and provide sustained therapeutic concentrations in addition to meeting the regulatory criteria for approval. The obstacles observed (or to be encountered) and possible considerations for delivering the APIs via topical instillation into eye are listed in Table 8.1.1.

The aim of this section is to highlight the developmental aspects of the cationic nanosized emulsion technology from simple lab work, scale-up, commercialization with regulatory approvals and finally to reach patients. The main obstacle of developing ophthalmic cationic emulsion at scale-up and regulatory clearance levels is the selection of suitable cation-conferring agent. Therefore, this section starts with the selection process of cation-conferring agent.

8.1.2. SELECTION OF SUITABLE CATION-CONFERRING AGENT

The surface charge of the nanosized emulsion is defined by the zeta potential. The zeta potential resembles the electric potential surrounding the dispersed oil droplets at the plane of hydrodynamic shear, which is measured by

TABLE 8.1.1. Non-comprehensive List of Obstacles Observed (or to Be Encountered) and Possible Considerations for Delivering the Active Pharmaceutical Ingredients (APIs) via Topical Instillation into Eye

Obstacles	Considerations
Absorption: only 3–4% ocular bioavailability of API after topical administration using traditional eye drops	*Anatomy and physiology of the eye*: mucus layer, eyelids, metabolism, blink wash-out, etc.
Poorly soluble APIs: conventional aqueous eye drops not suitable for lipophilic APIs (40–60% of new chemical entities)	*Tear composition*: lipid outer layer, stability of the tear film, enzymes, etc.
Patient compliance: multiple instillations are often needed with eye drops to reach desirable therapeutic levels in ocular tissues of interest	*Disease state*: impact of keratitis or inflammation on absorption and clearance, etc.
High comfort/tolerability requirements: limit to use different formulation options	*Ocular comfort*: drop size, osmolality, pH, tolerability of the formulation, viscosity, etc.
Excipient choice: few FDA-approved excipients (oils, surfactants, polymers, etc.) are listed to use in ophthalmic purpose	*Patient expectations*: type of packaging and squeezing ability impacting compliance, etc.
Posterior segment API delivery: no topical system for the posterior segment; invasive treatments are used due to lack of alternative approaches with commercial success	*API loading into the formulation*: impact on absorption, compliance, dosing regimen, efficacy, etc.

electrophoretic mobility. The latter depends on the nature of the cation-conferring agent used in the emulsion system, its concentration, and the electrolyte environment of the dispersed oil droplets. In Chapter 3, Section 3.2.6.3 devotes the measurement of zeta potential value using the instrumental techniques. The positive charge of the cation-conferring agent enhances the stability of the emulsion by creating an electrostatic repulsion between the dispersed oil droplets, side-by-side increasing the residence time on the negatively charged ocular surface (Benita 1999). Calvo et al. (1997) showed that two different types of cationic indomethacin-loaded nanocapsules [coated with poly-L-lysine (PLL) or chitosan] resulted in completely different API kinetics profiles, which provided evidence that the specific nature of the cation-conferring molecule may be responsible for improved uptake properties. Therefore, the cation-conferring agent selected needs to be carefully considered prior to starting pharmaceutical development as the success of the formulation is highly dependent upon the choice of the cationic agent as detailed below.

First of all, the o/w nanosized emulsions having a zeta potential value equal or below +10 mV could not be autoclaved without destabilizing the oil droplets. As a result, the first challenge of the cationic emulsion technology was to prepare a cationic emulsion with a zeta potential sufficiently high to stabilize the dispersed oil droplets of nanosized emulsion. Yet, it is important that the cationic surfactant concentration used is as low as possible to avoid compromising the safety of the nanosized emulsion. The optimal range for the zeta potential was found to be between +20 and +40 mV. Reviewing of literatures revealed a number of cation-conferring agents (as shown in Table 8.1.2) most of which are surfactants, indeed the positively charged region of the molecule does not enter the oil core of the droplet but instead remains at the surface, rendering them very useful for stabilizing the emulsions. It is most unfortunate that only very few cation-conferring agents are either listed in official pharmacopeias or accepted for ophthalmic products due to stability or toxicity issues.

When compared to anionic and nonionic surfactants, cationic surfactants are known to be the most toxic surfactants (Van Abbe 1973). Therefore, in order to develop the cationic emulsion technology, it was necessary to find an appropriate cationic surfactant, which would provide a sufficiently high cationic charge, have a low toxicity, and conform to regulatory standards.

Stearylamine is one of the most widely used cationic lipids in the academic world especially for the manufacture of cationic liposomes (Manosroi et al. 2002) or cationic emulsions (Abdulrazik et al. 2001). However, since this primary amine is very reactive toward other excipients including the APIs and even the monograph of stearylamine is not described in any pharmacopeias, it was not a reasonable choice for pharmaceutical development of o/w nanosized emulsions at the commercial level.

Oleylamine is another cationic lipid that has been used to manufacture ophthalmic emulsions (Rabinovich-Guilatt et al. 2004). But this lipid also has stability concerns due to its primary amine function and the presence of an unsaturated site in the aliphatic chain. Thus, oleylamine also possesses limitation to be used in the preparation of ophthalmic emulsions.

Other two cationic molecules such as poly(ethylenimine) (PEI) and poly-L-lysine (PLL) are frequently employed for DNA transfection and these two molecules

TABLE 8.1.2. Common Molecules Used for Cation-Conferration onto the Dispersed Oil Droplets of O/W Nanosized Emulsions

Cation-Conferring Molecules
Benzalkonium chloride
Cetalkonium chloride
Chitosan
Dioleoyl phosphatidylethanolamine
(*N*-(1-(2,3-dioleoyloxy)propyl)-*N*,*N*,*N*-trimethylammonium) chloride
Oleylamine
Poly(ethylenimine)
Poly-L-lysine
Stearylamine

are also tried to make the various nanoparticulate systems. For example, PEI is an organic polymer that has a high density of amino groups, which can be protonated. At physiological pH, the polycation is very effective in binding DNA and can mediate the transfection of eukaryotic cells (Boussif et al. 1995). It has been used as a cationic agent in micelles (Kuo et al. 2010), nanoparticles (Wang et al. 2010), albumin nanoparticles (Zhang et al. 2008), liposomes (Masotti et al. 2007), and nanosized cationic hydrogels (Vinogradov et al. 2002). However, while some authors claim this polymer to be safe, some others such as Hunter (2006) have reported PEI to be extremely cytotoxic. PLL is a polymer made of several lysines (amino acid). Lysine possesses an NH_2 function that is ionized at a physiological pH conferring several cationic charges to that polymer. It is sometimes used as a cationic agent in API delivery systems such as microparticles (Yeh et al. 2007). However, toxicity has been reported (Moreau et al. 2002), and this polymer is not authorized for use in ophthalmic formulations.

Cationic lipids, (*N*-(1-(2,3-dioleoyloxy)propyl)-*N*,*N*,*N*-trimethylammonium) chloride (DOTAP) and dioleoyl phosphatidylethanolamine (DOPE), represent another potential class of cationic agents. These are amphiphilic molecules with a fatty acid chain and a polar group bearing a cationic charge. Their main advantage is that they are biodegradable and well tolerated. DOPE, which also harbors a negative charge, is a neutral "helper" lipid often included in cationic lipid formulations like cationic nanosized emulsions (Kim et al. 2001). Cationic solid lipid nanoparticles were successfully made with DOTAP to transport DNA vaccines (Doroud et al. 2010). Hagigit et al. (2008, 2010) showed that using DOTAP was better than the seminatural lipid oleylamine to make stable cationic emulsions. Moreover, DOTAP cationic emulsion enhanced the penetration of antisense oligonucleotides after either topical ocular instillations or intravitreous injection. But like most of the seminatural lipids, these agents are chemically unstable and need to be stored at −20°C, thus drastically limiting their industrial use.

The primary limiting factors against the use of the previously cited cationic agents in the cationic emulsion technology, even though they showed potential in the formulation of cationic API delivery systems, is that (1) they are not listed in US and EU pharmacopeias or (2) their toxicity on the ocular surface has not been well documented, and (3) none of these cationic agents has been successfully commercialized in a pharmaceutical product. Consequently, the development of cation emulsion technology limits its search for the appropriate cationic agent among those already registered, used in ophthalmic products, or compliant to official pharmacopeias.

Other excipients previously accepted by health authorities were then considered. Quaternary ammoniums usually used as preservatives have surfactant properties and the potential to give a cationic charge onto the surfaces of dispersed oil droplets of o/w nanosized emulsions. The examples of quaternary ammoniums include benzalkonium chloride, benzethonium chloride, benzododecinium bromide, cetrimide, and cetylpyridinium. In their role as preservatives in pharmaceutical products, the quaternary ammoniums protect against infectious contaminants by electrostatically binding to the negatively charged surface of bacteria and mycoplasma and disrupting their cell membranes. The disadvantage of quaternary ammoniums is that their effect on cell membranes is not limited only to microorganisms but they are also capable of injuring epithelial cells lining the ocular surface by the same mechanism of action. It was consequently not obvious to foresee these molecules as cationic agents, therefore, quaternary ammoniums were not initially considered for use in emulsions. In 2002, Sznitowska, however, revealed in a report that the preservative efficacy of this class of surfactants was diminished or neutralized in the presence of emulsions (Sznitowska et al. 2002). Part of the quaternary ammonium is bound to the emulsion, resulting in the presence of less free surfactant molecules in the aqueous phase to exert their antimicrobial action, and, consequently, their toxic effect on the ocular surface epithelia. The cationic nanosized emulsion technology exploited this physicochemical property to make a new type of cationic nanovector using benzalkonium chloride (BAC) and cetalkonium chloride (CKC) as cationic agents. CKC is a highly lipophilic ($\log P = 9.5$) component of BAC.

It is hence BAC/CKC mostly included in the oily phase providing a higher zeta potential on surface of the oil droplets while leaving relatively no free molecules to induce ocular surface toxicity. BAC (and CKC as a component of BAC) has been routinely used as a preservative in other marketed eye drop solutions (e.g., BAC is used in Xalatan) and is accepted as compliant with regulatory requirements for ophthalmic products. These excipients used in lower concentrations as cationic agents in emulsions have been demonstrated to be safe for the eye. More importantly, the use of BAC and CKC as cationic surfactants only in emulsions are now protected by several granted and pending European and US patents [e.g., EP1655021 (Bague et al. 2008), EP1809237 (Bague et al. 2007b), EP1809238 (Philips et al. 2008), and EP1827373 (Bague et al. 2007a), which are granted].

The cationic nanosized emulsion technology finally finds specifications for commercially successful products (Table 8.1.3).

TABLE 8.1.3. Specifications for Commercially Successful Cationic Nanosized Emulsions

Parameters	Inference
Appearance (physical and visual)	Milky white to translucid
Mean oil droplet size	150–300 nm
Osmolality	180–300 mOsm kg^{-1}
pH	5.5–7
Sterility	Sterile
Surface tension	Similar to tears: 41 mN m^{-1}
Viscosity	1.1 $m^2 s^{-1}$
Zeta potential	+20 to +40 mV

8.1.3. SELECTION OF SUITABLE PRESERVATIVE IF NEEDED

Pharmaceutical and cosmetic products that contain the aqueous phase should be properly preserved against microbial contamination and proliferation during storage in normal conditions and proper use (Halla et al. 2018). The colloidal dispersions such as o/w nanosized emulsions if they are packaged in multidose containers, then, the necessity of adding preservative molecules become mandatory as the continuous medium is water. In this connection, there are few articles about the preservation of o/w nanosized emulsions with parabens (Jumaa et al. 2002; Han and Washington 2005; Watrobska-Swietlikowska and Sznitowska 2006). There is no publication about the preservation of colloidal dispersions using more safe preservative as BAC. Additionally, the preservation of o/w nanosized emulsion is a more challenging task due to a more complex internal structure, the existence of different phases, and the expanded interphase (Hippalgaonkar et al. 2010). Due to the complex internal structures, the preservative molecules added into the emulsion may not attain an effective concentration in the aqueous phase of the emulsion. Unfortunately, there is only limited research on this problem, which is important for the use of modern formulation in clinical practice. The preservative should protect the aqueous phase where the microbial growth occurs and it is important to choose the proper preservatives and their concentration for each formulation (Meyer et al. 2007; Puschmann et al. 2018). Unfortunately, most preservatives are lipophilic, so it is difficult to obtain the appropriate concentration in the aqueous phase of the emulsion. The BAC is a preservative and possesses properties of a cationic surfactant. It is very soluble in water, ethanol, and acetone (Rowe 2014). Upon shaking of BAC with water, a foam-like structure forms and this foam-like structure has low surface tension, detergent, and emulsifying properties. Chemically, the BAC is a quaternary ammonium compound used in pharmaceutical formulations as an antimicrobial preservative (Rowe 2014). The BAC, also known as alkyldimethylbenzylammonium chloride, is a mixture of alkylbenzyldimethylammonium chlorides of various even-numbered alkyl chain lengths. The BAC is one of the most widely used preservatives in ophthalmic preparations packed in multidose containers (Dao et al. 2017). It is also used in nasal and otic formulations (Marple et al. 2004) as well as in small-volume parenteral products. This preservative is additionally used as preservatives in cosmetics (Choi et al. 2018).

In recent years, there is a tendency to avoid preservatives in ophthalmic preparation because all preservatives, not only BAC, are not safe for the ocular surface. However, as it was mentioned, BAC is the most popular preservative in ophthalmic preparation (nearly 75% of ophthalmic preparation contains BAC) and if we say that the efforts have been put into preservative-free formulations that have replaced formulations that contain preservatives, meaning mainly BAC. It is noteworthy to mention that in parenteral small-volume multidose preparations, the presence of preservatives is essential and BAC characterizes effective bactericidal and fungicidal properties that help to minimize the growth of organisms in multidose containers.

Watrobska-Swietlikowska (2020) studied partitioning of BAC into the aqueous phases of colloidal dispersions such as nanosized emulsions, water lecithin dispersion, and nanostructured lipid carriers. The water phases of the investigated colloidal dispersions were obtained by ultracentrifugation and subsequently were subjected to ultrafiltration, which procedure allowed distinguishing between the fractions of free BAC in water and those incorporated in the liposomal and micellar region. The fractions present in the oil phase and in the interphase of nanosized emulsions were calculated. Despite the various compositions of the investigated formulations and the concentration of BAC added initially into them, the amounts present in water phase were very small at 0.2–8.0%. But the BAC amount at liposomal and micellar regions of respective colloidal dispersions was increased by increasing the total concentration of preservative from 29.0 to 42.0%. Using polysorbate 80 instead of lecithin resulted in a distribution of BAC to water-liposomal-micellar phase that was twice as high. The very low concentration of antimicrobial active form of BAC was analyzed in the water phase of emulsions stabilized with lecithin as well as in water lecithin dispersion and nanospheres (below 3%). Replacement of lecithin with polysorbate 80 in emulsions significantly increases (up to 8%) the fraction of BAC in the water phase where microbial growth occurs.

8.1.4. CLINICAL SAFETY ASSESSMENT

In Chapter 4, the commercially exploited o/w nanosized emulsions are displayed in Table 4.1. From Table 4.1, it can be seen that there are four ophthalmic emulsions (Cationorm, Catioprost, Cyclokat, and Vekacia) prepared using the cationic nanosized emulsion technology. The clinical safety assessment conducted in multicenter for Cationorm and Cyclokat is described below.

8.1.4.1. Clinical Evaluation of Cationorm

The dry eye syndrome (DES) is defined as a multifactorial disease of the tears and ocular surface that results in symptoms of discomfort, visual disturbance, and tear film instability with potential damage to the ocular surface. Currently, symptomatic treatment with artificial lubricants is the first line of treatment for patients with DES. However, the disadvantage of most conventional artificial tear solutions is that most of the instilled API is lost within the first 15–30 s after installation into eye, due to reflux tearing and the drainage via the nasolacrimal duct. The prolonged residence time of the cationic nanosized emulsion on the ocular surface due to the electrostatic attraction between the positively charged lipid nanodroplets and the negatively charged ocular surface and the augmentation of the tear film layers by the oil and aqueous phases of the emulsion suggested that the cationic nanosized emulsion technology could be inherently beneficial for the ocular surface even in the absence of any entrapped APIs.

By using cetalkonium chloride, glycerol, mineral oils, poloxamer 188, tris hydrochloride, tromethamine, tyloxapol, and water for injections, the Cationorm was prepared (Lallemand et al. 2012). The ocular tolerance and efficacy of Cationorm, an API- and preservative-free cationic emulsion, were evaluated and compared with Refresh Tears (Allergan) in a 1-month, phase II, multicenter, open-label, randomized, parallel-group study enrolling patients with signs and symptoms of mild-to-moderate DES. Adults with a history of bilateral DES were subjected to a washout period of prior DES treatments during which only artificial tears were allowed. At the inclusion visit, patients were randomized to treatment with either Cationorm (n = 44) or Refresh Tears (n = 35) in both eyes four times daily and evaluated at follow-up visits on Day 7 and Day 28. Ocular tolerance and efficacy were assessed at 1 month. Seventy-nine patients, 86% female with a mean age of 61.6 years, were enrolled in the study. At 1 week and 1 month, the mean reduction in individual symptoms of DES scores and total dry eye symptoms scores were greater in the Cationorm- than Refresh Tears-treated patients (36 vs. 21% at Day 7, and 49 vs. 30% at Day 28, respectively) demonstrating that the symptoms of DES improved better with Cationorm. While the global local tolerance was perceived similarly with both treatments, the study investigators rated the overall efficacy of Cationorm statistically significantly better than Refresh Tears (P < 0.001). Additionally, Cationorm-treated patients experienced greater improvements from baseline compared to Refresh Tears-treated patients for the Schirmer test (1.88 vs. 1.27 mm) and corneal fluorescein staining (−0.61 vs. −0.59) with statistically significant improvements in the tear film breakup time (2.00 vs. 1.16, P = 0.015) and lissamine green staining (−1.42 vs. −0.91, P = 0.046). The overall results showed that Cationorm was as safe as, but more effective than, Refresh Tears in patients with mild-to-moderate DES symptoms. A similar comparison using the Schirmer test and tear film breakup time was made for another cationic emulsion Cyporin N and Restasis in a clinical study (see the result in Table 3.15 of Chapter 3). Both Cyporin N and Restasis reduced the ocular surface staining scores with similar degrees of subjective symptom change, but Cyporin N produced faster improvement than Restasis due to its enhanced cyclosporin A (CsA) bioavailability.

In a subsequent 3-month, controlled, randomized, single-masked study conducted in Italy, the efficacy of Cationorm was evaluated in adults with moderate dry eye (Aragona et al. 2011). Seventy-one patients were randomized to treatment with Cationorm, Optive (Allergan), or Emustil (SIFI) four times daily, and efficacy assessments were conducted at 1 and 3 months. At 1 month, patients treated with Optive and Cationorm experienced a statistically significant improvement from baseline in their symptoms of DES, which was also evident for each of the three treatment groups at 3 months.

At 3 months, improvements from baseline in the tear breakup time and fluorescein staining were statistically significant for Cationorm and Optive but not for Emustil. While both Cationorm and Optive significantly reduced the tear film osmolarity, only Cationorm showed a statistically significant change compared to Emustil. In this study, Cationorm was clearly more effective than Emustil in patients with moderate DES and although not statistically better, the overall improvement in symptoms and signs of DES were greater in patients treated with Cationorm than Optive.

The results of the preclinical studies (corneal healing in alkali burn and de-epithelization rabbit models) and clinical trials evaluating Cationorm in patients with DES support its safety and efficacy for the treatment of symptoms of DES and showed the benefit of the cationic nanosized emulsion technology on the ocular surface independent of any entrapped APIs. However, it should be emphasized that the inherent efficacy of the preservative-free cationic nanosized emulsion on improving symptoms of ocular surface disease presented an unanticipated challenge when used as a vehicle in the evaluation of the efficacy of the preservative-free cationic emulsion loaded with CsA in patients with DES. The next section deals with the challenges of API-loaded cationic nanosized emulsion at the clinical study level.

8.1.4.2. Clinical Evaluation of Cyclokat

According to the definition of DES, the DES is accompanied by an increased osmolarity of the tear film and inflammation of the ocular surface. As such, DES can be considered a chronic, bilateral inflammatory condition for which appropriate treatment, particularly for patients unresponsive to symptomatic treatment with artificial tears would include an anti-inflammatory agent. While Restasis, an anionic emulsion of 0.05% CsA, is available for the treatment of DES in the United States, despite the widespread use of hospital compounded CsA formulations and even corticosteroids in the EU. However, there has been no approved pharmaceutical API indicated for patients with DES. Based on the preclinical data showing the potential advantages of a cationic nanosized emulsion over anionic emulsions and unmet medical need for an approved topical CsA formulation in the EU, the development of Cyclokat for the treatment of DES was undertaken by Santen Pharmaceutical, Osaka, Japan.

The initial clinical trial of Cyclokat was a phase II, 3-month, randomized, double-masked, placebo-controlled, dose-ranging study enrolling 53 Gougerot-Sjogren patients with moderate-to-severe DES. The primary objective of the study was to assess ocular tolerance and systemic safety of the cationic nanosized emulsion containing CsA at concentrations of 0.025, 0.05, and 0.1% compared with the cationic nanosized emulsion vehicle containing no API. An exploratory evaluation of efficacy was a secondary objective. At baseline, 62% of the enrolled patients had a Schirmer test score of ≤1 mm at 5 min and 49%

had a corneal fluorescein staining score of ≥3. Over the 3-month treatment period, there were no safety concerns and no evidence of systemic absorption of CsA following topical administration of either Cyclokat dose. Patients treated with the 0.1% Cyclokat formulation showed greatest improvements in corneal and conjunctival staining at 3 months and a dose–response effect was observed for the reduction of conjunctival HLA-DR staining (a biomarker for ocular surface inflammation) at month 3 compared to baseline (vehicle −10%; 0.025% CsA −8%; 0.05% CsA −23%, and 0.01% CsA −50%).

A second phase II, 3-month, double-masked placebo controlled study comparing Cyclokat 0.05 and 0.1% versus its cationic nanosized emulsion vehicle was conducted in 132 patients with mild-to-moderate DES utilizing the controlled adverse environment chamber. In this study, the efficacy and safety of Cyclokat was assessed by the evaluation of coprimary efficacy endpoints (corneal fluorescein staining as the sign and ocular discomfort as the symptom) at month 3 after and during exposure to controlled adverse environment chamber, respectively. Although superiority was not achieved for the coprimary endpoints, there was an overall favorable safety profile and efficacy was demonstrated for the improvement of several secondary endpoints addressing DES signs and symptoms with the results favoring the use of the 0.1% dose for subsequent clinical development.

The Siccanove study was a 6-month phase III, multicenter, randomized, controlled, double-masked trial of Cyclokat 0.1% administered once daily versus its emulsion vehicle in 492 patients with moderate-to-severe DES. The primary study objective was to demonstrate superiority of Cyclokat on both a DES sign [mean changes in corneal fluorescein staining (CFS) using the modified Oxford scale] and symptoms of DES (mean change in global score of ocular discomfort using a visual analogue scale, VAS). Following a washout period during which only artificial tears were allowed, patients were randomized at baseline to treatment with either Cyclokat ($n = 242$) or its cationic emulsion vehicle ($n = 250$) and evaluated at study visits at months 1, 3, and 6. As early as month 1 ($P = 0.002$), patients treated with Cyclokat showed a statistically significant improvement in the mean change in CFS grade compared with the cationic emulsion vehicle from baseline, which continued to improve from month 3 ($P = 0.030$) to month 6, the DES sign coprimary efficacy endpoint. The statistically significant improvements in CFS over 6 months ($P = 0.009$) were complemented by a statistically significant improvement in lissamine green staining ($P = 0.048$) and a reduction in HLA-DR expression ($P = 0.022$) (Lemp et al. 2011). Additional, post hoc analysis of the Siccanove study data showed that the benefit of treatment with Cyclokat was greatest in patients with the most severe keratitis (as defined by CFS) at baseline (delta in the mean change in CFS from baseline in CFS grade 2–4 = 0.22, $P = 0.009$; 3–4 = 0.32, $P = 0.005$; grade 4 = 0.77, $P = 0.001$) (Buggage et al. 2011). Although there was a clinically relevant improvement in DES symptoms from baseline the Cyclokat and cationic nanosized emulsion vehicle treatment arms, no statistically significant differences were observed at month 6 for the mean change in the global score of ocular discomfort, the DES symptom coprimary efficacy endpoint. However, there was a statistically significant improvement in symptoms for patients achieving a ≥25% improvement in the VAS score (50.21 vs. 41.94%, $P = 0.048$). The difficulty in demonstrating the benefit of Cyclokat over its cationic emulsion vehicle was in part attributed to the efficacy of the vehicle itself in improving the symptoms of DES as demonstrated in clinical trials for Cationorm. Additionally, the symptoms coprimary endpoint result can be related to poor correlation between dry eye disease signs and symptoms. At baseline in the Siccanove study, while the mean VAS scores increased with the severity of the CFS, the correlation between the VAS score, as an expression of DES symptoms, and the CFS grade, as an expression of a DES sign, at baseline was low (Spearman's correlation coefficient = 0.23) due to the wide variability in the severity of patient-reported symptoms. Similarly, at month 6, the statistical correlation between mean change in CFS grade and VAS score was low (Spearman's correlation coefficient = 0.094) with only approximately 68% of patients showing concordance in the direction of change in CFS grade and DES symptoms (Lemp et al. 2011). Although a poor concordance between dry eye disease signs and symptoms has been recognized in the literature, improvement in both signs and symptoms is an expected outcome in randomized clinical trials investigating new DES treatments. Hence, several APIs having shown promise for improving DES have failed due to the inability to demonstrate a statistically significant improvement in signs and symptoms of DES using coprimary efficacy endpoints.

8.1.5. CONCLUSION

A stepwise development of o/w cationic nanosized emulsions for ophthalmic topical use is narrated in this section. The main challenge is the selection of suitable cation-conferring molecules, which complies with the regulatory norms. The success of ophthalmic cationic nanosized emulsion should open new ventures for other route of administration.

REFERENCES

Abdulrazik, M., Tamilvanan, S., Khoury, K. et al. (2001), Ocular delivery of cyclosporin A II. Effect of submicron emulsion's surface charge on ocular distribution of topical cyclosporin A, *S.T.P. Pharm. Sci.*, 11 (6), 427–432.

Aragona, P., Spinella, R., Rania, L. et al. (2011), Assessment of the efficacy of Cationorm® in patients with moderate dry eye compared with Optive® and Emustil® eye drops, *Acta Ophthalmol.*, 89 (s248). doi:10.1111/j.1755-3768.2011.3125.x

Bague, S., Philips, B., Garrigue, J.S. et al. (2008), Oil-in-water type emulsion with low concentration of cationic agent and positive zeta potential, EP1655021, Novagali Pharma, Evry, France.

Bague, S., Philips, B., Rabinovich-Guilatt, L. et al. (2007a), Ophthalmic emulsions containing prostaglandins, EP1827373, Novagali Pharma, Evry, France.

Bague, S., Philips, B., Rabinovich-Guilatt, L. et al. (2007b), Ophthalmic oil-in-water type emulsion with stable positive zeta potential, US patent 20070248645, EP1809237, Novagali Pharma, Evry, France.

Benita, S. (1999), Prevention of topical and ocular oxidative stress by positively charged submicron emulsion, *Biomed. Pharmacother.*, 53 (4), 193–206. doi:10.1016/S0753-3322(99)80088-2

Boussif, O., Lezoualc'h, F., Zanta, M.A. et al. (1995), A versatile vector for gene and oligonucleotide transfer into cells in culture and in vivo: polyethylenimine, *Proc. Natl. Acad. Sci. U S A*, 92 (16), 7297–7301. doi:10.1073/pnas.92.16.7297

Buggage, R.R., Amrane, M., Ismail, D. et al. (2011), The effect of Cyclokat® (unpreserved 0.1% cyclosporine cationic emulsion) on corneal involvement in patients with moderate to severe dry eye disease participating in a phase III, multicenter, randomized, controlled, doublemasked, clinical trial, *Eur. J. Ophthalmol.* doi:10.5301/EJO.2011.7544

Calvo, P., Vila-Jato, J.L., and Alonso, M.J. (1997), Evaluation of cationic polymer-coated nanocapsules as ocular drug carriers, *Int. J. Pharm.*, 153 (1), 41–50. doi:10.1016/S0378-5173(97)00083-5

Choi, S.M., Roh, T.H., Lim, D.S. et al. (2018), Risk assessment of benzalkonium chloride in cosmetic products, *J. Toxicol. Environ. Health B Crit. Rev.*, 21, 8–23. doi:10.1080/10937404.2017.1408552

Dao, H., Lakhani, P., Police, A. et al. (2017), Microbial stability of pharmaceutical and cosmetic products, *AAPS PharmSciTech.*, 19, 60–78. doi:10.1208/s12249-017-0875-1

Doroud, D., Vatanara, A., Zahedifard, F. et al. (2010), Cationic solid lipid nanoparticles loaded by cysteine proteinase genes as a novel anti-leishmaniasis DNA vaccine delivery system: characterization and in vitro evaluations, *J. Control. Release*, 148 (1), e105–106. doi:10.1016/j.jconrel.2010.07.079

Hagigit, T., Abdulrazik, M., Orucov, F. et al. (2010), Topical and intravitreous administration of cationic nanoemulsions to deliver antisense oligonucleotides directed towards VEGF KDR receptors to the eye, *J. Control. Release*, 145 (3), 297–305. doi:10.1016/j.jconrel.2010.04.013

Hagigit, T., Nassar, T., Behar-Cohen, F. et al. (2008), The influence of cationic lipid type on in-vitro release kinetic profiles of antisense oligonucleotide from cationic nanoemulsions, *Eur. J. Pharm. Biopharm.*, 70 (1), 248–259. doi:10.1016/j.ejpb.2008.03.005

Halla, N., Fernandes, I.P., Heleno, S.A. et al. (2018), Cosmetics preservation: a review on present strategies, *Molecules*, 23 (7), 1–41. doi:10.3390/molecules23071571

Han, J. and Washington, C. (2005), Partition of antimicrobial additives in an intravenous emulsion and their effect on emulsion physical stability, *Int. J. Pharm.*, 288, 263–271. doi:10.1016/j.ijpharm.2004.10.002

Hippalgaonkar, K., Majumdar, S., and Kansara, V. (2010), Injectable lipid emulsions-advancements, opportunities and challenges, *AAPS PharmSciTech.*, 11 (4), 1526–1540. doi:10.1208/s12249-010-9526-5

Hunter, A.C. (2006), Molecular hurdles in polyfectin design and mechanistic background to polycation induced cytotoxicity, *Adv. Drug Deliv. Rev.*, 58 (14), 1523–1531. doi:10.1016/j.addr.2006.09.008

Jumaa, M., Furkert, F.H., and Müller, B.W. (2002), A new lipid emulsion formulation with high antimicrobial efficacy using chitosan, *Eur. J. Pharm. Biopharm.*, 53, 115–123. doi:10.1016/s0939-6411(01)00191-6

Kim, T.W., Chung, H., Kwon, I.C. et al. (2001), Optimization of lipid composition in cationic emulsion as in vitro and in vivo transfection agents, *Pharm. Res.*, 18 (1), 54–60. doi:10.1023/a:1011074610100

Kuo, W.-T., Huang, H.-Y., and Huang, Y.-Y. (2010), Polymeric micelles comprising stearic acid-grafted polyethyleneimine as nonviral gene carriers, *J. Nanosci. Nanotechnol.*, 10 (9), 5540–5547. doi:10.1166/jnn.2010.2454

Lallemand, F., Daull, P., Benita, S. et al. (2012), Successfully improving ocular drug delivery using the cationic nanoemulsion, novasorb, *J. Drug Del.*, 2012, 1–16. doi:10.1155/2012/604204

Lemp, M.A., Baudouin, C., Amrane, M. et al. (2011), Poor correlation between dry eye disease (DED) signs and symptoms in a phase III randomized clinical trial, *Invest. Ophthalmol. Vis. Sci.*, 52 (14), 3821.

Manosroi, A., Podjanasoonthon, K., and Manosroi, J. (2002), Development of novel topical tranexamic acid liposome formulations, *Int. J. Pharm.*, 235 (1–2), 61–70. doi:10.1016/s0378-5173(01)00980-2

Marple, B., Roland, P., and Benninger, M. (2004), Safety review of benzalkonium chloride used as a preservative in intranasal solutions: an overview of conflicting data and opinions, *Otolaryngol. Head Neck Surg.*, 130, 131–141. doi:10.1016/j.otohns.2003.07.005.

Masotti, A., Moretti, F., Mancini, F. et al. (2007), Physicochemical and biological study of selected hydrophobic polyethylenimine-based polycationic liposomes and their complexes with DNA, *Bioorg. Med. Chem.*, 15 (3), 1504–1515. doi:10.1016/j.bmc.2006.10.066

Meyer, B.K., Ni, A., Hu, B. et al. (2007), Antimicrobial preservative use in parenteral products: past and present, *J. Pharm. Sci.*, 96 (12), 3155–3167. doi:10.1002/jps.20976

Moreau, E., Domurado, M., Chapon, P. et al. (2002), Biocompatibility of polycations: in vitro agglutination and lysis of red blood cells and in vivo toxicity, *J. Drug Target.*, 10 (2), 161–173. doi:10.1080/10611860290016766

Philips, B., Bague, S., Rabinovich-Guilatt, L. et al. (2008), Ophthalmic emulsions containing an immunosuppressive agent, EP 1809238, Novagali Pharma, Evry, France.

Puschmann, J., Herbig, M.E., and Müller-Goymann, C.C. (2018), Correlation of antimicrobial effects of phenoxyethanol with its free concentration in the water phase of o/w-emulsion gels, *Eur. J. Pharm. Biopharm.*, 131, 152–161. doi:10.1016/j.ejpb.2018.08.007

Rabinovich-Guilatt, L., Couvreur, P., Lambert, G. et al. (2004), Extensive surface studies help to analyse zeta potential data: the case of cationic emulsions, *Chem. Phys. Lipids*, 131 (1), 1–13. doi:10.1016/j.chemphyslip.2004.04.003

Rowe, R.C. (2014), *Handbook of Pharmaceutical Excipients* [Internet], Pharmaceutical Press, London, Philadelphia; American Pharmacists Association, Washington, DC. Available from http://search.ebscohost.com/login.aspx?direct=true&?db=cat01789a&AN=GUM.vtls000317723&lang=pl&site=edslive (Accessed December 19, 2019).

Shell, J.W. (1984), Ophthalmic drug delivery systems, *Surv. Ophthalmol.*, 29 (2), 117–128. doi:10.1016/0039-6257(84)90168-1

Sznitowska, M., Janicki, S., Dabrowska, E.A. et al. (2002), Physicochemical screening of antimicrobial agents as potential preservatives for submicron emulsions, *Eur. J. Pharm. Sci.*, 15 (5), 489–495. doi:10.1016/s0928-0987(02)00034-9

Van Abbe, N.J. (1973), Eye irritation: studies relating to responses in man and laboratory animals, *J. Soc. Cosm. Chem.*, 24 (11), 685–692.

Vinogradov, S.V., Bronich, T.K., and Kabanov, A.V. (2002), Nanosized cationic hydrogels for drug delivery: preparation, properties and interactions with cells, *Adv. Drug Deliv. Rev.*, 54 (1), 135–147. doi:10.1016/S0169-409X(01)00245-9

Wang, J., Feng, S.S., Wang, S. et al. (2010), Evaluation of cationic nanoparticles of biodegradable copolymers as siRNA delivery system for hepatitis B treatment, *Int. J. Pharm.*, 400 (1–2), 194–200. doi:10.1016/j.ijpharm.2010.08.026

Watrobska-Swietlikowska, D. (2020), Distribution of benzalkonium chloride into the aqueous phases of submicron dispersed systems: emulsions, aqueous lecithin dispersion and nanospheres, *AAPS PharmSciTech.*, 21, 7. doi:10.1208/s12249-019-1540-7

Watrobska-Swietlikowska, D. and Sznitowska, M. (2006), Partitioning of parabens between phases of submicron emulsions stabilized with egg lecithin, *Int. J. Pharm.*, 312 (1–2), 174–178. doi:10.1016/j.ijpharm.2006.01.005

Yeh, M.K., Chen, J.L., Chiang, C.H. et al. (2007), The preparation of sustained release erythropoietin microparticle, *J. Microencapsul.*, 24 (1), 82–93. doi:10.1080/02652040601058533

Zhang, S., Wang, G., Lin, X. et al. (2008), Polyethylenimine-coated albumin nanoparticles for BMP-2 delivery, *Biotechnol. Prog.*, 24 (4), 945–956. doi:10.1002/btpr.12

CHAPTER 8.2

CASE STUDY 2 FISH OIL-BASED NANOSIZED EMULSIONS

SYED NAZRIN RUHINA RAHMAN* AND TAMILVANAN SHUNMUGAPERUMAL

Department of Pharmaceutics, National Institute of Pharmaceutical Education and Research (NIPER), Guwahati, Assam, India

*The funding provided by Department of Biotechnology, Govt. of India to Syed Nazrin Ruhina Rahman is greatly acknowledged.

Oil-in-Water Nanosized Emulsions for Drug Delivery and Targeting,
First Edition. Tamilvanan Shunmugaperumal.
© 2021 John Wiley & Sons, Inc. Published 2021 by John Wiley & Sons, Inc.

EXPANSION OF ABBREVIATIONS

AD	Alzheimer's disease
API	active pharmaceutical ingredient
BBB	blood–brain barrier
BMI	body mass index
BRCA1	breast cancer genes type 1
BRCA2	breast cancer genes type 2
CGB	caregivers burden
CIND	cognitive impairment no dementia
CKD	chronic kidney disease
COX	cyclooxygenase
CNS	central nervous system
CVD	cardiovascular disease
DHA	docosahexaenoic acid
DPA	docosapentaenoic acid
EC	European parliament and the council
EPA	eicosapentaenoic acid
EU	European Union
FAO	Food and Agriculture Organization
fMRI	functional magnetic resonance imaging
FO	fish oil
FOC	FO-containing
GPx1	glutathione peroxidase
HDL	high-density lipoprotein
HR	heart rate
HRV	heart rate variability
IgE	immunoglobulin E
IL	interleukin
IVLEs	intravenous lipid emulsions
LCTs	long-chain triglycerides
LOS	length of stay

LPS	lipopolysaccharide
ICU-LOS	length of stay in the intensive care unit
MAC	membrane attack complex
MCP	monocyte protein
MCTs	medium-chain triglycerides
MMP	matrix metalloproteinase
MMSE	mini mental state examination
NF-kβ	nuclear factor-kappaβ
NOD-LRR	nucleotide-binding oligomerization domain-leucine-rich repeat
OO	olive oil
ω	omega
O/W	oil-in-water
PCBs	polychlorinated biphenyls
PC	phosphatidylcholine
PEG	polyethylene glycol
PGE-2	prostaglandin E_2
PN	parenteral nutrition
PPAR	peroxisome proliferator activated receptor
PUFAs	polyunsaturated fatty acids
PYD	pyrin domain
RCT	randomized controlled trial
SCF	supercritical fluid
SO	soybean oil
SPI	soybean protein isolate
TAG	triacylglycerol
T2DM	type 2 diabetes mellitus
TLR	toll-like receptor
TNF	tissue necrotic factor
USFDA	United State Food and Drug Administration

8.2.1. INTRODUCTION

The sea is the interconnected system of all the Earth's oceanic waters, including the Atlantic, Pacific, Indian, Southern, and Arctic Oceans. Fishes are one of the major inhabitants of sea useful to human lives in many ways. In biosystem, the fishes belong to vertebrate animals (Phylum: *Chordata*) and having the living habits both in the fresh and salt waters of the world. By external appearance, the fishes are gill-bearing aquatic craniate animals that lack limbs with digits. The building components of fish are very useful to maintain and even supply many essential components such as amino acids, cholesterol, fats, minerals (calcium, iron, selenium, zinc, etc.), mono/di/triglycerides, free fatty acids (oils), phospholipids, proteins, sterylesters, vitamin A, B_3 (nicotinamide), B_6 (pyridoxine), B_{12} (cobalamine), D and E, etc., to the human body (Chee et al. 1990; Sidhu 2003; Porojnicu et al. 2008).

The objective of this section is to highlight the therapeutical utility of fish oil, fish oil supplement, fish oil-based emulsions, and regulatory aspects governing the fish oil-based products.

8.2.2. IMPORTANCE OF FISH OIL OR FISH OIL SUPPLEMENTS IN EVERYDAY HUMAN LIFE

Majority of the people love to eat the whole fish in their everyday life. Out of the many useful components of the fish, the fatty acids (oils) become the subject of health-care system due to their medicinal value able to manage variety of syndromes as shown in Table 8.2.1.

The fish oil (FO) accounts for about 2% of world consumption of fats and oils. Traditionally, the FO is obtained as a by-product of the fish meal industry. But currently smaller fishes (anchovies, eels, herring, and sardines) with a relatively high fat content are in the center of attention as a raw material in the FO industry. Historically, the FO has already played a significant role in the human diet, and currently, the demand for the FO is still growing, thanks to its curative and medicinal properties (Rizliya and Mendis 2014). Therefore, the FO and its products find places mainly in food and pharmaceutical industries as well as agriculture and aquaculture specializations as a feed additive to the fishes. Ranging from 25 to 30 million tons of healthy fish and fish cuttings consisting approximately 1.1 million tons of FO are produced around the world (Rizliya and Mendis 2014). However, only 5% of the produced FO is used to extract the Omega (ω)–3 fatty acids. The remaining 95% of FO is used in the aquaculture industry as feed additive to grow the fishes (Lembke 2013). Although in some regions of Europe and the rest of the world, fisheries sector still has a great place for growth and resource optimization (Xiao et al. 2017) and even the

TABLE 8.2.1. Non-comprehensive List of Syndromes Manageable by Fatty Acids (Oils) Present in Fish

Cancers
Compromised immunity
Hypertension, Cardiovascular Disease
Hypertriglyceridemia
Inflammatory diseases (arthritis, inflammatory bowel diseases, psoriasis, lupus, asthma, cystic fibrosis, dermatitis, neurodegeneration, etc.)
Weight gain, weight loss, obesity
Osteoporosis
Poor infant and childhood cognitive processes, learning, and behavior
Poor infant visual development (especially preterm)
Thrombosis
Type-2 diabetes

Adapted from Calder (2012).

analysis of the current situation shows that the FO production is relatively static. Because the future projections indicate that the available FO sources will not be able to provide the increasing demand from food and pharmaceutical industries as well as agriculture and aquaculture specializations (Rizliya and Mendis 2014). Therefore, the last decade has emphasized the research of a new source or species in different parts of the world, the environmental impact reduction of the extraction methods and the integration of green-extraction methods in an industrial scale (Ivanovs and Blumberga 2017).

The FO constitutes the fatty acids both in the forms of neutral lipid and free fatty acids in fish. The general physical properties of fish oil are displayed in Table 8.2.2.

In particular, the typical fatty acid composition of the FO extract can be divided into saturated fatty acids, monounsaturated fatty acids, and polyunsaturated fatty acids (PUFAs). The main saturated fatty acids found in FO include myristic acid (14:0), palmitic acid (16:0), stearic acid (18:0), and behenic acid (22:0) (Loftsson et al. 1995). The monounsaturated fatty acids in FO include myristoleic acid (14:1ω5), palmitoleic acid (16:1ω7), oleic acid (18:1ω9), eicosenoic acid (20:1ω9), gadoleic acid (20:1ω11), erucic acid (22:1ω9), and catoleic acid (22:1ω11). The major PUFAs in FO are presented as linoleic acid (18:2ω6), α-linolenic acid (18:3ω3), docosahexaenoic acid (DHA 22:6ω3), and eicosapentaenoic acid (EPA 22:5ω3). Both ω–6 and ω–3 PUFAs are essential fatty acids because the mammalian cells lack the desaturase enzymes capable of placing double bonds at the positions of ω–6 and ω–3. Both PUFAs should be obtained from diet and supplementation. The ω–6 and ω–3 fatty acids are needed for normal growth and health maintenance (Huang et al. 2018).

Going in depth, the fatty acids (oils) present in fish contain high, but variable, levels of ω–3 long chain PUFAs, including EPA (20:5ω–3) and DHA (22:6ω–3), as well as small quantities of docosapentaenoic acid (DPA; 22:5ω–3) (Ghasemi Fard et al. 2018). The commercially available FO supplements typically contain mixtures of EPA and DHA. The sources of fish from where the most abundantly produced and globally available fish oils are shown in Flowchart 8.2.1.

The most abundantly produced and available FOs globally are anchovy (*Engraulis ringens*) oil, capelin (*Mallotus villosus*) oil, menhaden (*Brevoortia tyrannus*) oil, and herring (*Clupea harengus*) oil, all of which contain variable

TABLE 8.2.2. General Physical Properties of Fish Oil

Property	Value
Appearance	Amber colored oil
Boiling point	>250°C
Flash point (fatty acids)	Approximately 220°C
Melting point	10–15°C
Odor	Fishy odor
Specific gravity (at 30°C)	0.91 (specific gravity of water at 30°C is 0.996)

Adapted from Rizliya and Mendis (2014).

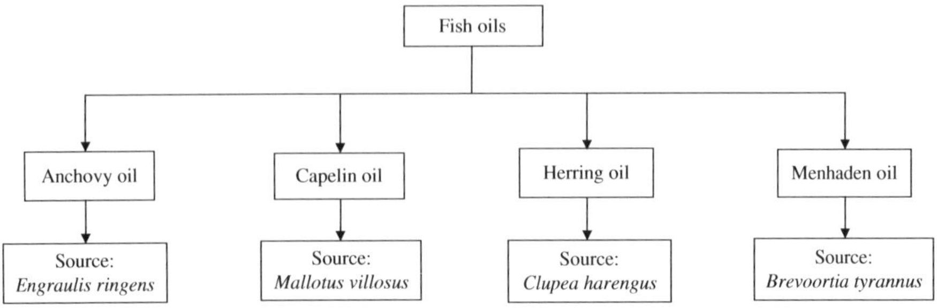

Flowchart 8.2.1. Commonly available various types of fish oil and their sources of fish. Modified from Turchini et al. (2009).

amounts of ω–3 long-chain PUFA, but are characterized by higher concentrations of EPA than DHA (Turchini et al. 2009).

The possible cardio-protective effect of dietary ω–3 long-chain PUFA present in oily fish has fueled numerous studies examining the health effects of dietary FO (Dyerberg et al. 1978). The Food and Agriculture Organization of the United Nations (2010) recommends a dose of 250 mg per day of EPA plus DHA for adult males and nonpregnant/non-lactating adult females, while the American Heart Association recommends about 1 g per day of EPA plus DHA for patients with known coronary heart disease, and 2–4 g per day for patients requiring triacylglycerol (TAG) lowering (Kris-Etherton et al. 2002). In human and animal studies, the most commonly studied FO is anchovy oil, which typically contains about 30% of EPA and DHA, in an 18/12 ratio (180 mg of EPA and 120 mg of DHA per gram of oil).

Most importantly, the endogenous biologic pathways through which ω–3 long-chain PUFA play essential roles include: (1) modulation of cell and organelle membrane structure and function, (2) modulation of ion channels and cellular electrophysiology, (3) regulation of nuclear receptors and transcription factors, and (4) production of ω–3 long-chain PUFA-derived bioactive metabolites (Mozaffarian and Wu 2011).

To get major understanding of FO after administration into the human body, the effect/performance of FO may be grouped into two types based on the therapeutic interventions: one with positive therapeutic intervention (Table 8.2.3) and another with negative intervention (Table 8.2.4).

8.2.3. OVERVIEW ON ENVIRONMENT-FRIENDLY GREEN-EXTRACTION METHODS OF FISH OIL FROM WHOLE FISH OR FISHERIES WASTE

After knowing the importance of fish oil and its products/supplements particularly to health-care system, the prime question arising in front of food and pharmaceutical industries as well as agriculture and aquaculture specializations is the extraction of essential fatty acids or simply FO from sources of fish. In principle, the FO

TABLE 8.2.3. Effect/Performance of Fish Oil Based on Positive Therapeutic Interventions in Humans

Disease	Effect	References
Cardiovascular disease, chronic obstructive pulmonary disease, myocardial infarct, neoplasia	Protective	Siscovick et al. (1995) and Daviglus et al. (1997)
Cardiac arrhythmias, graft rejection (transplantation)	Prevented	Leaf et al. (1998) and Grimml et al. (1998)
Chemotherapy	Supportive	Zadak and Cervinkova (1997)
Hypertriglyceridemia, peripheral arterial disease	Improved	Connor et al. (1993) and Tinker et al. (1999)
Psoriasis and dermatitis, systemic erythematosis, total oesophagus resection	Beneficial	Roulet et al. (1997), Grimminger and Mayser (1995), and Neuringer et al. (1988)

TABLE 8.2.4. Effect/Performance of Fish Oil Based on Negative Therapeutic Interventions in Humans

Disease	Effect	References
Antibody response and interferon-γ production, hypersensitivity, virus clearance	Delayed	Endres et al. (1993), Jolly et al. (1997), Taki et al. (1992), and Byleveld et al. (1999)
Asthma	Possibly harmful	Picado et al. (1988)
Cell mediated immune function, cytokine or eicosanoid production	Diminished	Wu and Meydani (1998) and Meydani et al. (1991)
Cellular immunity	Weakened	Meydani et al. (1993)
Incidence of tuberculosis in Alaskans	Increased	Comstock et al. (1967)
Upper respiratory tract and chest infections in Canadian Inuit	Recurring	Hildes and Schaefer (1984)

extraction methods from whole fish or fisheries waste can conveniently be classified into two categories, viz., traditional and nontraditional. This section briefly narrates especially the environment-friendly green-extraction methods (nontraditional) for FO extraction. Because the green-extraction methods utilize four different techniques, which involve enzyme extraction, microwave-assisted extraction, supercritical fluid (SCF) extraction, and ultrasound assisted extraction. Aside from the nontraditional green-extraction-based techniques, the FO extraction from whole fish or fisheries waste includes the traditional method like hydraulic pressing, heat extraction, and solvent extraction (Rubio-Rodríguez et al. 2010; Adeoti and Hawboldt 2014). The main drawbacks of traditional extraction methods in terms of the quality of the final products obtained are:

- the temperature-dependent degradation of labile/heat-sensitive natural compounds present in whole fish or fisheries waste and
- the appreciable amount of residual toxic solvents present in the final FO products.

Another important drawback of traditional methods is their greater impact on the environment. Because the traditional extraction methods either based on a significant amount of heat or utilized the high-risk organic solvents during the FO extraction and hence the chances of organic solvents leakage into the environment (Adeoti and Hawboldt 2014). In the last 20 years, the green-extraction methods are recognized as a promising alternative to the organic solvents and oil extraction grease. Mostly it is the SCF extraction using CO_2, but also other green methods coming up with the SCF-CO_2 regarding extraction yield, product quality, the content of ω–3 fatty acids, especially EPA and DHA (Sarker et al. 2012). Although the green-extraction methods can ensure the same quality of product, the green methods like traditional one also have their own drawbacks as listed in Table 8.2.5.

8.2.4. RATIONALE FOR DEVELOPING NANOSIZED EMULSIONS BASED ON FISH OIL

Despite its positive therapeutic interventions (as shown in Table 8.2.3), the fish oil usually loses potential health benefits during processing, transportation, and storage due to its fishy smell, low solubility in water, and oxidative instability. Thus, nanoencapsulated delivery systems such as polymeric nanoparticles, nanosized emulsions, and liposomes offer promising solutions to improve the processability, water solubility, and bioavailability of FO (Li et al. 2019). Keeping the points like low solubility and oxidative instability of FO in mind, many researchers prepared FO-based nanosized emulsions and these emulsions were stabilized by adding multiple emulsifiers at the varying concentration levels.

The first-generation oil-in-water (o/w) emulsions such as intravenous lipid emulsions (IVLEs) are an important component of parenteral nutrition (PN) as a source of essential fatty acids and as an energy source in patients who are unable to tolerate nutrition through other routes. Traditionally, IVLEs for PN were derived from soybean oil (SO) (Carpentier and Dupont 2000; Calder 2010). However, the SO possesses high concentration of ω–6 polyunsaturated long-chain triglycerides (LCTs). Many concerns have been expressed that administration of PN with the help of SO-based nanosized emulsions may have an immunosuppressive effect (Adolph 1999; Calder 2006). The SO may also exacerbate the release of pro-inflammatory cytokines and prostaglandin E_2 (PGE-2). This led for further contributing to an increased risk of complications in the human body (Adolph 1999; Calder 2006). To address this potential

TABLE 8.2.5. Brief Introduction, Advantages and Drawbacks of Different Green-Extraction Methods Used for Fish Oil Extraction from Whole Fish or Fisheries Waste

Green-extraction methods	Brief introduction	Advantages	Drawbacks	References
Enzymatic hydrolysis	Uses exogenous proteolytic enzymes to digest material for extracting the fish oil from whole fish or fisheries waste	No need for organic solvent The use of commercial low cost protease enzyme as an attractive alternative which leads to the overall reduction in extraction expenses	Scale up difficulty	Deepika et al. (2014)
Microwave-assisted	Uses microwaves to warm the solvents in contact with the solid matrix containing whole fish or fisheries waste to extract the contents from the sample solution	Reduction in extraction time and solvent usage Increased solvent permeation into solid matrix cellular material and thus escalated fish oil content release in the surrounding solvent medium Reduction in heat transfer into the surrounding environment Higher fish oil extraction yield Relatively lower temperature usage	High electric power consumption Heating affects only polar solvents and/or materials Scale up difficulty Undesirable oxidation of unsaturated fatty acids Limitation to use volatile solvent due to possible low fish oil extraction yield	Adeoti and Hawboldt (2014) Mercer and Armenta (2011)

TABLE 8.2.5. *Continued*

Green-extraction methods	Brief introduction	Advantages	Drawbacks	References
Supercritical fluid extraction using CO_2 (SCF-CO_2)	Uses supercritical fluids to separate fish oil contents from solid matrix containing whole fish or fisheries waste by means of SC-CO_2 as solvent.	Rapid extraction and higher fish oil extraction yield No need for organic solvent and hence extract is very pure Free of heavy metals and inorganic salts No chance of polar substances forming polymers. No further processing required for the obtained lipids Low operating temperatures (40–80°C)	Higher expenses and complex equipment operation at elevated pressures Suitable only for non-polar substances extraction as CO_2 is highly selective and thus not meant for polar substances Supply of clean CO_2 needed High electric power consumption	Adeoti and Hawboldt (2014) Sarker et al. (2012)
Ultrasound assisted	Uses ultrasound to promote the solvent permeation into solid matrix containing whole fish or fisheries waste to extract the fish oil contents	Rapid extraction and higher fish oil extraction yield Reduction in solvent usage Increased solvent permeation into solid matrix cellular material and thus escalated fish oil content release in the surrounding solvent medium	High electric power consumption Scale up difficulty	Abdullah et al. (2010)

Adapted from Ivanovs and Blumberga (2017).

risk, newer IVLEs have been developed to contain lower levels of long-chain ω–6 PUFAs to minimize the possible adverse effects associated with LCTs. One strategy in creating these new IVLEs has been to replace a portion of the SO component in an IVLE with medium-chain triglycerides (MCTs) and/or with olive oil (OO) and/or FO. Another strategy has been to use an IVLE supplement derived solely from FO, which is high in ω–3 PUFAs. Evidence from laboratory studies indicates that ω–3 PUFAs may result in lower inflammatory responses. The ω–3 and ω–6 PUFAs have competitive metabolic pathways. The metabolism of linoleic acid, a dominant ω–6 PUFA, produces arachidonic acid (C20:4ω–6), which is further metabolized to products with pro-inflammatory activity (Carpentier and Dupont 2000; Calder 2012). Therefore, in theory, PN high in ω–6 PUFAs may be detrimental for patients at risk of a pathologic inflammatory response. In contrast, metabolism of EPA (C20:5 ω–3) and DHA (C22:6 ω–3), which are the ω–3 PUFAs predominant in FO, may reduce the production of pro-inflammatory eicosanoids in a dose-dependent way:

a. by directly competing with the metabolism of arachidonic acid and
b. by the nature of EPA and DHA, which are metabolized to products that are less pro-inflammatory than arachidonic acid. Based on *in vitro* and animal experiments, the immune and pro-inflammatory properties of IVLEs rich in linoleic acid, a precursor of arachidonic acid, could have an adverse impact on clinical outcomes (e.g., by increasing risk of infection, sepsis, or systemic inflammatory response syndrome) (Calder 2007, 2010). Despite evidence from *in vitro* and animal models indicating that ω–3 PUFAs may have a dose-dependent beneficial effect on inflammatory responses, there have been conflicting reports from meta-analyses and systematic reviews regarding the clinical benefits of using fish oil-containing (FOC)-IVLEs over other lipids. While well-conducted meta-analyses can be key to demonstrating positive or negative effects, inappropriately conducted meta-analyses can lead to skewed results. Of the meta-analyses and systematic reviews conducted, some have reported a statistically significant clinical benefit for FOC-IVLEs (Wei et al. 2010; Pradelli et al. 2012; Li et al. 2014). However, one meta-analysis included only six relatively low quality trials (Wei et al. 2010), and another not only imputed up to 50% of standard deviations not reported by the original trial but did not correct for possible type 1 errors (Pradelli et al. 2012). Neither of these practices is recommended as good practice by the Cochrane Collaboration (Mascha 2015). Other reviews have reported either no significant clinical benefits with FOC-IVLEs or an overall poor quality of included studies with considerable heterogeneity, making it likely that any statistical benefit would be insufficient to recommend the use of FO over other IVLEs (Palmer et al. 2013; Tian et al. 2013; Langlois et al. 2017).

8.2.5. FISH OIL-BASED EMULSION PRODUCTS: COMPARATIVE LITERATURE SURVEY

Table 8.2.6 displays the already marketed intravenous fat emulsions and their compositions including definite ratio of ω–6 and ω–3 triacylglycerides (Vanek et al. 2012). When the intravenous fat emulsions prepared based on FO, the ω–6 and ω–3 triacylglycerides ratio obtained was 1:8 (Omegaven®) indicating the usefulness of FO to make emulsions.

To eradicate the fishy smell of FO, the FO was simply mixed with the aqueous solution containing stabilizers like soybean protein isolate (SPI), phosphatidylcholine (PC), etc. Furthermore, in order to show that the preparation of FO-based nanosized emulsions did not differ significantly from that of the preparation of fixed oil-based nanosized emulsions, Li et al. (2019) prepared FO-based nanosized emulsions stabilized by SPI-PC. The SPI (1, 2, 3, 4, 5%, w/v) was mixed with PC (0, 0.2, 0.4, 0.6, 1%, w/v) with varying ratios and they were dispersed in sodium phosphate buffer solution (0.05 M, pH 7.4), stirring continuously at room temperature (22–25°C) for 12 h. This mixture is considered as aqueous phase of the o/w emulsions. The FO (0.5, 1, 1.5, 2, 3%, w/v) was added to above SPI-PC solution followed by homogenization at room temperature by an Ultra-Turrax T18 homogenizer (ANGNI Co. Ltd., Shanghai, China) at 20,000 rpm min^{-1} for 3 min to form a coarse emulsion. Then, the coarse emulsion by high-pressure homogenization using a D-6L ultra-high-pressure homogenizer (PhD Technology Co., Ltd., Saint Paul, MN, USA) at different pressures was transformed into the nanosized emulsions. The FO-based nanosized emulsions can also be stabilized by Tween 20 wherein the composition contains 2% (w/v) Tween 20 and 1.5% (w/v) FO (Li et al. 2019).

TABLE 8.2.6. Marketed Intravenous Fat Emulsions Prepared Based on Fish Oil (FO), Long Chain Triglycerides (LCT), Medium-Chain Triglycerides (MCT), Olive Oil (OO) or Soybean Oil (SO) Along with ω–6:ω–3 Triacylglyceride Ratio

Product	Composition	ω–6:ω–3 triacylglyceride ratio
Clinoleic®	20% SO, 80% OO	9:1
Intralipid®, Intrafat®, Ivelip®, Liposyn®, Lipovenoes®, Soyacal®	100% SO	7:1
Lipoplus®/Lipidem®	40% SO, 50% MCT, 10% FO	2.7:1
Lipofundin MCT/LCT®, Lipovenoes MCT®	50% SO, 50% MCT	7:1
Omegaven®	100% FO	1:8
SMOFlipid®	30% SO, 30% MCT, 25% OO, 15% FO	2.5:1
StructoLipid 20%®	64% SO, 36% MCT	7:1

8.2.6. COMPLICATIONS OF FISH OIL AND FISH OIL-BASED PRODUCTS

The human body possesses the tendency of not accepting any exogenous substances that are administered through all possible routes. The entry of exogenous substances into the human body is usually marked with the generation of immune substances, opsonic molecules, and complement particles. The classical examples of opsonic molecules include various subclasses of immunoglobulins (Harnisch and Müller 1998, 2000), complement proteins like C1q and generated C3 fragments (C3b, iC3b) (Szebeni 2001), apos (Harnisch and Müller 1998, 2000), von Willebrand factor, thrombospondin, fibronectin (Price et al. 2001), and mannose-binding protein.

The complement system is comprised of more than 30 proteins in the plasma and cell membrane, organized through a hierarchy of proteolytic cascades that are triggered by initiators (Walport 2001). Following activation, the components of complement system assemble the membrane attack complex (MAC) to create pores in the cell membrane for destroying the target cells. During complement activation, several potent pro-inflammatory mediators, including complement proteins C3a and C5a, are simultaneously generated. These mediators further induce anaphylaxis, recruit infiltrating inflammatory immune cells and regulate T cell function (Arbore et al. 2016). It has also been reported that the complement activation has a role in reducing necrotic cell-induced inflammation by clearing dying cells and thereby aiding in the prevention of autoimmune diseases (Martin and Blom 2016). Furthermore, NLRP3 is a member of the nucleotide-binding oligomerization domain-leucine-rich repeat (NOD-LRR) protein family (Ting et al. 2006). It consists of a pyrin domain (PYD), a NOD (also known as NACHT domain), and a LRR domain. According to Suresh et al. (2016), the MAC formation can result in NACHT, LRR, and PYD domains containing protein 3 (NLRP3) inflammasome activation, which in turn activates caspase-1 and promotes secretion of the inflammatory cytokines interleukin (IL)-1β and IL-18. Thus, complement activation can affect inflammation through multiple mechanisms.

The liver is a unique organ that has important roles in both the digestive and immune systems. Xia et al. (2014) showed that a saturated fatty acid and ω–6 PUFAs-rich high-fat diet induced low-degree inflammation in the liver and promoted myeloid-derived suppressor cell differentiation, suppressed T-cell-mediated adaptive immunity, and promoted tumor growth through multiple mechanisms, including the alteration of hematopoiesis and the bone marrow microenvironment. Coming to the relationship between FO/FO-based products and complement substances, Jin et al. (2018) have recently studied the influence of high FO diet on complement activation and inflammation in the liver tissue of mice. Without adhering to an exact mechanism, the high FO intake induced inflammation and activated the complement system in the liver (Jin et al. 2018). Nevertheless, these authors concluded that ω–3 PUFA-rich

FO diet promoted the infiltration of CD11b+ Gr-1+ inflammatory myeloid cells into the liver, suggesting that the induction of liver inflammation is mediated specifically by ω–3PUFA-rich feeding, rather than by fatty acids and ω–6 PUFAs. However, the mechanism of a high-FO diet on the complement system, a key effector and regulator of innate and adaptive immunity, is not clear.

But the FO feeding into mice resulted in a modified bone marrow hematopoiesis and induced splenomegaly in treated animals (Xia et al. 2015). Interestingly, a high-FO diet not only increased the weight of the spleen but also increased the number of Gr-1+ myeloid cells in the liver (Jin et al. 2018). Considering the similar roles of the spleen and liver in extramedullary hematopoiesis, the increased myeloid cell infiltration into the liver was likely closely associated with the effects of FO on bone marrow hematopoiesis. In addition, the higher levels of MCP-1 in liver may have also contributed to this observed result in mice.

Wei et al. (2010) reported that the parenteral administration of FO-based nanosized emulsions could improve secondary clinical outcomes like length of stay (LOS). When compared with standard ω–6 fatty acids, PN adjustment with ω–3 fatty acids is associated with a statistically significant infection risk reduction in the elective postoperative patient. For outcome on LOS, the original meta-analysis found that heterogeneity existed among four included studies, and the random effects model was used for aggregation. The results showed a shortened trend associated with FO intervention but the difference was not significant. On stratified analysis of LOS, one study by Grecu and Mirea (2003) was found to have included in its sample patients with significant severe illness, and heterogeneity testing showed the source of variance from this study. This study was therefore excluded from stratified analysis. After excluding this study, the heterogeneity test result showed no heterogeneity among the remaining three studies. Hence, the fixed effect model could be used and a significant result was found. For the study by Grecu and Mirea (2003), the differences in length of stay in the intensive care unit (LOS/ICU-LOS) in the two groups were significant and FO intervention was associated with shortened LOS/ICU-LOS. The result of this single study is consistent with stratified analysis. Consequently, a conclusion was made that there is a significantly shorter hospital stay associated with FO-based lipid emulsions in elective postoperative patients.

However, for mortality, the primary outcome, there was no significant benefit after the data were aggregated. Because most of the included trials were conducted in elective postoperative patients (five of six), and the mortality risk was expected to be fairly low, the insignificant result is unsurprising. Considering the major therapeutic mechanism of FO is attenuating systematic inflammation, the possibility of benefit to patients with a higher mortality risk such as severe injury and sepsis existed. Non-randomized controlled trial (RCT) prospective studies presented by Heller et al. (2004, 2006) indicated that mortality in patients with severe head injury/multiple injuries or with abdominal systemic inflammatory response syndrome/sepsis may be decreased by emulsion therapy containing FO. However, in another RCT in patients with critical illness, no such effect was found (Friesecke et al. 2008). These controversial results indicate that the issue of mortality is not conclusive and indeed opens the possibilities for future more large-scale RCTs.

8.2.7. REGULATORY ASPECTS RELATED TO THE SAFETY OF FISH OIL/FISH OIL-BASED PRODUCTS

The fish catching and processing become a highly profitable business across different countries of the world. In an estimate given by Food and Agriculture Organization (FAO), the fish catching global fish production amounted to 175 million tons in 2017 and it is expected to reach 194 million tons by 2026 (FAO 2018). The FAO also said that the expansion of fish processing is creating increasing quantities of side streams and residuals, such as skins, heads, frames, viscera, and fillet cutoffs. These may account for up to 70% of fish used in industrial processing. Arason et al. (2010) have estimated that in cod processing only approximately 40% of fish raw material is used for food production, and side streams account for almost 60%.

If fish and its side streams are used for human consumption, then the quality and safety of the raw materials have to be ensured by handling and controlling side streams in the same manner as the primary product (fish). To this end, the application of food safety and quality control systems, including Hazard Analysis and Critical Control Point and Good Manufacturing Practice, are imperative for retaining the suitability of side streams as sources of human-grade food (Olsen et al. 2014). The hazards can be of biological and chemical in nature. Whatever it may be the nature of hazards, it should be appropriately controlled and even regulated through guidelines.

8.2.7.1. Biological Hazards Associated with Raw Materials of Fish

Following intentional or unintentional death of the fish, many physical and chemical changes (mediated by enzymes and microorganisms) begin to occur on the body of the fish leading to the complete decay of the fish. The fish decay process mediated by bacteria starts with protein decomposition and the proteolysis products such as amino acids or low molecular nitrogenous compounds provide better nourishment. The muscle tissue of live fish is generally sterile, but bacteria thrive in the alimentary tract and on the skin, and from there they penetrate into the muscles. Moreover, the by-products such as intestines and gills have also possessed high bacterial numbers. However, low temperatures strongly inhibit the activity of microorganisms in which case the fish decay/spoilage follows dominantly an autolysis or self-digestion process (Rustad et al. 2011).

Seafood allergies, including shellfish and fish, are typically lifelong affecting up to 5% of all children and 2% of all adults. A fish allergy is triggered by a specific immune response to fish proteins and is normally mediated by immunoglobulin E (IgE) antibodies. Most fish allergic patients are sensitized to β-parvalbumins, which is a highly stable protein. Fish allergic people often react to multiple fish species and therefore they are usually advised to avoid all kinds of fish. Also, gelatin and collagen derived from fish may be a health risk for persons suffering from a fish allergy (Kuehn et al. 2011; Sharp and Lopata 2014).

Adverse reactions to fish can also be mediated by non-immunological reactions as a result of exposure to the fish itself or various non-fish components in the product. These contaminants may include bacteria, viruses, marine toxins, and parasites. Biogenic amine histamine, formed from histidine by bacterial decarboxylation, may cause symptoms that seem like allergic responses. Fish of the scombroid family such as tuna, mackerel, and bonito contain naturally high levels of histidine. Scombroid fish poisoning is a clinically important type of fish intolerance (Kuehn et al. 2011; Sharp and Lopata 2014). Heating appears to increase the allergenicity of parvalbumin compared with raw parvalbumin and other allergens. Commercial heat processes used to generate canned fish seem to have a different effect (Sharp and Lopata 2014).

In the European Union (EU), the hygienic rules for fishery products are based on regulation no. 853/2004 of the European parliament and the council (EC 2004), as well as on regulation (EC) no. 2073/2005 on microbiological criteria for foodstuffs. The microbiological safety of fresh and prepared fishery products is controlled by the analysis of *Listeria monocytogenes*, *Salmonella* sp., *Escherichia coli*, and coagulase positive staphylococci. In addition, *Clostridium botulinum* needs to be controlled in some fishery products. Norovirus outbreaks associated with seafood have been reported worldwide. Contamination by the presence of naturally occurring pathogens (e.g., *Vibrio* spp., some species of *Aeromonas*) can result in unsafe seafood. Cross contamination may take place during harvesting, handling, preparation, processing, transportation, and storage (Elbashir et al. 2018). Fishery products that are obviously contaminated with parasites must not be placed on the market for human consumption [regulation (EC) 853/2004; EC 2004]. If parasites are detected by visual examination, such fish cannot be used as food and it must be treated according to regulation (EC) no 1069/2009 on animal by-products and derived products not intended for human consumption.

8.2.7.2. Chemical Hazards Associated with Raw Materials of Fish

Maximum levels of contaminants in fish and fish products have been set in regulation (EU) no 1881/2006. These contaminants include dioxins and dioxin-like compounds, heavy metals (cadmium, lead, and mercury), and polycyclic aromatic hydrocarbons (in smoked fish). The chemical hazards also include radioactive residues (e.g., cesium-137), histamine, allergens, and some biotoxins.

Almost every living creature has been exposed to dioxins or dioxin-like compounds. Dioxins are mainly by-products of industrial practices. They are produced through a variety of incineration processes, including improper municipal waste incineration and burning of trash, and can be released into the air during natural processes, such as forest fires and volcanoes. The modern life style of human beings allows the people to expose more on dioxins primarily by eating food, in particular animal products, contaminated by these chemicals. Dioxins are absorbed and stored in fat tissue and, therefore, accumulate in the food chain. More than 90% of human exposure is through food. Dioxin levels in fish depend on many factors, such as the species, fish age, fat content, type of tissues and organs tested, water pollution, fishing area, season, and habit migrations (Vuorinen et al. 2017). For instance, certain fish species originating from the Baltic region may contain high levels of dioxins and dioxin-like polychlorinated biphenyls (PCBs), which include a range of toxic and environmentally persistent substances. In aquatic environments, fish and other marine animals can absorb dioxins and dioxin-like PCBs (European Food Safety Authority 2010).

Whatever may be the nature of hazards either biological or chemical, the FO obtained from fish should be free from these hazards in order to ensure that the FO used to make the emulsions could be of safe raw materials. Therefore, the producers of FO-based nanosized emulsions must take an utmost care about the purity of FO before the start of the emulsion preparation process. That is why separate regulations are coming up to monitor the FO production from fish processing unit to avoid the presence of unwanted and most toxic side streams of fish in FO as residuals. It should be added that the composition of ω–3 PUFAs in the commercially available products depends upon the source of the fish, the body part of the fish, and the extraction methods employed. The United State Food and Drug Administration (USFDA) approved FO for the treatment of hypertriglyceridemia includes Epanova, Omtryg, Lovaza, and Icosapent ethyl (Vascepa). For pediatric patients with PN-associated cholestasis, the FO-based nanosized injectable emulsion is Omegaven (approval date 7/27/2018).

8.2.8. THERAPEUTICAL APPLICATION OF FISH OIL OR FISH OIL-BASED NANOSIZED EMULSIONS

Commercially, the FOs are available in both nonprescription and prescription-only varieties at different concentrations. However, the use of nonprescription FOs becomes a common practice by human beings for supplementing the diet with multivitamin regimen and healthy fatty acids, like EPA and DHA. Both EPA and DHA have the demonstrated abilities in minimizing or preventing hypertriglyceridemia when taken as an adjunct to a healthy diet. However, prescription-only FO products are sometimes prescribed for individuals who demonstrate severe ($\geq$500 mg/dl) hypertriglyceridemia (https://www.drugbank.ca/drugs/DB13961, accessed on 1/24/2020). This section narrates the medicinal uses of FO for managing various types of symptoms or syndromes such as Alzheimer's, cardiovascular, central nervous system, inflammation, malignancy (cancers), obesity, renal, sepsis, and type-II diabetes.

8.2.8.1. Alzheimer's Disease

Alzheimer's disease (AD) is a neurodegeneration disorder, with slow and continuous progression, characterized by progressive impairments of memory, language, reasoning, and other cognitive functions. To effectively prevent aging of brain tissues, the FO finds its position and thus helps in the management of AD (Kyle et al. 1999). The results obtained from three different RCTs self-exemplified the medicinal use of FOs in managing the AD totally in contrast way.

Hashimoto et al. (2016) conducted a RCT using 75 Japanese nursing-home residents with AD condition who received 1720 or 106 mg DHA per day (added in fish sausages) in a 12-month intervention study. The residents had substantial reduction of cognitive function value of 14 ± 1 at start [calculated from average Japanese Mini Mental State Examination (MMSE)] and were very old (approximate average age of 88 years). One year later when cognition, mood, depression, and caregivers burden (CGB) were assessed in 66 residents, depression and apathy were significantly less in those who received high DHA. Although no differences were observed for MMSE or CGB, some subitems of these two scales had improved for residents who received high dose of DHA.

A small British intervention study conducted by Phillips et al. (2015) in a RCT with the help of 76 participants [age 71 years, 57 with cognitive impairment no dementia (CIND) and 19 with AD] by using 1.225 g of EPA and DHA (1:1) or placebo for 4 months have, however, showed a contrasting observation (no effect of ω–3 fatty acids on several cognitive tests) in comparison to the RCT study conducted by Hashimoto et al. (2016) with 75 Japanese nursing-home residents.

A more experimental but interesting RCT was reported from the USA wherein the effects of 2.4 g ω–3 fatty acids per day (EPA:DHA 1:1) on cortical blood flow [measured by functional magnetic resonance imaging (fMRI)] when performing a memory task were assessed in 21 older adults with memory complaints but not dementia (Boespflug et al. 2016). After 24 weeks, those that received active treatment showed an improved fMRI signal in certain posterior areas of the brain during the memory task. The memory task was also performed better after ω–3 fatty acids intake. The results were interpreted as indications of enhanced neuronal responses to working memory challenges by providing ω–3 fatty acids.

8.2.8.2. Cardiovascular Disease

Most of the studies conducted to see the influence of ω–3 fatty acids (EPA and DHA) on the different parameters of cardiovascular system are usually performed in healthy human beings who regularly do the submaximal exercise.

Larsen (2006) initially indicated the usefulness of ω–3 fatty acids including EPA and DHA, may reduce the risk of mortality from cardiovascular disease (CVD) in people who have already experienced a cardiac event.

In a clinical study, the study design was made in such a way that the influence of high DHA intake by the human volunteers both at rest and in exercise conditions (O'Keefe et al. 2006). At rest, there was a reduction in heart rate (HR) value from 73 ± 13 to 68 ± 13 beats min^{-1} and after exercise, the value of HR recovery was improved by 1 min when the human volunteers took high DHA FO. Overall, heart rate variability (HRV) was unaffected by high DHA FO administration, though HRV in the high-frequency band increased significantly ($p < 0.02$). There were no significant effects on blood pressure, arterial compliance, lipids, or inflammatory markers reported between high DHA FO-treated and untreated human volunteer groups.

The administration of high DHA FO may also be considered as an adjunct to healthy volunteer exercise program. To emphasis this statement, Hill et al. (2007) reported that the daily intake of high DHA FO could be a useful adjunct to exercise programs aimed at improving body composition and decreasing CVD risk in overweight individuals. Furthermore, findings from this clinical study also suggested that the regular exercise coupled with a moderate dose of high DHA FO improves multiple cardiovascular risk factors, including plasma TAGs, high-density lipoprotein (HDL) cholesterol, and flow-mediated dilatation. To support the results of the above study (Hill et al. 2007), Buckley et al. (2009) showed that the erythrocyte EPA, DHA, and total ω–3 fatty acids content increased almost twofold, while serum TAG and HR reduced during submaximal exercise in athletic subjects receiving a high DHA FO supplementation, compared with the placebo. Despite no effect on resting HR, HR during submaximal exercise decreased by ~8 beats min^{-1} with high DHA FO, compared with only ~2 beats min^{-1} with sunflower oil, suggesting that high DHA FO improved CV efficiency during submaximal exercise. In this study, high DHA FO did not affect endurance exercise performance or recovery.

Ninio et al. (2008) reported that high DHA FO supplementation improved HRV in volunteers by increasing high-frequency power, and it reduced HR at rest and during submaximal exercise. The effects of high DHA FO supplementation on O_2 consumption during exercise was investigated in trained male cyclists (Peoples et al. 2008). No difference in O_2 consumption or peak workload after supplementation with FO was recorded.

The administration of normal meal and high DHA FO meal to the human volunteers resulted in producing variation in the enzyme level or enzyme-gene expression. To exemplify this argument, a comparative study was conducted by Armah et al. (2008). These authors found that after taking a high DHA FO meal by human volunteers, the isolated postprandial TAG-rich lipoproteins increased the endothelial nitric oxide synthase level but decreased the nicotinamide adenine dinucleotide phosphate oxidase gene expression in comparison to the enzyme level or enzyme-gene expression observed with human volunteers who took control meal only.

The values of indices such as augmentation and arterial stiffness were found to be attenuating/satisfying after the intake of high DHA meal by the human volunteers when compared with the augmentation and arterial stiffness index values observed following the intake of control meal by the human volunteers Chong et al. (2010).

McManus et al. (2016) reported that a single dose of DHA, but not EPA, significantly improved postprandial arterial stiffness compared with the control, as assessed by augmentation index, which, if sustained, would be associated with a significant decrease in CVD risk.

8.2.8.3. Central Nervous System

It is well known that the systemically administered active pharmaceutical ingredient (API)-loaded nanosized emulsions were able to improve the API exposer in the central nervous system (CNS) where the therapeutic intervention is needed in rodent models (Prabhakar et al. 2013; Đorđević et al. 2017; Tan et al. 2017). If the o/w nanosized emulsions are prepared by including the oils such as FO and flaxseed oil, then the ω–3 and ω–6 PUFAs have been shown to improve CNS API distribution and exposure in several preclinical experiments. Because the PUFAs regulate various activities in the brain, such as neurotransmission, cell survival, and neuro-inflammation despite the fact that the saturated and monounsaturated fatty acids can be synthesized *de novo* within the brain as well as the ability of PUFAs to enter into the CNS from blood (Bazinet and Layé 2014).

Various pathways have been proposed to describe the transport of PUFAs across the blood–brain barrier (BBB). While one theory proposes that fatty acids cross the BBB without specific transporters, crossing the luminal and transluminal leaflets of the endothelial cells by reversible flip-flop (Hamilton and Brunaldi 2007). Another theory indicates that fatty acids cross the BBB by facilitated diffusion wherein the fatty acid transporters attach to PUFA molecules on the luminal side of the BBB and diffuse across both the endothelial and neuronal cells (Liu et al. 2015). Additionally, the PUFAs can also cause membrane fluidization at the BBB interface, thereby impacting tight junctions and enhancing CNS absorption (Navarro et al. 2011). Although the precise mechanism responsible for CNS transport of PUFAs remains unclear, it has been hypothesized that encapsulating APIs in oils containing PUFAs may result in enhanced CNS API delivery. A major assumption of this hypothesis is that the API remains encapsulated in the PUFA containing oil phase of the nanosized emulsions, and the intact API containing oil droplet diffuses across the BBB. Despite the absence of experimental validation of this assumption, mainly due to lack of bioanalytical approaches to distinguish between the free and oil-encapsulated API in biological matrices, several studies claimed the improved CNS API delivery following systemic dosing of nanosized emulsions formulated with oils rich in PUFAs. Selected examples that illustrate the potential of such nanosized emulsions include the work of Vyas et al. (2008) demonstrating that flaxseed oil-based nanosized emulsions improve CNS exposure of saquinavir (an anti-HIV protease inhibitor) subsequent to oral dosing in mice and the results of Shah et al. (2014) illustrating improved CNS exposure and analgesic effects of a neuroactive peptide encapsulated in FO-based nanosized emulsions following systemic dosing in mice. Additionally, the *in vivo* properties of these nanosized emulsions and sensitive formulation characteristics that enhance CNS API delivery are not completely understood. Improved CNS API delivery from nanosized emulsions may be the result of several factors that may vary significantly between free and nanosized emulsion-encapsulated APIs. These factors include the characteristics of systemic pharmacokinetics and biodistribution and the interactions at the cell membrane of the BBB interface (Ganta et al. 2010). Increased circulation time of the nanosized emulsion-encapsulated API in the body often increases the time available for CNS penetration, and thus enhances the extent of CNS API delivery. Furthermore, the pharmacokinetic behavior of nanosized emulsions can be modified by varying the formulation approach involving the conjugation with polyethylene glycol (PEG) (Hak et al. 2015). In addition, unlike the free API molecule, which must cross the BBB via passive diffusion across the cell membrane, nanosized emulsion-encapsulated APIs may penetrate the BBB via different transport processes. One such transport process relies on endocytosis of the nanosized emulsion-encapsulated API following fusion of the oil droplet with cell membrane of the BBB, which depends not only on the composition of the nanosized emulsion but also on the size and charge of the nanosized emulsion oil droplet. Despite potential advantages, the role of different transport processes and the interactions between various nanosized emulsion formulation attributes that are relevant for regulating API delivery to the CNS in a systemic environment remain unclear. For example, the question of how the size of the oil droplet vs. the circulation half-life of the nanosized emulsion affects BBB permeability and resulting CNS penetration of the nanosized emulsion-encapsulated API. To elucidate relevant properties of CNS targeting nanosized emulsions when administered systemically, Kadakia et al. (2019) have recently formulated rapamycin containing FO-based nanosized emulsions with different formulation characteristics. It was found that the circulation half-life and particle size distribution did not impact the brain targeting efficiency of rapamycin containing FO-based nanosized emulsions. Furthermore, in the absence of any improvement in the systemic exposures of rapamycin, nanosized emulsions did not outperform their aqueous counterpart with respect to the extent of CNS API delivery (Kadakia et al. 2019).

8.2.8.4. Inflammatory Disease

The dietary pretreatment with ω–3 fatty acid favorably influences pathophysiologic response to endotoxins (Mascioli et al. 1988) and exerts important modulation on eicosanoid and cytokine biology. The most likely manner in which lipids might modulate pro-inflammatory cytokine biology is by changing the fatty acid composition in the cell membrane. As a consequence of the changes, two interrelated phenomena may occur, i.e., alteration in membrane fluidity and in products that arise from hydrolysis of membrane phospholipids (Grimble 1998). Indeed, inflammatory symptoms of rheumatoid arthritis, psoriasis, Crohn's disease, and ulcerative colitis are all ameliorated by FO-based preparations whether or not directly related to cytokine production. Consumption of EPA reduces the production of pro-inflammatory IL-1-α and -β, IL-6, as well as tissue necrotic factor (TNF)-α and -β in response to an inflammatory stimulus (Endres et al. 1989; Caughey et al. 1996). The anti-inflammatory effects of FO may also include decreased production of inflammatory substances like leukotriene B_4 and platelet-activating factors, released by the action of cytokines, as well as a large reduction of cytokine-induced synthesis of PGE-2 and thromboxane B_2 in the colonic mucosa (Endres et al. 1989; Engström et al. 1996). Other possible mechanisms include interaction with peroxisome proliferator activated receptor (PPAR)-α, which is a gene transcription factor that induces the breakdown of leukotrienes and thus has a role in limiting the duration and extent of inflammation (Devchand et al. 1996). After considering all the points discussed above, the FO has been used to replace SO for making lipid emulsions to reduce the possible risk of inflammatory complications (Abbasoglu et al. 2017).

8.2.8.5. Kidney/Renal Disease

Any deterioration in the normal functions of the nephron cells over instantly or chronically should affect the healthiness of human beings. The chronic kidney disease (CKD) affects around 10–15% of community-dwelling adults (Saglimbene et al. 2019). The ω–3 PUFAs supplementation less than 4 g per day might reduce proteinuria in patients with CKD and slow immunoglobulin A nephropathy (Donadio et al. 1994; Gopinath et al. 2011). It is very interesting to note that patients with advanced CKD have substantially lower blood levels of ω–3 PUFAs compared with the general population, probably due to lower dietary intake, inflammation, malabsorption, metabolic changes, and loss of ω–3 PUFAs during the dialysis process (Saglimbene et al. 2019).

8.2.8.6. Malignancy

Cancer is a collection of diseases involving abnormal cell growth with the potential to invade or spread to other parts of the body. The abnormal growth influenced by cancer can result in tumors, damage to the immune system, and other impairment that can be fatal. According to Ashfaq et al. (2019), the EPA and DHA can produce their antitumor activity against malignant cells by following four different pathways. The four different pathways include (1) carcinoma cell growth inhibition of apoptosis, (2) oxidative stress, (3) angiogenesis, and (4) cyclooxygenase (COX)-2/PGE-2 inhibition (Ashfaq et al. 2019). Either alone (EPA/DHA) or in combination with other chemotherapeutic agents, the therapeutic application of EPA and/or DHA for the management of various malignancy conditions (occurred in breast, lung, ovarian, and renal) are described below.

Breast cancer accounts for estimated 29% of cancer in women (Siegel et al. 2014) and the genetic factors such as breast cancer genes type 1 (BRCA1) and type 2 (BRCA2) are often the indicators for assessing the risk of malignant cells. As a single therapeutic agent, the influence of DHA on two different carcinoma cell line models such as MCF-7 (derived from human breast cells) and 4T1 (derived from mouse breast cells) was investigated. It was found that the DHA inhibitory effect was due to reduction of beta-catenin expression and causing apoptosis (Xue et al. 2014). In combination with other therapeutic agent, the DHA was able to produce/show synergetic effect in the treatment of metastatic breast cancer (Sophie et al. 2008; Philippe et al. 2009). During research on human breast cancer cell line (MDA-MB-231), it was suggested that DHA is responsible for increasing the reactive oxygen species level by inhibition of glutathione peroxidase (GPx1) activity (an enzyme for protection against hydrogen and lipid peroxides), which increases the treatment level of chemotherapeutic agent (Sophie et al. 2008). Recently, the effectiveness of DHA-enriched oil was investigated on surgically obtained tissue cultures of human malignant breast cancer. It was reported that anti-tumorigenesis properties of DHA is due to its ability to significantly promoting the expression of Toll-like receptor (TLR)-4 and PPAR-α, thus leading to DHA-induced breast tumor cells apoptosis (Geng et al. 2018).

According to statistic data provided by Siegel et al. (2014), the lung cancer and bronchitis constitute the top major cancerous cause of death (28% men and 26% of females). PUFAs, especially DHA, are thought to induce oxidative stress to inhibit tumor cell growth in lung cancer and the mechanism of tumor growth inhibition was studied in A549 lung adenocarcinoma cell line (Kikawa et al. 2010). Furthermore, the EPA and DHA supplementation may cause suppression of both primary tumor growth and lung metastasis occurrence (Nicolo et al. 2013). In addition, the EPA and DHA from FO have been demonstrated to affect the cancer cell replication and have shown significant suppression of lung adenocarcinoma proliferation by induced apoptosis (Yao et al. 2014).

The fifth leading cancer causing the death among women is ovarian cancer (Siegel et al. 2014). The best nontoxic way to prevent or suppress ovarian cancer is that the increased consumption of ω–3 fatty acids (Erfan et al. 2013a, b). It is generally believed that the anti-inflammatory effect of the EPA and DHA blocks the COX-2 pathway, which resulted in the reduced level of prostaglandins that leads to the prevention of ovarian cancer (Erfan et al. 2013c). The transforming growth beta-1 and p21 proteins levels was likely to be increased in ovarian cancer cell lines in the presence of ω–3 fatty acids in ovarian cancer cell lines, which in turn may result in the inhibition of growth of cancer cells

(Arun et al. 2009). The study by Mariko et al. (2010) indicated that the ω–3 fatty acids (EPA and DHA) effectively decreased the lipopolysaccharide (LPS)-induced nuclear factor-kappaβ (NF-kβ) activation and monocyte protein (MCP)-1 expression. Furthermore, the overexpression of PPAR-α activation by EPA and DHA resulted in LPS-induced apoptosis (Mariko et al. 2010). In fact, the inhibition of NF-kβ is one of the additional beneficial effects of FO.

Renal cell carcinoma originates in the lining of proximal convoluted tubules of the kidney and is one of the major cancer occurrences (Siegel et al. 2014). Renal cell carcinoma is characterized by enhanced levels of the metastasis promoting gene matrix metalloproteinase (MMP)-2 (Beardo et al. 2019). The ω–3 long chain PUFAs (EPA and DHA) from FO have shown to reduce of MMP-2 mRNA protein levels and killing of tumor cell line (Zhou et al. 2017). The DHA has shown to significantly reduce the growth of renal cell carcinoma, and that this reduction is regulated by levels of PGE-2.

8.2.8.7. Obesity

It is always good not to be obese for avoiding complications in the human body. In this respect, the obesity is a complex disease involving an excessive amount of body fat. In fact, the obesity is not just a cosmetic concern rather it is a medical problem that increases the risk of other diseases and health problems, such as heart disease, diabetes, high blood pressure, and certain cancers. There are many reasons why some people have difficulty in avoiding the obesity. Usually, the obesity results from a combination of inherited factors, combined with the environment and personal diet and exercise choices. In general, the obesity is diagnosed when the body mass index (BMI) is 30 or higher.

Munro and Garg (2012) reported a significant reduction in fat mass for the high DHA FO-consumed group after 14 weeks intervention in obese male and female subjects, but not for the control sunola oil group, and while the metabolic profiles of both treatment groups improved, overall there was no significant effect on weight loss or weight maintenance over the 14 weeks. Munro and Garg (2013) reported that obese females consuming high DHA FO over sunola oil exhibited a greater percentage decrease in weight (7.21 vs. 5.82%) and BMI (7.43 vs. 5.91%), respectively, but not in males. Therefore, the supplementation with high DHA FO had a time-dependent effect on weight loss in females.

8.2.8.8. Sepsis

Sepsis is a systemic inflammatory response syndrome that occurs during severe infection. Recently, it has been demonstrated that intravenous administration of fat emulsions rich in ω–3 PUFAs (6–6.5 g per day that is equivalent with 2.3 g EPA plus DHA per day) in patients with sepsis for 5 days from admission to the ICU can lead to their rapid incorporation into phospholipids of different cells, such as platelets, or monocytes, within the first 2 days of feeding, reducing serum pro-inflammatory cytokines over the next 7–8 days (Papaioannou and Pnevmatikos 2019).

8.2.8.9. Type 2 Diabetes Mellitus

Type 2 diabetes mellitus (T2DM) is one of the most common chronic diseases causing a dramatic public health burden. Poor insulin sensitivity is associated with T2DM and metabolic syndrome (Muley et al. 2014). The meta-analysis studies have shown that FO consumption and dietary ω–3 PUFAs decrease the risk factor of T2DM via enhanced insulin sensitivity (Gao et al. 2017). Gao et al. (2017) suggested that short-term FO supplementation was beneficial for insulin sensitivity than long-term intervention. The long-term treatment duration intervention was not effective on insulin sensitivity either. Therefore, the best intervention period might be short term, which is less than 12 weeks but at least over 4 weeks. Adiposity increases the risk of diabetes and reduces insulin sensitivity through the fat tissue release inflammatory markers, insulin-like growth factors, sCD163, and adipokines (Toupchian et al. 2016). Thus, it is biologically plausible that FO supplementation could improve insulin sensitivity among people with metabolic disorders. Using the diabetic retinopathy as a case study, the usefulness of FO in reducing the retinopathic ophthalmic tissues is shown below.

Diabetic retinopathy is caused by prolonged high blood glucose levels. Over time, high glucose levels can weaken and damage the small blood vessels within the retina. This may cause hemorrhages, exudates, and even swelling of the retina, which then starves the retina of oxygen, and growth of abnormal blood vessels takes place. The ω–3 fatty acids prevent its progression by inhibiting a wide range of inflammatory mediators, decreasing the formation of free radicals, and inducing the expressions of endogenous antioxidant enzymes. Also, they remarkably prevent the initiation of retinal angiogenesis by downregulating the expressions of various angiogenic agents such as COX-2, matrix metalloproteinases, and vascular endothelial growth factor (Behl et al. 2019).

8.2.9. CONCLUSION

Apart from considering the fish and its side streams as diet to human beings, the fatty acids present in fish possess the ability to intervene in many ailment conditions. Thus, the fish products such as FO containing the rich ω–3 fatty acids received renewed interest and therefore, the FO finds its place in modern medical care of many syndromes. The FO-based nanosized emulsions are slowly getting into the picture in curing, managing, and even preventing the syndromes that are dangerous to life.

REFERENCES

Abbasoglu, O., Hardy, G., Manzanares, W. et al. (2017), Fish oil-containing lipid emulsions in adult parenteral nutrition: a review of the evidence, *J. Parenter. Enteral Nutr.*, 43 (4), 458–470. doi:10.1177/0148607117721907

Abdullah, S., Mudalip, S.A., Shaarani, S.M. et al. (2010), Ultrasonic extraction of oil from *Monopterus albus*: effects of different ultrasonic power, solvent volume and sonication time, *J. Appl. Sci.*, 10 (21), 2713–2716. doi:10.3923/jas.2010.2713.2716

Adeoti, I.A. and Hawboldt, K. (2014), A review of lipid extraction from fish processing by-product for use as a biofuel, *Biomass Bioenergy*, 63, 330–340. doi:10.1016/j.biombioe.2014.02.011

Adolph, M. (1999), Lipid emulsions in parenteral nutrition, *Ann. Nutr. Metab.*, 43 (1), 1–13. doi:10.1159/000012761

Arason, S., Karlsdottir, M., Valsdottir, T. et al. (2010), Maximum resource utilisation-value added fish by-products, (Available at http://www.divaportal.org/smash/record.jsf?pid=diva2%3A707165&dswid=-6548). Accessed on 12/19/2019.

Arbore, G., West, E.E., Spolski, R. et al. (2016), T helper 1 immunity requires complement-driven NLRP3 inflammasome activity in $CD4^+$ T cells, *Science*, 352 (6292), aad1210. doi:10.1126/science.aad1210

Armah, C.K., Jackson, K.G., Doman, I. et al. (2008), Fish oil fatty acids improve postprandial vascular reactivity in healthy men, *Clin. Sci.*, 114 (11), 679–686. doi:10.1042/cs20070277

Arun, S., Jennifer, B., Joseph, E. et al. (2009), Effects of omega-3 fatty acids on components of the transforming growth factor beta-1 pathway: implication for dietary modification and prevention in ovarian cancer, *Am. J. Obstet. Gynecol.*, 200 (5), 1–6. doi:10.1016/j.ajog.2008.12.023

Ashfaq, W., Rehman, K., Siddique, M.I. et al. (2019), Eicosapentaenoic acid and docosahexaenoic acid from fish oil and their role in cancer research, *Food Rev. Intl.*, 1–20. doi:10.1080/87559129.2019.1686761

Bazinet, R.P. and Layé, S. (2014), Polyunsaturated fatty acids and their metabolites in brain function and disease, *Nat. Rev. Neurosci.*, 15 (12), 771–785. doi:10.1038/nrn3820

Beardo, P., Cacho, D.T., Izquierdo, L. et al. (2019), Cancer-specific survival stratification derived from tumor expression of tissue inhibitor of metalloproteinase-2 in non-metastatic renal cell carcinoma, *Pathol. Oncol. Res.*, 25, 289–299. doi: 10.1007/s12253-017-0339-7

Behl, T., Grover, M., Shah, K. et al. (2019), Role of omega-3-fatty acids in the management of diabetes and associated complications, in: *Bioactive Food as Dietary Interventions for Diabetes*, Academic Press, pp. 185–192. doi:10.1016/B978-0-12-813822-9.00012-6

Boespflug, E.L., McNamara, R.K., Eliassen, J.C. et al. (2016), Fish oil supplementation increases event-related posterior cingulate activation in older adults with subjective memory impairment, *J. Nutr. Health Aging*, 20, 161–169. doi:10.1007/s12603-015-0609-6

Buckley, J.D., Burgess, S., Murphy, K.J. et al. (2009), DHA-rich fish oil lowers heart rate during submaximal exercise in elite Australian Rules footballers, *J. Sci. Med. Sport*, 12, 503–507. doi:10.1016/j.jsams.2008.01.011

Byleveld, P.M., Pang, G.T., Clancy, R.L. et al. (1999), Fish oil feeding delays influenza virus clearance and impairs production of interferon-γ and virus-specific immunoglobulin A in the lungs of mice, *J. Nutr.*, 129 (2), 328–335. doi:10.1093/jn/129.2.328

Calder, P.C. (2006), Use of fish oil in parenteral nutrition: rationale and reality, *Proc. Nutr. Soc.*, 65 (3), 264–277. doi:10.1079/PNS2006500

Calder, P.C. (2007), Immunonutrition in surgical and critically ill patients, *Br. J. Nutr.*, 98 (S1), S133–S139. doi:10.1017/S0007114507832909.

Calder, P.C. (2010), Rationale and use of n–3 fatty acids in artificial nutrition, *Proc. Nutr. Soc.*, 69 (4), 565–573. doi:10.1017/S0029665110000157

Calder, P.C. (2012), Mechanisms of action of (n–3) fatty acids. *J. Nutr.*, 142 (3), 592S–599S. doi:10.3945/jn.111.155259.

Carpentier, Y.A. and Dupont, I.E. (2000), Advances in intravenous lipid emulsions, *World J. Surg.*, 24 (12), 1493–1497. doi:10.1007/s002680010026.

Caughey, G.E., Mantzioris, E., Gibson, R.A. et al. (1996), The effect on human tumor necrosis factor alpha and interleukin 1 beta production of diets enriched in n–3 fatty acids from vegetable oil or fish oil, *Am. J. Clin. Nutr.*, 63 (1), 116–122. doi:10.1093/ajcn/63.1.116

Chee, K.M., Gong, J.X., Rees, D.M.G. et al. (1990), Fatty acid content of marine oil capsules, *Lipids*, 25 (9), 523–528. doi.10.1007/BF02537158.

Chong, M.F.F., Lockyer, S., Saunders, C.J. et al. (2010), Long chain ω–3 PUFA-rich meal reduced postprandial measures of arterial stiffness, *Clin. Nutr.*, 29, 678–681. doi:10.1016/j.clnu.2010.02.001

Comstock, G.W., Ferebee, S.H., and Hammes, L.M. (1967), A controlled trial of community-wide isoniazid prophylaxis in Alaska, *Am. Rev. Respir. Dis.*, 95 (6), 935–943. doi:10.1164/arrd.1967.95.6.935

Connor, W.E., DeFrancesco, C.A., and Connor, S.L. (1993), N–3 fatty acids from fish oil, *Ann. N. Y. Acad. Sci.*, 683, 16–34. doi:10.1111/j.1749-6632.1993.tb35689.x

Daviglus, M.L., Stamler, J., Orencia, A.J. et al. (1997), Fish consumption and the 30-year risk of fatal myocardial infarction, *N. Engl. J. Med.*, 336 (15), 1046–1053. doi:10.1056/NEJM199704103361502

Deepika, D., Vegneshwaran, V. R., Julia, P. et al. (2014), Investigation on oil extraction methods and its influence on omega-3 content from cultured salmon, *J. Food Process. Technol.*, 5 (12), 140–159. doi:10.4172/2157-7110.1000401

Devchand, P.R., Keller, H., Peters, J.M. et al. (1996), The PPARα-leukotriene B 4 pathway to inflammation control, *Nature*, 384 (6604), 39–43. doi:10.1038/384039a0

Donadio Jr, J.V., Bergstralh, E.J., Offord, K.P. et al. (1994), A controlled trial of fish oil in IgA nephropathy, *N. Engl. J. Med.*, 331 (18), 1194–1199. doi:10.1056/NEJM199411033311804

Đorđević, S.M., Santrač, A., Cekić, N.D. et al. (2017), Parenteral nanoemulsions of risperidone for enhanced brain delivery in acute psychosis: physicochemical and in vivo performances, *Int. J. Pharm.*, 533 (2), 421–430. doi:10.1016/j.ijpharm.2017.05.051

Dyerberg, J., Bang, H.O., Stoffersen, E. et al. (1978), Eicosapentaenoic acid and prevention of thrombosis and atherosclerosis?, *Lancet*, 2:117–9. doi:10.1016/S0140-6736(78)91505-2

EC (2004), Regulation No. 853/2004 of the European parliament and the council of laying down specific hygiene rules for on the hygiene of foodstuffs, *Official Journal of the European Union L* 139/55, (Available at https://eur-lex.europa.eu/LexUriServ/LexUriServ.do?uri=OJ:L:2004:139:0055:0205:en:PDF). Accessed on 12/19/2019.

Elbashir, S., Parveen, S., Schwarz, J. et al. (2018), Seafood pathogens and information on antimicrobial resistance: a review, *Food Microbiol.*, 70, 85–93. doi:10.1016/j.fm.2017.09.011

Endres, S., Ghorbani, R., Kelley, V.E. et al. (1989), The effect of dietary supplementation with n–3 polyunsaturated fatty acids on the synthesis of interleukin-1 and tumor necrosis factor by mononuclear cells, *N. Engl. J. Med.*, 320 (5), 265–271. doi:10.1056/NEJM198902023200501

Endres, S., Meydani, S.N., Ghorbani, R. et al. (1993), Dietary supplementation with n–3 fatty acids suppresses interleukin-2 production and mononuclear cell proliferation, *J. Leukoc. Biol.*, 54 (6), 599–603. doi:10.1002/jlb.54.6.599

Engström, K., Luostarinen, R., and Saldeen, T. (1996), Whole blood production of thromboxane, prostacyclin and leukotriene B4 after dietary fish oil supplementation in man: effect of vitamin E, *Prostaglandins Leukot. Essent. Fatty Acids*, 54 (6), 419–425. doi:10.1016/S0952-3278(96)90025-3

Erfan, E., Carolynn, C.S., Stacey, R.M. et al. (2013a), Anti-inflammatory effects of fish oil in ovaries of laying hens targets prostaglandin pathway, *Lipids Health Dis.*, 12, 152–164. doi:10.1186/1476-511X-12-152

Erfan, E., Janice, M.B., and Dale, B.H. (2013b), Long term consumption of flaxseed enriched diet decreased ovarian cancer incidence and prostaglandins E2 in hens, *Gynecol. Oncol.*, 130, 620–628. doi:10.1016/j.ygyno.2013.05.018

Erfan, E., Karen, H., Yan, Z. et al. (2013c), Flaxseed enriched diet-mediated reduction in ovarian cancer severity is correlated to the reduction of prostaglandin E2 in laying hen ovaries, *Prostaglandins Leukot. Essent. Fatty Acids*, 89, 179–187. doi:10.1016/j.plefa.2013.08.001

European Food Safety Authority (2010), Results of the monitoring of dioxin levels in food and feed, *EFSA J.*, 8 (3), 1385. doi:10.2903/j.efsa.2010.1385

FAO (2018), *The State of World Fisheries and Aquaculture. Meeting the Sustainable Development Goals*, Food and Agriculture Organization of the United Nations, (Available at http://www.fao.org/3/i9540en/I9540EN.pdf). Accessed on 12/19/2019.

Food and Agriculture Organization of the United Nations (25–29 Jan 2010), Joint FAO/WHO expert consultation on the risks and benefits of fish consumption, Rome. http://www.fao.org/3/ba0136e/ba0136e00.pdf. Accessed 1/20/2020.

Friesecke, S., Lotze, C., Köhler, J. et al. (2008), Fish oil supplementation in the parenteral nutrition of critically ill medical patients: a randomised controlled trial, *Intensive Care Med.*, 34 (8), 1411–1420. doi:10.1007/s00134-008-1072-1

Ganta, S., Deshpande, D., Korde, A. et al. (2010), A review of multifunctional nanoemulsion systems to overcome oral and CNS drug delivery barriers, *Mol. Membr. Biol.*, 27 (7), 260–273. doi:10.3109/09687688.2010.497971

Gao, H., Geng, T., Huang, T. et al. (2017), Fish oil supplementation and insulin sensitivity: a systematic review and meta-analysis, *Lipids Health Dis.*, 16 (1), 131. doi:10.1186/s12944-017-0528-0

Geng, L., Zhou, W., Liu, B. et al. (2018), DHA induces apoptosis of human malignant breast cancer tissues by the TLR-4/PPAR-α pathways, *Oncol. Lett.*, 15, 2967–2977. doi: 10.3892/ol.2017.7702

Ghasemi Fard, S., Wang, F., Sinclair, A.J. et al. (2018), How Does High DHA Fish Oil Affect Health? A Systematic Review of Evidence, *Crit. Rev. Food Sci. Nutr.*, 1–44. doi:10.1080/10408398.2018.1425978

Gopinath, B., Harris, D.C., Flood, V.M. et al. (2011), Consumption of long-chain n–3 PUFA, α-linolenic acid and fish is associated with the prevalence of chronic kidney disease, *Br. J. Nutr.*, 105 (9), 1361–1368. doi:10.1017/S0007114510005040

Grecu, I. and Mirea, L.G. (2003), Parenteral fish oil supplementation in patients with abdominal sepsis, *Clin. Nutr.*, 22 (1), 23s. doi:10.1016/S0261-5614(03)80086-9

Grimble, R.F. (1998), Dietary lipids and the inflammatory response, *Proc. Nutr. Soc.*, 57 (4), 535–542. doi:10.1079/PNS19980078

Grimminger, F. and Mayser, P. (1995), Lipid mediators, free fatty acids and psoriasis, *Prostaglandins Leukot. Essent. Fatty Acids*, 52 (1), 1–15. doi:10.1016/0952-3278(95)90090-X

Grimml, H., Grimminger, F., Korom, S. et al. (1998), Use of fish oil to prevent graft rejection, *Proc. Nutr. Soc.*, 57 (4), 577–585. doi:10.1079/PNS19980084

Hak, S., Garaiova, Z., Olsen, L.T. et al. (2015), The effects of oil-in-water nanoemulsion polyethylene glycol surface density on intracellular stability, pharmacokinetics, and biodistribution in tumor bearing mice, *Pharm. Res.*, 32 (4), 1475–1485. doi:10.1007/s11095-014-1553-6

Hamilton, J.A. and Brunaldi, K. (2007), A model for fatty acid transport into the brain, *J. Mol. Neurosci.*, 33 (1), 12–17. doi:10.1007/s12031-007-0050-3

Harnisch, S. and Müller, R H. (1998), Plasma protein adsorption patterns on emulsions for parenteral administration: establishment of a protocol for two-dimensional polyacrylamide electrophoresis, *Electrophoresis*, 19 (2), 349–354. doi:10.1002/elps.1150190233

Harnisch, S. and Müller, R.H. (2000), Adsorption kinetics of plasma proteins on oil-in-water emulsions for parenteral nutrition, *Eur. J. Pharm. Biopharm.*, 49 (1), 41–46. doi:10.1016/S0939-6411(99)00064-8

Hashimoto, M., Kato, S., Tanabe, Y. et al. (2016), Beneficial effects of dietary docosahexaenoic acid intervention on cognitive function and mental health of the oldest elderly in Japanese care facilities and nursing homes, *Geriatr. Gerontol. Int.*, 17 (2), 330–337. doi:10.1111/ggi.12691

Heller, A.R., Rössel, T., Gottschlich, B. et al. (2004), Omega-3 fatty acids improve liver and pancreas function in postoperative cancer patients, *Int. J. Cancer*, 111 (4), 611–616. doi:10.1002/ijc.20291

Heller, A.R., Rössler, S., Litz, R.J. et al. (2006), Omega-3 fatty acids improve the diagnosis-related clinical outcome, *Crit. Care Med.*, 34 (4), 972–979. doi: 10.1097/01.CCM.0000206309.83570.45

Hildes, J.A. and Schaefer, O. (1984), The changing picture of neoplastic disease in the western and central Canadian Arctic (1950–1980), *Can. Med. Assoc. J.*, 130 (1), 25. PMID: 6317154

Hill, A.M., Buckley, J.D., Murphy, K.J. et al. (2007), Combining fish-oil supplements with regular aerobic exercise improves body composition and cardiovascular disease risk factors, *Am. J. Clin. Nutr.*, 85, 1267–1274. doi:10.1093/ajcn/85.5.1267

Huang, T. H., Wang, P. W., Yang, S.C., et al. (2018), Cosmetic and therapeutic applications of fish oil's fatty acids on the skin. *Mar. Drugs*, 16 (8), 256. doi:10.3390/md16080256.

Ivanovs, K. and Blumberga, D. (2017), Extraction of fish oil using green extraction methods: a short review, *Energy Proc.*, 128, 477–483. doi:10.1016/j.egypro.2017.09.033

Jin, H., Yan, C., Xiao, T. et al. (2018), High fish oil diet promotes liver inflammation and activates the complement system, *Mol. Med. Rep.*, 17 (5), 6852–6858. doi:10.3892/mmr.2018.8687

Jolly, C.A., Jiang, Y.H., Chapkin, R.S. et al. (1997), Dietary (n–3) polyunsaturated fatty acids suppress murine lymphoproliferation, interleukin-2 secretion, and the formation of diacylglycerol and ceramide, *J. Nutr.*, 127 (1), 37–43. doi:10.1093/jn/127.1.37

Kadakia, E., Harpude, P., Parayath, N. et al. (2019), Challenging the CNS targeting potential of systemically administered nanoemulsion delivery systems: a case study with rapamycin-containing fish oil nanoemulsions in mice, *Pharm. Res.*, 36 (9), 134. doi:10.1007/s11095-019-2667-7

Kikawa, K.D., Herrick, J.S., Tateo, R.E. et al. (2010), Induced oxidative stress and cell death in the a549 lung adenocarcinoma cell line by ionizing radiation is enhanced by supplementation with docosahexaenoic acid, *Nutr. Cancer*, 62, 1017–1024. doi:10.1080/01635581.2010.492084.

Kris-Etherton, P.M., Harris, W.S., and Appel, L.J. (2002), Fish consumption, fish oil, omega-3 fatty acids, and cardiovascular disease, *Circulation*, 106 (21), 2747–2757. doi.10.1161/01.CIR.0000038493.65177.94

Kuehn, A., Hilger, C., and Hentges, F. (2011), Fish allergy—a comprehensive review of recent developments in fish allergy diagnosis and fish allergen detection, in: Rodgers, P.M., Ed., *Food Allergies: Symptoms, Diagnosis, and Treatment*, Nova, Hauppauge, USA, pp. 51–78.

Kyle, D.J., Schaefer, E., Patton, G. et al. (1999), Low serum docosahexaenoic acid is a significant risk factor for Alzheimer's dementia, *Lipids*, 34, S245–S245. doi:10.1007/BF02562306

Langlois, P.L., Hardy, G., and Manzanares, W. (2017), Omega-3 polyunsaturated fatty acids in cardiac surgery patients: an updated systematic review and meta-analysis, *Clin. Nutr.*, 36 (3), 737–746. doi:10.1016/j.clnu.2016.05.013

Larsen, H.R. (2006), Fish oil news, (Available at http://www.oilofpisces.com/fishnews.html.) Accessed on 12/19/2019.

Leaf, A., Kang, J.X., Xiao, Y.F. et al. (1998), Dietary *n*–3 fatty acids in the prevention of cardiac arrhythmias, *Curr. Opin. Clin. Nutr. Metab. Care*, 1 (2), 225–228.

Lembke, P. (2013), Production techniques for omega-3 concentrates, in: *Omega-6/3 Fatty Acids*, Humana Press, Totowa, NJ, pp. 353–364. doi.10.1007/978-1-62703-215-5_19

Li, N.N., Zhou, Y., Qin, X.P. et al. (2014), Does intravenous fish oil benefit patients post-surgery? A meta-analysis of randomised controlled trials, *Clin. Nutr.*, 33 (2), 226–239. doi:10.1016/j.clnu.2013.08.013.

Li, Y., Li, M., Qi, Y. et al. (2019), Preparation and digestibility of fish oil nanoemulsions stabilized by soy protein isolate-phosphatidylcholine, *Food Hydrocoll.*, 105310. doi:10.1016/j.foodhyd.2019.105310

Liu, J.J., Green, P., Mann, J.J. et al. (2015), Pathways of polyunsaturated fatty acid utilization: implications for brain function in neuropsychiatric health and disease, *Brain Res.*, 1597, 220–246. doi:10.1016/j.brainres.2014.11.059

Loftsson, T., Gudmundsdottir, T.K., Fridriksdottir, H. et al. (1995), Fatty acids from cod-liver oil as skin penetration enhancers, *Pharmazie*, 50 (3), 188–190. PMID: 7732049.

Mariko, S., Shinji, K., Keiji, I. et al. (2010), Oleate and eicosapenataenoic acid attenuate palmitate-induced inflammation and apoptosis in renal proximal tubular cell, *Biochem. Biophys. Res. Commun.*, 402 (2), 265–271. doi:10.1016/j.bbrc.2010.10.012

Martin, M. and Blom, A.M. (2016), Complement in removal of the dead–balancing inflammation, *Immunol. Rev.*, 274 (1), 218–232. doi:10.1111/imr.12462

Mascha, E.J. (2015), Alpha, beta, meta: guidelines for assessing power and type I error in meta-analyses, *Anesth. Analg.*, 121 (6), 1430–1433. doi:10.1213/ANE.0000000000000993

Mascioli, E., Leader, L., Flores, E. et al. (1988), Enhanced survival to endotoxin in guinea pigs fed IV fish oil emulsion, *Lipids*, 23 (6), 623–625. doi.10.1007/BF02535609

McManus, S., Tejera, N., Awwad, D. et al. (2016), Differential effects of EPA versus DHA on postprandial vascular function and the plasma oxylipin profile in men, *J. Lipid Res.*, 57, 1720–1727. doi:10.1194/jlr.m067801

Mercer, P. and Armenta, R.E. (2011), Developments in oil extraction from microalgae, *Eur. J. Lipid Sci. Technol.*, 113 (5), 539–547. doi:10.1002/ejlt.201000455

Meydani, S.N., Endres, S., Woods, M.M. et al. (1991), Oral (*n*–3) fatty acid supplementation suppresses cytokine production and lymphocyte proliferation: comparison between young and older women, *J. Nutr.*, 121 (4), 547–555. doi:10.1093/jn/121.4.547

Meydani, S. N., Lichtenstein, A.H., Cornwall, S. et al. (1993), Immunologic effects of national cholesterol education panel step-2 diets with and without fish-derived *N*–3 fatty acid enrichment, *J. Clin. Investig.*, 92 (1), 105–113. doi:10.1172/JCI116537

Mozaffarian, D. and Wu, J.H. (2011), Omega-3 fatty acids and cardiovascular disease: effects on risk factors, molecular pathways, and clinical events, *J. Am. Coll. Cardiol.*, 58 (20), 2047–2067. doi: 10.1016/j.jacc.2011.06.063.

Muley, A., Muley, P., and Shah, M. (2014), ALA, fatty fish or marine *n*–3 fatty acids for preventing DM?: a systematic review and meta-analysis, *Curr. Diabetes Rev.*, 10 (3), 158–165. doi:10.2174/1573399810666140515113137

Munro, I.A. and Garg, M.L. (2012), Dietary supplementation with *n*–3 PUFA does not promote weight loss when combined with a very-low energy diet, *Br. J. Nutr.*, 108, 1466–1474. doi:10.1017/s0007114511006817

Munro, I.A. and Garg, M.L. (2013), Prior supplementation with long chain omega-3 polyunsaturated fatty acids promotes weight loss in obese adults: a double-blinded randomised controlled trial, *Food Funct.*, 4, 650–658. doi:10.1039/c3fo60038f

Navarro, C., Gonzalez-Alvarez, I., Gonzalez-Alvarez, M. et al. (2011), Influence of polyunsaturated fatty acids on cortisol transport through MDCK and MDCK-MDR1 cells as blood-brain barrier in vitro model, *Eur. J. Pharm. Sci.*, 42 (3), 290–299. doi:10.1016/j.ejps.2010.12.005

Neuringer, M., Anderson, G.J., and Connor, W.E. (1988), The essentiality of *n*–3 fatty acids for the development and function of the retina and brain, *Annu. Rev. Nutr.*, 8 (1), 517–541. doi:10.1146/annurev.nu.08.070188.002505

Nicolo, M., Lara, C., Laura, M. et al. (2013), Dietary ω–3 polyunasaturated fatty acid DHA: a potential adjuvant in the treatment of cancer, *Biomed. Res. Int.*, 2013, 1–11. doi: 10.1155/2013/310186

Ninio, D.M., Hill, A.M., Howe, P.R. et al. (2008), Docosahexaenoic acid-rich fish oil improves heart rate variability and heart rate responses to exercise in overweight adults, *Br. J. Nutr.*, 100, 1097–1103. doi:10.1017/s0007114508959225

O'Keefe, J. H., Abuissa, H., Sastre, A., et al. (2006), Effects of omega-3 fatty acids on resting heart rate, heart rate recovery after exercise, and heart rate variability in men with healed myocardial infarctions and depressed ejection fractions, *Am. J. Cardiol.*, 97 (8), 1127–1130. doi:10.1016/j.amjcard.2005.11.025

Olsen, R.L. Toppe, J., and Karunasagar, I. (2014), Challenges and realistic opportunities in the use of by-products from processing of fish and shellfish, *Trends Food Sci. Technol.*, 36, 144–151. doi:10.1016/j.tifs.2014.01.007

Palmer, A.J., Ho, C.K., Ajibola, O. et al. (2013), The role of ω–3 fatty acid supplemented parenteral nutrition in critical illness in adults: a systematic review and meta-analysis, *Crit. Care Med.*, 41 (1), 307–316. doi: 10.1097/CCM.0b013e3182657578

Papaioannou, V. and Pnevmatikos, I. (2019), Heart rate variability: a potential tool for monitoring immunomodulatory effects of parenteral fish oil feeding in patients with sepsis, *Nutr. Metabolic Insights*, 12, 1178638819847486. doi:10.1177%2F1178638819847486

Peoples, G.E., McLennan, P.L., Howe, P.R. et al. (2008), Fish oil reduces heart rate and oxygen consumption during exercise, *J. Cardiovasc. Pharmacol. Ther.*, 52, 540–547. doi:10.1097/fjc.0b013e3181911913

Philippe, B., Nawale, H., Ferrasson, M.N. et al. (2009), Improving outcome of chemotherapy of metastatic breast cancer by docosahexaenoic acid: a phase II trial, *Br. J. Cancer*, 101, 1978–1985. doi: 10.1038/sj.bjc.6605441.

Phillips, M.A., Childs, C.E., Calder, P.C. et al. (2015), No effect of omega-3 fatty acid supplementation on cognition and mood in individuals with cognitive impairment and probable Alzheimer's disease: a randomised controlled trial, *Int. J. Mol. Sci.*, 16, 24600–24613. doi:10.3390/ijms161024600

Picado, C., Castillo, J.A., Schinca, N. et al. (1988), Effects of a fish oil enriched diet on aspirin intolerant asthmatic patients: a pilot study, *Thorax*, 43 (2), 93–97. doi:10.1136/thx.43.2.93

Porojnicu, A.C., Bruland, Q.S., Aksnes, L. et al. (2008), Sun beds and cod liver oil as vitamin D sources, *J. Photochem. Photobiol. B Biol.*, 91 (2–3), 125–131. doi.10.1016/j.jphotobiol.2008.02.007.

Prabhakar, K., Afzal, S.M., Surender, G. et al. (2013), Tween 80 containing lipid nanoemulsions for delivery of indinavir to brain, *Acta Pharm. Sinica B*, 3 (5), 345–353. doi:10.1016/j.apsb.2013.08.001

Pradelli, L., Mayer, K., Muscaritoli, M. et al. (2012), *n*–3 Fatty acid-enriched parenteral nutrition regimens in elective surgical and ICU patients: a meta-analysis, *Crit. Care*, 16 (5), R184. doi:10.1186/cc11668

Price, M.E., Cornelius, R.M., and Brash, J.L. (2001), Protein adsorption to polyethylene glycol modified liposomes from fibrinogen solution and from plasma, *Biochim. Biophys. Acta (BBA)-Biomembr.*, 1512 (2), 191–205. doi:10.1016/S0005-2736(01)00330-3

Rizliya, V. and Mendis, E. (2014), Biological, physical, and chemical properties of fish oil and industrial applications, in: *Seafood Processing by-Products*, Springer, New York, NY, pp. 285–313. doi:10.1007/978-1-4614-9590-1_14

Roulet, M., Frascarolo, P., Pilet, M. et al. (1997), Effects of intravenously infused fish oil on platelet fatty acid phospholipid composition and on platelet function in postoperative trauma, *J. Parenter. Enteral Nutr.*, 21 (5), 296–301. doi:10.1177/0148607197021005296

Rubio-Rodríguez, N., Beltrán, S., Jaime, I. et al. (2010), Production of omega-3 polyunsaturated fatty acid concentrates: a review, *Innov. Food Sci. Emerg. Technol.*, 11 (1), 1–12. doi:10.1016/j.ifset.2009.10.006

Rustad, T., Storrø, I., and Slizyte, R. (2011), Possibilities for the utilisation of marine by-products, *Int. J. Food Sci. Technol.*, 46, 2001–2014. doi:10.1111/j.1365-2621.2011.02736.x

Saglimbene, V.M., Wong, G., Zwieten, A.V. et al. (2019), Effects of omega-3 polyunsaturated fatty acid intake in patients with chronic kidney disease: systematic review and meta-analysis of randomized controlled trials, *Clin. Nutr.*, 39 (2), 358–368. doi:10.1016/j.clnu.2019.02.041

Sarker, M.Z.I., Selamat, J., Habib, A.S.M. et al. (2012), Optimization of supercritical CO_2 extraction of fish oil from viscera of African catfish (*Clarias gariepinus*), *Int. J. Mol. Sci.*, 13 (9), 11312–11322. doi:10.3390/ijms130911312

Shah, L., Gattacceca, F., and Amiji, M.M. (2014), CNS delivery and pharmacokinetic evaluations of DALDA analgesic peptide analog administered in nano-sized oil-in-water emulsion formulation, *Pharm. Res.*, 31 (5), 1315–1324. doi:10.1007/s11095-013-1252-8

Sharp, M.F. and Lopata, A.L. (2014), Fish allergy: in review, *Clin. Rev. Allergy Immunol.*, 46, 258–271. doi:10.1007/s12016-013-8363-1

Sidhu, K.S. (2003), Health benefits and potential risks related to consumption of fish or fish oil, *Regul. Toxicol. Pharmacol.*, 38 (3), 336–344. doi:10.1016/j.yrtph.2003.07.002

Siegel, R., Naishadham, D., and Jemal, A. (2014), Cancer statistics, *Cancer J. Clin.*, 63, 11–30. doi: 10.3322/caac.21166

Siscovick, D.S., Raghunathan, T.E., King, I. et al. (1995), Dietary intake and cell membrane levels of long-chain *n*–3 polyunsaturated fatty acids and the risk of primary cardiac arrest, *JAMA*, 274 (17), 1363–1367. doi:10.1001/jama.1995.03530170043030

Sophie, V., Caroline, G., Philippe, B. et al. (2008), Sensitization by docosahexaenoic acid (DHA) of breast cancer cells to anthracyclines through loss of glutathione peroxidase (Gpx1) response, *Free Radic. Biol. Med.*, 44, 1483–1491. doi:10.1016/j.freeradbiomed.2008.01.009.

Suresh, R., Chandrasekaran, P., Sutterwala, F.S. et al. (2016), Complement-mediated "bystander" damage initiates host NLRP3 inflammasome activation, *J. Cell Sci.*, 129 (9), 1928–1939. doi: 10.1242/jcs.179291

Szebeni, J. (2001), Complement activation-related pseudoallergy caused by liposomes, micellar carriers of intravenous drugs, and radiocontrast agents, *Crit. Rev. Ther. Drug Carrier Syst.*, 18 (6), 567–606. PMID: 11789676

Taki, H., Morinaga, S.I., Yamazaki, K. et al. (1992), Reduction of delayed-type hypersensitivity by the injection of *n*–3 polyunsaturated fatty acids in mice, *Transplantation*, 54 (3), 511–514. doi:10.1097/00007890-199209000-00023

Tan, S.F., Kirby, B.P., Stanslas, J. et al. (2017), Characterisation, in-vitro and in-vivo evaluation of valproic acid-loaded nanoemulsion for improved brain bioavailability, *J. Pharm. Pharmacol.*, 69 (11), 1447–1457. doi:10.1111/jphp.12800

Tian, H., Yao, X., Zeng, R. et al. (2013), Safety and efficacy of a new parenteral lipid emulsion (SMOF) for surgical patients: a systematic review and meta-analysis of randomized controlled trials, *Nutr. Rev.*, 71 (12), 815–821. doi:10.1111/nure.12053

Ting, J.P.Y., Kastner, D.L., and Hoffman, H.M. (2006), CATERPILLERs, pyrin and hereditary immunological disorders, *Nat. Rev. Immunol.*, 6 (3), 183–195. doi:10.1038/nri1788

Tinker, L.F., Parks, E.J., Behr, S.R. et al. (1999), (*n*–3) Fatty acid supplementation in moderately hypertriglyceridemic adults changes postprandial lipid and apolipoprotein B responses to a standardized test meal, *J. Nutr.*, 129 (6), 1126–1134. doi:10.1093/jn/129.6.1126

Toupchian, O., Sotoudeh, G., Mansoori, A. et al. (2016), Effects of DHA-enriched fish oil on monocyte/macrophage activation marker sCD163, asymmetric dimethyl arginine, and insulin resistance in type 2 diabetic patients, *J. Clin. Lipidol.*, 10 (4), 798–807. doi:10.1016/j.jacl.2016.02.013

Turchini, G.M., Torstensen, B.E., and Ng, W.K. (2009), Fish oil replacement in finfish nutrition, *Rev. Aquacult.*, 1 (1), 10–57. doi:10.1111/j.1753-5131.2008.01001.x

Vanek, V.W., Seidner, D.L., Allen, P. et al. (2012), ASPEN position paper: clinical role for alternative intravenous fat emulsions, *Nutr. Clin. Pract.*, 27 (2), 150–192. doi:10.1177/0884533612439896

Vuorinen, P. J., Roots, O., and Keinänen, M. (2017), Review of organohalogen toxicants in fish from the Gulf of Finland, *J. Mar. Syst.*, 171, 141–150. doi:10.1016/j.jmarsys.2016.12.002

Vyas, T.K., Shahiwala, A., and Amiji, M.M. (2008), Improved oral bioavailability and brain transport of Saquinavir upon administration in novel nanoemulsion formulations, *Int. J. Pharm.*, 347 (1–2), 93–101. doi:10.1016/j.ijpharm.2007.06.016

Walport, M.J. (2001), Complement, *N. Engl. J. Med.*, 344 (14), 1058–1066. doi:10.1056/NEJM200104053441406

Wei, C., Hua, J., Bin, C. et al. (2010), Impact of lipid emulsion containing fish oil on outcomes of surgical patients: systematic review of randomized controlled trials from Europe and Asia, *Nutrition*, 26 (5), 474–481. doi:10.1016/j.nut.2009.09.011

Wu, D. and Meydani, S.N. (1998), *n*–3 Polyunsaturated fatty acids and immune function, *Proc. Nutr. Soc.*, 57 (4), 503–509. doi:10.1079/PNS19980074

Xia, S., Li, X., Cheng, L. et al. (2014), Chronic intake of high fish oil diet induces myeloid-derived suppressor cells to promote tumor growth, *Cancer Immunol. Immunother.*, 63 (7), 663–673. doi:10.1007/s00262-014-1546-7

Xia, S., Li, X.P., Cheng, L. et al. (2015), Fish oil-rich diet promotes hematopoiesis and alters hematopoietic niche, *Endocrinology*, 156 (8), 2821–2830. doi:10.1210/en.2015-1258

Xiao, L., Yu, J., Yu, Z. et al. (2017), Ultrasound-assisted extraction of bighead carp viscera oil and its physiochemical properties, *J. Jishou Univ. (Nat. Sci. Ed.)*, 38 (1), 49–55.

Xue, M., Wang, Q., Zhao, J. et al. (2014), Docosahexaenoic acid inhibited the Wnt/β-catenin pathway and suppressed breast cancer cells in vitro and in vivo, *J. Nutr. Biochem.*, 25, 104–110. doi:10.1016/j.jnutbio.2013.09.008

Yao, Q.H., Zhang, X.C., Fu, T. et al. (2014), ω–3 polyunsaturated fatty acids inhibit the proliferation of the lung adenocarcinoma cell line A549 in vitro, *Mol. Med. Rep.*, 9, 401–406. doi:10.3892/mmr.2013.1829

Zadak, Z. and Cervinkova, Z. (1997), PUFA *n*–3 lipid emulsion—a promising agent in ARDS treatment (editorial), *Nutrition*, 13 (3), 232–233. doi: 10.1016/s0899-9007(96)00408-x

Zhou, Y., Tong, L., Chu, X. et al. (2017), The adenosine A1 receptor antagonist DPCPX inhibits tumor progression via the ERK/JNK pathway in renal cell carcinoma, *Cell. Physiol. Biochem.*, 43, 733–742. doi:10.1159/000481557

INDEX

Oil-in-Water Nanosized Emulsions for Drug Delivery and Targeting,
First Edition. Tamilvanan Shunmugaperumal.
© 2021 John Wiley & Sons, Inc. Published 2021 by John Wiley & Sons, Inc.

Printed and bound by CPI Group (UK) Ltd, Croydon, CR0 4YY